THE
ROMANTIC
MACHINE

THE
ROMANTIC
MACHINE

*Utopian Science and
Technology after Napoleon*

JOHN TRESCH

THE UNIVERSITY OF CHICAGO PRESS
Chicago and London

The University of Chicago Press, Chicago 60637
The University of Chicago Press, Ltd., London
© 2012 by The University of Chicago
All rights reserved. Published 2012.
Paperback edition 2014
Printed and bound by CPI Group (UK) Ltd,
Croydon, CR0 4YY

23 22 21 20 19 18 17 16 15 14   3 4 5 6 7

ISBN-13: 978-0-226-81220-5 (cloth)
ISBN-13: 978-0-226-21480-1 (paperback)
ISBN-13: 978-0-226-81222-9 (ebook)
10.7208/chicago/9780226812229.001.0001

Library of Congress Cataloging-in-Publication Data

Tresch, John.
The romantic machine : utopian science and technology after Napoleon / John Tresch
    p. cm.
    Includes bibliographical references and index.
    ISBN-13: 978-0-226-81220-5 (hardcover : alkaline paper)
    ISBN-10: 0-226-81220-0 (hardcover : alkaline paper)   1. Technology—Social aspects—France–19th century.   2. Utopias—France—History—19th century.   3. Machinery—Social aspects—19th century.   4. Science—Social aspects—France—19th century.   5. Technology—Philosophy—19th century.   6. Romanticism—France.   7. France—History—February Revolution, 1848.   I. Title
T26.F8T74 2012
509.44'09034—dc23
                                                                                                2011038172

♾ This paper meets the requirements of ANSI/NISO Z39.48-1992 (Permanence of Paper).

To all of my teachers, and most
of all to Simon Schaffer, Bruno Latour, and
George Stocking

Every epoch has its technology, and this technology has the style of an epoch: a style that demonstrates to what extent everything is connected, and to what extent everything interferes with everything else.

LUCIEN FEBVRE

You are not the same as or different from conditions on which you depend; you are neither severed from nor forever fused with them.

NAGARJUNA, translated by Stephen Batchelor

We shape our tools and they in turn shape us.

MARSHALL MCLUHAN

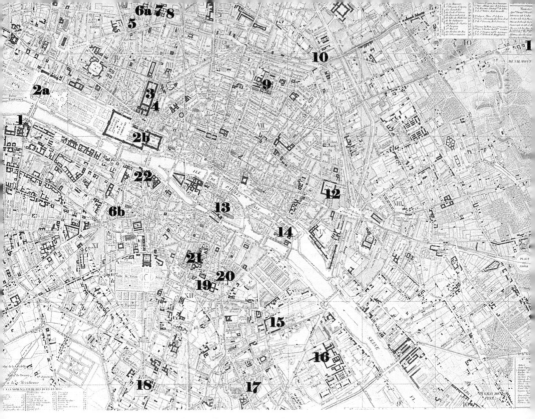

Detail from Girard Xavier, *Plan de la ville de Paris divisé en 12 arrondissements, en 48 quartiers indiquant tous les changemens faits et projétés* (Paris, 1843). Hand-colored engraved map, 53 × 96 cm. John Hay Library Maps Collection.

1. Chambre des Députés
2. Sites of Expositions Nationales: *2a*, Concorde; *2b*, Louvre
3. Wooden Galleries and Palais Royal
4. Athénée Royal de Paris
5. Robert-Houdin's theater for *Soirées Fantastiques*
6. Saint-Simonian lecture halls: *6a*, Salle Taitbout; *6b*, Rue Taranne
7. Opéra de Paris, Salle Le Peletier
8. Mairie du IIème Arrondissement (Comte's Astronomy Lectures)
9. Conservatoire Nationale des Arts et Métiers (CNAM)
10. Daguerre's Diorama
11. Retreat of Saint-Simonians at Ménilmontant
12. Hugo's house at Place des Vosges
13. Notre Dame; Saint-Simonian Temple
14. Hôtel Pimodan (Club des Hachichins)
15. Muséum d'Histoire Naturelle, Jardin des Plantes
16. Salpêtrière (Esquirol, Georget, magnetism research)
17. Gobelins manufactory (Chevreul)
18. Observatory of Paris
19. Panthéon
20. Ecole Polytechnique
21. Sorbonne (Cousin and company)
22. Academy of Sciences

Contents

Preface: Absolute and In-between xi

1 Introduction: Mechanical Romanticism 1

PART 1 * DEVICES OF COSMIC UNITY

2 Ampère's Experiments: Contours of a Cosmic Substance 29
3 Humboldt's Instruments: Even the Tools Will Be Free 61
4 Arago's Daguerreotype: The Labor Theory of Knowledge 89

PART 2 * SPECTACLES OF CREATION AND METAMORPHOSIS

5 The Devil's Opera: Fantastic Physiospiritualism 125
6 Monsters, Machine-Men, Magicians: The Automaton in the Garden 153

PART 3 * ENGINEERS OF ARTIFICIAL PARADISES

7 Saint-Simonian Engines: Love and Conversions 191
8 Leroux's Pianotype: The Organogenesis of Humanity 223
9 Comte's Calendar: From Infinite Universe to Closed World 253

10 Conclusion: Afterlives of the Romantic Machine 287

Acknowledgments 313
Notes 317
Bibliography 377
Index 431

Preface

Absolute and In-between

The kinds of machines we use are bound up with the ways we think about nature and the ways we know it. When our machines and our understandings of them change, so does nature, and so does our view of knowledge. The dominant image of modern science has depended on certain exemplary "classical machines": balances, levers, and clocks. These imply a nature that is stable, fixed, and made of discrete points of matter, subjected to a limited number of forces in equilibrium; they also suggest a view of knowledge as a detached, impersonal, and emotionless objectivity. This image was solidified in the seventeenth century and continues to exert a powerful hold on our ideas of what science is and ought to be.

This book examines a period in which a different image of science—as a theory of nature and a theory of knowing—appeared, at the same time as a new set of machines came on the scene: steam engines, batteries, sensitive electrical and atmospheric instruments, improved presses, and photography. These were "romantic machines." Unlike "classical machines," they were understood as flexible, active, and inextricably woven into circuits of both living and inanimate elements. These new devices accompanied a new understanding of nature, as growing, complexly interdependent, and modifiable, and of knowledge, as an active, transformative intervention in which human thoughts, feelings, and intentions—in short, human consciousness—played an inevitable role in establishing truth.

This alternative scientific tradition rose to prominence in the 1820s, 1830s, and 1840s amid the upheavals of early industrialization. I focus on its manifestations in Paris, the city to which many looked for signs of things to come. After 1850 the classical image of science again took the upper hand; even today, we largely take for granted that real knowledge is possible only where there is a radical divide between subjects and objects and where nature is reduced to discrete, predictable mechanisms. Yet this other view, which I call mechanical romanticism, made major contributions to physics, evolutionary theory, the social sciences, mass entertainment, modern transport and communications, and precipitated the Revolution of 1848. This book reconstructs this neglected theory of nature and knowledge, one that may be a resource for those who seek to redraft the relationships between machines, knowledge, and the earth.

No one better captured the drama and the high stakes involved in this new way of thinking than Honoré de Balzac, who called *The Human Comedy*, his interlocking series of novels, a "natural history of society." His novel of 1834, *The Quest for the Absolute*, featured a character, Balthazar Claës, who trained in Paris with the chemist Lavoisier. After returning to comfortable domesticity in the northern province of Douai, Claës received a visit from a shadowy Polish scientist—modeled on the real-life Count Hoëne-Wronski, a mathematician, inventor, and mystic—who revealed to him the existence of *the Absolute*. This was the underlying principle of matter and life: "the First Cause, the key to all the phenomena of nature . . . the last word of creation."[1]

Claës was enthralled. He immediately threw his energies and his fortune into the experimental isolation and control of this principle.[2] He spent his days and nights in an attic laboratory, surrounded by ominous machines of metal and glass, conducting strange experiments with rare substances delivered from Paris, powered by voltaic batteries and the sun's light and heat. When his wife pressed him about the vast sums literally going up in smoke, he assured her that once he grasped the Absolute, wealth would be an afterthought, because power over matter and life would be his.

Much like the *prima materia* of the ancient alchemists, the Absolute held matter together, fueled animal life, and lay behind human thought. Mankind, Claës explained, was the most perfect device that nature had created to wield this force:

> Each time that nature has perfected an apparatus, into which for some unknown reason she has added the three distinct degrees in the system of organic nature—feeling, instinct, or intelligence—these tend towards a combustion whose action is in direct proportion to its results. Mankind, which represents the highest level of intelligence and which is the apparatus that demonstrates a power of near-creation—the faculty of *thought!*—is also the zoological creation in which combustion is most intense. Its effects are revealed by the presence of phosphates, sulphates, and carbonates that can be analyzed in man's body. What are these substances but traces of the action of electric fluid, the life-giving principle? Would not electricity manifest itself in mankind through more varied combinations than in any other animal? Is it not to be expected that man would possess

greater faculties for absorbing a larger share of the absolute principle, and would he not assimilate it in order to compose his strength and his ideas into a more perfect machine? I am sure of it.[3]

Claës's experiments aimed at replicating and improving the electrochemical, metabolic, intellectual processes of the human apparatus. His pious wife was shocked by his definition of human life as a kind of machine, a material process that could be technologically reproduced and controlled: "Accursed science! Accursed fiend! You are forgetting, Claës, that this is the sin of pride by which Satan fell! You are encroaching on God!" Yet her tears, for Claës, were nothing but a chemical compound.[4]

In other tales, Balzac expressed similar theories that linked electricity, heat, light, and magnetism to life and thought; he claimed that the will could redistribute, amplify, and direct these "imponderable" fluids.[5] Dramatized by Claës and other characters, including the precocious Swedenborgian philosopher Louis Lambert and the doomed composer Gambara, Balzac's theories were closely connected with the scientific thought of his time. There were good reasons for this. While he was composing *Quest for the Absolute*, Balzac lived next to the Observatory of Paris and frequently dined there with its director, the astronomer François Arago; his brother Etienne Arago, who was a playwright and a politician; the physicist and astronomer Félix Savary, who read and corrected his drafts; and opticians who built the novelist a "divine pair of opera glasses."[6] Claës has been seen as an amalgam of the character and physiognomy of François Arago, with other traits—including his obsessiveness and absentmindedness—borrowed from the pioneer of electromagnetism, André-Marie Ampère, whose son was a leading literary critic.

Like many other scientists of this time, Arago and Ampère were fascinated by the relationships among the "substances known only by their effects," electricity, magnetism, heat, and light. How could their physical force be quantified or put to use? How did they affect one another? Were they distinct fluids, or were they vibratory motions in some underlying medium or ether? If so, what was the nature of that surrounding circumstance, or milieu—the *mi-lieu*, the halfway place, the zone in-between—and how did it interact with less subtle matter? It was not only physicists and chemists who were captivated by these questions. In the 1820s and 1830s, animal magnetism, or mesmerism, was once again being investigated by physicians and anatomists. Some attributed the power that magnetizers had over their patients to the transmission of a fluid, others to a vibration in the ether, some to a change induced in the organization of the vessels and fluids of the brain. Subtle fluids were important for anatomists, as the life of an organism depended on the balance of gases, light, heat, electricity, and magnetism that made up its milieu. Those who followed Lamarck understood these phenomena as keys to the generation and metamorphoses of living things; changes in their quantities, or the decline of other sources of nutrition, would force organisms to acquire new habits in order to survive. Exercising certain organs, and using others less, would modify a species' overall character, in some cases even bringing new organs into existence; these changes would be inherited by offspring.

By calling his work a "natural history of society," Balzac drew attention to his descriptions of the elements that made up the human milieu: the settings, accoutrements, passions, strategies, and arrays of forces that shaped human lives. *The Quest for the Absolute*, for instance, carefully analyzed the balance of social relations and material elements (tulips, paintings, silver) that made up the society of Douai, and it showed the ways Claës's obsessions upset that balance. Likewise, Balzac's Parisian novels are a precious document of the era's salons, boudoirs, offices, concert halls, and workshops and the varieties of social species that inhabited them. His concept of the milieu extended beyond wallpaper, fabric, backdrop, and mise-en-scène to include those chemical and physical components that were the focus of scientific fascination. In *Le père Goriot* he spoke of the precise combination of gases that produced the "boarding house smell," olfactory evidence of the nutritive envelope that determined the life course of its inhabitants.[7] His introduction, or "Avant Propos," to *The Human Comedy* approvingly cited the theories of Lamarck's supporter Geoffroy Saint-Hilaire to describe the role of fluids in bringing about modifications and improvements in organisms.

More generally, Balzac's "Avant Propos" asked how an organism might rise on the scale of beings toward greater perfection. The projects of his heroes—for social distinction, wealth, love, artistic achievement, and enlightenment—were expressions of a universal aspiration. In natural-historical terms, Balzac's central question was one of adaptation and control. Through what means does one come to master one's environment? How does one alter one's relation to one's milieu? For Balzac, everything depended on the ability to focus and direct the will. As shown in his belief that his own frenzied productivity depended on sexual abstinence and very strong coffee, Balzac saw vital and spiritual forces as subject to physical modification. His Parisian novels spelled out the care with which characters composed their thoughts, gestures, speech, and attire to gain advantage in political and social intrigues. Similarly, in *Gambara* he described the precise experimental setup required by the composer, addressing both matter and mind—custom-made musical instruments, a precise quantity of alcohol—to achieve transcendent musical effects.

In the *Quest*, Claës's scientific instruments allowed him to access and manipulate the hidden springs of nature and gave him hope of complete mastery. Yet just as many of Balzac's other schemers endured tragic reversals, Claës's investigations abounded in ironies. When in the clutches of his mania, Claës's vision became fixed and he was said, repeatedly, to move "mechanically"—an adverb applied to no other characters in the book.[8] Claës was enchanted, mechanized, by the very scientific machines that have been held responsible for the disenchantment of the world. In his laboratory, his daughter encountered him "almost on his knees before his machine" as it channeled both sunlight and the electricity of a voltaic battery—a scene presented as an insane and idolatrous attempt at mechanical transubstantiation.[9] Most cruelly, the book's final scene showed Claës seized by a sudden inspiration, finally grasping the secret for which he had sacrificed his life: "Eureka! I have found it!" he exclaimed. He died before he could share his discovery. His fate compares to that of Gambara, whose performance of his grandiose and sublime opera was a pure cacophony to his listeners, or of the painter Frenhofer, whose "Unknown Masterpiece," a lifetime in the making, was seen by everyone else as an inscrutable mass of opaque, meaningless brushstrokes.[10]

Balzac feared a similar fate: that his own gloriously conceived works might, through excess of will, through obsession gone wrong, through loss of control over his means, lead to nothing but incomprehension or silence. Balzac's masterpiece *Lost Illusions*, written from 1837 to 1843, spelled out the ambivalent power of his own chief instrument, the press. At one level the book presented the press as a material technology, through a detailed history of paper making, descriptions of editorial offices, and a recounting of the moment-by-moment decisions of authors; at another, the press appeared as a transformative force, capable of turning a showgirl into a legend and quickening the development and decline of the naïve and irresolute hero Lucien de Rubempré. Balzac affectionately followed this "great man in embryo" from the provinces, as he attempted to use all the tools available to him to conquer the capital—a quantum of talent, an innocent charm, the armor of the dandy, the aid of fickle allies, and above all, his words, as they were multiplied and transmitted by the press—only to find himself drawn into the sophistry-strewn swamp of journalism. Lofted by the forces of publicity, hype, and criticism, Lucien remade himself by making these tools his own, then found himself ruined when they were directed against him. Again and again Balzac showed how the apparatus, instruments, or "organs" that his characters merged with to master their environments—intoxicants, the tools of art, social stratagems, or strangely active musical and scientific instruments—might turn upon their users.

Significantly, *Lost Illusions* also offered a brief, admiring view of a handful of minor characters who were just as scientifically, philosophically, and artistically fixated as Balzac's self-destructive major protagonists, yet whose projects radiated idealism. These were the members of the "cenacle," an informal club that met regularly near the Sorbonne; its members at first welcomed Lucien to their high-minded society, though they foresaw his descent into the hell of journalism. Two of the cenacle's members were clearly based on figures known to Balzac in the 1820s and 1830s. They were both former followers of Saint-Simon, the prophet of industry: one resembled Balzac's friend Philippe Buchez, a physician and republican who founded a new "science of history" that culminated in a Catholic form of socialism; another appeared to have been modeled on Pierre Leroux, a printer and editor who adapted Geoffroy Saint-Hilaire's natural history into a new "religion of humanity," weaving science, technology, and the arts into a vision of a democratic, self-organizing society.[11]

This group of idealists lacked the self-interest and narrow-mindedness that led many of Balzac's other heroes to their tragic ends. But just like them—and like Balzac himself—they sought to create and control the tools that would lift them higher on the scale of being. For these reformers, this aspiration meant putting feeling, activity, and intelligence to work to bring humans into closer, less competitive association, to guarantee their subsistence and the flourishing of their environment. Their tools included the physical sciences, instruments of production, and the media of communication. Although the Absolute might lie beyond the possibility of direct experience, by reorganizing the space between the mind and the world—the milieu, the mediations between humans, and nature—humans might draw nearer to it.

* * *

The dates of Balzac's activity, from the last years of the Restoration through to the collapse of the July Monarchy, match those of the present book. He knew many of its central characters and mingled with them in Paris's salons, offices, printed pages, and streets. His quest and the quests he described closely resembled and intersected with the projects described here. His protagonists, like those of this book, were seized by the passions, mystic flights, uncertainties, and obsessions of romanticism even as they sought new machines and rational sciences to let them know and act upon their environment. These individuals' attempts to orchestrate combinations of tools to reshape their milieu transformed their society and themselves. In various ways they kept the Absolute in sight—whether as an underlying principle, an encompassing totality, or the union of mind and world—even while many of them noted the intrinsic limitations of knowledge and action and the impossibility of reaching a point external to the circumstances into which humans are thrown. As his friend the early feminist writer George Sand wrote, "Balzac, who searched so much for the absolute in any number of discoveries, nearly found in his very own work the solution to a problem unknown before him: all of reality in a complete fiction."[12] Balzac, Sand, and their fellow travelers aimed to weave together all of reality in fabrications that were at once scientific, artistic, technological, and political. To these ends they imagined, built, and joined themselves with romantic machines.

I

Introduction

Mechanical Romanticism

HOPELESS ROMANTICS AND SOULLESS MECHANISTS

Both romanticism and mechanism have defined the modern world. Creators, dreamers, and nature-lovers have confronted—or run from—hard-headed realists and rationalizers. Technophiles, bureaucrats, and scientific fundamentalists, on the other hand, have tried to ban emotion, individuality, and fantasy from the serious work of learning about nature. In this constantly reformulated battle, perhaps you have joined your voice to those who call out for more creativity and spontaneity, raging against mechanization and standardization. Or you may have nodded your assent to those who plead for better data and technological fixes, shinier equipment and starker rationality, less superstition and feeling.

Or perhaps you see this as a false opposition. If so, you might have found yourself at home among the protagonists of this book, in Paris before 1848: a time and place when romantic aspirations shaped mechanical sciences and industry and when new discoveries and devices intensified organic and artistic visions. In diverse but overlapping projects, emotion and aesthetic experience were valued on a par with technical and rational mastery; individuals and entire movements set themselves the task of remaking society and the natural world with the help of new machines.[1]

The opposition between romanticism and mechanism is abundantly familiar, thanks in part to histories of ideas from the first half of the twentieth century, such as Lovejoy's *Great Chain of Being* and Whitehead's *Science and the Modern World*, with its key chapter "The Romantic Reaction."[2] According to these influential works, romanticism brought with it a new conception and experience of the self, accentuating emotion, expression, aesthetics, and purposeful action. Romanticism saw all of nature as united through an underlying living force and through archetypal forms. Humans themselves were part of nature's incessant growth and development, which manifested in creations of artistic and philosophical genius. The human imagination and senses played an active role in shaping phenomena and creating a "second nature." Art and philosophy sought to raise audiences beyond individual consciousness into participation in a wider whole. Politics itself appeared as an art: land, language, and birth were melded together, laying the ground for both nationalism and a vibrant cosmopolitanism.[3]

The varied, at times contradictory, tendencies of romanticism have frequently been summed up in a single term: *organicism*. Romantics used notions associated with the processes of life—growth and holism, circular causality, and the productive struggle and harmony between polarities—to characterize poems, persons, nations, and the cosmos. More bluntly, the term *organicism* has been defined as the opposite of *mechanism*.[4] Put forth influentially by German thinkers clustered in Jena at the end of the eighteenth century, this opposition oriented much of the intellectual life of the subsequent two centuries. As a refusal of the rationalism of the Enlightenment and as a protest against the emergence of a cold and fragmented scientific and industrial order, romanticism has been defined as "that attempt, apparently doomed to failure and abandoned by our time, to identify subject and object, to reconcile man and nature, consciousness and unconsciousness."[5] "Hopelessly romantic," we say, of someone who wants life to be like a novel, a *roman*: someone childishly addicted to intense passions, unable to yield to hard realities and common sense. If romanticism rallies around hearts, spirits, and the vital pulse of living things, it is at the same time the enemy of reason, matter, and machines.

In contrast to romantics, promoters of machines and mechanism have been seen as pursuing detached, rational analyses of nature's objective properties. Against the celebration of individual freedom, they hold to an all-embracing determinism; against the nostalgic flight to nature, they relentlessly construct artificial improvements; against organic holism, they analyze and reduce. According to many interpreters, this has been all to the good. In the history of science, for example, there is a recurrent assumption that a field's progress can be measured by the extent to which its phenomena are treated "mechanically," implying a removal of teleology and metaphysics, and the extent to which its procedures are performed by mechanistic algorithms or repeatable experiments. This narrative of scientific progress overlaps with the view that measures a society's degree of civilization by its use of machines.[6]

Yet the very "inhumanity" that has made machines a standard for progress and the model for the well-governed state—their supposed uniformity, efficiency, and lack of emotion—has also made them targets of hostility. A machine has no feelings, no soul; it lacks the growth, flexibility, and freedom attributed to life.[7] Mechanistic science proceeds by decomposition and analysis, separation and distinction; it kills what it studies. A set of oppositions has thus emerged in which those who mourn the disenchantment of the world and the unweaving of rainbows confront those who embrace the adult, real world of facts, laws, and objects:

Romanticism	Mechanism
emotion, will, passion	reason
spirit	matter
sensation, color, feeling	mass, motion, number
moral, personalized nature	amoral, impersonal nature
freedom	determinism
wholes (synthesis)	parts (analysis)
retrograde, nostalgic	progressive, forward-looking
organisms	machines

Any number of cultural and disciplinary clashes—from the "dissociation of sensibility" and the "two cultures," to the "science wars"—have taken place along this fault line.[8] Yet as accurate as this opposition may be as a description of certain intellectual skirmishes and posturings, it has obscured important features of the intellectual and political landscape of modernity, leading us to misunderstand both romanticism and mechanism as well as the manifold cases in which they were entwined.

This book explores numerous such cases in one pivotal setting: Paris in the uncertain interval between the fall of one Napoleon and the rise of another. At that time, debates about the impact of technology were at the center of cultural and political life. While some expressed wariness or even hostility toward machines, others embraced them; and many did so with attitudes and ideas usually associated with romanticism. Taken as aids for externalizing and expressing the self, machines drew forth virtual powers and brought about conversions among hidden forces; they could be used to create new wholes and organic orders, remaking humans' relationship to nature and renewing nature itself. Romantic themes guided research across scientific fields, as shown by recurrent scientific interest in development, conversion, and metamorphosis; in reflections about phenomena as the interface between the mind and external objects; and in a new attention to the aesthetic, emotional, and subjective aspects of knowledge. At the same time, artworks and popular spectacles used new and elaborate techniques to produce powerful emotions and lifelike effects and often took the demiurgic powers of science and technology as their central themes.[9] In many cases, the encounter between mechanism and romanticism fed into a thorough re-imagining of the system of government, the distribution of the fruits of labor, and the proper relationship between humans and the earth. For the loose association of individuals whom I call the mechanical romantics, science and technology appeared not as the enemies of the human, but as integral components—both tools and actors—in the creation of a "second nature."

One point shared by these diverse projects was an interest in protean, weightless, invisible fluids. These "imponderables" were in turn closely related to the concept of the "milieu," or the totality of substances surrounding an organism and forming its "conditions of existence." In English, the term refers primarily to a social setting, but in the French writings discussed here, it also had connotations relating it to biology, spatial location, and physics. In the late eighteenth century, Jean-Baptiste Lamarck—who coined the term *biology*—spoke of milieus, referring to the material elements that sustained and contributed to the transformation of living things, including light, heat, electricity, and magnetism. Milieu was a central concept both in Alexander von Humboldt's global mapping of environments and in Balzac's "natural history of society," with its "physiology" of salons, newspaper offices, and opera loges.[10] Despite the central role the term *milieu* came to play in biology—as a precursor to the twentieth-century concepts of "ecological niche" and *Umwelt*—Georges Canguilhem traced its origins to classical mechanics, specifically to Newton's concept of the *ether* as the medium of transmission for light. Milieu thus had a doubled spatial implication: it referred to the life-sustaining envelope surrounding an individual (like its synonym *atmosphere*) as well as, going back to Newton, a sense of linear emanation, as the space connecting two entities: the "milieu," or place in between. It is thus a term of both connection and separation. Further, because it was applied both to the fluids and ethers of physics and to the nutritive environment of organisms, it was a link between scientific fields that were being pulled apart by specialization. In addition, the milieu was increasingly the level at which projects of reform and transformation were aimed. Rather than acting directly upon an organism, an individual, or a society, change was sought through alterations to its surrounding environment.

Lamarck also famously argued that changes in the milieu led organisms to develop new habits to satisfy their needs. The use and disuse of organs changed their size and configuration and even produced new ones—a view elaborated in the 1820s and 1830s by Geoffroy Saint-Hilaire, with support from the *Naturphilosoph* Lorenz Oken's concept of organogenesis.[11] This book shows that a decisive shift in the conception of nature in the first half of the nineteenth century expanded this view of an ongoing process of natural production and adaptation to include human technology: machines and tools were seen as new organs modifying humans' relation to their environment. At its grandest scale, what we might call a technological Lamarckism saw human industry as a natural expression of the development of the earth itself.[12]

A law of progress was seen to be at work in shaping not only organisms but also geologic formations, governments, and ideas; the widest frame for this historical, even evolutionary understanding of nature and society was set by the "nebular hypothesis" that described the origin of the solar system as the successive condensation of a cloud of nebular gas. On earth, humans could contribute to this development at multiple levels: by remaking the landscape and altering nature's material order; by framing and arranging phenomena and concepts; and through the activity of perception, conceptualization, and imagination. At each of these levels, the modification of nature was aided by machines, eroding the dichotomy between natural and artificial. New instruments and machines were theorized as extensions of human senses and intentionality, as fluid mediators between mind and world, and as the ligaments of society; they appeared as transformative, even sublime devices. These devices were not the basis for a polarization between the sciences and the arts; they inspired instead a strong sense of commonality and even identity among artists, philosophers, and scientists. In the 1830s and 1840s, romantic genius was alive and well and living in Paris—surrounded by machines.[13]

The chapters that follow thus trace a set of closely related transformations. There was a shift in the image of the machine from an idea of balanced, inhuman clockwork to a "romantic machine" exemplified by the steam engine and other technologies of conversion and transmutation. Concepts of mechanism and organism merged in several ways: mechanical processes were seen as the instruments of organic teleology; human technical innovations expressed nature's development; devices and machines fused with human actions, intentions, and perceptions. More broadly, a new concept of nature emerged, with the recognition that nature not only has a history but is subject to alteration by human technology.[14] A new theory of knowledge also emerged: the sense organs and inner faculties of observers were seen to play an active role in the constitution of phenomena, as were devices of observation and experiment. As has been argued in recent discussions of scientific ideals, epistemology has an ethical dimension: many of the thinkers considered here advanced one form or another of an ethics that aimed at freedom through connection—with other humans, with the rest of nature, and with machines.[15] Finally, these metaphysical, epistemological, and ethical shifts helped inspire and guide a new political orientation: they formed the background both for a radicalized republicanism and for the birth of modern socialism.

In short, what was at stake in these diverse projects was the emergence of a new cosmology: a new conception of the relationships between the domains of nature, social order, and human activity. To track these diverse shifts, my account focuses not on a vague *Zeitgeist* or episteme, but on the concrete and specific means through which actors presented the order of the cosmos to themselves and to their fellows. To do so they frequently created *cosmograms*—material assemblies using words, images, numbers, songs, stories, or monuments to convey the order of the universe as a whole. These were artifacts of different scales and genres, made of different materials; some aimed at faithful representation of the world as it was, and others were intended as propositions, guideposts, anchors, or even satirical jests indicating how the world ought to be. Among the examples I discuss are Ampère's classification of the sciences, Humboldt's views of nature, Grandville's *Un autre monde*, the Saint-Simonian temple, Leroux's *L'humanité*, and Comte's calendars. Although the cosmograms of this period reflected their creators' idiosyncratic life histories and intentions, and although they varied according to disciplinary and generic requirements, each of them insisted, in its own way, that nature was subject to human modification and that machines were going to play a central and beneficial role in its development; and each of them reflected an urgent, historically specific sense that all parts of the cosmos had to be brought together and represented in a single site, in order to focus and organize human activity and remake the world.[16]

PARIS AS A CENTER OF CONVERSION

France might appear the least likely place to study the interweaving of romanticism and industrialization. After all, it was here that the most extreme tendencies of the Enlightenment had taken hold, in a hyperbolic rejection of faith and tradition that led to the bloody excesses of the Revolution; its neighbors often saw it as the center of materialism and rationalism. Such revolutionary attitudes were only intensified under Napoleon, an artillery engineer who arranged a chair for himself in the mathematics section of the National Institute of Sciences and Arts. In contrast, Germany is often seen as the source of both the romantic movement and crucial scientific and technical developments; likewise, England played the leading role in the industrial revolution, and its romantic poets have shaped our view of the movement. The intersections of romanticism, technology, science, and social reform in this period could be fruitfully explored in other national contexts, focusing, for instance, on Johann Ritter, Lorenz Oken, Novalis, Wilhelm Weber, and Friedrich Schelling; on Humphry Davy, Samuel Taylor Coleridge, Michael Faraday, Thomas Carlyle, and Mary Shelley; or, perhaps, on Joseph Henry, Ralph Waldo Emerson, Samuel Morse, and Edgar Allan Poe.[17]

Yet the mergers of romanticism and mechanism that appeared in Paris were both distinct and distinctly influential. For Karl Marx, for Walter Benjamin, for any number of historians and historical actors, the arts, politics, and intellectual life of Paris were the template and the vanguard of modernity. It was here that key modern political categories and concepts were established, from individualism and socialism to "the working class"; where the notion of an artistic avant-garde was formulated; where new mass spectacles, forms of commercial consumption, and popular publishing in literature, journalism, and science took root; and where epoch-defining sciences and technologies arose, including electromagnetism, photography, quantitative experimental physics, and biological transformism, along with the landmark philosophy of positivism.

In France by 1820 the awakening of the romantic consciousness was exactly contemporaneous with a dawning awareness of the political importance of industrialization and the role science would play in it. As the defeat at Waterloo showed, the excellence of French mathematics and engineering was no guarantee against English naval supply routes and production capacities. The widespread application of mechanism and steam technology to industry that was already under way in late-eighteenth-century England did not begin in France until after the fall of Napoleon, and it did not take off in earnest until the late Restoration and the July Monarchy. Napoleon's censors had also largely blocked the entry of foreign writings. Thus German literary and philosophical developments after Immanuel Kant flooded into France with the fall of the empire in 1815. The Parisian intellectual scene rushed to catch up with its neighbors, both technologically and intellectually, and strived self-consciously to put its distinctive mark on the modern sensibility. New journals appeared, devoted to digesting and reviewing events from around the globe, reframing them for both national and international consumption.[18]

Two other features encouraged a uniquely explicit and extensive cross-fertilization of romanticism and mechanism in France. First, French cultural and political life revolved around the capital of Paris. This exceptional centralization ensured frequent face-to-face encounters among innovators across fields and fostered dense, overlapping social networks. The worlds of literature, politics, science, and industry were in constant contact and exchange in the vibrantly theatrical social life of Paris and on the pages of its journals. In the period treated in this book, Paris was growing, yet still a small city by late-nineteenth-century standards, with a population of about 750,000 in 1820 and just over 1 million by 1851.[19] This was Paris before Haussmann, the planner of the Second Empire who prepared the wide boulevards, glittering shops, restaurants, and cafés that now define the city's physiognomy.[20] A potent traditional social geography still marked the city: the Faubourg de Saint-Germain as primarily the home of the wealthy; the Latin Quarter as the place of students, scientists, and owners of small shops; and the area around the Louvre, and increasingly the Bourse, as the site of power in politics and finance. On the Right Bank, the Faubourgs of Saint-Denis and Saint-Antoine were packed with narrow, often unnamed streets with unnumbered dwellings: here, and in the smaller communes at the city's outer limits, annexed over the course of the century, were the homes of workers and the poor. Small workshops in leather, woodworking, and metallurgy occupied these regions, with a few sites for state-sponsored production of luxury goods such as the Gobelins' tapestry work; in this period, many of these industries began to introduce new chemical processes and heavy machinery.[21] Yet the city was still small and dense enough that members of diverse social strata could meet in salons, reading rooms, balls, and the streets. Further, in times of discontent, the revolutionary tradition was still strong: popular assemblies could directly influence policy, and barricades could stop government troops. As suggested by the dominance of theatrical metaphors in literature, criticism, politics, and epistemology at this time, Parisian life appeared to many as a vast play in which all performed, on the mobile stages of streets, salons, and journals.

Second, France's recent history gave the nation's historical trajectory a significance that went well beyond its borders. Following the universalizing tendency of French orators in the Revolution, works like Jules Michelet's *Le peuple* presented French history as the history of the world.[22] The period's instability was presaged by the earth-shattering events of the Revolution: the storming of the Bastille and the execution of Louis XVI, the lurching transitions from Convention to Directory to Consulate and Empire, and Napoleon's defeat at Waterloo in 1815. After a brief return to power that year (the Hundred Days), Napoleon was exiled, and the Restoration of the decapitated king's younger brother, Louis XVIII, was imposed by allied powers whose troops remained for several years in Paris. When

Louis XVIII died in 1824, a still younger Bourbon brother, the ultraroyalist Charles X, was crowned. His refusal to yield to the restraints imposed by the constitution (the Charte) triggered the largely peaceful Revolution of 1830, which took only three days in July (*Les trois glorieuses*). The new monarchy was headed by King Louis-Philippe, a scion of the Orleans branch of the royal family. Although freedom of the press briefly increased, the "July Monarchy" showed little concern for the social circumstances of the 1830s, in particular the poverty and uncertainty faced by the growing working class. Strikes and riots took place throughout the period, followed by repressive acts: these tensions culminated in the Revolution of 1848, which resulted in the brief, unstable Second Republic. Universal suffrage led to the election in 1849 of a new president, Louis-Napoleon Bonaparte (the first Napoleon's nephew). In 1851 this new Napoleon staged a coup d'état, declared himself emperor, exiled or imprisoned his opponents, and proclaimed the Second Empire.

This social and political turmoil fostered an awareness among inhabitants of Paris between 1815 and 1851 that they were living in extraordinary times. The physical sciences had recently made massive strides: in terrestrial physics, a combination of precise experimentation and mathematization was bringing about a "second scientific revolution" in the study of heat, light, electricity, and magnetism. Georges Cuvier's classification of the animal kingdom and medical advances by Bichat, Broussais, and Etienne Serres suggested a more profound understanding of the motors of life. At the same time, steam technology and its economic and social effects were recognized, for the first time, as creating an "Industrial Revolution."[23] The political scene was also unprecedented. The memory of the Revolution was still fresh. Its historical significance was still up for debate. Was it an aberration to be wiped from memory (which the very term *Restoration* seems to imply)? Or was it the precursor of a coming liberal state, or of a "juste milieu" between monarchy and liberalism? Which aspects of the eighteenth-century legacy of political, scientific, and artistic experimentation should be retained, and which should be left behind? Republicanism thrived as an underground movement that frequently erupted in the light of day: radical conspirators, or Carbonari, plotted new revolutions faithful to the ideals of liberty, fraternity, and equality—even as these terms were themselves reinterpreted. Did equality imply the redistribution of property and wealth? How would the freedom of owners to pursue their interests be balanced with that of workers? Could brotherhood overcome emerging class divisions?

These were historical questions of daily concern for the "children of the century," those individuals born around 1800—a group whose members increasingly imagined themselves as a "generation." In the opening pages of *La confession d'un enfant du siècle* (1836), Alfred de Musset lamented: "We were born into a world in ruins, with no glory or wars to die for. Despair was our only religion." The Restoration was a false, partial solution to the destruction left by the Revolution, a retrograde grasping for the lost certainties of king and church. The anomie that de Musset described as characteristic of his generation has been attributed, in sociological terms, to the absence of institutions to house the new generation: for the educated offspring of the commercial or professional classes, opportunities for advancement and sources of patronage were suddenly blocked as a result of the Restoration's institutional reshuffling in favor of the nobility and the clergy.[24] This was a generation that saw itself as having been thrown back on its own resources.

Yet what truly marked the mechanical romantics as a group was not their similar social origins, nor a fixed institutional context—in fact they tended to intersect only episodically, in varied combinations and diverse sites—but instead their closely overlapping horizons. They shared a keen sense of the need to unify the domains of knowledge, society, and nature itself. Society was yet to be made; new kinds of social relations and institutions had to be invented. To be sure, an analyst committed to a classic sociology of ideas would consider their ideas as reflections of their social location. What I wish to emphasize instead is that against the backdrop of a highly unstable social order, these individuals shared a forward-looking intention to change the social playing field; and to an extent that has been little appreciated, these "romantics" saw science and technology as means of building a more just, free, and harmonious society.

Much of their work was stimulated by the apparent moral vacuum and disunity of the present, and it led to reflection on the nature and basis of the social bond. A source for many of these reformers, perhaps paradoxically, was reactionary thinkers who emphasized a need to return to Catholicism and to submit the "temporal power" to the "spiritual power" as these categories were theorized in the Middle Ages. For some this meant recovering and reinterpreting the Catholic tradition or rediscovering mystical, illuminist traditions.[25] But the key idea of the neo-Catholic Joseph de Bonald—that there is an inevitable unity between ideas and institutions—was also taken up by those with no sympathy for the dogmas of the Vatican. A new spiritual power was needed, but one that incorporated the changes in worldview brought about by the previous two hundred years of discoveries in science: religion had to make a place for science. Paul Bénichou's *Le Sacré de l'écrivain* depicted romantic literature as an attempt to find a new affective and intellectual basis for social unity. There was also a *"sacré du savant"*: scientists took on the task of providing guiding methods and concepts for social transformation, both through public presentations of nature and through the emerging social sciences. In the religious imagination of this period, machines of various kinds were presented as sacraments for spiritual reunification and renewal.[26]

These projects of metamorphosis drew inspiration from the new ideas arriving from Germany, as presented by de Staël, Cousin, Humboldt, and Sainte-Beuve; aspects of Enlightenment-era thought returned to France in a German wrapping.[27] German romanticism also gave a boost to a newly emotional Catholicism and a mystical religiosity visible in both the revival of mesmerism and the "mystical school of Lyon," many adherents of which made their way to Paris. At the same time, considerable attention was paid to English social thought. The English constitution was discussed as an example of how to balance central power, tradition, and popular will. Even more influential was British political economy, as in Smith, Malthus, or Ricardo—a set of theories that seemed well suited to guide a society on the path toward industrialization.

It would be a mistake, though, to read this book as simply the history of the influence of German romanticism or English industrialism on France. Parisian traditions, modes of sociability, and historical concerns provided a unique and enabling atmosphere for the works of Kant, Goethe, Schiller, Schelling, and others, as well as for Watt's steam engine and the political economy of Malthus or Babbage. This alchemical conjunction of romanticism and mechanism depended on ingredients peculiar to France: its Catholic past, revolutionary history, and recent Restoration; Napoleon's technocratic empire; the nation's extreme centralization; Paris's intense and stratified social life of salons, cenacles, and cliques; and the prevalence of the theater as an organizing trope and exemplary experience, in contrast to the pastoral emphasis found in German and English romanticism.[28] The aspects of Parisian culture of this period that made it the emblem and pivot for modernity—the "capital of the nineteenth century"—arose from this singular combination of circumstances.

ROMANTIC MACHINES

In familiar images of romanticism, plants or animals frequently played the role of cosmic symbol, representing the universe as a growing unity. This organicism contrasted with the notion of the "clockwork universe" that was made prominent in the mechanical philosophy of the seventeenth century. This philosophy had its own cosmic symbol: an ideal type of mechanism we might call the "classical machine." Exemplified by the clock, the lever, or the balance, the classical machine was identified with primary qualities of mass, position, and velocity; it was seen as a passive transmitter of external forces, as a symbol of balance and eternal order. It implied a stable, determinist nature and was associated with an epistemology of rational, unfeeling detachment.[29] It was against this classical machine—and its implication of a lifeless, unchanging agglomeration of points and forces—that romanticism exalted the spontaneity and holism of the organism.

Yet in the early nineteenth century, the distinction between a machine, as that which is moved by an external force, and an organism, as a system whose motive force is internal, often broke down. The exemplary machines of the romantic era, powered by steam, electricity, and other subtle forces, could be seen to have their own motive force within them; they were presented as ambiguously alive.[30] Such mergers were suggested by the concepts of *forces vives* and "potential energy," by the fluids shared by physics and physiology, and by a growing recognition of the fusions and exchanges of properties that took place between humans and their tools. In contrast to the clock, the lever, and the balance, the devices that were most visible and celebrated during the Restoration and the July Monarchy—steam engines, electromagnetic apparatuses, geophysical instruments, daguerreotypes, and industrial presses—suggested a new ideal type of mechanism: the "romantic machine." In the early nineteenth century, machines were not exclusively associated with detachment, rationality, and fixity, but with the conversions, imaginative flights, and metamorphoses of the fantastic. They drew forth invisible powers, converted them, and put them to use. Unlike the classical machine, the romantic machine did not stand alone; it involved the active participation of the observer and articulated a spontaneous, living, and constantly developing nature; it produced aesthetic effects and emotional states.[31] From a certain angle, the romantic machine might still be seen as the embodiment of instrumental rationality, as an agent of the deadening and alienating routines of modern life. Yet from other angles—as explored throughout this book—it was imbued with the aesthetics and the affects of the organic, the vital, and even the transcendent.

Because machines and mechanism have frequently been seen as expressions of rationality and even modernity, the shifts in emphasis brought by the romantic machine also imply shifts in key arguments of critical theory, including those of Marx, Weber, and Benjamin. In their writings after 1848, Karl Marx and Friedrich Engels defined their own scientific socialism as the opposite of the reveries of "utopian socialists." In *The Manifesto of the Communist Party*, they ridiculed the schemes of the Saint-Simonians and the Fourierists as "fantastic," using the word five times in just two pages; these predecessors failed to live up to the standards of a rigorous social science. Similarly, in *Capital*, Marx sought to debunk the "phantasmagoric" illusions of "commodity fetishism": consumers attributed magical powers to inert objects only because the real material and social relations that produced them were hidden.[32] Yet Marx's thought was shaped not only by German idealism and English political economy but by the riotous intellectual ferment of Paris in the 1840s; and many of his arguments, not least those in his 1844 writings—on the human "species being," which externalizes itself by transforming its environment and realizes its essence through varied creative activity—closely resemble those of the mechanical romantics discussed here. Furthermore, this period's fetishization of new industrial objects, whether novel commodities or the machines that made them, did not necessarily entail the concealment and forgetting of labor. As is shown on the following pages, the fetishism directed at technology and its products in the 1830s and 1840s in fact accompanied a heightened visibility of labor: magical powers were attributed to machines even as the points connecting them to the earth and humans were highlighted and celebrated. The visibility of machines, of work, and of workers in producing wealth, knowledge, and sublime effects did not diminish the luster of their products. A celebration of the fantastic aspect of machines went along with the recognition of the concrete social relations they expressed and maintained.

Furthermore, although "fetishism" was frequently discussed at this time as an error to dispel, it was also presented as a valuable social force. Auguste Comte's writings in the 1840s took a deliberate turn backward to fetishism and the view that inert objects have moral and spiritual powers, as a means of humanizing science and unifying society.[33] Similarly, Barthélémy-Prosper Enfantin argued that the origin of the arts and sciences lay in certain sensitive individuals' ability to empathize with the feelings and desires of nonhuman life and inert matter, a view he summarized by declaring: "We must follow the lead of the imaginary; there is a lesson to be drawn from the fantastic." The investment of inert matter with sensation and spiritual powers resonated with romanticism's enthusiasm for folk tales and symbolism that imbued everyday life with a mythical dimension, orientations that contributed to what Cathy Gere calls "prophetic modernism," a recapitulative mode of history that looked to the past as a means of anticipating a more wholly human future.[34] Of course, this was also the moment of the solidification of a faith in progress through science and industry. Although usually studied as opposites, these exactly contemporary cultural formations—a return to a mythical past and faith in a rational future—intersected in the figure of the romantic machine: a concrete, rational, often utilitarian object that was nevertheless endowed with supernatural, charismatic powers.[35] Just as the romantic-era concept of the self championed the imagination as a power that could transform the world, these charismatic technologies were seen as capable of awakening obscure forces of nature and fixing the coordinates of technical systems that grew into living, dynamic wholes.[36] To mangle together two of Weber's slogans, this period witnessed a *routinization of enchantment*; in its fascination for the uncanny aspects of machines and mechanical procedures, there was a complementary *enchantment of routine*. Science and technology had arrived at means of manipulating nature's hidden powers so successful that they no longer needed to be kept secret.[37]

The response to technology in this period thus confounded familiar oppositions: fetishism and scientific truth; magic and mechanization; charisma and instrumental rationality. Walter Benjamin's discussion of "the aura" of a work of art offers insight to such doublings. In "The Work of Art in the Age of Mechanical Reproduction," he spoke of the aura as a "nearness in a distance," explaining the concept with reference to a poem of Novalis that described a landscape that seemed to look back at a human spectator. For Benjamin, such an encounter was the paradigmatic experience of aura: "the transposition of a response common in human relationships to the relationship between the inanimate or natural object and man." In other words, "To perceive the aura of an object we look at means to invest it with the ability to look at us in return."[38] Benjamin read this experience dialectically, as the fusion of two contradictory movements. On one hand, the technologies of the early nineteenth century hastened the "decline of the aura," as exemplified by the case of staring at a camera, an inert object that cannot return the viewer's gaze. Likewise, the violent shocks brought by the new urban environment—noise, crowds, construction, and speed—deadened sensibilities, dimming even the intersubjective aura created by glances between humans: just as the increasing mechanization of industrial production transformed humans into machines, so did the city dweller come to see fellow humans as mere objects. Yet on the other hand, in the "Work of Art" essay, Benjamin held out a hope that certain uses of machines might actually produce auratic experience: in the cinema in particular, with its scenes of crowds and its close-ups, where the viewer is seen and sees herself as part of the scene, without hiding the artifice involved in producing such images. Here he referred again to Novalis, whose unfinished bildungsroman *Heinrich von Ofterdingen* described a young man obsessed with "the blue flower," an obscure quarry associated with an earlier state of humanity, one to which he longed to return: "I have heard it said that in ancient time beasts, and trees, and rocks conversed with men. As I gaze upon them, they appear every moment about to speak to me; and I can almost tell by their looks what they would say."[39] Benjamin wrote of the cinema as "the blue flower in the land of technology"; the early romantic desire for union with nature might be attained by machines.[40]

Similar notions appeared in the *Arcades Project*, Benjamin's fragmentary work on nineteenth-century Paris.[41] While attentive to the ways in which human values and experiences were transformed and often anesthetized by mechanization, he also recovered less familiar developments, such as Charles Fourier's cosmological predictions—seas turned to lemonade, new helpful species coming to the aid of mankind, copulating planets—which Benjamin read as distorted, dreamlike perceptions of the powers unleashed by scientifically directed industry. This was a version of the dream, expressed in Baudelaire's "Correspondances," of nature as a temple whose living pillars murmur at us, a forest of symbols watching us with "familiar glances," one also evoked in Gérard de Nerval's pantheistic *Vers dorée*: "A mystery of love rests in metal; 'Everything senses!' And everything acts on your soul."[42] For Benjamin, the experience of inert matter as alive and conscious was fed by memories both from the promiscuous identifications of individual childhood and from the childhood of humanity, when the "primitive horde" experienced itself as part of a divinely animated nature. Such memories might also look forward, fueling revolutionary desire for a future state of wholeness; they were awakened by the new powers of industry. Even as industrialization destroyed the aura, it offered new possibilities for its return.

Echoes of Benjamin's readings of this period should be audible throughout this book, and not only where it deals with topics Benjamin made his own such as Grandville's chimeras and Balzac's "physiologies." Nevertheless, an important nuance must be noted. Benjamin found the central element of auratic experience—the sense of an inert object coming to life—in the nostalgic ruminations of poets and in the neglected ravings of visionaries; the possibility of its wide-scale reactivation had to wait until the early twentieth century and the arrival of the cinema. This book's primary focus, in contrast, is on scientists, engineers, and social philosophers and their attempts to understand and manage the complexities of nature. Rather than marginal speculations, idle musings, or idiosyncratic dreams buried in the abandoned shops of dusty arcades, theirs were projects at the center of widespread social concerns, projects that contributed directly and concretely to the scientific and technical infrastructure of the modern world. In these works, as well, we find a recurrent fascination with animated matter, lifelike machines, the creative forces of "naturing nature," and even pantheism. Benjamin's detection of a dream image of nature alive with dynamic, transformative powers was confirmed not just by the works of extravagant speculators like Fourier or artists like Grandville, nor only in works of fantastic literature, visual arts, and music, but even—and most consequentially—in the writings of prominent physicists, astronomers, biologists, and engineers.

At the dawn of industrialization in Paris, the notion of using technology to allow humans to participate intimately with the nonhuman world was widespread both in the arts and in the sciences. Fantastic longings comparable to those of Novalis's "magical idealism" thrived in Paris: in the sciences of dynamic and convertible powers associated with electricity and steam, in the philosophy of positivism, and in the demiurgic aspirations spurring both research and industry. The external world was transformed as if by magic into a domestic space, a temple for worship, a reflection of the mind.[43] In these writings, machines themselves—rather than destroying aura or hastening the disenchantment of the world—were granted an uncanny power to animate the inanimate, to emancipate and spiritualize "vibrant matter."[44] The powers of new technology triggered aspirations toward an intersubjectivity that would embrace more than just humans; they lent support to the view that all elements of the world participate in a single living, intelligent, and perhaps divine substance.

PROGRESS OR DECLINE?

This book's case studies demonstrate that the stark divide usually assumed to exist between the mechanical and the romantic—between mind and nature, objects and subjects, science and humanities—is not the only possible way to organize a world in which science and industry play a central role. Many recent scholars have likewise enriched our awareness of the contradictions and paradoxes inherent in romanticism. Although romantic poets and artists often featured pastoral and rural scenes, remote cliffs, sublime mountains, and mazy rivers, they were immersed in contemporary developments in politics and science; they did much of their business in the city, and the nature over which they rhapsodized had been long occupied and shaped by humans.[45] Further, historians of science have noted how romantic themes, approaches, and attitudes contributed to the life sciences and debates about the possible harmony between teleological and mechanical views of nature. The connection between the romantic quest for unity and nineteenth-century interrogations of work, conversion, and energy has been a recurrent theme in the history of physics. Concepts from industrial economics have been recognized as the basis for holistic thinking about organisms, as in Milne-Edwards's "division of organic labor."[46]

This book builds on such work, extending it into previously neglected domains. In doing so it also directly confronts an obstacle to the study of "romantic science." Because the progress of science and of civilization more generally has been measured by mechanization, scientific projects that smacked of romantic aesthetics, emotion, and epistemology—assumed to be antimechanical—have been largely seen as a dead end. One exception has been the life sciences, where romantic conceptions of organism and environment are increasingly recognized as crucial for evolutionary theory. Yet the central role played by romantic views of the convertibility of forces and the underlying reality of "energy" in the development of nineteenth-century physics remains underappreciated, and especially so in studies of French science of this time. Equally neglected are the connections between these romantic concepts and Enlightenment thinking about imponderable fluids and milieus and debates about animal magnetism.[47] This book aims to fill those gaps. It also traces the impact of romantic philosophy on epistemology, especially the growing recognition of the role played by human senses and activity in producing phenomena, and broadens the concept of "romantic science" to include technology.[48] It shows that in France, romantic themes were not retrograde tendencies working against the progress of science and industry. Instead, romantic ideas provided collective goals, conceptual resources, and an emotional intensity that proved decisive for building France's technical infrastructure. In concrete and significant ways, romanticism structured, inspired, and gave direction to mechanical science and industrial development.

These points run contrary to received wisdom not only about romanticism's impact on science but also about the progress of French science in this period. According to an influential thesis of sociologist Joseph Ben-David, Napoleon's empire was a high point for French science; its mathematical and engineering training was the envy of Europe and a contributing factor to military dominance. Then, during the Restoration and the July Monarchy, French science went into decline, as scientists' attention was drawn to politics; the promising yet somehow disappointing career of astronomer-turned-politician François Arago is often cited as a case in point.[49] From this point of view, science declined because it was politicized—a predictable result of the violation of the norm of disinterestedness laid out by sociologist of science Robert Merton. Yet one problem with this explanatory scheme is that under Napoleon, science was at least as politicized as it was in the Restoration and the July Monarchy; the difference, of course, was that there was no successful opposition to its imperial co-optation: this was science in the service of a centralized, rigorously hierarchical, top-down state. Again, in the aftermath of the period we are discussing, in the Second Empire, science was again politicized: the choice to replace the republican Arago with the man of order, Leverrier, as director of the Observatory was a deliberate decision to elevate savants sympathetic to the emperor. Thus the problem for proponents of the "decline thesis" seems not to be with politicized science, but only with science's association with political views that oppose centralized hierarchies.

Further, historian Matthias Dörries has shown how the argument of a "decline of French science" had been loudly proclaimed in the wake of the French defeat by Prussia in 1871: the nation's military failure was attributed in large part to Germany's successful mobilization of science and technology.[50] Yet this insistence on the "decline thesis" was itself a repetition of one advanced immediately after Waterloo, although the primary rival at that time was England; reforms in scientific education and state support for industry were launched to close this gap.[51] A similar argument was soon made in England about British science, by Charles Babbage, who saw the Society of Naturalists and Natural Philosophers founded by Humboldt and France's Ecole Polytechnique as models that Britain would do well to imitate.[52] Indeed, claims of "decline" and "backwardness" bounced around Europe and across the Atlantic throughout the nineteenth and twentieth centuries, contributing heavily to the international arms races to which so much of "the progress of science" is due. The close connection between claims of scientific decline and the fear of military defeat should make us pause before taking the decline narrative at face value; at the very least, it should force us to ask just what we mean by scientific progress and to ask how it ought to be measured.

Yet even if we remain within the usual measures of scientific progress—for example, the number and significance of discoveries and the fecundity of research programs—a strong case could be made that French science, far from declining in this period, remained robust and influential. The approaches in physics that contradicted Laplacean mechanism (Fresnel's optics, Fourier's thermology, Ampère's electromagnetism) contributed massively to the syntheses of the second half of the century; the fact that the young William Thomson and James Maxwell read deeply in French mathematics and physics (and that Thomson took a Parisian pilgrimage to the laboratory of Regnault) testifies to the continued high esteem in which French mathematics and physics were held in the 1840s.[53] The connections made in Paris among steam engines, work, and political economy were decisive for the subsequent development of thermodynamics and, more generally, for the concept of energy. Further, the impact of Etienne Geoffroy Saint-Hilaire's transmutationist comparative anatomy on German and British medicine, physiology, and evolutionary thought has begun to be recognized; Charles Lyell likewise sat at the feet of Elie de Beaumont before assembling his geologic tracts; Comte's initial success outside of France can be attributed in large part to the high value placed abroad on the Ecole Polytechnique.[54]

A rich secondary literature exists on science in the Revolution and the empire; yet the cultural history of French science in the three decades before 1848 remains obscure. To individual studies of scientists and institutions of the late Enlightenment and the "second scientific revolution" have been added studies of moral codes of scientific institutions and the rhetoric of patronage, revealing science's role in the establishment of a stable, disciplined, "bourgeois" order.[55] These accounts, although illuminating, often imply continuity between the revolutionary and imperial drive for new institutions and the mechanistic materialism and rigid disciplinary divisions that took command after midcentury in the Second Empire. What often falls out, or becomes a mere curiosity, are developments in the sciences— particularly during the Restoration and the July Monarchy—that deviate from a smooth transition from the late Enlightenment and the Revolution to the state-centered, university-based science that flourished under the "high capitalism" of the second half of the century. There has been a tendency to follow the historians of the Second Empire—with their admiration for hierarchy and tradition—in writing the republican and romantic opposition out of this history, making it difficult to see the intellectual, social, and political coherence that existed among those scientists who opposed the established order. Entranced by the self-perpetuating significance of the institutions of royal science, we have overlooked the nomadic social organization of oppositional scientists—the cliques and *cénacles*, the cells, committees, and banquets—that fostered the reformist and revolutionary spirit and opposed monarchies throughout this period.[56]

For example, we know a great deal about Cuvier, whose drive for stable order in classification systems, in organisms, and in governments accompanied a dexterous affiliation with whatever government happened to be in power; we are much less familiar with his opponent, Geoffroy Saint-

Hilaire, whose archetypal, historical, and materialist approach resonated with romantic philosophy and republican politics, in the press and in salons outside of the central institutions of science—and whose transcendental anatomy was arguably more influential for Darwin than Cuvier's new classification system was. Likewise, studies of this period have focused on Pierre-Simon Laplace and his followers (notably Poisson and Biot), who aligned their physics with the stabilities of imperial order. While it is clear that the unified program of Laplacean physics disintegrated after the empire, we know little about the broader directions of research shared by "anti-Laplacean" physicists or the ties that bound them to romantic-era philosophy, natural history, literature, and politics. In philosophy, we have recently learned a great deal about the ways in which Victor Cousin made himself indispensable to the pedagogical institutions of the July Monarchy and beyond, offering a philosophy based in large part on German idealism and a conception of the self suited to the ruling party and its ideology of individualism; yet less is known of the ways in which his philosophy was opposed outside of state institutions, not only by phrenologists and positivists, but also by the influential and neglected romantic socialist Pierre Leroux.[57] Further, it is only very recently that historians have begun to recognize the vitality and political importance of popular science in the Restoration and the July Monarchy.[58]

A reason for this neglect is that conflicts between the scientists who affiliated themselves with the established order and those who challenged it did not always take place within the familiar halls of science. Their conflicts in fact sought to redefine where it was that science took place and who could participate. Crucially, they also tried to redefine science's connections with other parts of society. The period covered by this book reveals huge deviations from the assumption of an uninterrupted transition between empires. We will see that, instead, reformist, revolutionary, and "romantic" turmoil encouraged important scientific and technical advances. Romanticism in France around 1830 became obsessed with social reform, depicting the new situation of industrial workers in the rhetoric of republicanism and revolution. Scientific and technical fields touched by romanticism were put to use for projects of political transformation, whether republican, as in the case of Arago, or socialist, as with the Saint-Simonians. The movement of such ideas was facilitated by the close personal and social relations among scientists, journalists, poets, composers, and, increasingly, workers. Romantic science and technology contributed to the utopian visions of the 1830s and 1840s and to the workers' revolution of 1848.

The failure of these utopias led to their historical erasure.[59] The same fate was often met by the scientists most closely allied with these utopians. To grasp the significance of the scientific and technological developments before 1848, therefore, we need to consider action taking place outside the central institutions of the Restoration and the July Monarchy; we have to heed voices that challenged the center from the outside as well as those that challenged the political order and conventional wisdom from within. These voices of opposition briefly became the center during the Revolution of 1848 and the ephemeral Second Republic.

PLAN OF COMPOSITION

This book unfolds in a range of settings, including classic sites of French science: the academies of science and medicine, the Ecole Polytechnique, the Muséum d'Histoire Naturelle, and the Conservatoire National des Arts et Métiers; the salons of various literary cenacles and social cliques, from Nodier's reunions at the Arsenal to Madame de Krudener's mystical gatherings and Julie Récamier's convocations of the romantic and liberal literati. Many of its characters were part of what has been called the *France des notables*: a microsociety of about ten thousand people making up the elite of finance, politics, and culture, combining members of the traditional aristocracy with the new aristocracy of wealth and accomplishment. Yet, thanks in part to the dense social ecology of Paris, these largely closed, elite settings were permeable to other spaces and populations.[60] In the capital, people from all classes, all regions of France, other nations, and overseas territories were thrust into constant contact. In the early nineteenth century, the city grew and changed, increasing the number and variety of encounters on the streets and in restaurants, cafés, and reading rooms; in public lecture halls; in the offices and workshops and on the pages of new journals; at the Paris opera and the nearby vaudeville theaters; and in the wooden galleries near the Palais Royal (home of publishing, inexpensive entertainment, and prostitution). The city was the setting for an endless series of paradoxical encounters: Arago meets Hugo on the Allée de l'Observatoire, while Balzac eats lunch with Arago's son on the other side of the building; the Saint-Simonians move between the Ecole Polytechnique and the Palais Royal; George Sand frequents the opera, the vaudeville, and the salon of Geoffroy Saint-Hilaire near the museum. Paris itself, with its complex and changing social ecology, must be a central character in any investigation into the intersections among French science, arts, and politics of the nineteenth century.[61]

Against this shared urban background, the case studies of this book are divided into three parts. The first deals with physical scientists obsessed with the imponderable fluids of physics, the second explores technology's impact on theories of the self and the human by concentrating on the fantastic arts and public spectacles, and the third shows how these themes informed religiously inflected projects of social and natural transformation. In each part, a profound aspiration toward unity is evident: developments in the natural sciences were recognized as having social and aesthetic import; the arts depended on the sciences and were given a political vocation; and projects of social reform began by systematizing the sciences and recruiting arts to their causes. Further, each chapter shows how the romantic aim of integrating all aspects of humanity into a living synthesis depended upon technology. As the chapter titles suggest, each concentrates on one machine, instrument, device, or system that either appeared in this period or was newly recognized as decisive: electromagnetic experimental apparatus and geophysical instruments, daguerreotypes, steam presses, orchestras, operas, automata, steam engines, typesetting machines, and calendars, as well as the politically charged spectacles of the Second Republic. Although there is considerable temporal overlap, the chapters follow a roughly chronological order, from Ampère's research on electromagnetism in 1820 and his classification of the sciences, to the cultures of popular and scientific spectacle around 1830, to the utopian projects that followed the turn toward social and political reform leading up to the Revolution of 1848.

The first part looks at scientists known for their contributions to physics, geophysics, and astronomy; for each, the model of precision experiment and mathematics associated with Laplacean mechanism was enlarged and transformed by an encounter with romantic philosophy and aesthetics. Physicist André-Marie Ampère, the focus of chapter 2, developed experimental devices to demonstrate the equivalence of electricity and magnetism and to measure their dynamic force; adopting notions from the *Naturphilosophie* of Hans Christian Oersted, he conceived of electromagnetism as an ambient ether that propagated heat and light in waves. He set his research within a philosophy of knowledge formed in dialogue with the introspective philosopher Maine de Biran. For Ampère, knowledge was rooted in feeling around in the invisible: it arose from the regular, and frequently technically mediated, encounter of the active subject with "noumenal" points of resistance. As we see in chapter 3, interest in the environmental effects of electricity and magnetism were central to the geophysical research of Alexander von Humboldt, the Prussian explorer who made his home among Parisian savants. Humboldt's geophysical research, as presented in his carefully staged images of natural landscapes, borrowed heavily from Kant's and Schiller's conceptions of beauty. The geophysical instruments he used to measure the global distributions of heat, humidity, electricity, and terrestrial magnetism were animated, personified extensions of himself—willing assistants in the production of an uplifting image of the cosmos. A notion of knowledge as the outcome of combining human activity with precision instruments also informed the "labor theory of knowledge" embodied by the subject of chapter 4, the astronomer and politician François Arago, Humboldt's close friend. Unlike his predecessor Laplace, whose view of mechanism emphasized detachment, instantaneity, and obedience, Arago presented both instruments and citizens as dynamic mediators who modify the forces they transmit. His expressive personal style and his advocacy on behalf of workers and inventors—as shown in his support of the daguerreotype, an instrument that harnessed the active force of light—offered a bold contrast with Laplace's imperial mechanism.

The second part focuses on popular spectacles in which new scientific discoveries in optics, mechanics, and natural history occupied center stage. Chapter 5 considers new technologies for visual and auditory illusions: panoramas, dioramas, the "fantastic" symphonies of Berlioz, and Meyerbeer's hallucinatory opera *Robert le diable*. Scientists were often consulted for assistance in staging these works, and philosophers' reflections on their uncanny effects contributed to new theories of perception in which the perceiver's memory, habits, and sense organs were seen to play an important role in "creating" the experience. Maine de Biran was the source for many of these "physiospiritual" theories; the chapter ends with Balzac's reading of *Robert* as a reflection of technology's power to produce the transcendent. Chapter 6 examines issues raised by two popular displays in this period—the Museum of Natural History and the Expositions of the Products of National Industry—where the encounter with orderly displays of animals and machines prompted reflections on the origin, behavior, and capacities of humans. Debates about the fixity or mutability of species at the museum were inseparable from metaphysical and theological debates over materialism, vitalism, and the existence of the soul; these controversies also attached to images of machine-humans, or automata. Intersections between biological metamorphosis and demiurgic engineering coalesced in Robert-Houdin's magic shows and in Grandville's *Another World*, fantastic entertainments whose serious play superimposed technology and life.

Part 3 deals with the ways in which reflections on the proper relations between humans and machines fed into the utopian and revolutionary imagination of new social philosophies. In chapter 7 we see how imagery from the physical sciences and engineering—productive forces, conversions, active fluids, and networks—furnished the followers of Saint-Simon with material for their visions of a new world in which rewards and power would depend on individuals' contributions to society. A close connection can be seen between the Saint-Simonians' theories of conversion in industrial mechanics and their efforts to bring followers to their "New Christianity." Chapter 8 presents the philosophy of the former Saint-Simonian Pierre Leroux, one of the coiners of the term *socialism*. A printer and a literary critic, Leroux saw the printed word as a technological communion; the "pianotype" he invented allowed typographers to read while they set type, making authors and printers the engineers of a new collective consciousness. His view of "Humanity" as an ideal, virtual being realizing itself in increasingly perfect social and intellectual orders was an application of the philosophical anatomy of Geoffroy Saint-Hilaire to the realm of society. Chapter 9 focuses on Auguste Comte, founder of positivism and sociology. Rather than a sterile denial of emotion and imagination in favor of facts, Comte's philosophy was one of the most inspired, fantastic solutions to the political crises of the postrevolutionary period. A new "spiritual power" in the hands of scientist-priests would present the heterogeneous fields of knowledge as a coherent dogma and provide a framework for the technical development of the human milieu. Comte's "hierarchy of the sciences" and "Positivist Calendar" were mnemonic devices to guide the temporal coordination of the diverse time scales of nature and society.

The protean energies tapped and amplified by new machines were woven into utopian visions of abundance, freedom, and harmony; these aspirations exploded with the workers' revolution of 1848. The conclusion examines the impact of 1848 and the Second Republic's paradoxical presentations of machinery and labor. These include the Festival of Industry of 1849, the pendulum experiment of Léon Foucault, and Meyerbeer's opera *The Prophet*. In each of these spectacles, machines continued to serve as the bearers of prophetic hopes for an egalitarian society. Yet they were overshadowed by a coming reversal. The view of machines as liberating and integrating could easily give way to a more familiar and ominous view in which machines were instruments of murderous analysis and suffocating repression.

These chapters suggest routes that the industrial West might have taken if not for the political reaction and the hardening of social, institutional, and disciplinary boundaries signaled by Louis-Napoleon Bonaparte's coup d'état of 1851, which launched the Second Empire. Yet the book ends on an optimistic note with a discussion of notions from this period that are relevant for today's still unsettled relations with technology: the vision of nature as mutable, but only within limits; the recurring notion of the human as "the technological animal"; the embrace of projects at the medium scale; and the importance of putting human consciousness back into the world picture. Without denying the risks posed by unchecked technological development, the message of balance between human activity and the demands of nature found in the forgotten futures of the first half of the nineteenth century offers a helpful point of comparison for today's ambivalent—often apocalyptic—views of science and technology.

RECYCLING ROMANTICISM

The principal human subjects of this book were well known in their lifetimes, and their accomplishments had a significant impact on subsequent generations. This book is not a nostalgic record of the dreamtime of a modernity whose brightly lit day was later defined by materialism and reason. Taken individually, many of the innovations studied here helped establish the scientific, technical, artistic, and conceptual landscape of modernity; taken together, the projects with which they are associated add up to a coherent if neglected orientation toward the industrial world.

As we will see, however, the prophecies of the 1830s and 1840s were thwarted by the authoritarian reaction of Louis-Napoleon Bonaparte in 1851.[62] It might therefore be possible to read mechanical romanticism as the soft preparation for subsequent hard realities, easing the passage from Enlightenment in its most blatantly militaristic, totalitarian form (Napoleon) into rigidly hierarchized, mass-scale industrialization (Napoleon III). Yet this book is not primarily a story about good intentions gone wrong, or about the sinister underpinnings of seemingly liberatory projects. It does not simply unveil the mechanical aspects of what we think of as inspired or the irreducibly speculative, perspective-bound and affect-laden aspects of what we take as objective and deterministic. Instead, this book offers the mixtures of the 1830s and 1840s as anticipations of, and perhaps as propositions for building on, developments of the present. The railroad, the steam press, the telegraph, and the daguerreotype were welcomed with fears and hopes akin to those we have seen intensified in recent years. We live in a world of technically enabled connectedness that offers both nightmares (enhanced surveillance, relentless global warfare, accelerated consumerism, and environmental destruction) and utopian aspirations (new alliances among global multitudes, new techniques of expression and discovery, biotechnology, and green development). The constellation before 1848 in many ways prefigures today's "new" mergers of nature and machine. As at that time, many of today's approaches to technology, although acutely aware of the havoc wrought by unchecked industrial production and consumption, go beyond the self-defeating naïveté, Luddism, or nostalgia of earlier liberation movements by building technology into their utopias.[63]

Yet current forms of technophilia often come in a package of consumerism and individualism: an activist impulse is routed into stocking nucleated domiciles with elegant, minimalist technologies and "whole" foods of questionable origin. In contrast, the mechanical romanticism of inter-Napoleonic France kept constant sight of the changes that technology brought not only to nature but to the entire social fabric and sought to guide these transformations toward the good of all. In many cases this meant coupling an ethics of individual self-determination with the recognition of individuals' dependence on society and nature, with mutual respect between humans and nonhumans. Rethinking technology meant rethinking the basis of the social bond and the order of the universe and, potentially, living very different lives. Updated to the present, mechanical romanticism suggests that even if solutions must be small and local, they require a conceptual and aesthetic frame that is deep and wide.

The following pages trace the outlines of alternative modernities in which scientists, engineers, poets, painters, composers, philosophers, and politicians cast the powerful new technologies of the early industrial age as the basis for social arrangements that differed from both the retrospective organicism of the parties of order, tradition, and stability and the possessive individualism of the liberals. They proposed paths beyond the stalemate between hopeless romanticism and soulless mechanism, making technology and science into instruments of inspiration and even salvation, in which experiments with external nature and internal subjectivity offered practical sources of hope. These modes of thinking through the proper division of goods, spaces, and actions offer critique as well as serious alternatives to the fragmentation, destruction, and alienation found on the paths actually taken by modern society in the past 150 years.

What follows, then, is not just an account of what happened in the course of a few decades nearly two centuries ago, but the detailing of possible worlds which may be of use—in a new historical juncture marked again by transformative technologies, along with the failures of high modernity and an atrophied sense of progress—in conceiving of futures still to come.

PART ONE

DEVICES OF COSMIC UNITY

2

Ampère's Experiments

Contours of a Cosmic Substance

ANOTHER GERMAN REVERIE?

In 1820 a scientific discovery was made that eventually transformed worldwide communication and industry, making telegraphs and dynamos possible. It also appeared to prove the romantic speculations of *Naturphilosophie* right. This was the dynamic force of electromagnetism.

In the Paris Academy of Sciences on September 9, 1820, François Arago reported a finding of the Danish natural philosopher Hans Christian Oersted, a follower of Schelling: when a magnetic compass needle was moved above or below an electrified wire, the needle would turn, at right angles to the current. This discovery directly contradicted a fundamental assumption of "the standard view" in physics at the time: that magnetism and electricity, along with the other imponderable fluids, light and heat, were distinct and independent. Distrust of the romantic science with which Oersted was associated led many in France to dismiss the report as "another German *rêverie*."[1] But the mathematician, chemist, and philosopher André-Marie Ampère, with help from his collaborators Arago and Augustin Fresnel—the leading supporters of the wave theory of light, against the standard, Newtonian view that light was made of tiny particles that traveled in a straight line—immediately set about reproducing and elaborating Oersted's unexpected effect. Over the next months Ampère conducted experiments that secured the basic principles of electrodynamics, culminating in a mathematical expression of the force of attraction and repulsion between two electrical currents. He continued to elaborate and defend his theories until his death in 1836.[2]

The interaction between electricity and magnetism has been seen as a "monster" in the sense discussed by anthropologist Mary Douglas: an unclassifiable anomaly falling outside of cultural categories.[3] Electromagnetism threatened to undermine important distinctions not only within physics but between the kingdoms of nature: between matter, life, and the mind. In Gaston Bachelard's expression, electricity in the early nineteenth century was a cosmic substance: the locus of moral, emotional, and symbolic values, forged in a cauldron of intuitions, attitudes, and expectations derived from the culture at large. For Bachelard a cosmic substance was necessarily an irrational object. Yet for Ampère, the figure who brought electrodynamics closer to what Bachelard would call the "rationally purified" form of a mathematical law, electricity's cosmological associations were not an irrational residue but a constitutive frame.[4]

Ampère's electromagnetic research has been understood in the context of the "second scientific revolution," in which the study of light, heat, electricity, and magnetism was given precise, mathematical expression through the use of delicate experimental devices and precise measurement. A central figure in this development was Pierre-Simon Laplace, the leading astronomer and physicist under Napoleon's empire, although a more recent strand of scholarship connects Ampère's work to the anti-Laplacean physics of ethers.[5] For many of Ampère's closest friends and collaborators, an interest in the ether as the medium of propagation of light and heat went even further, leading to hypotheses of the ultimate unity and convertibility among electricity and nervous, vital, or spiritual fluids. Those associates of Ampère viewed electricity, like Balzac's hero in *Quest for the Absolute*, as linked to life, thought, and spirit.[6]

Furthermore, a connection has been suggested between Ampère's research and German *Naturphilosophie*. Kenneth Caneva has shown Ampère's debt to Oersted, through both personal contact and the central importance both gave to electrochemistry and the conception of an "electrical conflict," a constant process of composition and decomposition between negative and positive electricity that lay behind the electric current. Oersted, several of whose works were collected under the title *The Soul in Nature*, embraced the project Schelling introduced in his *Ideas for a Philosophy of Nature*. Schelling urged scientists to recognize and demonstrate experimentally the "absolute identity" of mind and nature: "For what we want is not that Nature should coincide with the laws of our mind *by chance* (as if through some *third* intermediary), but that *she herself*, necessarily and originally, should not only *express*, but even *realize*, the laws of our mind. . . . Nature should be Mind made visible, Mind the invisible Nature. Here then, in the absolute identity of Mind *in us* and Nature *outside us*, the problem of the possibility of a Nature external to us must be resolved."[7] According to Schelling, the absolute identity of mind and nature could be proved, because mind and nature had the same origin: the underlying principle of the "World-soul." In striving to know itself, this primordial power divided into two opposed principles that Schelling called the infinite ego and the absolute ego; in a recurrent dialectical process of division and limitation, the infinite ego continued to divide itself to form the entities we recognize as the empirical world of nature, while the absolute ego unfolded itself until it took the form of discrete, individual minds.[8] The progressive elaboration of empirical nature took the form of increasingly specific and complex phenomena: from the dynamic opposition between attraction and repulsion that Kant, in his *Metaphysics of Nature*, had suggested was the foundation of matter, into light, vital air, magnetism, electricity, chemical cohesion, organic life, and eventually consciousness.[9] Because the dialectical development of empirical nature mirrored the dialectical development of the human mind—both in individual maturation and in the development of human civilization— it was possible for the artist, philosopher, or scientist to retrace material nature back to the underlying, primordial spirit that humans shared with it.[10] The physical sciences drew nearer to this identity by decomposing the observable world into increasingly fundamental principles. For Schelling, the recent discovery of electrolysis (the decomposition of compounds by the application of electricity) pointed in this direction, as did Oersted's discovery of the interaction of electricity and magnetism.[11]

It is generally taken for granted that Ampère saw electromagnetism as a material phenomenon distinct from mind and soul. Although this is indeed the case, this chapter suggests how strange this is. Like his contemporaries, Ampère sought to extend the study of electricity into geophysics, physiology, comparative anatomy, and psychology: it was for him a site of unification of wildly diverse phenomena. His refusal to identify electricity with thought or the action of the mind should not be taken for granted as scientific common sense; instead it is a stance that needs to be explained.

Ampère's theoretical and experimental explorations of electricity took place against the background of his lifelong philosophical obsessions, which also placed him centrally within the currents of romantic, post-Kantian science and philosophy. Much of his philosophical work was inspired by his exchanges with the "spiritualist" philosopher Maine de Biran. Yet in important respects he went beyond Maine de Biran's favored method, introspection. Like Oersted, Ritter, and other *Naturphilosophen*, Ampère designed ingenious and delicate experimental devices that allowed him to feel his way (*tâtonner*) in the realm of the invisible. These machines were continuations of the central notion of his epistemology: that knowledge of both the self and the world emerges through the experience of resistance. Yet in line with the norms of French physics, Ampère did not consider his research complete until the relations it established had been formulated as a general mathematical law.

For *Naturphilosophen* like Schelling and Oersted, electricity was a bridge between matter and mind, a manifestation of the soul that humans and nature shared. Although this was a bridge that Ampère did not cross, his work on electromagnetism, and the broader philosophical and epistemological project of which it was a part, was as romantic as it was mechanical. His science and philosophy offered a distinctive and historically illuminating set of means for understanding the soul's action in the material world. The conceptual malleability and ubiquity of the "cosmic substance" of electricity made it analogous to his efforts to form a classificatory system that would contain the entire cosmos: a single taxonomy uniting all fields of knowledge. The unifying power of electricity enacted in the order of things what his unified philosophy aimed to accomplish in the order of ideas.

THE "STANDARD VIEW" AND ITS DISCONTENTS

Parisian physics was dominated in the first fifteen years of the nineteenth century by Laplace and his supporters, notably Jean-Baptiste Biot and Siméon-Denis Poisson. One of the most powerful figures in French science, Laplace sought to extend the Newtonian regularity he had demonstrated in the solar system to the innermost recesses of matter, and he played a key role in "mechanizing" the practices of science, engineering, war, and the administration of the modern state. Under the empire, the physical and mathematical sciences were so highly prized that Napoleon had himself elected to the Institute (the body that replaced the Academy of Sciences during the Revolution); the geometers' emphasis on reason, number, and uniformity were imposed across Europe through the Continental System's axiomatic, centralized bureaucracy. Just as his colleague Lagrange had claimed that mathematics was complete, Laplace's *Mécanique céleste* was seen to have perfected Newton's system of the heavens. He also worked as a dutiful instrument of the state, creating a uniform system of measures and establishing national standards in the teaching of mathematics and engineering. Napoleon made him a marquis, and he maintained his pedagogical and institutional supremacy in a number of ways: as a teacher and administrator he selected candidates, set topics for prize competitions, and served as a patron for the highly influential Society of Arcueil at the home of his neighbor the chemist Berthollet.[12] Laplace's students combined virtuosic mathematics, precise experimental machinery, and practical applications, laying the theoretical and technological infrastructure for wide-scale industrialization and imperial expansion.[13]

The "standard view" on the imponderable fluids held by Laplace and others in the early nineteenth century, as described by John Heilbron, held that light, heat, electricity, and magnetism were independent, weightless fluids, which, despite their lack of mass, were subject to Newton's laws of attraction. They were assumed to consist of microscopic particles that repelled each other (hence the tendency of light to diffuse when unfocused) but which at a macro level had attractive powers acting at a distance. The imponderable fluids were assumed to behave in analogous ways: Coulomb's discovery of mathematically similar force laws for both electricity and magnetism was one example, as was the parallel between latent heat and electric potential. Despite these analogies, these fluids were understood to be independent and incapable of affecting each other. Further, following the model of Coulomb, it was assumed that the direction of the forces they exerted was rectilinear.[14] By 1815 Laplace's ideas were being developed by his followers, most notably Biot, who presented the Laplacean program systematically in his *Précis élémentaire de physique expérimentale*.[15]

Thus, for those who held to the "standard view," Oersted's announcement of an interaction between electricity and magnetism, and an unexpected perpendicular motion in the compass needle, was indeed monstrous. Yet the "standard view" was not universally shared either among physicists or among others concerned with the nature of matter. For many, Oersted's discovery suggested a confirmation of a relationship already assumed if not yet proved: that light, electricity, magnetism, and caloric might be simply modifications of some single, underlying principle.

The leading apostate from the gospel of Arcueil was François Arago, about whom we will learn more in chapter 4. Arago had been an outstanding student at the Ecole Polytechnique; he benefited heavily from Laplace's patronage in his early career, yet after Napoleon's fall, Arago led the charge to dismantle Laplace's program, championing the cause of physicists who suggested explanations of phenomena without reference to linear action at a distance. Strong encouragement in this direction had been offered by Joseph Fourier, whose analysis of the diffusion of heat, attained by an innovative mathematical summation of sine waves, the Fourier series, began with a refusal to make any claims about the ultimate nature of heat. While not an outright denial of Laplace's caloric theory, Fourier's work was pointedly not a confirmation of it. In 1819, the polytechnicians Dulong and Petit made a study of known elements and the ratios in which they combined;

their results, which revealed fixed proportions across quantities, were seen as an argument for the atomic theory of matter, the idea that heat is a vibratory motion, and the rejection of the caloric theory.[16] More open hostilities were provoked by the optical research of Augustin Fresnel, who was Ampère's lodger in Paris.[17] Fresnel's experiments on light and its diffusion directly clashed with Laplace's assumptions. He argued that light traveled in waves and was propagated through an elastic ether, rather than in rectilinear streams of particles. Ampère and Arago aided Fresnel in his experiments and conclusions and helped sharpen their argumentative thrust into an attack on the Laplaceans.[18]

So what came next? French physics immediately after Laplace may be understood not as a story of disintegration and decline (as much of the historiography has suggested), but as a network of technical practices, personal liaisons, and shared enmities, with a loosely shared if not uniform conceptual horizon. The anti-Laplaceans did not entirely renounce "mechanistic" explanations in terms of action at a distance between central forces; nor did they abandon mathematical analysis or the development of precise experimental devices. Yet they often made use of physical conceptions and methods at odds with those of the Laplaceans. The wave theory of light, the vibrational theory of heat, and Ampère's understanding of the nature of electromagnetism relied on the idea of an underlying, undulatory ether.[19] In each case, the emphasis shifted from substances to processes; experimental apparatus produced effects that made it possible to trace the movement of phenomena through a milieu. Further, while Laplace and Lagrange avoided diagrams and geometric solutions, Arago and his allies employed rival mathematical methods linked to those taught by Gaspard Monge, whose courses in descriptive geometry at the Ecole Polytechnique (as we will see in chapter 4) inspired a generation of research in fluid dynamics, topography, and practical engineering.

It was shortly after Ampère's publications on electromagnetism that Fourier replaced Delambre in the position of permanent secretary of the Academy of Sciences (renamed in 1814), a position that was soon awarded to Arago. Ampère's work, after the successful treatments of heat and light by Fourier, Dulong and Petit, and Fresnel, has been seen as the death blow to the Laplacean program in terrestrial physics.[20]

A DIVERSITY OF UNIFICATIONS

Ampère's willingness to believe Oersted's report and explore its ramifications has been attributed to his marginal, "outsider" status in the world of

Parisian science; a social misfit notorious for his distraction and obsessiveness, he has even been identified as a possible model for Balzac's Balthazaar Claës.[21] The comparison is not without justification. After learning of Oersted's discovery, Ampère launched his investigations with single-minded intensity; a letter from this period shows him falling asleep in the midst of writing a sentence. Anecdotes recounted by Arago depict him as becoming so lost in thought that he forgot where he was: while dining at the home of a politician, he suddenly exclaimed, "But this meal is truly loathsome; why did my sister hire this useless cook?" On another occasion he walked away from a lively theological discussion, taking as his own the hat of another guest—a bishop's mitre. His tendency toward distraction was in proportion to the breadth of his interests. Sainte-Beuve recounted: "One night with his friends he set out to expound the system of the world . . . and as the world is infinite, and everything is linked, and since he knows the world from circle to circle in all directions, he did not stop [for thirteen hours], and if he hadn't been halted by fatigue I believe he would still be talking."[22] Further, Ampère's scientific monomania, like that of Claës, led to a neglect of social duties. Compulsive purchases including the largest voltaic battery available and experimental equipment left him with debts making it difficult for him to provide materially for his children, sister, and mother.

Yet Ampère was not intellectually isolated. He had strong institutional support from Arago and Fresnel, and his interests in philosophy, psychology, chemistry, and natural history tied him to other French thinkers and drew him into the orbit of post-Kantian *Naturphilosophie*. Born in 1775 near Lyon and given a Rousseauian, self-directed education, Ampère read early in his life Buffon, Thomas à Kempis's *The Imitation of Christ*, and the *Encylopedia* of Diderot and D'Alembert. After the death of his father at the Jacobin scaffold in Lyon in 1792, Ampère became mute, gradually regaining speech thanks to classical nature poetry and solitary promenades around his family's home.[23] He wrote poetry and his emotional life swung between exaltation and despair, feelings prompted by his first idyllic love affair and his wife's death at the birth of their son; his second marriage to a woman who refused contact after their daughter's conception; tearful arguments and reconciliations over his son's missteps; his daughter's disastrous marriage to a drunken maniac and her eventual death in an asylum; his recurrent respiratory illness; and his solitary death in Marseille.[24]

The fact that Ampère had not been trained at the Ecole Polytechnique set him apart from most of his colleagues in physics, yet his ability to speak the language of mathematics—in early essays on the theory of probability and on the analysis of functions—earned him a teaching position at the Ecole and eventually a chair at the Academy of Sciences.[25] His closest friends in Paris came from Lyon, a city known as a center for illuminist thought, home to disciples of Saint-Martin and the spiritualist Allan Kardec. Members of the Société Chrétienne of Lyon, which he helped form, included the veterinarian and anatomist Bredin and the moral philosopher Degerando, who brought him into the circle surrounding Maine de Biran once Ampère arrived in Paris. This circle included the *idéologue* Destutt de Tracy, the physiologist Cabanis, the doctrinaire political advocate of constitutional monarchy Royer-Collard, and the young Victor Cousin, who later built a new and influential philosophy on the rock of Maine de Biran's introspective method. Another of Ampère's close friends from Lyon was the celebrated religious writer Pierre-Simon Ballanche, the inventor of a typographical machine and a heat-based engine and eventually the author of influential texts expressing a prophetic and emotional Christianity.[26] Ballanche was a fixture on the Parisian social scene, the confidant of the Restoration hostess Julie Récamier, in whose salon oppositional politics and liberal Catholicism mingled with romantic literature; he also acted as patron and chaperone for Ampère's son, Jean-Jacques, who became a literary critic for romanticism's key journal, the *Globe*. Ballanche also frequented salons in which the Prussian Dr. Ferdinand Koreff reintroduced the Parisian public to animal magnetism.[27]

Interest in animal magnetism intersected with a long-standing cultural fascination for electricity and magnetism. In the Enlightenment, the forces unleashed by magnets, friction machines, and the voltaic pile had been seen by many as manifestations of active powers inherent to matter. In France, electricity—along with light, heat, and magnetism—also played a critical role in materialist natural history and, eventually, the transformist biology developed by Lamarck.[28] Volta's electric battery reinforced this connection: in a report to the Academy of Sciences, Volta's close ally Etienne-Gaspard Robertson, promoter and innovator of the sound and light spectacle of the *Fantasmagoria*, asked about electricity: "Couldn't this extraordinary fluid be the first of the acids available in nature? Couldn't it be the first agent of living movement, that the ancients called *nervous fluid*?"[29] Relays between physics and the sciences of life were nothing new: the concept of "combustion" was first developed in studies of animal respiration; Galvani, Humboldt, and others had explored circuits of animal electricity; Lamarck argued that interactions among matter, ambient vapors, and the imponderables could prompt the development of new organs and even new organisms.[30] Through short steps, it was possible to move from etherian (and anti-Laplacean) theories of the underlying identity of the imponderable fluids to an identification between these fluids and the source of life and thought.

Such ideas were bound up with the fortunes of mesmerism. At the end of the eighteenth century, Mesmer and his disciples had proclaimed the existence of a single universal fluid cascading between the earth and the stars, uniting mind and matter, a fluid whose manipulation could unlock individual and social well-being. Using passes of the hands, wands, and magnetized objects such as tubs and trees, the magnetizer brought about changes in internal states by redistributing the flow of this fluid in patients' bodies. Despite the conclusions of the Commission of the Academy of Sciences of 1784, to which Laplace and Benjamin Franklin contributed—widely read as denying the existence of a single "mesmeric fluid"—Mesmer's disciple Puységur continued teaching, as did J. P. F. Deleuze at the Muséum d'Histoire Naturelle; Deleuze wrote a series of cautious books on magnetic phenomena. What, Deleuze asked, is the nature of this fluid, whose motions appear to be the cause of astonishing psychological states and miraculous cures? He inventoried the possibilities: "Is it the same as light? Is it a single thing variously modified by the channels that it runs through? Is it composed of many different fluids? Electricity, caloric, mineral magnetism, the nervous fluid, etc. ... are they its modifications? Is it subject to the law of gravity? What is its movement, and what causes direct its movement? We do not know."[31] A friend of Ampère, the chemist Eugène Chevreul—director of dyeing at the Gobelins tapestry factory and author of an influential work on color perception—had been initiated into animal magnetism by Deleuze in 1812. Fascinated by the interaction between psychological, physiological, and physical phenomena, Chevreul conducted experiments on magic wands and pendulums; he later concluded that their effects were brought about by the interaction between the mind and the muscles. Ampère witnessed Chevreul's experiments, and at his urging, the chemist eventually published his reflections.[32]

Imagery taken from animal magnetism was a frequent feature of French romanticism.[33] Beyond the mesmerists' suggestion of a unity of physical powers and demonstration of the powers of the mind, many of Ampère's other interests and personal contacts drew him close to romanticism and *Naturphilosophie*. He was one of the first French readers of Kant (in Latin), and he harassed Maine de Biran until he also read the *Critique of Pure Reason*. During Napoleon's empire, Ampère told a Swiss correspondent, "All the hope that remains in Europe is in Germany." He requested from a traveling friend "portraits of Schiller, Klopstock, Goethe, and others if possible" and praised London for having "a school very near to that of Kant and Schelling, whom it admires a great deal."[34] Ampère had a long-standing interest in natural history, as shown in his correspondence with the anatomist Bredin, and he was a strong supporter of the transcendental anatomy of Goethe and of Geoffroy Saint-Hilaire (see chapters 6 and 8 on Geoffroy). Although at various points he appeared to be on friendly terms with Cuvier, the latter joined Laplace in shooting down Ampère's first candidacy to the Institute, in favor of Cauchy. In 1824 Ampère published articles in the *Annales des sciences naturelles* on the relationship between vertebrates and insects, in support of Geoffroy's theory of the unity of animal types, and after 1830 he openly attacked Cuvier's view of the fixity of species.[35] He was also fascinated by chemistry, which he discussed with Nicolas Clément, who collaborated with Sadi Carnot on attempts to measure the motive force of heat (see chapter 4), with Chevreul, and with Oersted when the latter visited France before 1820.[36] Independently, Ampère identified chlorine and iodine as elements; and under the influence of Linnaeus and Jussieu, he constructed a table of elements, a "natural classification for simple bodies," on the assumption that molecules must be shaped like polyhedrons.[37] In addition to coming up with a chemical theory of the earth's formation, he corresponded with Humphry Davy, whose discovery of new elements, experiments with electrolytic decomposition, and philosophy of living nature—cultivated in dialogue with Samuel Taylor Coleridge—also drew upon *Naturphilosophie*.

Ampère showed a lifelong interest in the diverse topics dear to romantic scientists: the unity of physical phenomena, energy conversions, chemical transformations, the development of organisms, and the unity of species. We can also align his methods, at least in part, with those of the *Naturphilosophen*. L. P. Williams has argued that a trait shared by "post-Kantian researchers"—among them, "in the background," Ampère—was a method of discovery that proposed hidden relationships and then attempted to create the circumstances or conditions to make them manifest. In the case of Oersted's electromagnetic research and the investigations leading up to Faraday's "effect" (the deviation of a light ray by a magnetic field), Williams wrote: "No number of unsuccessful experiments could prove anything except that the right conditions for the production of the expected effect had not yet been achieved."[38] A similar conception of experiments was found in Goethe's evocatively titled essay "Experiment as Mediation between Subject and Object." Goethe contended that long series of variations on experiments were needed to grasp the concept that lurked within the concrete: "The greatest accomplishments come from those who never tire in exploring and working out every possible aspect and modification of every bit of empirical evidence, every experiment." Goethe put this method to use in experiments that he conducted on electricity and light in the company of both Ritter (one of Oersted's mentors), who was a follower of Schelling, and Alexander von Humboldt.[39] In discussions of early electromagnetic research, a similar notion of "exploratory experimentation" has been developed by Friedrich Steinle: this is research that does not seek to confirm or reject hypotheses, but simply to become familiar with a phenomenon and its extensions.

In a lecture on scientific methods, Ampère discussed such an exploratory process as intrinsic to research; James Hofmann has argued for this text's importance for understanding Ampère's method. Ampère argued that complicated physical phenomena often made it impossible to apply the method he called direct synthesis—formal presentation in terms of axioms. Further, in such cases, his preferred method of direct analysis— deriving a fundamental law from a statement of approximate empirical regularities—was also impossible. Thus when faced with complexity, the researcher turned instead to the open-ended process of indirect synthesis. He explained: "The necessity of changing methods is all the more obvious when it is a question of finding the explanation of a phenomenon that nature offers in all of its complication. There, where the givens are by their very existence more complicated than the results we seek, direct synthesis becomes inapplicable, and it is necessary to take recourse either to direct analysis if possible, or to indirect synthesis, to feeling around (*tâtonnement*) and explanatory hypotheses."[40] While *tâtonnement* may be translated as "trial and error" and may include speculative reasoning, the word's root meaning of tactile testing—feeling one's way around—resonates, as we will see, with Ampère's notion of knowledge as arising from resistances between the will, the muscles, and external objects. Another comparable notion was enshrined in Ampère's *Essay on the Philosophy of Sciences*, in which every "first-order" science included four "third-order" subsciences, one of which he called "*troponomique*." This subfield considered "the successive modifications that a single object undergoes, whether in what is immediately observable or in what one may discover by the analysis or interpretation of the facts, with the aim of finding the laws that these modifications follow."[41] Every science, therefore, included a method of examining and exploring its phenomena through varied circumstances or "successive modifications." Armed with a range of concepts from *Naturphilosophie*, chemistry, and animal magnetism about nature's unity, a hostility toward Laplacean physics, and a method that embraced exploration as a means of getting a handle on uncertain phenomena, in 1820 Ampère was ready to take Oersted's discovery and run.

FEELING AROUND IN THE INVISIBLE

From October 1820 to January 1821, Ampère presented papers nearly every week at the Academy, many of which were printed in the journal Arago edited, *Annales de chimie et de physique*. The exact sequence of his experiments and thoughts has been the topic of considerable discussion, provoked in part by his later self-presentation as moving seamlessly from observations to mathematical theory, but his actual course has been seen to be much more haphazard and serendipitous.[42] He began with an open-ended testing of material objects and their relations—wires, batteries, glass tubes, mercury—both to produce expected effects, such as the attraction between currents, and to see what effects might be produced by varying the circumstances. He was feeling his way around in the invisible, devising experimental apparatus to capture, redirect, and thus come to know the properties of electromagnetism.[43]

The first of these romantic machines was a metal frame in which a mobile rectangular wire was suspended above a fixed wire. Both the frame and the wire could be made to carry a current in either direction, and the wire could be replaced by an ordinary magnet (see fig. 2.1A). The apparatus could thus test what happened when currents were juxtaposed flowing in either the same direction or in opposite directions and when the fixed wire was replaced by an ordinary magnet. Some sense of the character of this object can now be obtained at the Musée Ampère—located in his childhood home outside of Lyon and sponsored by the French national electric company—where working reconstructions of his experimental machines are on display. Kept under glass cubes, they appear both rigid and delicate, like wire hangers bent into rectangles of various configurations, not much larger than one foot in any direction; when you press a button next to them, with a great humming noise, they are roused sleepily into motion, their slender bars either pulling together or gently propelling apart and coming to rest; when the button is released, cutting the current, they swivel back to their starting points. The first of his experimental machines, once set in motion and run through a set of permutations, confirmed that an electrified current and a magnet behaved in the same ways: an electrified wire would have the same effect as a magnet. He also found that two parallel wires with currents traveling in the same direction would attract each other, while if one of the currents was reversed, they would repel each other.[44]

In this first phase, Ampère sought to come to grips with the surprising movements initiated by the juxtaposition of voltaic currents and magnets; he sought a workable "construal" (David Gooding's term), an expression that captures experimental regularities but is not yet a full-blown theory.[45] One of the strangest aspects of Oersted's effect was that the compass needle moved in different directions according to whether it was held above, below, or to one or another side of the wire. Ampère quickly arrived at a standard terminology for the direction of the current, the "*bonhomme d'Ampère*": in an imaginary man traversed by an electric current from head to feet, the outstretched left arm indicates the direction of the magnetic force rotating around the wire.[46] Another conceptual innovation of this phase was in his use of the term *circuit* to describe not only the wire carrying electricity but the battery itself, a significant displacement from previous electrostatics. He also arrived at several technical innovations, including the "astatic needle," a magnetic setup arranged to cancel out the effects of terrestrial magnetism, and a "galvanometer," a magnetized needle used to detect a current.

This somewhat chaotic phase of elaborating the phenomenon was not undertaken alone. Working with Arago, in an attempt to imitate what he imagined to be the structure of a natural magnet, Ampère experimented with wires coiled into a spiral, and later into helices: these resulted in the first electromagnet, a powerful attractor. He also imagined a system whereby temporarily magnetized copper, attached to a needle pointing at the letters of the alphabet, could be used to transmit messages over as long a distance as a wire could reach; as Samuel Morse later acknowledged, it was a discussion with Alexander von Humboldt about Ampère's electromagnetic discoveries that led him to invent the technical and semiotic system of telegraphy. Further, Arago noted the pattern formed by iron filings around an electric current; he also observed that a compass needle suspended above a spinning copper disk would rotate. Faraday later extended these effects with his notion of "lines of force" and with his discovery of electrical induction, or the generation of an electric current by moving a magnet back and forth through a coiled wire.[47]

The fact that electrodynamic attraction acted perpendicularly to the direction of the current made the angle at which wires met a relevant variable. Ampère planned an apparatus, again working with wire frames with some parts fixed and some parts movable, that would allow him to vary the angle at which the currents approached each other; his hope at this point was to attain an absolute measure of electromagnetic force. Yet the extreme variability of the electrical charge of the battery, as well as the minute movements of the wires, made direct measurement of these angles difficult.[48] His solution arose from an apparatus he designed in which an expected motion failed to appear, as it was canceled out by an equal and opposite force. This was the origin of the "equilibrium" or "null" experiments that formed the basis of the final presentation of his theory.

Ampère's *Théorie mathématique des phénomènes électromagnetiques*, read to the Academy in 1825, was based on four separate equilibrium experiments using apparatus, like that described above, in which an unfixed current-carrying wire was juxtaposed with another fixed wire: if there had been a difference in the force between these two wires, the first would move.[49] By showing, however, that these movements did not occur—that the interaction between the currents was null—Ampère was able to deduce the major claims that went into his mathematical theory. The claims were (1) the electrodynamic action works with equal and opposite force between infinitesimal points in the current elements; (2) it acts in a direction perpendicular to the flow of the current; and (3) its force is independent of the overall length of the circuit or its configuration. This last was shown by his second equilibrium experiment, where one of the wires was bent into a zigzag, and in the fourth, in which loops of different circumferences were shown to have the same effect (see fig 2.1B).

In the *Théorie* these experiments were framed as successive points in his argument, culminating in the derivation of a mathematical equation expressing the force acting at a distance between two infinitesimal points of two electrified wires, when the strength of the current, the distance between them, and a magnetic constant were known; this was a Newtonian force acting at a distance.[50] The fact that the fourth experiment was never conducted (as he admitted) underlines that in this magisterial presentation of his mathematical theory, his electromagnetic machines changed their status: his "equilibrium experiments" had ceased to be tools of exploration, becoming instead axioms in an argumentative structure. They had gone from tools of discovery to tools of demonstration. He said, "[These experiments] immediately furnish the many laws that lead directly to the mathematical expression of the force that two elements of voltaic conductors exercise on each other."[51] His calculation of this formula involved the integration of discrete elements of a current into a plane; although he himself did not use the term *field*, and although he was primarily interested in the interaction between two currents, his equations were adapted by James

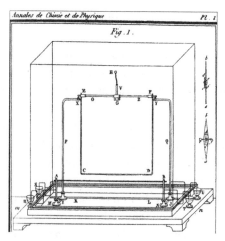

FIG. 2.1A. Apparatus for demonstrating the equivalence of electric and magnetic currents, from André-Marie Ampère, "Mémoire présenté." AB is an electric conductor that, when replaced by a magnetic bar, produces the same rotary motion in the suspended electrified wire CD (Planche I, discussed on 216).

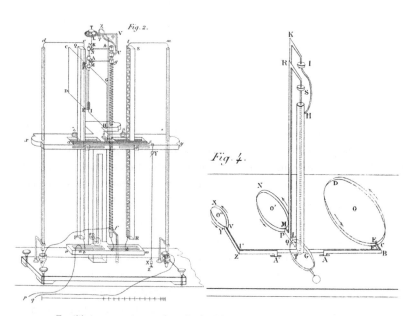

FIG. 2.1B. Equilibrium experiments from André-Marie Ampère, *Théorie mathématique des phénomènes électro-dynamiques*. Experiment 2 shows that the force exerted by a straight electrified wire and another one in a zigzag shape cancel each other out and are thus equivalent; experiment 4 likewise shows that electrified wire loops of different sizes have the same effects. Both experiments were offered as demonstrations that electrodynamic force acted perpendicularly to the direction of the current and was independent of the length or configuration of the circuit.

Clerk Maxwell into the formula describing the rotation of a magnetic field around an electric current.[52]

Ampère contended that the magnetic effects produced by a current were the results of electricity in movement. This meant that he was dealing with a new scientific domain, one distinct from the electrostatics studied by Coulomb and later by Poisson. It also marked a fundamental difference from the research that the Laplaceans Biot and Savart conducted on electromagnetism, which assumed that magnetism and electricity, even if they could interact, were both "primitive facts" unrelated to each other. Ampère thus aimed at a more general theory: he saw magnetic effects as simply a more limited case of electrical interactions. Yet in several important respects, Ampère's presentation followed the norms of Laplacean physics. In the first pages of his essay, he cited Newton as an authority for his method, which he presented as "to observe the facts first of all, to vary their circumstances as much as possible, and to accompany the latter with precise measures in order to deduce from them the general laws"; he later alluded to Fourier's study of heat, which deliberately avoided making claims about the nature of heat, as an example for his own reticence—or positivism—about the underlying physical causes of electrodynamic effects.[53] This profession of methodological rigor and his explanation of the action between the currents in terms of linear central forces acting between infinitesimal points at a distance earned him Maxwell's epithet "the Newton of electrodynamics."[54]

Nevertheless, in both private and public communications, Ampère contradicted the Laplacean view of magnetism as a "primitive fact." Like Oersted, he saw electrodynamic phenomena as the result of two opposed fluids brought into conflict in electrical currents: "Electromagnetic effects of attraction and repulsion can be produced by the rapid motion of the two electric fluids passing in opposite directions in the conductor by a series of almost instantaneous decompositions and recompositions.... In attributing the attractions and repulsions of the conducting wires to this cause, it is impossible not to admit that the motion of the two electricities in these wires is propagated in the neutral fluid which is formed by their union, and which fills all space."[55] In 1832 Ampère proposed that this electric ether served as the medium for propagation of the vibratory phenomena of heat and light. Writing of himself, he said, "It is to molecular vibrations and their propagation through the ambient milieus that M. Ampère attributes all phenomena of sound; it is to atomic vibrations and to a propagation in the ether that he attributes all [phenomena] of heat and light."[56] He ratified a theory of the unity of electric, magnetic, and luminous phenomena, a central tenet of *Naturphilosophie*: "In the work of M. Oersted ... we already find that all heat and all light result, as he says, from the electric conflict."[57] In these presentations, the electric ether was a dynamic connective glue, the principle underlying the formerly distinct imponderable fluids that were seen to travel, like sound, as a vibration through a milieu.

Ampère's use of Laplacean and Newtonian explanations and rhetoric, according to Kenneth Caneva, was primarily window dressing, a means to "force acceptance of his views upon his reluctant and silent opponents"; it was a strategic adjustment to local norms, when in fact his physical intuitions derived from a worldview "in which [light, heat, electricity, and magnetism] were seen as depending on the vibrations and combinations of an all-pervading ether."[58] From this perspective, Ampère was a romantic who pretended to be a mechanist. Christine Blondel has stressed the genuine incompatibility between Ampère's use in the *Théorie* of rectilinear forces acting at a distance and his arguments elsewhere on behalf of an ether propagating phenomena in waves by direct contact; she argues that Ampère entertained a "double discourse," holding "at the same time two rival irreconcilable languages."[59] Ampère's research and theories of electromagnetism thus inscribed the tension between the linear determinism of Newtonian mechanism and the vibrations, emanations, and conversions of the *Naturphilosophie* endorsed by Oersted and Schelling; it also embodied the tension between an image of science as a combination of methodical, rule-bound observation and pristine calculation and that of science as an open-ended, tactile, and spontaneous interaction between material objects and human actions, perceptions, and symbols. Ampère was able to hold these seemingly contradictory strands together in a quiet synthesis — at once mechanistic and romantic — that proved decisive for the development of subsequent physical theory and for the electrification of modern industry, communication, and everyday life.

AMPERE'S EXTENSIONS OF ELECTRICITY — AND THEIR LIMITS

Although Ampère's electromagnetic research is often presented as one of the nineteenth century's victories of reason and empiricism over speculation and superstition, the reception of Ampère's experiments, and his own presentation of them, were inseparable from reflections on the unity of nature. He sought to extend electromagnetism into fields beyond physics, an urge that ran counter to the tendency toward specialization and fragmentation that are often presented as definitive of this period.

Such reflections were spurred both by *Naturphilosophie* and by renewed enthusiasm for animal magnetism, being explored now in the terms of the emerging sciences of the mind.[60] In 1816 Ampère expressed interest in meeting Mesmer's disciple Puységur. During the Restoration, a number of striking new phenomena were added the repertoire of mesmeric performances, including somnambulism (trance states) and seeing and reading at a distance; among the new enthusiasts exploring these phenomena in practice and in print were the doctors Jules Dupotet at the Hôtel Dieu and R. J. Georget and Léon Rostan at the Salpêtrière.[61] New societies and journals appeared, devoted to the study of the medical uses of magnetism and reflecting on its principles of action. The renewal of public interest led to the appointment of a new Commission on Animal Magnetism by the Academy of Medicine in 1824.[62] Investigators continued to express the idea that the principles underlying animal magnetism were connected to imponderable and vital fluids. The *Archives du magnétisme animal* stated, "The more research we do, the more we discover that the means of nature are simple. The *electric, magnetic-mineral, or organic* fluid, consists perhaps in a single elementary fluid, which is modified in different ways."[63] The analogy that Mesmer and his disciples had made between magnets and the effects of animal magnetism was now reinforced by new experimental facts. One of the commission's distinguished savants, Bailly, confessed that "*at one moment he was strongly shaken in favor of the belief in an animal or organic magnetism*: it was when he received knowledge of the experiments by means of which M. Arago managed to impress a rotary movement onto a copper needle with a piece of magnet which he made to turn at some distance."[64] The experiments carried out by Arago and Ampère were read as arguments on behalf of animal magnetism.[65]

In the popular press, post-Laplacean physicists were seen as marching toward a unified theory of forces, one that replaced previous dogmas and crossed over into the study of living things. Among their strongest supporters was also one of the period's great experts on animal magnetism: Alexandre Bertrand, a physician who had trained at the Ecole Polytechnique and who helped invent modern science reporting in articles for the *Globe*, the Restoration's leading journal of romantic criticism and liberal politics. In an article reviewing recent research at the Academy, Bertrand described Ampère's experiments as a "revolution in the high regions of physics." Linking Ampère's discoveries to Fourier and Fresnel's work, he announced, "The universe appears to us now as if entirely plunged into an infinite ocean of imponderable matter, in the midst of which weighted (*ponderable*) matter is merely an accident."[66] Bertrand further suggested that the analogies between heat, light, electricity, and magnetism might be extended to chemistry and to physiology. In his works on animal magnetism, Bertrand argued against the interpretation of animal magnetism as a fluid: instead, he claimed, although the mesmerists' effects were obtained through psychological suggestion, they nevertheless induced genuine organic changes in the brain and nerves of susceptible patients: the resulting states of "ecstasy," he claimed, were at the root of religions.[67] Ampère and his mother both received magnetic cures from Bertrand; Ampère's "Classification of Sciences" acknowledged Bertrand's theories by including the field of *phrénygiétique*, which studied phenomena "due to the exaltation of the sensitivity of certain of our intellectual faculties, which have been designated by the names of ecstasy, somnambulism, and animal magnetism."[68] Like Ampère, Bertrand's interest in the relations between mental and physiological states was encouraged by exchanges with the philosopher Maine de Biran.

Bertrand also shared Ampère's support for the theories of Geoffroy Saint-Hilaire, whose debate with Cuvier was stoked by Bertrand's coverage in the *Globe*. In an early philosophical fragment, Ampère had compared thought to electricity: "The relationship of causality, resulting necessarily from organic circumstances, is given in combination with muscular action or sensation, which is relayed [*rapporté*] to the *sensorium commune*, in the same way as if this result were produced by galvanism."[69] Likewise, Geoffroy's philosophical writings of the 1830s featured not only the concept of the unity of animal plan, comparable to Oken and Goethe's notion of "archetype" (as discussed in chapter 6 below), but also flamboyant speculations on the causes of matter, life, and death that merged physics, chemistry, and life science: "Combustion of a solid body restores it to its original form, which is a state of extreme attenuation of its molecules. Electrification takes up these molecules, proceeding by a progressive chain of translations, and remakes bodies." Geoffroy's use of the terms "combustion" and "electrification" suggested that electromagnetic research as well as the study of heat engines had implications for the understanding of life. Ampère's vocal support of Geoffroy Saint-Hilaire against Cuvier may well have been motivated in part by their shared emphasis on the importance of mobile fluids in life processes and his sense that electricity might be the key to understanding relations between mind and body.[70] Discussing medical experiments in which electrical currents were used to reestablish digestion, Ampère argued as follows: "These experiments will show that it is again the same cause which presides over vegetative [autonomic] life and over the execution of our wills by our members."[71] On several occasions, Ampère reported in the Academy of Sciences on research about medical uses of electricity and the connections between electricity and nervous reactions; he also reported on the medical efficacy of acupuncture, which assumed circuits of bodily energy.[72]

Other "cosmic" extensions of electricity caught Ampère's attention. He rarely missed the opportunity to point out electricity's relation to terrestrial magnetism, an enthusiasm he shared with Humboldt.[73] Humboldt was one of the first scientists invited to observe Ampère's experiments; he helped coordinate the "magnetic crusade" to measure worldwide distributions of terrestrial magnetism, making the Paris Observatory one of its key nodes.[74] Ampère also considered the possibility that "the heat and light of the sun are due to electricity," and in the *Revue des deux mondes* he put forward a theory of the electrical origin of the earth.[75] Following a suggestion of Fresnel, Ampère went on to explain magnetism in general according to his theory of an "electric molecule," in which electricity flowed in circular currents around every molecule of a body: magnetization aligned these currents to flow in a single direction. By analogy, this theory offered an explanation of terrestrial magnetism: the earth's magnetic currents resulted from the stacking of different layers of metals, and their direction and strength was a result of their alignment. In September 1822 Ampère enthused, "Never have the séances of the Institute had such important reports so frequently. Here we see electric currents which produce, on one hand, chemical actions, on the other hand, digestion and the contraction of muscles. All is brought back to the same principle of action in nature by experiments and observations which are its necessary consequence."[76] Ampère saw the study of electricity as unlocking the secrets of the universe.

In all of these examples, Ampère extended the boundaries of electromagnetism beyond the domain of physics as it was then defined by Laplace and Biot: not only did he seek to link it to heat and light, he pursed its implications for astronomy, geophysics, physiology, and psychology. In this light, it is tempting to suspect that the electromagnetic fluid appeared to Ampère, as it did to so many of his contemporaries—Schellingian nature philosophers, enthusiasts of animal magnetism, philosophical anatomists such as Geoffroy, and even Balzac's Balthazaar Claës—as a bridge between the world of matter and that of spirit: a step toward an "absolute" overcoming of the gap between mind and nature. Tempting, but ultimately disappointing. Ampère's thinking went right up to this brink before it stopped. For Ampère, mind and matter remained two unalterably different kinds of thing. Yet given his abundant social and intellectual contacts with animal magnetizers, promoters of "nervous fluids," and romantic authors—many of whom saw spirit and matter as inseparable—how can we explain his refusal to use the universal solvent of electricity to erase the boundary between the mind and the world?

On one reading, Ampère's restriction of the "cosmic substance" of electromagnetism to its physical dimensions might be understood as sober-minded restraint, the reflection of a new spirit of positivism and objectivity. But Ampère was not one to shy away from cosmic speculation; and he was far from irreligious. The more likely source for his refusal to contemplate a spiritual or mental interpretation of electromagnetism is precisely his Catholic faith. In an autobiographical sketch, Ampère noted the importance of his early reading of a précis of the life and thought of Descartes that emphasized not only the latter's ontological separation of extended substance and thinking substance but also his Catholicism. Ampère shared both of these convictions.

His faith was pronounced if erratic. He frequently saw a conflict between his scientific and philosophical research and his duties to God. Doubts assailed him after his father's execution at the hands of the Jacobins, but his first wife brought him back into the fold; he wrote to her in 1802 about a morning when "ideas of God, of eternity, were dominant among those which floated in my imagination." Her illness shook him, but the "cruel test" of her death brought another renewal.[77] In 1804 he helped to found the Christian Society of Lyon, where he presented an essay titled "On the Historical Proofs of the Existence of the Divinity of Christianity." Another essay offered a natural theological anticipation for his research in physics: only a conception of the underlying unity of light, heat, electricity, and magnetism was adequate to God's power.[78] He and his friends counseled each other in letters, recommending readings that included Jakob Boehme, the theosophical linguist Fabre d'Olivet, and the socialist priest Lamennais. His friends warned him against the temptations of metaphysics. Ballanche—no stranger to speculative extravagance—wrote: "I know that you cannot put brakes on your brain. Will not this *idéologie* in some ways contradict your religious sentiments?"[79] They worried about the influence of the Parisian salons and his successes: "Last year, he was a Christian; today, he is nothing but a man of genius, a great man!"[80] Heeding their warnings, Ampère later interpreted the sufferings that followed his disastrous second marriage as a message from God: "Why, in giving myself over to so many vain occupations, have I given myself over to this unpardonable laziness where the things of heaven are concerned? According to the world, today, I have attained the wealth, the reputation, which many men might desire. But you know, dear Bredin, God has sought to prove to me that all is vain, all except loving and serving Him."[81] Ampère portrayed his faith as demanding a singular concentration and focus; he viewed reason and its worldly rewards with a deep suspicion, as in the following note to himself, which reads like a scientist's edition of *The Imitation of Christ*:

> Be wary of your mind; it has often betrayed you! How can you still rely on it? . . . My God! What are all these sciences, all these reasonings, all these ingenious discoveries, all these vast conceptions which the world admires and to which curiosity gives itself so avidly? In truth, *nothing* but pure vanities.
>
> Study, meanwhile, but without any hurry.
>
> Take care not to allow yourself to be preoccupied by the sciences as in recent days.
>
> Work in the spirit of a prayer (*une oraison*). Study the things of this world; such is the duty of your station; but look at them only with one eye. May your other eye be constantly fixed upon the eternal light.[82]

These self-admonitions show his thoughts in perpetual oscillation between God and the "vanities" of philosophy and science. Despite what a biographer has called a "complete and definitive return to faith" in 1817 (suggested by the quote "Today, it is only in the Catholic Church that I find faith, in the gradual accomplishment of the promises which God only made to her"[83]), he continued to struggle to balance the demands of heaven and earth. In 1824 he was appointed to the Collège de France; on learning that this "honor" meant doubled teaching obligations with no increase of income, he wrote to his son: "Here you see this frenzied scientific passion which is punished by events as it deserved to be, since it was one of the causes that have taken me away from that which I should never have abandoned by concerning myself uniquely with the discoveries of dynamic electricity."[84] Ampère experienced worldly setbacks as divine punishment for letting down his inner vigilance; the temptation, the attraction that pulled him away from God, was electrodynamics itself.

The reality of the eternal light was as certain for Ampère as was the light studied by opticians. Yet these were necessarily separate realities: his Catholicism harmonized with a Cartesian dualism that made it impossible for him to embrace any theory of electromagnetism that identified it with thought, spirit, or divine activity. Nevertheless, the question of how these two domains related—how thinking substance, the mind or the soul, comes to know and act upon extended substance—was the central issue of his psychology and epistemology. He pursued these questions with the support of his conversations with the Catholic philosopher Pierre Maine de Biran.

FROM SOUL IN NATURE TO WILL IN ACTION

The profound impact on French philosophy of Maine de Biran (1766–1824), an aristocrat engaged in civil service under the empire and the Restoration, was effected less through publication than through the conversation circles he hosted weekly. Among those who attended were Ampère, the "doctrinaire" Royer-Collard, Georges Cuvier and his brother Frédéric, and the wunderkind of Restoration philosophy, Victor Cousin, whose philosophy of the "moi" was to dominate national education for decades.[85] In the late twentieth century, it became a commonplace to oppose two traditions of French philosophy, one concerned with introspection and subjectivity, the other with the analysis of scientific concepts and rationality; the source of this opposition has been seen as the opposition between "spiritualism" and "positivism," between Maine de Biran (and after him, Victor Cousin) and Auguste Comte.[86] Yet we will see in chapter 9 that although Comte never embraced a methodology of introspection, his positivism was by 1848 deeply invested in personal experience and subjectivity; likewise, as is developed further in chapter 5, Maine de Biran influenced a number of thinkers who, while inclined toward introspection and concerned with the variability of experience, were fascinated by the effects of physical activity, repetition, and technology on internal states as well as external facts.

One of the few works Maine de Biran published in his lifetime, *On the Influence of Habit*, analyzed the processes by which discrete elements of light and sound (as well as the sense givens of taste, smell, and touch) are synthesized, through the "internal play" of rapid, barely noticeable organic movements and mental judgments, into recognizable and familiar forms. According to Maine de Biran, active impressions, or perceptions, always involve the movement of muscles, nerves, and the organs of the brain. Each modification of thought comes from an alteration in the "sensory system." When the same stimuli are repeated, these movements become weaker and are accomplished automatically by habits, which rapidly and unnoticeably convert stimuli first into perceptions and then into ideas.[87] Discussing sight, he wrote, "As habit renders judgments, like movements, always more prompt and less noticeable, the activity of the individual ends by transporting itself entirely into the exterior object; color, figure, form, distance, all accumulate on the solid kernel, and melt together in an impression, in an indivisible *sensation* which the eye seems to receive *naturally* in opening itself to the light."[88] Thus repetition trains us to add our own habitual judgments and motions to exterior objects — in fact to *transport our own activity into them*. As a result, we live in a world shaped by our routines of perceiving and judging, most of which escape conscious control.[89] For Maine de Biran, habit was neither purely good nor bad. "Mechanical judgments" may lead to belief in false notions that cannot be shaken, despite changes in external stimuli. At the same time, habit was needed for both memory and knowledge.[90]

Like his ideologue predecessor Destutt de Tracy, Maine de Biran rejected the "sensualist" philosophy of the eighteenth century associated with Locke and Condillac. They saw empiricism as promoting a view that thought was the passive reception of external sensations.[91] In contrast, Maine de Biran considered the willful "self" (*le moi*) to be the noncorporeal starting point for all experience.[92] Nevertheless, this "primitive fact" can be known only through the "inner sense" of muscular resistance that one feels in trying to move some part of the body. The resistance of the body is needed in order to learn about the existence and primacy of what Maine de Biran called the "hyperorganic," the nonmaterial self.[93] With the encouragement of Ampère, he read Kant, who confirmed him in the conviction of the foundational activity of the self, although he qualified Kant by insisting that this activity could be known only through the activity of the body, the sensations experienced by internal and external organs.[94] According to his interpreter François Azouvi, Maine de Biran's philosophy involved a perpetual search for a "resting point" to end the incessant movement he observed in his thoughts and feelings. In the years before his death in 1824, Maine de Biran's introspection became a quest for direct knowledge of God, a reality that surpasses individual identity: a philosophical method became a spiritual practice.[95]

Although he published little, Maine de Biran left behind him a set of influential questions about the variability of consciousness, the relationship between physical states and states of mind, and the effects of habit on perceptions. His works arrived at a view of the mind as shaped by the body and its experiences, as well as of a mind that can select its actions and experiences and thereby in a sense shape and craft itself through a kind of mental calisthenics. Physical activity—of the muscles, nerves, or the more obscure "movements" of the organs of thought—was an essential component of such changes in experience and thought. While his thought has long been associated with Cousin and "spiritualism," its emphasis on embodiment and compatibility with technology suggests that his philosophy, and that of many of his successors, would be more fittingly characterized as a physiospiritualism.[96]

Ampère was thrilled with Maine de Biran's philosophy. He coined a term, *emesthèse*, for the sensation of the activity of the will upon the muscles (recall his interest in the possibility of electricity as the vehicle of this sensation). *Emesthèse* was the basis of all knowledge, as it provided the foundational experience of causality: "The composite modification of muscular sensation and causality is named resistance.... [Causality] only exists and in consequence can only be perceived in the deployment of our activity."[97] Activity was thus the primitive fact out of which experience of causes, and eventually knowledge, was composed, through "commemoration," in which images are gathered together, and "concretization," where they are combined with present sensations.[98]

Yet Ampère worried that Maine de Biran's philosophy failed to give external reality its due: "It's an entirely spiritual metaphysics like that of Kant, perhaps even more distant from any which holds to materialism." Using Kantian language, he argued that it lacked "a criterion to distinguish between those notions which depend on the nature of our organs, and which cannot without absurdity be applied to noumena independent of us, and those which, being absolutely independent of the nature of our organs, can on the contrary be attributed to noumena themselves."[99] Psychology had to be capable of grounding knowledge on something more than human senses and faculties: "Without this theory, psychology becomes the enemy of the sciences and of all the consoling ideas on which morality and virtue rely."[100] A guarantee of adequate reference was needed to keep psychology on firm epistemological—and moral—ground.

Ampère extended his theory of resistance from the limited case of what he called *emesthèse*, or the feeling of resistance between the will and the muscles, to include a notion of external resistance: "Movements predetermined in phenomenal extension must be halted at certain points, free in others, and ... these movements lead little by little to a recognition of the limits of the field of activity; because the fundamental idea of matter is impenetrability."[101] In other words, the self comes to know itself by the internal sensation of resistance in the muscles; it then comes to know extended matter only by means of resistance (or its absence) in interaction with other phenomena. Ampère completed this "tempered realism" with a further step. Within the realm of thought, it is possible to establish relations among these sites of resistance: "Objective truths, the only ones which deserve the name of truth, consist in the agreement of the *real relations of beings* with *those relations that we attribute to them* in the conceptions that we form of them to ourselves."[102] Thus science established relations among points of resistance, which allowed, he claimed, knowledge of "noumenal" causes. Refusing such a possibility would be as absurd as asserting that "there is no enormous, eternal, foreseeing, powerful and free cause, but that an unknown cause makes us believe these things about the Divinity," just as absurd as denying that our thinking survives after death. In other words, knowledge of the relations among external noumena must be as certain as our knowledge of God and the soul; leaving aside Ampère's crises of faith, these must be the most certain of facts.[103]

This philosophy resembled Descartes's (and later, Cousin's, Husserl's, and Merleau-Ponty's) in that it began with the investigation of internal experience as a means of building up knowledge about the external world. Unlike Descartes, however, Ampère saw the founding instant of certainty in physical activity, not thought; we know we exist because our muscles resist the impulse of our will. Instead of *cogito ergo sum*, both Ampère and Maine de Biran began with *volo ergo sum*, I will, therefore I am. Upon this foundational friction between soul and internal resistance, Ampère built an activist epistemology. We know external objects exist because we sense their resistance as we feel our way around them; out of this highly tactile process of *tâtonnement*, we recognize regular relations among phenomena. These laws guide our further efforts at the physical modification of objects.

Ampère incorporated his thinking about the distinct phases of knowledge and their respective points of view into his *Essay on the Philosophy of the Sciences*, which he worked on for years and saw as the culmination of his life's work.[104] This classification of the sciences was founded on the opposition between world and mind; all knowledge belonged either to *les sciences cosmologiques*—concerning material things—or *les sciences noologiques*, a coinage from the Greek "nous," meaning understanding (see fig. 2.2). Ampère explained that the book emerged from his course on physics at the Collège de France, as a means of clarifying to his listeners the difference between physics and other fields. Following the example of the Museum of Natural History's taxonomist Bernard de Jussieu, he claimed this was a "natural classification"; if a field didn't exist yet, he invented it, with a Greek-based neologism.[105]

Each of the two kingdoms, the "cosmological" and the "noological," was divided into four *embranchements*, echoing Cuvier's four *embranchements* of organisms; each embranchement was further divided into four "sciences of the first order" (see fig. 2.3). These were in turn divided in two, and in two again, to form the four "sciences of the third order": the four possible points of view on any object. Each of the four subfields considered one of "the diverse points of view under which we know" objects: external properties, hidden properties, variations according to circumstances (*la troponomique*, as discussed above), and underlying causes.[106] Thus, any given phenomenon could be understood through its immediately visible

FIG. 2.2. Ampère's classification of sciences. The first table divides the sciences into two kingdoms: "cosmological" sciences dealing with external objects and "noological" sciences concerned with the understanding. Each of these is divided into four "embranchements" (an echo of Cuvier's classification of the animal kingdom), which further break down (in the second table) into sub-*embranchements* and then into four first-order sciences for each kingdom. Finally, in the third table, these first-order sciences are divided into second-order and third-order sciences (including, for example, "la technesthétique"). Ampère, *Essai sur la philosophie des sciences*. Credit: CNRS, www.ampere.cnrs.fr/, with thanks to Christine Blondel.

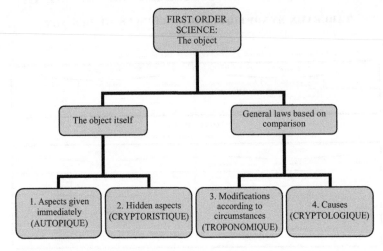

FIG. 2.3.

qualities, those revealed by further experiment and investigation, those that emerged by varying its circumstances, and finally, according to its underlying causes. As confirmation of the "naturalness" of his classification, Ampère noted that the order of his four subfields corresponded to the development of the human understanding: at first, we know only phenomena, but in a second stage we come to have ideas of matter and substance, which arise from moving our bodies—at which point we also come to know the existence of both self and outside objects. In the third "epoch," we use words to compare and judge, and in the fourth we are able to analyze from a variety of cases and, using signs, to proceed to the real causes of things. Ampère's order of classifications, therefore, followed the natural order of human development.

Of the trees of knowledge created by thinkers across the centuries, many feature a fundamental division between theory and practice—a division going back to the classical division of *episteme* and *techne* and the medieval division of liberal and mechanical arts.[107] Instead of making this the foundational division, Ampère replicated it over and over throughout his system. Many of the first-order sciences formed pairs of theory and practice: geology was matched with "Oryctotechnie," the science of extracting minerals from the earth; medicine was paired with hygiene; zoology was followed by "Zootechnie," dedicated to "the utility or the agreement which we draw from animals, the labors and the care by which we procure raw materials from the animal kingdom";[108] social economics was followed by politics (which included the neologism "La Cybernétique," from the

Greek for "governor"), concerned with "the maintenance of public order"; general physics was matched with technology, a field coming into its own at this time.[109] Far from neglecting "applied" sciences, Ampère granted a prominent place to those sciences that aimed at modifying objects. The technological orientation of the system was also visible in the sciences concerned with human thought, the "noological" half of the table. Ampère dedicated "Les Sciences Nootechniques" to "the means by which man acts on the intelligence or the will of other men": the relevant categories were "La Technesthétique," or the fine arts in their effects on the sentiments and thoughts of others; "Glossogie," the power of words; and pedagogy, including "Mathésiologie," which aimed at the amelioration of the minds of other men. Ampère's own "Classification" fell recursively into that class; it was conceived as an instrument of "nootechnique."[110]

Ampère understood knowledge neither as purely intuitive participation nor as detached, calculating reason. Instead, knowledge involved modifications to phenomena that were brought about by humans' willful interventions. Across the chain of causes leading to knowledge, he identified modifications of different sorts: from the experience of resistance the will encountered as it set the muscles into motion; to the resistance of external objects; to the framing of regularly appearing relations among these phenomena; to, ultimately, the extraction, care, transformation, and exploitation of the goods of the earth. For Ampère, knowledge and its tools were technologies that modified both inner sensations and the external world. Just as he followed the effects of the "cosmic substance" of electricity from molecules to organisms to the earth and the sun, his philosophical work underlined the importance of projects of ordering the cosmos and the variety of forms they took in the postrevolutionary period. Ampère sought a general theory of the processes of knowing—from feeling one's way around immediately in apparent phenomena to identifying their hidden aspects, their modifications, and their underlying causes—as the basis for a taxonomy of sciences, which in turn was a map of all that can be known. In every domain of both the "cosmological" and "noological" sciences, Ampère showed how activity and technical intervention gave shape to the mind's encounter with the world.

UNIFICATIONS OF DIVERSITY

Once we get past the idea of Ampère as a misfit, as an isolated and unrepresentative thinker, his work suggests important nuances to our understanding of French physical sciences of this time. The main epistemological reference point for early-nineteenth-century French physics has often been assumed to be Condillac and his project of reducing ideas to their basic sensory elements.[111] Ampère's double interest in the physical sciences and in "spiritualist" philosophy is one place in which Maine de Biran's development of ideology's analysis of ideas had a decisive impact. Ampère's philosophy of science, and many of the concepts that guided his exploration of electromagnetism, also showed the influence of Kant's philosophy and its romantic aftershocks. Many of the *Naturphilosophen*, and Ampère along with them, merged Kant's transcendental critique with an attention to physiology and the impact of instruments. Knowledge was shifted from pure reflection and unchanging categories into sites that were open to observation and physical manipulation, whether in the internal realm of introspection (as exemplified by Maine de Biran's increasingly mystical meditations) or in bodies, signs, instruments, and machines, as in the case of Ampère and several of his fellow anti-Laplaceans. Merging Kantian critique with Maine de Biran's emphasis on the active will and French physics' obsession with precision experimentation, Ampère's philosophy of science saw knowledge as the result of an active, embodied encounter between mind and matter.

Furthermore, at this time the disciplines were taking their modern institutional forms, defined by specialization and competition; early-nineteenth-century physics has often been taken as exemplary in this sense.[112] Yet, as we see with Ampère, and as we will see in different ways with Humboldt and Arago, post-Laplacean physicists often reached for a general conception of nature that was closely attuned to the unifying, cosmological concerns of more traditional understandings of natural philosophy—or *Naturphilosophie*, the German translation of the term. While recognizing differences among the sciences, for example, Ampère nevertheless sought to bring all fields of knowledge together into a single "natural" classification, discovering a fourfold set of "points of view" for each. The "Classification of Sciences," by inscribing a basic soul-body dualism and giving a prominent place to natural theology, was also a pious affirmation of the place of God and humans in any system of nature. As an attempt to unite all domains of existence, one of many such projects that attempted to reorder the system of knowledge and society, his "Classification" was a cosmogram that merged aspects of Enlightenment mechanism, romantic organicism, and a technologically informed epistemology; it was part of Ampère's distinctive set of attempts to set the mind within nature.

One of Ampère's closest friends from Lyon, the romantic printer, poet, and prophet Pierre-Simon Ballanche, whose influence will be glimpsed again in subsequent chapters, offered a suggestive philosophical complement to Ampère's work.[113] Ballanche modeled his lyrical statement of divinely ordained social progress and regeneration, *Palingénésie sociale*, on the *Palingénésie naturelle, ou idées sur l'état future des êtres vivants* of the Leibnizian naturalist Charles Bonnet, one of the first to use the term *evolution* in the study of living things. Bonnet relied on the notion of a graduated series of animals stretching from the simplest creatures up to God. Famously, he "temporalized the great chain of being" by predicting the gradual perfection of the entire animal series all together, in tandem with preordained changes in the environment. The relative positions of all organisms would remain the same, as would the "unalterable germ" that defined each of them, to which they would be restored at the Apocalypse.[114] He called this preestablished, chrysalitic process *palingenesis*, or rebirth. Ballanche spelled out a vision of history as a series of providential destructions and reconstructions: ruptures like the Revolution and the Terror were preordained by God in order to prepare the perfection of civilization and, through it, of mankind. A new religion would arrive through the prophetic works of artists and poets; enlightened social ideals—charity, equality, and association—would become guiding social principles.[115] As he put it in *Palingénésie sociale*, the highest mission of man was the development of thought: "to exercise the intellectual magnetism which tends to spiritualize matter."[116]

The electromagnetic research and philosophical writings of André-Marie Ampère, Ballanche's close friend and spiritual confidant, vibrated in sympathy with these prophecies. While refusing to see the material world as the incarnation of a *naturphilosophisch* absolute ego or "world-soul," Ampère nevertheless showed how the individual soul was at work in the material world: through the primary interaction between the will and the body, through the extension of the understanding into experimental devices, and through the work, both practical and conceptual, of bringing order to the relations among the "resistances" offered by the world. Through technology and activity, the mind was put into nature, just as nature was composed in the mind.

3

Humboldt's Instruments

Even the Tools Will Be Free

It was an ineffable and profound music,
A fluid oscillating incessantly around the world.
And in the vast skies, renewed by its tides,
It rolled onward, expanding its infinite orbits
Till it reached the depth in which its flux was lost in shadow,
Along with time, with space, with number and form.

Like a second atmosphere, scattered and overflowing,
The eternal hymn cloaked the whole flooded globe.
The world, enveloped in this symphony,
As it sailed through the air, sailed through harmony.
Lost in thought, I listened to those harps of ether,
Immersed in this voice as if in a sea
....

And in this great concert, which sang day and night,
Every wave had its voice and every man had his cry.

VICTOR HUGO, "What One Hears on a Mountain," 1829

AN AESTHETIC ANXIETY

For all his idiosyncrasies and internal turmoil, Ampère was firmly embedded in the social and intellectual milieu of romantic-era Paris: he frequented

Epigraph: C'était une musique ineffable et profonde,/Qui, fluide, oscillait sans cesse autour du monde,/Et dans les vastes cieux, par ses flots rajeunis,/Roulait élargissant ses orbes infinis/Jusqu'au fond où son flux s'allait perdre dans l'ombre/Avec le temps, l'espace et la forme et le nombre.//Comme une autre atmosphère épars et débordé,/L'hymne éternel couvrait tout le globe inondé./Le monde, enveloppé dans cette symphonie,/Comme il vogue dans l'air, voguait dans l'harmonie./Et pensif, j'écoutais ces harpes de l'éther,/Perdu dans cette voix comme dans une mer./[...]/ Et dans ce grand concert, qui chantait jour et nuit,/Chaque onde avait sa voix et chaque homme son bruit. Victor Hugo, "Ce qu'on entend sur la montagne," in *Les feuilles d'automne* (Paris: Hauman, 1832), 19–22.

salons and journals and was in constant contact with his Lyonnais friends, the thinkers surrounding Maine de Biran, and anti-Laplacean scientists. One of his colleagues, Alexander von Humboldt, who settled in Paris after exploring the Americas, was both more of an outsider and more of an insider: though a Prussian, he was an honorary member of Laplace's Society of Arcueil, a close confidant and ally of Arago, and, eventually, one of the most celebrated and sociable savants in the world. Despite living under the official hostility of Napoleon and the suspicion of the Restoration government, Humboldt was a fixture in Parisian social and intellectual scenes from 1804 to 1827, and he visited Paris regularly over the next two decades as well, serving as a diplomatic liaison between the Prussian court and King Louis-Philippe.[1] Humboldt was also, like Ampère, an ambassador for German romanticism and *Naturphilosophie*. He went out his way to praise Schelling's conception of "the holy, eternally creative, primary force of the world, who actively generates and produces all things out of herself," and he strove to unite the increasingly specialized domains of physics and natural history within this living, ever-widening whole.[2]

The culmination of Humboldt's lifework, based on lectures he gave in 1827–28 in Berlin immediately after his return from France, was *Cosmos: A Sketch of a Physical Description of the Universe*, an overview of all past and present knowledge of the earth and the heavens. In taking on such a staggeringly vast topic, he confessed in his first paragraph to an anxiety with a "two-fold cause": "The subject before me is so inexhaustible and so varied, that I fear either to fall into the superficiality of the encyclopedist, or to weary the mind of my reader by aphorisms consisting of mere generalities clothed in dry and dogmatical forms." The solution he found was an aesthetic one: "Nature is a free domain," he wrote at the end of the introductory paragraph, "and the profound conceptions and enjoyments she awakens within us can only be delineated by thought clothed in exalted forms of speech, worthy of bearing witness to the majesty and greatness of the creation."

Many have dwelled on the fact that Humboldt's "exalted forms of speech" in *Cosmos*, like the beautiful colored prints that accompanied his books, appealed to the senses and attempted to convey, indeed to produce, the "different degrees of enjoyment presented to us by the aspect of nature and the study of her laws," which *Cosmos* announced as one of its topics.[3] Yet the concept of "the aesthetic" brought with it philosophical and political implications that went well beyond the sense pleasure and formal harmony of artworks.[4] The tension Humboldt expressed above between two opposed tendencies—one leaning toward the limitless play of particular empirical phenomena, the other drawn to dogmatic, formal generalities— lined up precisely with the oppositions that troubled Friedrich Schiller in his *Letters on the Aesthetic Education of Man* (1794), a text as important for German romanticism as it became in France, thanks to de Staël, Cousin, and Jouffroy.[5] Schiller, the poet, philosopher, and playwright who was closely associated with Goethe and the Humboldt brothers, found a balance between these oppositions in an intermediary "aesthetic state." Between the barbaric formal emptiness of excessive refinement and the slavish attachment to the senses of savagery, there was a middle ground in which the two fundamental urges of humanity—the formal drive and the sense drive—could be reconciled. This third drive, the "play drive," roved freely between form and sense and expressed itself in the beautiful appearances of fine art. Schiller's view of the aesthetic drew on Kant's third critique, in which aesthetic judgment was examined as a "subjective universal" intimately connected with the goal of culture and *Bildung*.

From his earliest youth, Alexander von Humboldt was in the midst of the tempest that was German romanticism; he personally knew Schiller, Goethe, and most of the major figures of the movement. Yet he was equally steeped in French mathematical physics. Humboldt brought about a revolution in the organization and direction of the sciences throughout Europe and the Americas in the first half of the nineteenth century. Like his brother Wilhelm, whose study of languages strove for "a conception of the world in its individuality and totality,"[6] Alexander described the different spaces of the globe in their singularity and their interactions, studying the reciprocal combinations of forces on a cosmic scale. He used a vast range of instruments to measure geophysical phenomena; in addition to gathering data on his own voyages, he helped establish international networks of observers to collect similar measurements from around the world. New techniques of combining data into striking visual images allowed him to chart the changes in average temperature across the globe and changes in vegetation in different milieus. With the beautiful images accompanying his texts and the evocative language that filled them, he sought to format and diffuse the sciences to the widest audience possible (see fig. 3.1). His hope was that exposing readers to the pleasures of nature and the methods of the sciences would contribute to moral and political reform.[7]

In Humboldt's life and work, the aesthetics and holism of German natural philosophy merged with the precision and "quantitative spirit" of French physics. Humboldt's partner for his journey to the Americas was the French

FIG. 3.1. Humboldt and Bonpland share a theodolite with an Ecuadoran. *Alexander von Humboldt and Aimée Bonpland at Mount Chimborazo, Ecuador* (1806), by Friedrich Georg Weitsch. Credit: Hermann Buresch, Schloss Bellevue, Stiftung Preussische Schlösser & Gärten Berlin-Brandenburg, Berlin, Germany; Bildarchiv Preussischer Kulturbesitz; Art Resource, New York, NY.

botanist Aimé Bonpland; he frequented Laplace and Berthollet's Society of Arcueil and later undertook expeditions—including one in a hot-air balloon to evaluate the atmosphere at higher elevations—with a student of Laplace, the chemist Gay-Lussac. Yet Humboldt's closest Parisian confidant was the republican astronomer and physicist François Arago, whom we saw leading the charge against Laplace. While Humboldt's travels set a heroic example, his nomadic geophysical wanderings were always tethered to astronomical observatories, and Arago's Observatory was a central node in the global network he assembled. Like Arago's, Humboldt's interest in popularizing science, and the means he used to do so, reflected the theatrical sociability of romantic Paris.[8]

Humboldt lived the slogan of Friedrich Schlegel: "All art should become science, and all science should become art."[9] This chapter explores the relation between the "aesthetic" dimension of Humboldt's work and his instrumental practices. In particular, I seek to link Humboldt's new mode of conducting the sciences to two sources for the outpouring of romantic-era science and art: Immanuel Kant and one of his major interpreters, Friedrich Schiller.[10] Kant and Schiller's conception of the aesthetic shone through Humboldt's scientific work, not only in his evocative writing style and his visual representations, but in the very content and aims of his "physical description of the universe" and in the theory it provided for his practice of observation and efforts to organize the sciences. Discussing the way in which precision instrumentation and its numerical data helped create a common language for an international community of researchers, Marie-Noëlle Bourguet has written of Humboldt's "republic of instruments."[11] Among the leading "citizens" of this republic were the instruments themselves. This conception was strengthened by the attitudes and methods of Parisian science, along with France's republican heritage.

A guiding question of this chapter's genealogy of scientific morals is the relation between objectivity, as both theory and practice, and the views of freedom that were associated with it. Humboldt's focus on instruments and technology—as well as his strong support for the daguerreotype, discussed in the next chapter—might conjure up the ascetic denial of individuality, desire, and unreliable senses that has been identified with "mechanical objectivity."[12] Yet his version of objectivity, building on Kant's philosophy and its transformation by Schiller, was not an ascetic theory of escape, negation, or disconnection. Humboldt did not perceive the machines he used to make readings of natural phenomena as the antithesis of the human, organic, or natural; the best way to know the world was to multiply mediations and observers, not to eliminate them. The kind of "freedom" pursued in the Humboldtian regime of observatory science, and the version of "objectivity" it realized, implied interdependence, shared labor, mediation, and community—notions closely aligned to the republican ideology of his French colleagues.

Humboldt's phrase mentioned above, that "nature is a free domain," nonchalantly restated the central problem of Kant's *Critique of Judgment* and Schiller's *Aesthetic Education*.[13] Despite the occasionally dry and "anaesthetic" nature of such a discussion, a presentation of central themes in Kant and Schiller is essential if we want to understand the particular tensions that Humboldt's science confronted and sought to overcome.[14] Beyond philosophical interpreters such as Maine de Biran, Victor Cousin, and de Staël, Humboldt's regime of instruments was one of the most important channels through which Kantian and post-Kantian thought arrived in

Paris, and it reveals one of the ways in which romanticism did not merely accommodate but also domesticated and even liberated the machine.

FROM NATURE TO FREEDOM

Kant's influence on late-eighteenth- and nineteenth-century philosophy, literature, and science in German-speaking lands was unrivaled. His famous distinction between the phenomenal (sensible) and noumenal (supersensible) aspects of nature was both a statement of the limits of human knowledge and an enticement to surpass them: equally provocative was that he defined the knowable world by mechanical determinism yet saw ethics, art, and organisms as expressions of freedom.[15] For Kant, humanity had a doubled nature. As an empirically observable, determined phenomenon within the system of physical, mechanical causes, man "belongs to the world of sense" and is "subject to laws of nature." But, said Kant, "Insofar as he belongs to the intelligible world subject to laws which, independent of nature, are not empirical but are founded only on reason," man also possesses a supersensible nature: a free will.[16] Kant's *Critique of Pure Reason* dealt with laws of nature, in which sensory givens were brought into conformity with the "schematism" of time and space and the a priori categories, the concepts of the understanding. These categories, provided spontaneously by the mind, were the necessary conditions for experience: they presented phenomena as occurring within uniform time and space, subject to mechanical causality.[17] The judgments made by the understanding were "objective" because of the categories' universal applicability and necessity.

The "objectivity" that pertained to laws ruling over desire (and not knowledge), however, took a different form: instead of laws of nature, here it was a question of laws of freedom.[18] In *The Critique of Practical Reason*, Kant argued that these laws were legislated not by the faculty of the understanding, but by reason. Reason dealt with questions about which certain knowledge was impossible: the origins of the universe, the existence of God, the freedom of the soul.[19] The one "Idea of Reason" about which humans could be certain, however, was that rational beings were capable of freely choosing their actions. Yet for Kant—as for a long tradition of political and religious thinkers—ethical "freedom" did not mean doing whatever one wanted. Kant's morality has been rightly seen as having an ascetic bent: moral choice demanded a purely "good will" that acted only from duty. Kantian freedom was a curiously doubled notion: one was free to act immorally, but only truly free when one chose to obey a rational law. To learn this law, one abstracted from the specific circumstances of a moral dilemma and expressed the action as a general maxim: only if no logical contradiction followed from the idea of a world in which everyone would follow such a maxim was it acceptable. The freedom of a rational being meant that one could choose not to behave in such a way, choose instead to follow desires, the "hypothetical" motivations for doing what one wanted. Yet this very freedom of will—the essential characteristic of the "supersensible" side of our nature—demanded that we hold ourselves to a *categorical* law, that is, one that held with the same kind of determining, universal, and thus "objective" necessity as the a priori categories (or concepts) of the understanding did. The *categorical imperative* required that "I should never act except in such a way that I can also will that my maxim should become a universal law." A will that chose to follow its duty by behaving as if under the compulsion of a universal law was a will that possessed autonomy, or obedience to the law one gives oneself—which was for Kant, as for Rousseau, synonymous with freedom.[20] To act morally, according to Kant, one had to act toward others as though they also had the freedom to act morally. This meant acting as if a "kingdom of ends" existed, treating all other rational beings with the dignity they deserved, as ends, not as mere means or instruments.

Although laws legislating over knowledge and laws legislating over desire were both "objective"—both took the form of a universally valid law prescribed by the intellect—their modes of objectivity were completely different: the one concerned objects determined by mechanical causality, and the other concerned rational subjects who were free to obey or not, and whose freedom consisted of choosing to follow their duty. But because humans were simultaneously empirical beings in the system of causes and free subjects of reason, they were subject to both kinds of law. The vast difference between these two natures led to the central question of Kant's later philosophy. How could the empirically perceivable, determined system of causes that was nature be brought into line with the ideal law of the kingdom of ends? In other words, how could freedom act within nature?

The Critique of Judgment (1790) sought to bridge this divide between the understanding, which legislated over knowledge, and the reason, which legislated over desire, by examining a third faculty: judgment. Judgment legislated over feeling—for instance, the pleasure that came from the "free play" of our faculties when confronted with a beautiful scene of nature, or the pleasure derived from the harmony between our faculties and the objects of the world. In neither of these cases was it possible to pass from subjective pleasure to a universally valid, "objective" intellectual concept— such as the judgment that "this is beautiful" or that "the natural order of nature must have been designed by a creator." While the system of a priori categories involved *determinant* judgments—they were the necessary forms that any judgment about natural phenomena had to take in order to be considered knowledge—judgments of taste and judgments that involved the postulation of final ends made use of the *reflective* judgment. The evaluation that some object was beautiful was based not on a concept, but on feeling: the observer had to feel that everyone would assent to the same view in matters of taste, yet this feeling was merely a "subjective universal," not authorized by an objective a priori concept.

Similarly, in teleological judgments, the observer attributed a final end or a purpose to natural entities and to nature as a whole. Such finality could not be observed directly in the entities being considered; nor did it involve the kind of mechanical necessity of causes and effects that was cognized as one of the a priori categories of the understanding. Nevertheless, teleological explanations of the reflective judgment were required, Kant argued, as *regulative principles* for the understanding. They made it possible to grasp the intrinsic interrelations of parts within living organisms (and eventually, within nature as a whole), providing a guiding principle for conceiving of *generation*, a phenomenon that purely mechanical explanations could not grasp.[21]

Teleological explanations brought with them a version of "instrumentality" that highlighted the difference, running throughout Kant's system (and reclaimed by familiar strands of romanticism) between a machine and a living being. In both a watch and an organism, the motion of each part caused and was caused by every other, working in tandem as an organized whole.[22] Yet it was only in the case of organisms that the different parts actually engendered each other, due to living things' capacity for self-organizing development, growth, and self-repair: "We must think of each part as an organ that *produces* the other parts (so that each reciprocally produces the other.)"[23] The immediate (or in Aristotle's language, efficient) cause of a watch was the person who made it, while the efficient cause of an organism was identical with its final cause, the reason it was made: itself. Each part of the watch was at most an instrument of art, potentially designed and built by an external agent and set mechanically to perform some task; in the organism, however, each organ was an instrument of nature, formed by the organism itself. The part was both means and end.

Further, Kant argued that extrinsic relations among organisms, such as the complementarity of the sexes, or the interactions between an organism and its habitat, encouraged the search for interconnections among the various domains and empirical laws of nature; along these lines, as we will see, Humboldt's *Cosmos* was guided by the idea that diverse empirical laws formed a single system. Whether dealing with an isolated organism or a wider system, researchers had an obligation, Kant said, to pursue interconnections among phenomena and explain them according to mechanical principles of cause and effect as far as possible. At the limit of these mechanical explanations, however, our thought necessarily pushed beyond the realm of appearances, seeking the supersensible cause and goal of the system.[24] At this limit, enquirers encountered themselves. "Finally the question is: What are the predators good for, along with all the other natural kingdoms? For man, for the diverse uses to which his understanding teaches him to put all those creatures; man is the ultimate purpose of creation here on earth, because he is the only being on earth who can form a concept of purposes and use his reason to turn an aggregate of purposively structured things into a system of purposes."[25] Man's *understanding* allowed him to clear forests, build houses, dam rivers, and bring the empirical givens of nature into conceptual order as sciences. Yet such tasks addressed only man's habitat and material needs; they did not draw upon his unique possession, the free use of *reason*. This aspect of his nature beckoned him to a goal superior to "his own happiness on earth," a goal that goes beyond making him "merely the foremost instrument for establishing order and accord in nonrational order outside him."[26]

Humans were for Kant the "lords of nature" not simply because of their intelligence or technical skill but because they could conceive and create a "kingdom of ends." This highest human calling took the form of *culture*, which means "producing in a rational being an aptitude for purposes generally (hence [in a way that leaves] that being free)"; hence, "only culture can be the ultimate purpose that we have cause to attribute to nature with respect to the human species."[27] Nature exists to produce human culture. In the nation, this meant creating "that constitution of human relations where the impairment to freedom which results from the mutually conflicting freedom [of the individuals] is countered by lawful authority within a whole called *civil society*. For only in this constitution of human means can our natural predispositions develop maximally." At the next level up was a *cosmopolitan whole*, a law-bound "system of all states that are in danger of affecting one another detrimentally."[28] Although such a system might be possible only as an ideal, much like the kingdom of ends in *The Critique of Practical Reason*, activities undertaken toward this goal had immediate benefits. Kant's privileged examples were "fine art and the sciences," which "involve a universally communicable pleasure as well as elegance and refinement, and through these they make man, not indeed morally better for [life in] society, but still civilized for it: they make great headway against the tyranny of man's propensity to the senses, and so prepare him for a sovereignty in which reason alone is to dominate." Art and science, activities undertaken in the world of sensuous particularity but whose results were universally communicable, prepared humanity for the cosmopolitan rule of free reason and perpetual peace.[29]

In a short occasional essay entitled "What Is Enlightenment" (1799), the progress Kant sketched as the gradual appearance of "freedom" in the realm of causally determined "nature" was presented as a metamorphosis from mechanical determinism to the rule of freedom. In current society, Kant wrote, the Enlightenment's audacious quest for knowledge was possible only in the "public" side of man's life; only there could the scholar and cosmopolitan "world citizen" question old dogmas and advance new doctrines. On the "private" side, however, duties to the state had to be fulfilled; "some sort of mechanism" was needed to ensure obedience to communal laws. Eventually, however, the pursuit of freedom would allow the supersensible to act upon the mechanical: "At last," he foresaw, "free thought acts even on the fundamentals of government and the state finds it agreeable to treat man, who is now *more than a machine*, in accord with his dignity."[30] The causality of freedom would eventually act upon and transform mechanical causality: the result would be a polity in which all could freely discern and obey the law that determined their duty, investigate the empirical relations of the phenomenal world, and create and enjoy works of beauty. In their function as instruments of nature, such citizens were liberated: they were both means and ends.

The *Critique of Judgment* united the two sides of mankind by the discovery of a supersensible goal that must nevertheless be realized within nature. Culture (or its cognate, *Bildung*)—the creation of institutions and practices for the development of the aptitudes of mankind in keeping with reason—fused the laws of freedom and nature, as well as their distinct modes of objectivity. Yet despite these optimistic sketches of human destiny in his later works, many found in Kant's system—especially his ethics and its apotheosis of the "purely good will"—only austerity, self-denial, and a grim refusal of pleasure. His works seemed to demand a superhuman refusal of natural inclinations while expressing a disdain for temporal change, human variability, and sensory enjoyment.[31] It was this impression that Schiller sought to combat by refashioning Kant's philosophy, in *The Aesthetic Education of Mankind*, in a way that made objectivity not ascetic and passive but sensuous and active.

SCHILLER'S AESTHETIC STATE AND ITS CITIZENS

When the poet and critic Schiller penned his treatise on the philosophical and political importance of the arts, he acknowledged the influence of "Kantian principles" yet paid little heed to Kant's neat divisions and rigorously patrolled boundaries.[32] *The Aesthetic Education* melded together the two types of objectivity in Kant—moral and theoretical, laws of nature and laws of freedom—in Schiller's notion of the aesthetic state. His discussion's central term was *Selbständigkeit*, "self-standingness," translated and reappropriated as "autonomy" in the critical tradition that has followed him.[33] This did not, however, mean a pure, godlike state of disengagement and self-sufficiency. Schiller took up and developed Kant's reflections on instrumentality and means and ends in practical terms; his view of autonomy implied connection with other humans and linked "pure form" with sense and inclination. In focusing on the movement, exchange, and activity taking place in this intermediate realm, he also brought readers' attention to concrete practices and materials. As much as it was a theory of an intermediate state between universality and particularity, Schiller's *Aesthetic Education* was a theory of the action of material mediation.

In *The Aesthetic Education*, the tension in the architecture of Kant's theory between nature and freedom became the engine for an unfolding argument whose harmonic elaboration mirrored the dialectical development of humanity from savagery through barbarism to culture and morality. Schiller set out binary pairs that were sublimated, overcome, or reconciled in a higher unity or dialectical reconciliation (to speak as did Hegel, who learned much from him). The essay's form numerologically restated its content: the dyads that became triads, in the classic symphonic form of A, B, A′, were echoed throughout his twenty-seven (3^3) letters. Where Kant laid out a static map of the regions of the mind, Schiller portrayed the faculties as drives and forces in active conflict. The development of the individual, like the development of society, involved a dramatic movement and struggle among its parts. He asserted a fundamental tension between the passive "sense drive," "life," the empirical world with its desires and impressions, on one hand, and, on the other hand, the active "form drive," which concerned abstraction and eternal principles in science or in art. Between the two was the "play drive," which relished the mere appearance of things—the autonomous [self-standing] "*schöne Schein*," or beautiful appearance, detached from the desire either to possess an entity or to freeze it in the form of timeless knowledge. Writing in 1794, after the storming of the Bastille and the Terror, Schiller traced the root of contemporary political imbalance—

the viciousness of the mob as much as the arrogance and indifference of the rulers—back to a fundamental imbalance within mankind. If he could cure the latter, he could cure the former. Thus his solution demanded a transformation at once artistic, emotional, political, and moral.

Kant had disparaged those who believed that moral action could result from following subjective inclinations or pursuing rewards; they were chasing "a dream of sweet illusions (in which not Juno but a cloud is embraced)." Early in *The Aesthetic Education*, Schiller rejected the self-abnegation implied by Kant's claim that the pure and dutiful will was the only possible basis of morality. "If then, man is to retain his power of choice and yet, at the same time, be a reliable link in the chain of causality, this can only be brought about through both these motive forces, inclination and duty, producing completely identical results in the world of phenomena." Duty and inclination, which for Kant were divided by the same gulf that separated man's spiritual nature from his physical nature, had to fuse; law and sense had to harmonize. Like Kant, Schiller aimed to unite the two systems of causality—moral and natural, free and mechanical—laid out in the first two critiques (and their respective modes of objectivity) by means of culture, art, and science. Yet for Schiller, for freedom to play a role in nature, reason and desire had to agree.[34]

An instrument [*Werkzeug*] was required to bring about this transformation. He proposed fine art, the analysis of which recapitulated the fusion of duty and inclination in the realm of morality: form and matter had to be balanced in a dynamic equilibrium. His key example was the famous Roman bust *Juno Ludovisi*, a recurrent topos of classicist and romantic criticism, thanks to Winckelmann. In what can be read as a reply to Kant's dismissal of the delusions of consequentialist ethics by referring to a mirage of Juno, Schiller described this concrete divinity as the embodiment of the ideal physiognomy: the statue balanced physical and spiritual perfection, grace and dignity, woman and God, the sensible and the supersensible. In it, desire longed for formal purity and formal perfection guided desire. The statue led its observers into the *aesthetic state*, one of "both utter repose and supreme agitation, and there results that wondrous stirring of the heart for which mind has no concept nor speech any name." Here the aesthetic was a specific psychological state balanced between opposed forces, combining "melting beauty" with "energizing beauty." In contradiction to Kant, an experience of freedom no longer required humanity to depart from the world of the senses. "We need, then, no longer feel at a loss for a way which might lead us from our dependence upon sense towards moral freedom, since beauty offers us an instance of the latter being perfectly compatible with the former, an instance of man not needing to flee matter in order to manifest himself as spirit." In works of beauty, man was "already free while still in association with sense"; the "objective" moral law was joined with the "stuff" of sense and matter.[35]

This psychological state paved the way for a new political state, in which the desires of the individual would harmonize with the demands of civil society: "Once man is inwardly at one with himself, he will be able to preserve his individuality however much he may universalize his conduct, and the State will be merely the interpreter of his own finest instinct." Schiller again recounted a history of humanity, this time in terms of the development of an individual's consciousness: the appearance of the play drive took the subject from a first state of immersion in mere sense to an awareness of the distinction between self and object and a delight in everything that offered "material for possible shaping." This new relation with things accompanied a new relation with people. Relations between men and women went from mere satisfaction of physical desire to an intellectual and existential exchange of recognition: "From being a force impinging upon feeling, [man] must become a form confronting the mind; he must be willing to concede freedom because it is freedom he wishes to please." This reciprocity spread throughout "the complex whole of society, endeavouring to reconcile the gentle with the violent in the moral world," opening up a new, autonomous realm between "the fearful kingdom of forces" and "the sacred kingdom of laws." He called this autonomous zone "a third joyous kingdom of play and of appearance, in which man is relieved of the shackles of circumstance." This aesthetic realm was characterized by a specific notion of freedom.[36]

For Schiller, freedom was not merely a state but an activity, one that could be brought into being only by the presence and reciprocal involvement of others: "To bestow freedom by means of freedom is the fundamental law of this kingdom."[37] This conception of freedom—and its vicissitudes—was a central theme in his writing, from *Wilhelm Tell* and *Don Carlos* to the poem he penned for the chorale of Beethoven's *Ninth Symphony*. Although it is known as "Ode to Joy," the original title was "Ode to Freedom":

> Thy magic reunites those
> Whom stern custom has parted;
> All men will become brothers
> Under thy gentle wing.
>
> May he who has had the fortune
> To gain a true friend

And he who has won a noble wife
Join in our jubilation!

Yes, even if he calls but one soul
His own in all the world.
But he who has failed in this
Must steal away alone and in tears.

Even the last two lines' stinging *Schadenfreude* captured the essential point of Schiller's notion of freedom. In Kant's *Critique of Practical Reason*, freedom was the property of an individual in his innermost isolation, abstracted from all social ties and fellow-feeling, ascetically bound to duty. For Schiller, freedom (and the morality that was founded upon it) could not exist in the absence of active relations with others. In this exchange of recognizing gazes, a phenomenology of mutual and reciprocal self-possession, freedom was given to others and received back from them, implying ownership and being owned—as in Rousseau's social contract, where each gave himself to all and received the others back.

Schiller's conception of the state that incorporated this freedom involved a change in the definition of and relation between the "objective" and the "subjective." In Kant, the "subjective" pertained to the sensuous and changing, both empirical sense givens and judgments not regulated by a universally valid concept; "objective" judgments involved a priori concepts, thereby constituting an object of knowledge. In Schiller, the terms began to take on more familiar contours. Again, before Hegel, Schiller called the state "the objective and, as it were, canonical form in which all the diversity of individual subjects strives to unite." But as if in anticipation of those who have seen, in the line of political thought that grew out of this statement (from Hegel to Marx and their followers), the possibility of a quasi-totalitarian uniformity imposed from above, Schiller took pains to note that "the State should not only respect the objective and generic character in its individual subjects; it should also honour their subjective and specific character." Again he sought an autonomous middle ground: the good society would preserve what he called here the "subjective" character of a citizen who was "inwardly one with himself" and was therefore able "to preserve his individuality however much he may universalize his conduct." Thus, said Schiller, the "State will be merely the interpreter of his own finest instinct." In the case of morality, Schiller mixed the "objectivity" of the good will that acted in accordance with reason alone with the "subjective" desires, inclinations, and experiences of the individual. The political sphere demanded a comparable adjustment of the formal laws regulating conduct to the character and circumstances of the individuals who made up a society. "Autonomy," *Selbständigkeit*, was the name he gave to this balance between the universal or objective and the particular or subjective.[38]

Moving from morality to aesthetics and politics, Schiller's arguments also extended to the natural sciences. His first letter decried the narrow aridity of the sciences and the murderous consequences of their investigations: "Truth is a paradox for the analytic thinker; analysis dissolves the very being of that which is analysed." The same was true for the analyst himself. Humanity paid a heavy price for the extreme specialization of the modern disciplines: "Once the increase of empirical knowledge, and more exact modes of thought, made sharper divisions between the sciences inevitable, and once the increasingly complex machinery of State necessitated a more rigorous separation of ranks and occupations, then the inner unity of human nature was severed too, and a disastrous conflict set its harmonious powers at variance."[39] Here science was linked to the "machinery" of the state and its regulation of the division of labor, the "objective form" of human interactions. Yet the solution to science's intensification of social and internal divisions was not an escape from all social bonds, nor a flight from science or technology into either an idealized state of nature or a subjective state of reverie. Instead, the aesthetic state would take science out of its austere and self-enclosed abstraction; science had to humanize itself.

As in his discussion of politics, Schiller continued to use "objective" in a sense like Kant's: as validation by an abstract, formal law. But his depiction of the "free" practice of science, like that of politics, demanded a mixture between the pure formalism of Kantian objectivity and the variable, sensuous "life" with which it engaged.[40] Once more, this state of interdependence between two poles was called autonomy. Scientific specialists had to share their knowledge; with diffusion came transformation: "From within the Mysteries of Science, taste leads knowledge out into the broad daylight of Common Sense, and transforms a monopoly of the Schools into the common possession of Human Society as a whole." By means of the aesthetic, freedom would saturate the realm of science as surely as it did the realm of politics. The result would be the metamorphosis of the lifeless machinery of the state, and of the mere instruments of art, into a free, cosmic republic: "At the touch of the wand [of taste], the fetters of serfdom fall away from the lifeless and the living alike. In the Aesthetic State, everything—even the tool which serves [*auch das dienende Werkzeug*]—is a free citizen, having equal rights with the noblest; and the mind, which would force the patient mass beneath the yoke of its purposes, must here first obtain its assent."[41]

The mind, whose province is truth, "object, pure and simple," could rule only if it had the assent of its subjects. All individuals over whom the laws of form held sway—the laws of the objective polity, the objective law of morality, or the objective categories of science—had to be liberated, allowed to choose to obey the law, all the way down to the "lifeless" instruments that were previously only means to ends. The result would be a kingdom in which the means (the instruments) were also ends; this kingdom included humans and all of nature, along with the material mediations—art, tools, language—that shaped and articulated the relations among its members.

In Schiller's aesthetic state, knowledge was no longer the relation between a transcendental subject and sense data subsumed under universal categories; it was now a liberating relation of mutual respect among users, tools, and their objects. For the pedagogical or the political artist, "Man is at once the material on which he works and the goal toward which he strives. In this case the end turns back upon itself and becomes identical with the medium; and it is only inasmuch as the whole serves the parts that the parts are in any way bound to submit to the whole." Schiller had moved from a vertical relation to a horizontal one; from a hierarchical to an egalitarian model; from a linear, mechanical causality to a reflexive, circular, organic causality. The subjective individuality of all entities had to be preserved within the beautiful appearance of the work of art, the logical interconnections established in the scientific "tableau," the objective apparatus of the state.[42]

While Schiller's main goals were artistic and political, his arguments, as we have seen, were directly concerned with natural science, offering an antidote to the deadly consequences of fragmentation and mechanization. *The Aesthetic Education* also presented the theory of how to get from an intellectual, immaterial, and individual view of objectivity and "autonomy" to one in which objectivity and autonomy were external, embodied, and collectively validated. The work of Alexander von Humboldt, culminating in his *Cosmos*, offered the practice.

A POLITY OF FREE INSTRUMENTS

Cosmos was a publishing sensation, one of the best-selling books of its time. The renown of this work, just like François Arago's popular astronomy lectures at the Observatory of Paris (see chapter 4), was to a great extent due to its pleasing style, vivid descriptions of natural phenomena, and liberal references to literature. Humboldt quoted, for instance, "the immortal poet" Schiller that mankind, "amid ceaseless change[,] seeks the unchanging pole." But the "aesthetic" concerns that underwrote Humboldt's project were not just window dressing.[43] The task Schiller announced of fusing particular, "ceaselessly changing" sensory givens with "the unchanging pole" of abstract principles was of a piece with Humboldt's overall project. The naturalist consciously worked in the intermediate zone between extremes that Schiller opened up and with much of the same theoretical apparatus. While Humboldt set himself the task of actively submitting observations of nature "to the test of reason and intellect," he also paused to note how a "romantic landscape" could serve as "a source of enjoyment to man, by opening a wide field to the creative powers of his imagination." Just as Schiller found in the *Juno Ludovisi* a combination of melting and energizing beauty, so did Humboldt praise the "soothing yet strengthening influence" of natural observation. The guiding idea of *Cosmos*—to present the universe as a law-bound, unified whole while respecting the specificity and "freedom" of each individual part—was a scientific realization of Schiller's reconfiguration of Kantian autonomy.[44]

In *Cosmos*, Humboldt made his endorsement of post-Kantian epistemology plain: "Science is the labor of mind applied to nature, but the external world has no real existence for us beyond the image reflected within ourselves through the medium of the senses." The appearance of the term "labor" in what might otherwise be construed as an expression of idealist faith went beyond Kant's view of cognition as "transcendental" and universal; some kind of *activity* was demanded to shape what passed through the "medium" of the senses. Discussing progress made in the theory of matter, Humboldt noted improvements in natural philosophy over earlier speculations and haphazard observations "by the ingenious application of atomic suppositions, by the more general and intimate study of phenomena, and by the improved construction of new apparatus." Contemporary science was marked by its refinements in theory, method, and, crucially, in apparatus: the technical configuration of the middle ground between mind and nature.[45]

Like other scientists and engineers of this period, Humboldt placed a heavy emphasis on the development of new instruments and observational apparatus to measure and reduce differences in perception. His research depended on an embarrassment of devices for registering a huge range of phenomena: chronometers, telescopes, quadrants, sextants, repeating circles, dip needles, magnetic compasses, thermometers, hygrometers (based on a strand of human hair that grew longer or shorter depending on the moisture of the air), barometers, electrometers, eudiometers (to measure the air's chemical composition).[46] These instruments were not understood as transparent means of registering nature "in itself." Like Schiller's "fine art," Humboldt's instruments were the concrete media occupying the milieu, the "halfway-place" between the mind and the world: the concrete locus for the fusion of sense and intellect.

In other words, Humboldt's regime of instrumentation externalized, temporalized, and "communalized" Kant's categories. Jonathan Crary has argued that in this period perception was increasingly theorized as a phenomenon of particular bodies disciplined by external devices and practices. Humboldt's work with instruments showed that in the same moment that perception was somatized and particularized—thereby dismantling the Kantian notion that the categories of experience were universal properties of the transcendental ego—these categories were *made into universals in practice*, in the technical apparatus of the observatory sciences. The processes of instrumental calibration, standardization, and coordination externalized the process of the understanding described in the *Critique of Pure Reason*. By ensuring that instruments shared a standard scale of measures for quantities of time and space and degrees of magnitude, that they were subject to identical thresholds for determining the presence and composition of substances, and that they possessed the same sensitivity to specific causal relations, the work done in observatories to bring instruments into agreement literally built the concepts of the pure understanding—the basis for the communicability of knowledge—into the physical apparatus.[47]

The new instruments of nineteenth-century science were often seen to embody the qualities of the ideal human subject.[48] For Humboldt, this symbolic identification was rooted in *interchangeability* and *intimacy*. In his extensive research on galvanism and animal magnetism in the 1790s, he constructed elaborate circuits for galvanic electricity in which different metals, chemical solutions, and frogs had equal status as instruments. He presented prepared muscle and nerve fibers as a "living anthracometer" capable of detecting the presence of carbon. In several experiments, the main site of inscription and observation, another link in the chain, was his own body—with welts and blisters as proof of his wholehearted commitment to science.[49] Furthermore, his correspondence repeatedly testified to the extraordinary care he took with his instruments. He typically identified them by the patronymic of their makers and went to great lengths to assure their well-being; his most cherished compasses, barometers, and sextants were discussed with the same enthusiastic affection as his dearest friends. His letters of introduction written on behalf of his human protégés often followed equally solicitous letters on behalf of instruments. Even in the midst of inquiries about friends and their families, there was hardly a single letter in his correspondence which did not mention an instrument or a meteorological observation. Gleefully describing a phase of his expedition in the Americas in which "all of my instruments were in action," he followed an apparently rhetorical question, "But how can I tell you about that?" with thousands of words describing the behavior of each member of his brood. Observations, he said, must be made "with exactitude and '*con amore*'" when tropical heat makes them burn his hands. Instruments that made good traveling companions—those that were small, light, and versatile—were favored, like the portable barometer that could be fitted onto the head of Humboldt's walking stick. The tools arrayed in the famous painting of Humboldt and Bonpland during their voyage—visually uniting the two researchers at the center of the painting—were totems and extensions of the researcher's self.[50] Similarly, the image of the explorers sharing a theodolite with a Native American at the foot of a mountain (fig. 3.1) was a scene of contagious enlightenment and mutual delight. Humboldt's instruments not only extended his senses, heightening his perceptual faculties and submitting sensory phenomena to mathematical scaling; they were embodiments of his relations with others and his place in the natural and social world.[51]

This mode of sociability was central to Humboldt's conception of scientific knowledge. In much late-Enlightenment republican thought, including Schiller's, scientific "objectivity" was linked to ideals of moral and political freedom. But Schiller's assimilation of the laws of nature to laws of freedom in the aesthetic state points toward a distinct moral meaning that was encoded in Humboldtian instrumentation. Daston and Galison's influential discussions of mechanical objectivity consistently stress its negative, self-abnegating aspect, comparing it to the hollow remainder of wax impressed by the more robust and positively defined seal of subjectivity, defining it as an "escape from perspective" and from intention: "Instead of freedom of will, machines offered freedom from will."[52] Elsewhere, Daston describes the moral economy of science in the mid-nineteenth century in terms recalling Kant's second critique: "The self-restraining and self-effacing counsels of mechanical and aperspectival objectivity reverberate with the stern voice of moral duty: the self-command required in both cases to suppress the merely personal is indeed the very essence of the moral."[53] In these arguments, freedom is associated with escape, suppression, and denial. Yet for Humboldt, it was not simply the negative virtues of tirelessness, restraint from intervention, and lack of bias that commended machines as observers and scribes for natural observation. The freedom and objectivity of Humboldtian instruments were positive and active virtues.

The well-tempered instrument, like a reliable but spontaneous human, oscillated within a specific range of values, passive in receiving, active in transmitting its phenomena. The process of standardization and calibration built an a priori principle into the instrument, a categorical imperative ruling over its "desire"—like Schiller's "objective man" who is "ennobled into participation with the law." Each instrument responded "freely" to its milieu and its particular circumstances; but like human laws and the regularities of human language and practice, the agreement produced between the field instrument and the master instrument against which it was calibrated—often located in an urban observatory, like that of Paris—fixed the former's action within a defined range of values, providing the shared and stable background needed to make local difference communicable.[54] This was not, however, an automatic process or a unilateral application of force. Just as freedom, for Schiller, could only emerge in reciprocal exchange with other beings, so the objectivity of the Humboldtian tool demanded cooperation with a highly skilled and patient human. The observer had to gain the instrument's assent by entering into a dialog, "playing" with it, becoming familiar with its limits and habits. Humboldt's letters and travel reports were filled with accounts of awkward moments at the beginning of his relationship with an instrument and his joy at learning to cooperate successfully with it. Making good measurements meant knowing and adjusting to an instrument's particularities. For instance, he preferred a chronometer that lost time gradually at a regular rate to one that kept time perfectly yet was subject to unexpected stops. Precisely the same logic underwrote the device François Arago introduced at the Paris Observatory to measure observers' "personal equation"—an individual's regular lag time in registering a star moving across a fixed space. The law might be the same for all; yet as Schiller argued, the law was only a law of freedom if it adjusted to the living particularity of the individual, whether a person or a machine.[55]

Indeed, for Humboldt there was a strong sense in which the distinction between the two lost significance. Tool and human became a single unit: the instrument was humanized, and the human incorporated the machine. The final sentences of *Cosmos* read:

> The creation of new organs (instruments of observation) increases the intellectual and not infrequently the physical powers of man. More rapid than light, the closed electric current conveys thought and will to the remotest distance. Forces, whose silent operation in elementary nature, and in the delicate cells of organic tissues, still escape our sense, will, when recognized, employed, and awakened to higher activity, at some future time

enter within the sphere of the endless chain of means which enable man to subject to his control separate domains of nature, and to approximate to a more animated recognition of the Universe as a Whole.[56]

Humboldt was one of the first to make explicit a theme that came to have a lasting impact on thinking about the relationship between humans and technologies: that of the tool as the extension of human faculties, or of the human and the instrument forming a "machinic assemblage," as in Deleuze and Guattari's *man-horse-stirrup*.[57] Instruments would seamlessly merge with "forces peacefully at work in elementary nature," allowing humans to recognize and make use of them, elevating and animating our knowledge.

Humboldt's instruments were privileged members of a cosmopolitan society. All instruments of a single type constituted a globally distributed class or function registering the same phenomenon; this was one stratum of a larger society into which humans, machines, and certain sites—notably, observatories—were woven. Humboldt's preeminent role in the internationalization and institutionalization of science has long been recognized. He encouraged younger scientists, giving advice and training and supporting their candidacies; he facilitated contacts among scientists and secured government support. Along with the transcendental anatomist Lorenz Oken, he was a stalwart administrator of the German Society of Naturalists and Natural Philosophers. For its 1828 meeting in Berlin, just after his return from Paris, he commissioned music from Felix Mendelssohn-Bartholdy and orchestrated scientific sing-alongs; his opening speech inspired Charles Babbage, in attendance, to form the similar British Association for the Advancement of Science.[58] Even more, Humboldt's incessant correspondence allowed him to cultivate and tend a worldwide network of natural researchers. The global network of investigators Humboldt helped coordinate directly recalled his early experiments in animal magnetism, in which animals, metals, batteries, and the experimenter formed an energetic circuit.

The aim of Humboldt's physics of the earth was to map the patterns of global systems of natural forces, charting in its local detail and its particular interactions "a general equilibrium which reigns among disturbances and apparent turmoil," the result of an infinity of mechanical forces, vital powers, and chemical attractions balancing each other out. By charting the values registered by instruments distributed around the globe, global patterns became manifest. The clan of thermometers and their readers, for instance, made it possible to trace *isothermal lines*, regions of shared average temperature in bands across the earth. An early article he wrote for the Society of

Arcueil contained a tableau with the positions in latitude and longitude, the average temperatures over the year, and the maximum and minimum temperatures of forty-eight different locations around the globe, taken from reports at all the sites. The table was also a "who's who" of observers, field stations, and observatories: Humboldt in Curana, Saussure in Geneva, Dalton in Kendall, Arago in Paris, Euler in Petersburg, Young in London, and Playfair in Edinburgh.[59] This study, in which half of the sites' averages were calculated on the basis of around eight thousand observations, prepared the way for the "magnetic crusade" that traced the distribution of magnetic intensity, inclination, and declination around the globe. As John Cawood has shown, this campaign—for which Arago's Observatory of Paris served, briefly, as the clearinghouse—was a major step in the establishment of a global network of scientific observation, highlighting the need for communication and shared standards; the multiple readings taken in observatories in both metropolitan centers and colonial outposts played a key role in the maintenance and implementation of standards for time, space, and other measures.[60] It was also a massive operationalization of a fundamental question of *Naturphilosophie*, as pursued by Ritter, Oersted, and, as we saw in chapter 2, Ampère: the electromagnetic properties of the earth and their geographic modifications.

The range of local variations of phenomena across the planet were depicted in the synoptic tableaux that Humboldt invented. In addition to his global maps of isothermic lines, Humboldt also mapped the distribution and relative change of numerous phenomena within a single limited region. In the painted tableau that accompanied his *Essai sur la géographie des plantes*, each instrument and the phenomenon it registered belonged to one vertical column; the y-axis represented altitude (see fig. 3.2). The temperature, air pressure, degree of magnetic phenomena, light quality, blueness of the sky, moisture, and boiling temperature of water for each stratum could thus be seen and compared at once. In a given tableau, each instrument (like each citizen) performed its Kantian duty, but not in isolation and not in pure, abstract relation to the law. When each did its duty, the whole system was described—a balancing act among opposed forces. In the tableau, Humboldt rendered the image of a vast natural chorus, expressing itself freely through its liberated (and at the same time lawbound) instruments. If we rotate the image ninety degrees to the right, so it stands on its side, with the columns reading across horizontally, we can read each column as tracing the part played by each phenomenon, isolated and recorded by its respective instrument. The tableau followed the structure of an orchestral score, with progression over time replaced

FIG. 3.2. *Géographie des plantes équinoxales* (1805), after a sketch by Alexander von Humboldt; engraving by Lorenz Schönberger and Pierre Turpin. Ibero-Amerikanisches Institut, Stiftung Preussischer Kulturbesitz, Berlin, Germany; Bildarchiv Preussischer Kulturbesitz; Art Resource, New York, NY.

by ascension in altitude. The observers, instruments, and phenomena united in Humboldt's tableau sang the same song, one variation of which could be expressed, again, in Schiller and Beethoven's "Ode to Joy"/"Ode to Freedom":

> All the world's creatures
> Draw joy from nature's breast;
> Both the good and the evil
> Follow her rose-strewn path.
> Be embraced, Millions!
> . . . .
> Can you sense the Creator, world?
> Seek him above the starry canopy.[61]

As in Kant, a view of the agreement of empirical laws within the realm of appearances suggested a supersensible basis for this harmony and splendor.

The shimmering intermediate zone of appearances, with each phenomenon isolated, brought into a form that allowed for comparison, and joined with the others in a global system of energetic forces in constant interplay, gave testimony to a higher order of eternal things. The numerous connections Humboldt wove between the starry canopy and the sublunary globe were indices of a transcendent reality that he sought to bring as near as possible to presence. Each particular element and each sweeping line that joined it to wider phenomenal currents, each note, each melody, each movement pointed to the dynamic whole it was part of and the sublime principle of order beyond it.

A "symphony" is so named because it contains a range of sounds working together at once. Humboldt composed something altogether new, combining the surge and shades of feeling tapped by romanticism with the mechanical detail and social coordination of the emerging worldwide regime of the precision sciences. Humboldt was composing a cosmic *symphenomenony*—and helping to assemble the worldwide orchestra needed to perform it. An image of this society, in which humans and natural phenomena create vast, harmonious patterns of order by following their own intentions, had already been suggested by Schiller:

> I know of no better image for the ideal of a beautiful society than a well executed English dance.... A spectator located on the balcony observes an infinite variety of criss-crossing motions which keep decisively but arbitrarily changing directions without ever colliding with each other. Everything has been arranged in such a manner that each dancer has already vacated his position by the time the other arrives. Everything fits *so skillfully, yet so spontaneously*, that everyone seems to be following his own lead, without ever getting in anyone's way. Such a dance is the perfect symbol of one's own individually asserted freedom as well as of one's respect for the freedom of the other.[62]

The harmonies of the gently tamed English garden became the easy choreography of the English dance; Humboldt suggested the entire world might celebrate together in a humanized landscape, in a skillful yet spontaneous ensemble.

The notion of a polity of scientific instruments and users might prompt the questions of who makes the laws, who enforces them, and how. Such concerns, of course, recall the enduring political issue of the late Enlightenment and the French Revolution: how to determine and enforce the laws of a republic whose legislators were in principle exactly as numerous as its subjects. It would be difficult to identify one single mode of arbitrating such questions in the physical sciences. Instead, diverse solutions emerged in the early nineteenth century for establishing the accuracy, reliability, objectivity, or truth of claims about physical phenomena: the construction of increasingly precise instruments, relying on trust and experience in makers and users; instruments and methods to account for and to correct individual error; Gauss's mathematical method of least squares to reduce the errors in a large number of observations; the creation and maintenance of standard values. From the late 1790s to the 1830s, Humboldt was centrally involved in attempts to perfect observation, measurement, and representation according to every one of these methods. While these modes of verification were diverse, a common thread ran through them. In each case, objectivity was recognized as the product of communal activity, a process of exchange and coordination taking place within a single sphere—a common ground between mind and nature, shaped by the social instruments that articulated it.[63]

At all levels, Humboldt's works played a key role in the shift in the meaning of "objectivity" from an internal and rationalist sense in Kant to an external, communal, and emergent model, one that relied increasingly on machines.[64] The development of scientific associations, journals, and vast collective research projects in the nineteenth century, which Daston has offered as evidence of an ascendant ideal of "aperspectival objectivity," might thus in Humboldt's case be thought of as an ideal of *multiperspectival* objectivity. While Humboldt's science relied thoroughly on mechanical devices, these were not seen as the negation of individual perspective and the embodiment of values of restraint and denial; they aspired to an ideal of communal, active, productive, and spontaneous mediation. Humboldt's instruments were free citizens in a cosmic polity: flexible individuals that nevertheless obeyed laws. As long as their "users"—whose actions they regulated in turn—understood and adjusted to the instruments' individual qualities and temperaments, they served as autonomous go-betweens, fulfilling their duty in the liberated universe of humans and nature depicted and, Humboldt hoped, realized by *Cosmos*.

REFRAMING THE MODERN WORLD PICTURE

Humboldtian science begat the laboratory science of the rest of the nineteenth century, especially in Germany. Du Bois-Reymond wrote, "Every industrious and ambitious man of science . . . is Humboldt's son," and Helmholtz modeled his own persona as humanist spokesman for German science on Humboldt.[65] For French scientists—thanks to a large extent to Arago—Humboldt offered a compelling example of a holistic, aesthetic science, one permeated by republican sensibilities and aimed at wide diffusion, and one in which the activity of both observers and their "new organs," scientific instruments, were given starring roles.

Cosmos was a decidedly romantic attempt to harmonize the forces of nature and human industry; unlike otherworldly and antirationalist strands of romanticism, the work incorporated the natural sciences and their machines into a vision of harmony. In addition to reporting the current state of all the natural sciences, the book discussed the history of representations of the cosmos, from the ancients through to the present, and reflected on the best means of encouraging a taste for and diffusion of the knowledge of nature. These reflections on the ends and means of scientific popularization shaped the work itself. The book, as much as Humboldt's *Tabelau of Equatorial Plants*, was a cosmogram: he intended its image of a stable yet dynamic, diverse but integrated universe to raise awareness of the interconnections of nature and to place human endeavors within a wider frame of meaning. While *Cosmos* undeniably participated in its era's projects of imperial and industrial expansion, it can also be read as an argument against the madness of seeing human autonomy as liberation from all restraint. Instead it advanced a view of freedom-in-connection that applied to all beings, the effects of whose actions reverberated throughout the system.[66]

The new mode of science launched by Humboldt may seem paradoxical. It required a huge range of sensory instruments—human and artificial—distributed across the globe, each working independently and immersed in its specific circumstances; at the same time, the many voices and individuals brought into this circle of exchanges had to be calibrated within known parameters, measured by shared standards, and held to universal principles. What was the best means of creating a national or global community ordered by laws that preserved and respected phenomenal variety, individual spontaneity, and freedom? While it took a variety of forms, this problem was faced in the same terms in the sciences, the arts, and the politics of the first half of the nineteenth century. Humboldt wrote to Arago in May 1848, after the uprising for a constitutional monarchy in Berlin: "My ardent hopes for democratic institutions, hopes which date back to 1789, have been fulfilled." The next year, after the king dissolved the assembly and repudiated the constitution, he was less sanguine: "I am reduced to the banal hope that the noble and ardent desire for free institutions is maintained by the people and that, though from time to time it may appear to sleep, it is as eternal as the electromagnetic storm which sparkles in the sun."[67] For Humboldt, the work of freedom—and the discipline it demanded—combined hope, effort, and artifice to bring forth dormant natural potentials. To awaken, guide, and frame this unity in the natural world and among humans, Humboldt served as a charismatic and gregarious instrument. Humboldt fused the passions and projects of unity found in post-Kantian philosophy and German romanticism with the precision mathematics and mechanisms of French science; thanks to his friend Arago, his example helped reshape astronomy and physics and put romantic machines at the center of republican politics.

4

Arago's Daguerreotype

The Labor Theory of Knowledge

BALLOONS, BIRDS, AND MECHANICAL PROPHECY

The predominant image of French astronomy in the early nineteenth century has featured Laplace. His five-volume *Celestial Mechanics*, published from 1799 to 1825, with its analysis of a stable cosmos of points, masses, and forces, was presented as a confirmation of Newton's mechanics and a vindication of the differential calculus. His use of Newtonian force laws, action at a distance, and precise observation and calculation, applying them to the weightless fluids of light, heat, electricity, and magnetism, has also dominated discussions of the physics of this period. Equally prominent has been Laplace's physical determinism, vividly conveyed by his thought experiment of an intelligence capable of calculating all future states of the universe on the basis of its initial conditions; even his speculations about the origin of the solar system in a cloud of nebular gas that slowly condensed was an expression of Laplace's own drive to provide a mechanistic and deterministic explanation for the celestial clockwork.[1] The deterministic, classically mechanistic aspects of Laplacean physics are reinforced by the heavy support he received from the artillery engineer Napoleon Bonaparte, who was made a member of the Institute in 1798; Laplace's tightly managed scientific regime harmonized with the emperor's emphasis on universal law and the mastery of the microscopic details of military and political domination.

Given these images, we might expect the leading astronomer in the years after Laplace, François Arago—a graduate of the Ecole Polytechnique who earned his scientific stripes on a state mission to measure the earth during the wars of the Revolution—to present a cold, analytic demeanor in step with scientific and military regularity.[2] We might also expect that the invention with which he is closely associated today, the daguerreotype, would be best understood as a "classical machine" akin to Laplace's cosmic clockwork. Yet consider Victor Hugo's recollection:

> One evening I was walking in the Allée de l'Observatoire with that great pioneer thinker, Arago. It was summer. A balloon that had ascended from the Champ de Mars passed over our heads in the clouds. Its rotundity, gilded by the setting sun, was majestic. I said to Arago: "There floats the egg waiting for the bird; but the bird is within it and it will emerge." Arago took both my hands in his, and fixing me with his luminous eyes exclaimed: "And on that day, Geo will be called Demos!" A profound remark. The whole world will be a democracy.[3]

A profound remark, perhaps. A strange one? No doubt. But it was hardly an uncharacteristic one. Instead of ascetic, unfeeling rationality, Arago spoke here with mesmerizing verve and passion; in place of the unchanging, idealized, and imperious clockwork of Newton, the sight of an "aerostatic machine" provoked a vision of global democracy. Arago's emotional, aesthetic, and prophetic enthusiasms set the tone for his science.[4]

As our discussion of Ampère showed, the shadow of Laplace's achievements has obscured understandings of the epistemological and political commitments of post-Laplacean physics. This is particularly the case for Arago, the leader of the anti-Laplaceans. Arago was the public face of science in the July Monarchy; his renown was worldwide. He assisted and championed the major developments in physics after Laplace, within France and internationally; his own achievements included discoveries in astronomy, optics, and physics, and the development of many new instruments. But his influence went well beyond the sciences: he represented the interests of workers in the Chamber of Deputies and briefly headed the national government after the Revolution of 1848.

Arago is now best known for the extraordinary role he played in 1839 in helping introduce an invention with a very bright future: the daguerreotype, one of the first effective forms of photography. The term "daguerreotype" refers at once to the device used to produce images, the images produced, and the process of creating them. The device was a camera obscura reduced to a small size—and eventually made portable, thanks to the improvements of the polymath Pierre-Armand Séguier; it was a wooden box, approximately 18 inches on each side, with a variable lens set into one of its faces (see fig. 4.1). In the back of the box was placed a silver-coated copper plate that had been treated with iodine to make it sensitive to light. The lens was directed at external objects, and the light reflected off them was captured and focused by the lens and projected on to the plate inside the box. When the plate was removed, exposure to a mercury vapor revealed the image, and immersing it in a saline solution fixed the image. The lustrous images

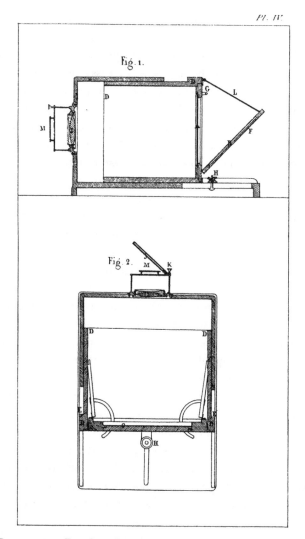

FIG. 4.1. Daguerreotype. From Louis Daguerre, *Historique et description des procédés du daguerréotype et du diorama*, 1839.

that resulted—the size of a small postcard—were strikingly sharp and had a detailed resemblance to the scenes they captured both in their outlines and in their shadings. In 1839, to great fanfare, Arago—who had many close friends in the worlds of romantic arts, literature, and stage craft, as well as in politics and the press—announced the invention, made it available to the public, and secured a lifetime government pension for the families of Daguerre and Niépce.

The fact that this mechanical process of representation was introduced to the world by a student of the arch-determinist, Laplace, makes it easy to interpret the automatic light-writing of the daguerreotype as an expression of the ideals of the classical machine. Developments in the 1860s and 1870s encourage this association: the emergence of mass industry and technology of unprecedented scale and power helped solidify an image of machines as antithetical to human individuality, emotion, and spontaneity. According to Daston and Galison, nineteenth-century scientists were wracked with anxiety about the influence of individual will, interpretation, and emotion on observation and analysis; to avoid the unwanted interference of personality into observation, they relied increasingly on mechanical techniques of observation and inscription. The resulting scientific ideology, "mechanical objectivity," took the classic conception of the machine—as automatic, unfeeling, and impersonal—as the model for knowledge. Photography has been presented as the "essence and emblem" of this new scientific asceticism, in which unalterable, transparent machines were preferred to humans.[5]

Yet Arago introduced the daguerreotype as part of a set of projects that were in many ways at odds with the image of the machine as uniform, impersonal, and automatic; these projects ran counter to Laplace's emphasis on primary qualities, equilibrium processes, and military discipline. As Hugo's anecdote suggested, in the period between the apex of Laplace's influence and the late nineteenth century's faith in impersonal mechanism, a quite different style of physics prevailed, led by Arago—a savant closely tied to romantic artists and republican activists. Arago's astronomy and physics revealed the influence of Humboldt and, through him, the currents of *Naturphilosophie* that also shaped Ampère's conception of electromagnetism. Further, Arago was part of an important school of researchers whose methods and assumptions contrasted with those of Laplace: these were the "scientist-engineers" associated with engineering schools, mostly notably the Conservatoire National des Arts et Métiers, who studied steam engines and developed the sciences of routes and work.

The daguerreotype stood at the intersection of the diverse social alliances, scientific interests, and political hopes that Arago braided together. Before examining the meaning and use of this romantic machine for Arago, we need to understand Arago's Humboldtian astronomy, his republican politics, and first of all, the theories and attitudes towards knowledge of the scientist-engineers. Along with Arago, they developed an epistemology focused on activity and real-world machines—what I call the "labor theory of knowledge." They were also deeply involved in the social and technical dimensions of industrialization. Their work—and that of the artists, poets, stage designers, and republican politicians in Arago's milieu—sheds light on the views of nature and of how we know it that were framed by the daguerreotype, the romantic machine Arago introduced to the world.

AN ASTRONOMER AMONG ENGINEERS

The history of French physics and engineering in the nineteenth century is inseparable from the development of the Ecole Polytechnique. Founded in 1794 as the Ecole Centrale, it was intended to prepare engineers for the military challenges faced by the revolutionary government. In 1795 it was moved to the Latin Quarter.[6] Its two founders, Gaspard Monge and Lazare Carnot, had distinguished themselves as scientists, citizen-politicians, and soldiers, and Napoleon converted it into a military institution (see fig. 4.2). A rigorous code of discipline encouraged a lifelong esprit de corps; yet the training went beyond the administration of docile bodies.[7] Well into the 1830s, the school aimed at inculcating a specific ethical and political ideal, one that echoes Rousseau's and Kant's views of autonomy: the ideal of the republican citizen-soldier who is free because he chooses to obey the demands of the nation. Lazare Carnot, in his "Discourse against Passive Obedience" of 1792, argued that "an army which obeys by reason will always defeat an army which acts like a machine (*machinalement*), because the free soldier is better than the slave."[8] Obedience that comes from free, rational choice, making the soldier "more than a machine" in Kant's sense, is not only morally preferable to blind obedience, but it also makes an army more effective.[9] Carnot and Monge's school remained a hotbed of republican agitation in the Restoration; it was briefly shut down by the government in 1816 when student protests were led by Auguste Comte against a royalist professor.[10]

Throughout the nineteenth century, the school was accused of preferring theory over practice, an accusation that was increasingly justified after Laplace reorganized its curriculum in 1816.[11] Laplace, along with Lagrange, favored theoretical finesse and rewarded research into abstruse regions of the theory of functions. Their approach to analytic mechanics privileged static and reversible systems over dynamic processes; its exemplary cases were systems in equilibrium in which combined forces were in balance and thus canceled each other out. In his introduction to *Mécanique analytique*, Lagrange had pointed out with pride that his method avoided visualizations.

Yet despite Lagrange and Laplace's tendency toward abstraction, the dominant ethos of the school, as set by Monge and Carnot, aimed at practical knowledge.[12] As suggested by the etymology of *poly-technique*, the school

FIG. 4.2. *Visit by Napoleon to the Ecole Royale Polytechnique, April 28th, 1815*, drawn and engraved by Péronard (1815–20). Credit: Copyright Collection École Polytechnique.

provided a heterogeneous tool kit that could be applied to any imaginable setting.[13] Its pedagogical cornerstone was Monge's descriptive geometry, which taught students to render objects swiftly in their essential relations regardless of the perspective of the viewer, representing three dimensions in two. Future military engineers learned to formalize and map a situation through a variety of transformations: Monge's legendary lectures taught students to build up figures by "folding" and "unfolding" them, rotating them and considering the forms their shadows cast. Such manipulations were at the heart of "projective geometry" and led to diagrams whose internal relations revealed relations in the original object.[14]

The son of a knife sharpener, Monge held that theoretical knowledge depended on direct experience and activity; he stressed that the sciences were valuable to the extent that they could be adapted to practical use. The Ecole's chemistry laboratory was one of the first to combine lectures with experiments and research, using equipment originally requisitioned from Lavoisier; the young Justus Liebig modeled his own landmark teaching laboratory on this pedagogical experiment.[15] For the fiercely devoted disciples of Monge and Carnot, experience in the field or the laboratory and work with concrete machines—as opposed to mere rational calculation or symbolic manipulation—was foundational: theory followed from practice. In addition, these engineers developed an alternative approach to mechanics by emphasizing nonequilibrium situations and dynamic processes that changed over time, as in studies of conversions between linear and circular motions, fluid mechanics, elastic solids, and pressure thresholds; in contrast to Lagrange and Laplace, they favored geometrical analysis and diagrams.[16]

This group of mathematicians formed a distinct professional identity; Ivor Grattan-Guinness has dubbed them the "engineer-scientists." After graduating from the Ecole Polytechnique, they frequently entered the state engineering corps, continuing their education at the school of Ponts et Chaussées, the engineering school at Metz, or the Ecole des Mines, then conducting engineering missions for the state. In their research, they sought general explanations for evaluating and increasing the efficiency of mechanical constructions and production processes. Grattan-Guinness classes Arago among the rational physicists whose key exemplars were Laplace and Lagrange—no doubt because he received early patronage from Laplace and because of his accomplishments in astronomy, the model science for rational mechanics.[17] Yet, as we have seen, Arago was the driving force in the dismantling of the Laplacean program in terrestrial physics; he supported Fresnel, Fourier, and Ampère in their alternative studies of the microphysics of particles and imponderables; he also aided Dulong and Petit's studies of the expansion of gases and the capacity of steam engines. Further, Arago taught the introductory engineering course at the Ecole Polytechnique and was closely engaged with the Conservatoire National des Arts et Métiers, the central Parisian institution for engineer-scientists. Arago was an engineer-scientist who studied the skies.

In the 1820s and 1830s, the research of many of the engineer-scientists concentrated on routes and work. The methods of Monge's descriptive geometry, foundational for hydraulics and the study of the resistance of vessels through water, were transferred to the planning of canals, sewage pipes, roads, and railroads for optimal efficiency. What applied in the water could apply on the land; according to naval engineer Charles Dupin, "When one compares the tactics used by the different armed services, the infantry, the cavalry and even the navy, one is always surprised to come across the same system of evolutions."[18] The science of routes also involved the comparison of the costs of different modes of transport and thus was used to demonstrate the economic advantages of new canals and railroads. Work on optimization also led to the crystallization of a central concept in French engineering theory, that of the network. The ideal network was to be an efficient and rapid "reticular system," in which any point could connect or "communicate" with any other without having to pass through a single center, keeping military supply routes open even if the capital was occupied. Network theory combined military fortifications planning, the science of routes and water systems with the dream of crossing France by boat. It coalesced the concept of the nation as a united economic territory defined by the extension and density of industry, transport, and communications.[19]

Engineer-scientists also developed the field of industrial mechanics. An obligatory starting point for this field was the Ecole Polytechnique's required course on machines, given by Carnot's disciple Hachette, whose textbook examined the construction and effects of a variety of simple machines: pullies, gears, and waterwheels, as well as the means of transforming circular force into linear force.[20] The Ecole's chemistry classes, with their analyses of specific heats of gases and expansion processes, were another point of reference. Yet the central focus of industrial mechanics was the measurement and comparison of the productive force of different machines. This question had emerged in the eighteenth-century debate over "vis viva" or "forces vives" in which D'Alembert faced off against Leibnizians over the quantification of the forces acting upon two weights hanging from two sides of a pulley.[21] The dispute was quieted with Lagrange's concept of "virtual velocities," which expressed the motion of a system in terms of a static equilibrium.[22]

In contrast to this highly idealized solution, Lazare Carnot's *Essay on Machines in General* sought to measure and compare the "quantity of action" of actual machines. Carnot concentrated on waterwheels and their relative efficiency at transmitting the force provided by a waterfall, and he analyzed the loss of "moment-of-activity" from violent shakes. Sudden changes of speed and shocks, he demonstrated, greatly reduced the overall effect of the machine. Another early contribution to industrial mechanics came from Charles Augustin de Coulomb—famous for establishing the laws of electrostatics—who compared the action of machines and human labor. As in the phrase of the balloon and engine expert Joseph Montgolfier, "live force is that which must be paid for," Coulomb developed an economic approach to "determining the quantity of action that men can provide by their daily work according to the different manners in which they employ their forces," including the gradual decline caused by fatigue.[23]

In the Restoration, diverse attempts to measure and compare productive force of all kinds were consolidated in the concept of "work," defined as force times distance in the direction of the force. The engineer Claude-Louis Navier saw this as a concept that could apply to all machines, regardless of their specific mode of action; he spoke of the need "to establish a kind of *mechanical currency* [monnaie mécanique], if one can express it thus, with which one may estimate the quantities of work employed in order to effect all kinds of manufacture."[24] At Arago's urging, Navier created a course on applied mechanics at the Ecole Polytechnique. Another key theorist was Navier's successor, Coriolis, who distinguished between the total work of an engine, the "lost work" needed to overcome resistance, and the final amount of "useful work" that remained; he later focused on kinetic energy and its effects on systems in rotation.[25] To quantify this force, measuring instruments were invented; the Prony brake, placed on an axle to measure the force of the engine that turned it, was improved by Dupin's "dynamometer." Dupin also announced a new field of "dynamie" to study work of all kinds. Likewise, Jean-Victor Poncelet, professor at the Ecole de Génie at Metz (and after 1848 the commander of the corps of the Ecole Polytechnique), made "work" central to his course on industrial mechanics, a field he defined as the study of "the diverse transformations or metamorphoses which the work of engines can undergo by the means of machines or of tools, to compare between them the quantities of work, to evaluate them in money or in production of such and such a kind, etc.," highlighting both work's "transformations or metamorphoses" and the close connection between engineering science and political economy.[26]

POPULAR MECHANICS

In Paris, the activities of the engineer-scientists were focused on the Conservatoire National des Arts et Métiers (CNAM), an institution which stood, significantly, halfway between the Hotel de Ville and the workers' quarters of northern Paris. The CNAM was founded in 1794, the same year as the Ecole Polytechinque and the Ecole Normale Supérieure, and located in a former priory; it was intended to offer instruction as well as to hold a collection of technical objects for training, a mission advanced thanks to a bequest from the legendary automaton-maker Vaucanson that consisted of his scientific instruments, androids, and mechanical animals. The CNAM's goal was to stimulate the technical arts in the service of the state. Its first director, Gérard-Joseph Christian, composed a work that has been recognized as crucial in the history of the concept of technology: *Vues sur le système industriel, ou Plan de technonomie*, an overview of the state of the arts, the political economy of production, and strategies for the "perfection" of all of the mechanical arts.[27] Christian's "technonomie" was a blueprint for the CNAM, as a clearing house of industrial inventions from which national production could be coordinated.

The prominence of the CNAM in the political economy of machines was amplified by its reform in 1819, under the initiative of Dupin, a disciple of Carnot. After training at the Ecole Polytechnique, Dupin contributed to the study of tangents, networks, and routes; he was also obsessed with the industrial "evolution" of France. Attributing the defeat at Waterloo to England's superior industrial power, he traveled to England in 1818–19 to inspect its manufactures; his published report held up the communications and transport infrastructure, steam power, and educational institutions in England as indicating the road to progress. Dupin was also a driving force behind French statistics. He founded an Institute of Statistics and published statistical pamphlets and articles on issues of national concern; he helped popularize statistical writings as a form of political writing in the 1820s, earning praise from Goethe.[28] At the central auditorium for Restoration-era popular science, the Athénée, near the Palais Royal he gave lectures, later published as *Forces productives de la France*, in which he presented for each region of the national territory tables of its natural resources, manufactures, distribution centers, and supply routes.[29] Dupin also counted schools, museums, and lecture halls. Thus intangibles like education and literacy were placed on a par with coal mines, forests, waterfalls, and factories as contributors to national productivity. *Forces productives* included a map depicting a region's degree of instruction according to a gray scale: the regions below the line from Saint-Malo to Geneva appeared alarmingly dark, vividly displaying how far southern departments had to go on the road to Enlightenment.[30]

On his return from England, Dupin campaigned to turn the CNAM from a highly specialized school with a formal enrollment into a site for open-access lectures on general topics in industry and arts. The committee that decided on the issue, which included Arago, agreed, naming Dupin to the Chair of Mechanics.[31] Following the model of Andrew Ure's Andersonian Institute in Glasgow, Dupin lectured to audiences of artisans about geometry and the mechanics of industrial production; his aim was both to raise their state of learning and to encourage invention by showing the underlying principles of their craft.[32] The lectures attracted audiences as large as two thousand; his yearly inaugural lectures, reprinted and distributed as inexpensive pamphlets, stressed the importance of education for improving workers' conditions, as well the importance of the working class for the well-being of the nation as a whole. These courses overlapped with those given by the Association Polytechnique, started by Arago and Auguste Comte and also hosted at CNAM, in which polytechnicians gave lectures on mechanics, mathematics, and geometry.[33]

Another of the CNAM's three chairs belonged to J. B. Say, founder of the French school of political economy. Say departed from the physiocrats' view that land was the source of all value. Like Adam Smith, he held that a commodity's value came from the quantity of work that went into making and delivering it—what economists know as the labor theory of value. An advocate of free markets, he is known for "Say's Law," which has been parsed as "supply creates demand": for example, although the introduction of machines into printing threw many printers out of work, eventually it led to a greater demand for books, and thus new jobs. Like Dupin, Say argued that despite some initial disturbances, in the long run mechanization was beneficial for workers.[34]

The third professorial chair was occupied by Nicolas Clément-Desormes, a chemist who determined experimentally the ratio of the specific heats of gases and was one of the first to use the term *calorie* as a unit of heat. His lectures were attended by a young polytechnician named Sadi Carnot, Lazare's oldest son. The two collaborated on experiments, including one that established how much work could be produced by burning one kilogram of coal—an early inquiry into the latent productive force or potential energy in matter.[35] According to François Vatin, the study of "work" took mechanics up to its conceptual limits; beyond them lay thermodynamics, which studied the transformation not just of one kind of force into another, but of heat into productive force. Sadi Carnot's *Reflexions on the Motive Power of Heat* (1824)—to which we will return in chapter 7 in relation to his brother Hippolyte, the director of propaganda for the Saint-Simonians—arose from the mixture of technical, economic, and social concerns under discussion in this site.

The CNAM was thus a meeting point between polytechnicians, liberal political economists, state-level planners, and members of the working class. It concentrated on work in its distinct and overlapping senses: "le travail" as the quantifiable effect of machines, labor as the basis of economic value, and work as the defining activity of an emerging class of artisans and laborers—a class whose relationship with the professional scientists, managers, and industrial planners at CNAM in the 1820s, 1830s, and 1840s was an increasingly pressing concern.[36]

THE LABOR THEORY OF KNOWLEDGE

Le travail was built into the scientist-engineers' views of how we learn about nature. Many recent discussions about the epistemologies associated with industrialization have focused on a group of reform-minded and entrepreneurial British scientists who played key roles in the British Association for the Advancement of Science, including Charles Babbage, John Herschel, and David Brewster. These "gentlemen of science" and their associates believed that the sciences would aid industry by discovering and developing new production processes, by measuring and rationally improving efficiency, and by applying mathematics and statistics to political economy.[37] Although many of these authors were in close contact with the engineer-scientists in the orbit of the CNAM, British industrial epistemology—concerned with the question of who knew what about industrial processes and how they knew it—must be distinguished from comparable developments taking shape in France.

Primarily remembered now for his proto-computer, the calculating machine called the difference engine, Charles Babbage was also celebrated in his lifetime for his *Economy of Machines and Manufactures*. The book advocated a constant increase in the division of labor by developing and applying machinery. It depicted a strict demarcation and hierarchy among unskilled laborers, who were easily replaced (often by machines, at a savings for owners); skilled laborers, who were more highly valued and difficult to replace but whose work also could also be increasingly mechanized; and managers, who were able to see production as a unified system, to remember, to predict, and to anticipate alterations in the production chain. This great chain of labor and of classes was also a great chain of knowledge; it corresponded to a hierarchy that elevated mind over body.[38]

Above all of these stood the natural philosopher, capable of discovering the natural laws that would then be applied by managers. Babbage's separation of a faculty of reason—capable of comparison, recollection, and prediction—from the thoughtless body was derived from a long-standing tradition reinforced by the philosophical successors of Christian and Platonic dualism, which remains an important resource for images of knowledge. In its nineteenth-century renderings, the ideal knower was often one without a body or distinguishing characteristics; pure knowledge transcended the idiosyncrasies of individual personalities and bodies. The ideal of knowledge as performed by detached, abstracted observers was the epistemological correlate of the inhuman "classical machine," an ideal easily transferred, late in the nineteenth century, to machines themselves, which were valued to the extent that they provided transparent access to phenomena.[39] Within England's strict class hierarchies, the division between the laboring body and the knowing mind was mapped onto the division between a class of laborers and a managerial and ownership class, a division that "dehumanized" laborers by associating them with and replacing them by machines.

However, a different epistemological tradition also runs through the history of science, one that came back into focus in the age of the romantic machine.[40] According to this tradition, the actions and experience of knowers, along with their tools, were celebrated as the necessary conditions of knowledge. In place of disembodiment and transparency, this craft-centered view emphasized the body of the artisan and the transformations that labor and instruments brought forth. An important moment in this tradition was Renaissance alchemy; as Pamela Smith has argued, Paracelsus legitimated his claims by referring to his long experience and his patient labor with herbs, metals, and instruments.[41] Craft knowledge and embodied experience were given even greater legitimacy in the programmatic works of the keeper of the seal, Francis Bacon, who, in his protracted crusade against the "Idols of the Theater" worshipped by scholastic systems-mongers, argued repeatedly that sensory experience and active testing of the world by means of perfected tools of observation and experiment were the only ways to advance learning. Bacon was a hero of eighteenth-century *philosophes*, and the practical orientation of his science—both its real-world applications and its emphasis on practice, craft, and technology as the basis of knowledge—ran through the *Encyclopédie*. Diderot's recurrent critique of the sterility of geometrical knowledge went hand in hand with his promotion of the knowledge of skilled craftsmen.[42]

These attitudes shaped the orientations of the engineer-scientists at the Ecole Polytechnique—thanks in part to the examples of Carnot and Monge—and at the Conservatoire National des Arts et Métiers. The activities of the engineer-scientists and the discoveries of workers brought a new attention to the "work"-based aspects of knowledge production. Engineers frequently set theoretical works on mechanics in the context of the machines and manufactories where this science developed and was applied.[43] Similarly, Jean-Baptiste Dumas's lectures on chemistry stressed the origins of his science in craft, including a history of alchemy.[44] Between 1820 and 1850, this epistemology resonated with the rising visibility of labor and the working class. Arago, for instance, recalled how Benjamin Franklin, the "son of a poor artisan," invented the lightning rod; he listed the looms of Lyon, steam engines, and the clocks and spyglasses used by sailors as inventions from the "class of artisans." Arago's formal and popular scientific reports featured discussions of the creation and use of new instruments; he framed the labor and tools of the scientist as intrinsic elements of the knowledge produced.[45]

This was a social epistemology suited to the places in which it was taught. More than simply technical training institutes, the Ecole Polytechinque and the CNAM were hothouses of republican and liberal sentiment, suffused with the revolutionary-era ideals of liberty, fraternity, and equality. The emphasis on labor compounded the activities of unskilled laborers, skilled "sublimes," managers, planners, and theorists into a single kind of thing, as so many different species of transformative work.[46] All were joined in the transformative national project of industry. In France, Babbage's hierarchical view of the relation between industrial technology and knowledge was countered by this alternative tradition, in which the skilled and active body was the bearer of knowledge and which asserted an identity of work of all kinds. At the CNAM—whose mission was the development of industry through the spread of knowledge—the epistemology based on craft, technology, and labor was in constant contact with the theorization of the bases of an industrializing society in the political economy taught by Dupin and J. B. Say. In line with the *labor theory of value* promoted by these political economists, this epistemological orientation can be seen as *the labor theory of knowledge*.[47]

Just as industry and labor were seen as the ultimate source of value, with technical developments increasing the productivity of an individual's labor, so too, for many French engineer-scientists—prominent among them Arago, Dupin, and, in a more complex way, as we will see, Auguste Comte—labor, and the intensification it underwent thanks to new scientific instruments and machines, was seen to lie at the root of knowledge. This epistemology underwrote the introduction of new scientific devices and corroborated the notion that the progress of technology, the progress of knowledge, and the progress of society were inseparable. Recall Poncelet's emphasis, in his textbook of industrial mechanics, on "the diverse transformations or metamorphoses which the work of engines can undergo by the means of machines or of tools": machines increase or decrease, redirect, or otherwise modify the forces to which they are submitted.[48] In economics, the labor theory of value implied a comparable metamorphosis: matter became valuable according to the effort that was mixed with it. Similarly, a metamorphosis was involved in producing epistemological "surplus-value": naturally appearing particulars were converted into general theories; embodied knowledge or rules of thumb were converted into explicit, mathematical formulas.[49] To hasten this process of conversion—bringing knowledge that was hidden, or merely local and tacit, into a space of explicit universal communication and equivalence, and reinvesting that knowledge in local practice—was precisely the aim of Dupin's lectures and those of the Association Polytechnique. This ideology was as central for the liberalism of Dupin as it was for the republicanism of Arago and, as we will see, for the socialism of the Saint-Simonians. All of them advanced the notion that the interests of scientists, managers, and entrepreneurs were ultimately the same as those of workers, thanks to the classes' shared participation in the general project of "industry." This was, however, a fragile identification; the revolution of 1848, framed by its more radical instigators as the revolt of the working class against the capitalists, was a sign of its breakdown.[50]

The new epistemological importance that the economists' labor theory of value placed on technically assisted physical effort paralleled the replacement of the balanced clock, lever, or balance as symbol of order and knowledge by the productive steam engine. Tracing the implications of this transition, Norton Wise has argued that in the late eighteenth century the balance—as concrete instrument, method, and cosmic symbol—played a mediating role across distinct intellectual fields, from astronomy, physics, and chemistry to agronomy and political economy.[51] Balances and balance sheets were tools of discovery and proof, as in Lavoisier's and Laplace's work on the balance between inputs and outputs of heat and weight in chemical reactions, studied with the aid of the calorimeter; balanced situations of equilibrium likewise served as the model for Lagrange's mechanics and virtual velocities; physiocratic political economy traced the variations in productivity around a steady state. In each of these fields, researchers described situations in which variations over time counterbalanced and canceled each other out. To the importance of the balance as both a technology and a model of physical systems, we might add an epistemological dimension as well: Wise argued that despite the ubiquity of the balance, the instrument (and its related technologies) was largely invisible in accounts of knowledge.[52] Well-made scientific instruments also followed the model of the balance: instruments of measure and observation were transparent, adding or subtracting nothing from the phenomena they measured or observed.

But what happens when we move forward into the period when—as Wise, Michel Serres, Anson Rabinbach, Isabelle Stengers and Ilya Prigogine, and others have suggested—the central cosmological model ceased to be the classical machine, the enlightenment balance (or its relatives the clock and the lever) and was replaced by the steam engine? At the conceptual level, the consequences were many. Linear temporality became crucial: instead of the reversibility and equilibrium of Lagrange's virtual velocities, the concept of potential energy became ubiquitous, and physical systems were understood as undergoing irreversible processes. At the same time, political economy came to trace limitless, or asymptotic, developments, instead of deviations around a fixed point. In theories of the solar system, cosmologies of balance and stability gave way to narratives of evolution and collapse. In short, the steam engine, as a cosmic symbol, supplanted a universe of timeless balance and equilibrium with one defined by history, activity, conversions, combustion, waste, and eventually entropy.[53]

Beyond these conceptual changes, I suggest that the steam engine—and its related romantic machines—was also implicated in the shift in epistemological emphasis we have identified as the labor theory of knowledge. If Enlightenment-era instruments were seen as passive and transparent, neutrally balancing inputs and outputs, the early nineteenth century's labor theory of knowledge acknowledged and sought to account for the modifications introduced by the physical properties of instruments and the modifications brought about by the activity of observers. As we have seen in the cases of Ampère and Humboldt, Maine de Biran and Kant's view that individual knowers actively construct their experience provided the background for epistemologies in which experimental devices and instruments were understood as extensions of the observers' faculties—tools for feeling around in the invisible and new organs of perception. The idea, familiar from romantic poetics and aesthetics, of the power of the imagination to make and remake the world was merged here with a profound experience and reflection on technology: machines were assistants in the process of making the invisible visible, constructing objects of knowledge, and framing the image of the world.

François Arago's science—his theories about the structure of nature, as well as his conceptions of how we know it and how that knowledge should be used—overlapped with the mechanical romanticism of his colleagues, yet added another element that reflected his engagement with Parisian republican politics. Far more than either Ampère or Humboldt, he linked questions of technology and activity to the heated political question of the changing status of workers in the face of widespread industrialization. Arago's scientific and popular writings made constant allusion to the importance of the activity of observers and instruments. The polyvalent term *labor* linked the content of his science (his interest in the dynamic effects and conversions of heat, work, light, and electromagnetism), his epistemology (the labor theory of knowledge and an emphasis on "interference" and "modification"), and his politics (his support of workers and inventors). These commitments combined with themes from the romantic and fantastic arts to form the background for his presentation of one of his age's most remarkable machines, the daguerreotype.

A HUMBOLDTIAN SCIENCE OF THE HEAVENS

By allying himself with industrial mechanics and the CNAM's engineer-scientists, Arago departed from his predecessor Laplace. The sources of his opposition ran deep. He recalled dining at Laplace's home as a student: "My mind and heart were highly disposed to admire everything, to respect everything, at the home of the man who had discovered the cause of the secular equation of the moon.... But what then was my disenchantment when, one day, I heard Madame de Laplace approach her husband and say to him, 'Would you entrust me with the key to the sugar?'"[54] Reminding us of the embargo that isolated France's sugar colonies during the Napoleonic wars, Arago stirred in a soupçon of accusation, presenting his mentor as a petty domestic tyrant who kept sweetness under lock and key—a caricature also inspired, perhaps, by the contrast between Laplace's origins among the tight-fisted Norman gentry and Arago's reputation as a hot-blooded son of France's Spanish border.

Arago's intimation of a lack of warmth and generosity applied to his presentation of Laplace's science as well. In a biographical sketch, he praised Laplace for demonstrating how much "an observant geometer who, from the moment of his birth, never left his work cabinet, who never saw the sky except through the narrow opening running from north to south in which the principal astronomical instruments move in the vertical plane" might discover. This stationary observer would learn that "his humble and narrow dwelling was part of a flattened, ellipsoidal globe.... He would have found as well, still without moving, his true distance from the sun."[55] This is backhanded praise: throughout his career Arago distanced himself from the image of the astronomer fixed in the observatory analyzing a limited quantity of observations. Arago depicted Laplace's approach to knowledge (and to sugar) as ruled by rarity, arbitrary authority, and enclosure. He looked back on Laplace's coterie in the Society of Arcueil as both deficient and excessive: "Preconceived ideas, to which the best minds succumb more easily in a group which is, so to speak, intimate, than before a larger public, could result in stifling the spontaneity of genius and restrain research to a conventional level"; the group mentality inhibited spontaneous genius and promoted obedience to expectations.⁵⁶ After Laplace's death in 1827, Arago presented Laplace's system of science as a moribund machine turning endlessly in circles.

Arago used Laplace's own methods for securing control over the sciences against him. Taking over a number of powerful positions—director of the Bureau of Latitudes, director of the Observatory in 1843, and in 1830, permanent secretary of the Academy of Sciences, which made him editor of its *Comptes Rendus*—Arago redirected research, controlled publications, and promoted candidates opposed to Laplace and his allies; he also publicly sided with Geoffroy Saint-Hilaire against Cuvier, who had been Laplace's ally under the empire. From these institutional strongholds, Arago delivered a jolt of romanticism to precision physics, taking inspiration from the personal and scientific example of his friend Humboldt and advancing an astronomy shaped by the experience of working in the field. By making positional astronomy the model for all sciences, Laplace ignored the aesthetic and affective aspects of natural knowledge and reduced its experiential dimension. In contrast, Arago's picaresque *Histoire de ma jeunesse* recounted his near-death encounters while measuring the meridian through France; his later scientific reports frequently included first-person narration describing his preferences and emotional responses.⁵⁷ Like Humboldt's *Cosmos*, Arago's lectures on astronomy, published as *Popular Astronomy*, were full of evocative rhetorical flourishes and sensuous descriptions aimed to engage the listener's imagination. One chapter, for example, inquired about the habitability of comets, while others related anecdotes from Greek mythology, the history of science, and Arago's own adventures.⁵⁸

Arago's speeches as permanent secretary and deputy also testified to a poetic sensibility. He quoted Goethe as an authority on the use of numbers in politics; an epigram from Byron colored his "Eloge d'Ampère," which mentioned verses by Ampère that could "figure in the debate, if it was renewed," over whether or not scientific studies "dry out" the intellect.⁵⁹ In his obituary for the deputy Eusèbe Salverte, reprinted as the preface to the latter's history of the occult sciences, he argued, "Yes, *messieurs*, he had a warm heart."⁶⁰ For Arago, a *coeur chaud*, the quality that Laplace lacked, was *de rigueur*. Professions of strong feeling were frequent: in his discourse on the daguerreotype, he pleaded, "We are forced to share our convictions with you because they are lively and sincere"; elsewhere, out of his "intimate conviction" of "the great, the majestic figure of Condorcet," he combated the philosopher's detractors "with visor raised."⁶¹ Such linguistic forms and gestures typified the "expressive" romantic subject. Yet against assumptions about romanticism as subjectivist, asocial, or merely fantastic, Arago, the "Jupiter of the Observatory," applied these conventions of unconventionality to projects marked by patriotism, mathematics, and machines.[62]

Arago's science made common cause with the sciences of the field. The primal scene of Mongean engineers was Napoleon's expedition to Egypt, where Fourier and Geoffroy Saint-Hilaire had participated in geographical surveys and collections that extended the French empire in space as well as in time. Likewise, Humboldt's global science was based in travel and field science and offered an alternative cosmological horizon for both terrestrial and celestial physics.[63] At the Observatory, Arago developed the astronomical wing of Humboldt's geophysics, presenting a Humboldtian science of the heavens.[64] He cultivated the same descriptive, aesthetic approach to the skies as his Prussian ally developed in the study of the variations of the globe; like Humboldt, he presented science as an open-ended exploration and dialogue. While Laplace's reports of his findings were largely mute about the role of specific apparatus, Arago's *Astronomie Populaire* abounded in descriptions, images, and analyses of the functioning of new scientific devices. As for Humboldt, the play of instruments was part of the scientific drama. Arago celebrated the mechanical inventiveness of researchers and artisans, citing the work of lens grinders, clockmakers, and his preferred instrument makers, the Soleil family. Against the back wall of the lecture hall he had built for his popular lectures, he affixed large, realistic paintings of a telescope, an astronomical clock, and a meridian circle, totemic medallions that gazed out at his rapt crowd.

Arago has also been shown to be a forerunner of astrophysics by devising instruments to measure light and color intensity as a way to ascertain the physical makeup of celestial bodies: the sun's gaseousness, the origin of moonlight, the physical nature of comets and planets. His ingenious experimental setups and instruments combined and redirected discrete techniques to fix on new phenomena: his cyano-polarimeter, a hybrid of telescope and polarimeter, quantified the degree of blue in the sky. Furthermore, the roof of Arago's Observatory was crowded with apparatuses for meteorological and atmospheric measures (fig. 4.3). Colbert and Cassini's royal watchtower and Laplace's imperial fortress became, under Arago, a cosmopolitan outpost in the midst of the metropolis, a crucial node in Humboldt's global network tying together observers, instruments, and the natural world.[65] Again and again the object of Arago's study was the space between the object observed and the observer: "the nature ... of the milieu crossed."[66] This interest can be found from his early studies with Biot on

FIG. 4.3. Meteorological instruments on rooftop of Observatory of Paris. Copyright Observatoire de Paris.

the effect of different gases on the transmission of light, through to his later explanation of the scintillation of stars by their transmission through atmospheric layers of varying degrees of temperature and humidity.[67] To study these milieus, one needed to interrupt, divert, or capture light with the regular, repeatable "interference" provided by instruments; by studying the properties of these media, the researcher took account of the ways in which they structured the milieu between the object studied and the observer.

According to the image Arago built up in his writings, Laplace had presented the universe as a balanced system unhindered by resistance or friction; likewise, heavenly bodies could be known as points of light, observed by an abstracted eye in the astronomer's cabinet using obedient, unproblematically "transparent" instruments. Arago had no illusions that such an "unmediated" vision was possible. The human eye, like other optical devices, was an instrument with specific sensitivities and limitations. His writings referred to a variation on the daguerreotype as an "artificial eye," and he stated elsewhere that "the eye may be treated as a lens having for its 'focus' a screen of nerves named the *retina*."[68] Arago made it clear that *something happens* at the interface between the object and the observer. He presented his instruments as specific modes of mediation and interference, which introduced friction and thus, in many respects, a deformation between the observer and the natural world. The deviation introduced by the instrument was not a hindrance to knowledge but rather its necessary condition: to learn about the primary object of study, one also had to study the modifications effected by this interface. Arago set his model of scientific sociability against the image of an echo chamber of frictionlessly obeyed commands and of a science that progresses through limited observations processed by unfeeling machines of reason.

INSTRUMENTS OF THE GENERAL WILL

Arago's reign at the Observatory and the Academy of Sciences was marked by spontaneity, dialogue, and openness—positive qualities shared by humans and machines. Following the revolution of 1830, in which he took a leading part, Arago served in the Chamber of Deputies as a representative of his native Pyrénnées Orientales. He was quickly identified with the republican opposition to the plutocracy running rampant under Louis-Philippe. Similarly, he sought to replace Laplace's imperial control over scientific knowledge with a "republican" drive to make science available to the widest audience possible for the benefit of the nation as a whole. Along with his lectures for the Association Polytechnique, for years he gave lectures on popular astronomy at the Observatory to a full house.[69] Further, he reshaped the representation of science in the press: he took over the editorship of various journals and changed the format of the reports of the Academy of Science to make them more accessible, with complete accounts in the form of scientific articles, rather than schematic records based on oral presentations. He also opened the Academy's gallery to reporters and the public, encouraging newspapers to publish a weekly account of its scientific news and debates: this became the "feuilleton scientifique"—with Alexandre Bertrand as a forerunner, and eventually Léon Foucault as contributor—which occupied the same space at the bottom of page 1 as the weekly installments of romantic novels. These provocative changes aimed at increasing public participation and "transparency" in the conduct of science, as well as making the Academy a stage for Arago's republican politics.[70]

The job of permanent secretary required Arago to write memorial obituaries for departed scientists.[71] Arago deepened his connection to the revolution through this role as historian of science. Retrieving the examples of his predecessors, including, notably, Condorcet and Carnot, he was a preeminent inventor of a tradition of Enlightenment that connected the philosophes and the Revolution to the political struggles of industrialization.

Condorcet had been demonized by reactionary writers of the Restoration; they blamed him for the Terror, portraying him as a ruthless advocate of egalitarian uniformity. Arago showed Condorcet's main preoccupation to be the protection of personal freedom and democratic process in an orderly but progressing society, one strengthened by vibrant civic participation and state-funded education.[72] Rather than the automatic obedience demanded by a despot, or the transparent realization of the general will envisioned by Rousseau, Arago recalled the structures that Condorcet proposed for painstaking democratic process.[73]

In his *Eloge de Carnot*, Arago noted the "analogy that can almost always be pointed out between scientific theories and the rules of conduct of their authors." Lazare Carnot revealed to him that he had maintained a steady course through the upheavals of the Revolution by recalling his own theory of machines, which demonstrated that a sudden change of speed caused a greater loss of momentum than a gradual one.[74] Arago identified not only with Carnot—the founder of the Ecole Polytechnique, the exemplary citizen-soldier, the "organizer of victories"—but with his mechanics: in a speech in the Chamber in 1840, Arago argued that gradual measures of reform were required to ensure a progress that would be "constant, regular, without shakes, without violence"—precisely Carnot's language for efficient accelerations. Elsewhere, Arago argued that there was no difference in kind between an "instrument," a "tool," and a "machine": "Such a distinction is puerile: [it would be] impossible to say with precision where the tool ends and the machine begins."[75] We might thus hazard an identification between Arago and both the explosive and dynamic power of the steam engine and Humboldt's flexible and gregarious instruments. Arago endorsed an image of machines that combined disciplined regularity with spontaneity and freedom.

Steam engines were more than a symbolic resource for Arago. We have mentioned his studies of engines with Dulong and Petit, conducted at the Observatory. Taking the political stage in the late 1820s and the revolution of 1830, he made himself one of the leading public advocates on the question of the social impact of steam technology, transforming revolutionary rhetoric about "the people" into a concern for "the working class."[76] Arago saw the fates of machines and of workers as inseparably entwined. In his biography of James Watt from 1834—much reprinted in both France and Britain—the central section, "Machines Considered in Their Relations with the Well-Being of the Working Classes," combated "the opinion that machines are harmful to the workers," calling it "an old prejudice without any current value, a true phantom."[77] He cited Say's argument that increases in productivity and the subsequent lowering of prices result in a demand for more varied machine-produced goods, leading to the expansion of markets, and eventually to a net increase in the number of workers employed and an improvement in their quality of life. Further, he pleaded for state recognition and patent protection for inventors, lamenting the British government's neglect of Watt (the essay was swiftly and appreciatively noted and republished in British journals).[78] He concluded with a prophecy: "A time will come when the science of destruction will bow down before the arts of peace"; steam engines will let humans "penetrate into the entrails of the earth"; railroads will connect distant regions, and steamboats will cross the seas; all branches of each domain of production will be joined under a single roof. Even the "steppes of Europe" will be covered with "elegant habitations"; "the population, well fed, well dressed, well heated, will quickly grow."[79] This was a utopia of communication, circulation, and production where technological advances would hasten the flow of goods and ideas through reticulated networks, drawing forth nature's wealth and renewing the earth itself—a vision exactly aligned with those we will see expressed by his socialist contemporaries.[80]

Arago's discourse "On Electoral Reform" of 1840 likewise incorporated the science of machines and advocacy on behalf of inventors into a political program for improving the lot of workers. It warned about the political threat posed by the miserable conditions of the poor: their "cruel sufferings" were the result of the control by a "très-petit" number of "capitalists" over mechanical industry. He denounced his fellow deputies for tolerating these egoists; the "monopoly Chamber," as it had been called, had "duped and blinded the people."[81] To defuse this explosive situation, Arago passionately argued for expanding the electorate to include all males. As proof of workers' political merit, he listed the inventions they had contributed to the commonweal, mobilizing the labor theory of knowledge in the politics of representation of the laboring classes. In recognition for Arago's strong support of the movement of workers' liberation, he was named one of the heads of the Provisional Government that was installed after the workers' revolution of 1848, and he is still celebrated for issuing the law abolishing slavery in the colonies.

In Arago's view, instruments could be "autonomous" in a specific sense: they were disciplined and interconnected but at the same time spontaneous, active, and free. His own conduct exemplified these values. As a deputy, Arago stood in an intermediary position between responsiveness to the interests of those who elected him, the needs of the nation, and the demands of parliamentary procedure. Yet he was not a passive channel for these disparate demands. As seen by his often explosive performances in the Chamber and in public meetings, fulfillment of his duty did not mean the suppression of his individuality, emotion, or personality. He presented the republican image of the obedient but enthusiastic citizen-soldier in a key of romantic individualism.[82] As a dissenting deputy, he played the role of a "useful interference": he captured the ordinary course of government and redirected it in a useful way.

"THE DREAM HAS BEEN REALIZED"

The threads of Arago's commitments with regard to science, machines, representation, and politics all wove together to form the background for the text most often associated with his name today, his "Discourse on the Daguerreotype" of July 1839. Interest in photosensitive materials had been growing in previous years, with experiments in Britain by Erasmus Darwin, John Herschel, and William Fox Talbot. In France, the collaboration between the stage designer Louis-Jacques-Mandé Daguerre—well known for his diorama, an audio-visual illusion discussed in chapter 5—and the inventor Nicéphore Niépce resulted in a successful and repeatable process for fixing the images in the camera obscura. They sought in vain for investors until Arago agreed to promote the technique and to secure them a lifetime pension, in exchange for making it available to the public. After a long campaign of tantalizing leaks to the press, Arago at last announced the discovery to the world in the Chamber of Deputies, to great fanfare. Arago's speech, quickly reprinted, has become a classic reference in the history of photography.[83]

Aesthetic, moral, and political contexts wove through his presentation. He detailed Daguerre and Niépce's innovations, focusing on their experiments with chemicals before arriving at the combination of silver, iodine, and mercury. The images that they produced appeared against a silver background and were visible only from certain angles and distances, giving them a spectral and ghostlike quality (see fig. 4.4).[84] Yet as suggested by the images that Daguerre originally offered Arago (a satyr and a nymph by a fountain, an image of Eve and Venus, and between them a view from the artist's atelier—a series that juxtaposes classical, biblical, and contemporary figures of creation), the daguerreotype was immediately recognized as more than a technical object.[85] Accordingly, Arago's speech was much more than a technical report.

By setting the speech in the Chamber of Deputies, not the Academy of Science, Arago consecrated the daguerreotype in a ritual of the state,

FIG. 4.4. "Bust of Homer," daguerreotype (gift from Louis Daguerre to François Arago). Credit: Collection Musée des Beaux-Arts Hyacinthe Rigaud, Perpignan. Service Photo, Ville de Perpignan.

"in the name of national glory," to recognize achievements that occupied a "very elevated region."[86] Making the daguerreotype the final entry in a history beginning with "the Neapolitan physicist Jean-Baptiste Porta," the Renaissance natural magician whom he identified as a founding figure of his own profession (physicist), he presented discoveries concerning metals and chemicals made by early alchemists as in continuity with the more recent explorations of silver chloride by poetic chemist Humphry Davy. Della Porta and others had longed to see the lines that the camera obscura drew on a wall become permanent; but this was "a dream that was destined to find its place among the extravagant conceptions of a Wilkins or of a Cyrano de Bergerac." With the daguerreotype, however, Arago proclaimed, "the dream has been realized."[87] Having established its novelty, he moved to its utility: he foresaw its use in inventorying monuments of France and regretted its absence during Napoleon's survey in Egypt. As this expedition had stoked a widespread Egyptomania, with frequent speculations about hieroglyphics and the secrets of the pharaoh's magicians, Arago associated the daguerreotype with the strangely complementary projects of mapping the national territory, imperial expansion, and an illuminist decoding of nature.[88]

The device, Arago noted, preserved "nearly mathematical" relations, making it useful for drafts or for surveying.[89] Yet his observations on the technique's geometric fidelity vied with comments about its subjective, affective, and physiognomical impressions. In Arago's speech the daguerreotype's photochemical effects were bound up with its theatrical, dramatic, aesthetic effects. He cited testimony of the contemporary painter Paul Delaroche to allay the fears of those who saw the technique as a threat to artists and engravers. Anticipating Baudelaire's famous pronouncement of 1859 that photography should restrict itself to the role of a servant, Delaroche claimed, "The admirable discovery of M. Daguerre is an immense service rendered to the arts," as it provided them more exact models.[90] Yet Delaroche granted that these images possessed striking aesthetic qualities of their own: "The correctness of the lines, the precision of the forms is as complete as possible in the designs of M. Daguerre, and we see in them at the same time a large, energetic model, and an ensemble as rich in tone as in effect." Delaroche, whose own paintings reconciled classicism and romanticism, saw the geometrical precision of the image as a complement to its dynamic impact. Along these lines, recent critics have suggested that formal aspects of daguerreotypes were continuous with the ambitions of romantic painting; meticulous observation was not antithetical to meaning or effect.[91]

At last Arago spoke of specifically scientific uses. The daguerreotype would aid photometry, the measurement of intensities of light. Instead of comparing, with difficulty, the brightness of a star to an artificial light, the physicist now would "compare lights by their effects," measuring the rate at which they developed a silver plate.[92] In this use, the daguerreotype had to be combined with telescopes and clocks; the new instrument did not stand alone but was immediately understood as a member of the community of the observatory's tools. Other scientific uses—topographical surveys and a map of the moon—were firmly set within the Humboldtian field sciences. Another application he proposed bypassed the image's visual content entirely. Noting that the plates develop at different rates depending on the time of day and the location, he suggested that "the meteorologist would have one more element to include in his *tableaux*, and to the previous observations of the state of the thermometer, the barometer, the hygrometer and of the transparency of the air, he will have to add an element that the other instruments do not grasp, and he will have to take account of a particular absorption, which cannot be without influence on many other phenomena, even on those touching on physiology and medicine." Here Arago was not interested in the objects depicted on the silver plate, but rather in what the process and rate of its development revealed about the invisible atmospheric phenomena that made them appear. Rather than reproducing what the perfect, unbiased human eye would see, in this instance the daguerreotype registered *invisible* phenomena unfolding over time.[93] The daguerreotype was presented by its first public supporter as another member of the family of Humboldtian geophysical instruments—another temperamental, site-specific, and networked tool for registering and mapping a specific range of phenomena, manifesting the invisible, dynamic connections among them.

REFLECTIONS ON TRANSPARENCY AND RECIPROCITY

Arago's quickly reprinted and much-cited discourse on the daguerreotype invites a more detailed reflection on the place he assigned this technique in science and society. In line with the labor theory of knowledge, Arago foregrounded the technical, labor-intensive, and transformative aspects of scientific research; this image of scientific labor contrasted with a rival tendency to hide or minimize labor and artifice so that experimental facts spoke as the unmodified voice of nature.[94] Arago's emphasis on process and transformation also applied to the daguerreotype—a peculiarly dense, active, and idiosyncratic mediation between natural objects and human viewers.

Recent writings on early photography have, however, stressed its putative transparency. As part of a rich study of Arago's battles with the Laplacean physicist Biot, Theresa Levitt argues that Arago in 1839 took the daguerreotype as "an unproblematic representation of the thing it depicted," an image that "could easily be used to stand in for the world."[95] When speaking of a technique that presented images of natural objects, two distinct though frequently combined notions of "transparency" may be implied. First is that the technique produced faithful, unmodified visual images of the objects of the world. A frequent claim about photography is that the image of nature it produces is in principle verifiable by simple comparison with the object as we see it with unaided eyes; yet such resemblance is necessarily incomplete, and strikingly so in daguerreotypes. Colors are altered, movement is lost, and a landscape is reduced to a few square centimeters. Though obvious, the point bears mentioning, since the fascination for daguerreotypes in the 1840s derived in part from their difference from the objects they represented: the difference between a three-dimensional, full-color setting and a scaled-down, static, reflective, two-dimensional image. Even if certain proportions were maintained, the image was in fundamental ways a deformation.[96] Similarly, Arago's hypotheses on its mode of action suggested not transparency but transmutation.[97] To describe the operation, he used language of activity, transformation, and even vitality: he spoke of "the action of light," "penetration," and a "sensing substance"; "the most feeble rays of light modify the substance of the Daguerreotype." During development, the "operator" of the technique could see "the mercury vapor, like the most extremely delicate pincer, mark each part of the plaque with the appropriate tone."[98] Far from a passive transmission of visible properties, Arago described an active modification. The idea of transparency as visual resemblance was further undermined by Arago's reference, discussed above, to the "particular absorption," where the relevant observation was the time taken for exposure, not the image developed. As with the photometric observations of Arago's assistants Foucault and Fizeau, this "photogenic effect" was indifferent to mimetic representation; it was a photochemical process whose interesting feature was its duration.[99]

Alternatively, transparency can mean that a technique or instrument's use has become so much a part of routine activity that its mode of action is taken for granted: it has been "black-boxed." Disputes about it have reached "closure": "Closure makes instruments into what are seen as uncontestable transmitters of messages from nature, that is, it makes them 'transparent.'"[100] As long as outputs follow inputs in a regular way, unresolved questions about an instrument's actual operations may be ignored. In this sense, the process of fixing the images in the camera obscura was indeed a black box. According to Arago, this "mysterious" process displayed "many curious phenomena"; his hypotheses about the interaction between the forces of iodine, silver, mercury, and light were so many submolecular fantastic symphonies; he concluded, "We will make perhaps thousands of beautiful drawings with the daguerreotype before its mode of action will have been completely analyzed."[101] However, the daguerreotype was definitely not a black box, if by that we mean a rapid, automatic, and completely predictable passage from input to output. Many steps were required before photography could be seen as a reliable, instantaneous technique in astronomy, before stars could "register themselves."[102]

In 1839 the time for development was still thirty to forty-five minutes, varying widely according to atmospheric conditions. Although Arago insisted on the ease of the process, he was immediately rebuked in reviews in the *Journal des débats*, which asserted a much steeper learning curve.[103] Moreover, Arago deliberately downplayed the possibility of making portraits, the most hotly anticipated use, because of the difficulty of getting subjects to sit immobile for long periods in bright sunlight. Finally, mass reproduction was out of the question because of the "delicacy" of the process: it would be impossible, Arago said, to use the silver plates themselves as stamps for lithography, as the rolling press would destroy them. "But," he wondered, "would anyone imagine giving a strong pull on a band of lace, or of scrubbing the wings of a butterfly?"[104] The process was not automatic; nor was it instantaneous; nor was it well understood. Rather than emphasize the rigid repeatability of the process or the sturdiness of its products, he compared the daguerreotype to lace and butterfly wings, the most fragile of artisanal and natural creations.

For Arago, the daguerreotype fused the aesthetic and the cognitive, the fleeting and the fixed. Like the citizen who is at once autonomous and completely dependent on the nation, the unique properties of the daguerreotype allowed it to participate in an existing network of instruments and researchers and, at a higher level, in the "moral economy" that Arago worked to establish in French science after Laplace.[105] The term *moral economy*, retrieved in the history of science by Robert Kohler and Lorraine Daston, was coined by Marxist historian E. P. Thompson, for whom it meant the assumed obligation of the wealthy to guarantee the basic subsistence of the poor. As Arago warned in "Electoral Reform," violation of this tacit agreement would justify revolt.[106] In keeping with this original use of the term, the moral economy established by Arago encompassed a much broader social space than laboratories, universities, and qualified audiences.

According to anthropologist Marcel Mauss, all economies are moral economies: material exchanges are a vessel for the redistribution of status and esteem and a means of structuring social relations.[107] Arago's presentation of the daguerreotype was an overdetermined move along Maussian lines. It manifested a complex system of reciprocal relations—at once symbolic and material—linking inventor, science, government, and people. In sponsoring the pension for Niépce and Daguerre, Arago, the broker of this exchange, guaranteed financial recompense and gave lasting credit to Daguerre by immortalizing his name. Similarly, in exposing the device to the broadest possible public, he acknowledged the debt of an intellectual elite to the productive labor of the rest of society—a debt he also repaid with his promotion of electoral reform and his popular lectures and lessons on mechanics for workers. Finally, by ensuring the priority of Daguerre and Niépce and recognizing the invention as French, Arago's intervention became a patriotic act of republican piety, a gift for the glory of the nation, one whose future rewards could only be imagined: "When observers apply a new instrument to the study of nature, what they have hoped for is always insignificant compared to the succession of discoveries which the instrument orginates. In this manner, it is on *the unexpected* that one must particularly count."[108]

TECHNOLOGIES OF TRANSMUTATION

The daguerreotype was presented by Arago as a productive, active interface, introducing a productive friction between light, chemicals, and silver. Despite its eventual ability to produce images that appeared "free from human intervention," this earliest instance of photography was explicitly and ineluctably recognized as a process dependent on variable settings and skills of individuals to make it work. We might extrapolate from this example to the other techniques introduced at this time by scientist-engineers, including dynamometers and, as discussed in chapter 7, indicator diagrams. These instruments have been seen as forerunners of the impersonal, ascetic, and automatic techniques celebrated later in the century; the machine-based ideal of objectivity has been read back onto engineers' obsession with precision, which in turn has been presented as an ascetic negation of the human. Yet as suggested by the labor theory of knowledge championed by Arago—as well as his flamboyant public persona—the early nineteenth century's pursuit of precision, even or especially when it involved mechanical techniques, was acknowledged as inexorably bound to its fleshy, temporal, "laboring" aspects. Machines were seen not as the negation of the human but as vital extensions of human activity, and thus they could be celebrated as part of labor's foundational role in a society increasingly organized around industry.

Further, Arago's emphasis on the opacity and mystery involved in the photographic process—qualities accentuated by Balzac and Nerval, who feared the process would capture the outer layer of their spirit—suggests a kinship between its reception and the metaphysical excitement generated by the contemporary experiments of Oersted, Ampère, and Arago on the convertibility of electricity into magnetism; as we will see at greater length below, similar commotion met technologies that converted fire into motion through the mercurial medium of steam.[109] Much of the fascination for these romantic machines lay in their status as processes, as technologies of time and transformation.[110] To make sense of these novelties, precedents were found in the past. Just as Arago saw the natural magician Della Porta as a colleague, the sculptor Dantan Jeune caricatured Arago as a wizard holding a telescope and a compass in place of a magic wand, perched on his observatory-castle tower, commanding a globe crisscrossed by railroads

FIG. 4.5. Statuette of Arago as a sorcerer, by Dantan Jeune. Credit: Copyright Musée Carnavalet; Roger-Viollet. With thanks to Monique Comminges.

and illuminated by the lighthouse lenses he helped Fresnel invent (fig. 4.5). The statue was not intended as a slight against Arago, nor did it imply that Arago's science was identical to magic. Nevertheless, it is one of many cases in which modern mastery over the secrets of nature was portrayed as a fulfillment of Renaissance natural magic.[111] This sense of the recurrence of a magical past could also be heard in the quote from Victor Hugo at the start of this chapter, in which an aerostatic machine was transmuted into a cosmic egg—a symbol whose alchemical resonances are hard to miss when read alongside Hugo's hermetic reveries about the stars seen at Arago's Observatory.[112] At the time of its introduction, photography was received as an instrument of science and art and as a technology of rationalized magic.

Arago did not always discourage fantastic rapprochements between science and magic. Just as he had called Della Porta a physicist, his allusion to "myriad new worlds" recalled pantheist strands of Renaissance cosmology and its promises of the metamorphosis and conversion of a *prima materia* through secret techniques. Just as Dumas retraced the history of alchemy in his lectures on chemistry, Arago included discussions of astrology and multiple worlds in his *Popular Astronomy*. Though apparently satisfied with the report of the Commission on Mesmerism of 1784, Arago also undertook his own investigation of the new claims of mesmeric clairvoyance in the 1820s and 1830s, declaring that "the refusal of a possibility is not a scientific attitude" and urging continued investigation of animal magnetism.[113] While pursuing techniques of precision and mechanical analysis, and in certain respects remaining committed to the Newtonian "celestial mechanics" of Laplace, Arago nevertheless drafted mechanism—both as physical theory and as a tool—into the service of higher goals of transformation and unity. In significant respects, the realism frequently attributed to photography was entwined in its first presentation with concerns more often associated with the fantastic arts. The daguerreotype balanced delicately among diverse segments of society and the ever-growing family of geophysical instruments, as well as between humans and the shimmering phenomenal surface of nature. In its first frame, it offered an image—precisely scaled down, yet singular, flexible, and autonomous—of its presenter and the milieu out of which it developed.

MECHANICAL ROMANTICISM FOR THE PEOPLE

While the work of Laplace's successors is often characterized by a lack of unity, the last three chapters have shown that more than generational politics was at stake in the new approaches to physics in the Restoration and July Monarchy. That it was Arago who promoted the daguerreotype, throwing his considerable institutional and cultural weight behind it, is just one proof of the vitality and the lasting impact of post-Laplacean physics. Other testimony is provided by Fresnel's wave theory of light and Pierre Dulong and Alexis Petit's contributions to the theory of specific heats and the elastic force of steam (conducted with Arago's help at the Observatory) as well as the theoretical, practical, and institutional accomplishments of Ampère and Humboldt. The anti-Laplaceans shared the aim of unified explanations of terrestrial forces, pursuing the connections and possible conversions among them. They worked with theories of undulatory motion for heat, light, and electromagnetism and had frequent recourse to ethers—whether Ampère's constantly composing and decomposing electromagnetic fluid or Fresnel's largely immobile solid, parts of which were dragged along by matter as it moved though space.[114] These ideas intersected with an interest in the vital and nervous fluids of physiology, psychology, and animal magnetism. Thanks in part to Humboldt's influence, as well as their allegiance with Geoffroy Saint-Hilaire (as we will see in chapters 6 and 8), the post-Laplaceans contributed to the development of evolutionary theories of the organism and the emergence of ecological thought.

Their emphasis on conversion, modification, and mediation suggested a newly developmental, historicized conception of nature. Although Arago and Humboldt were wary about the theory that the solar system originated out of the condensation of nebular gas—due, no doubt, to the association of the nebular hypothesis with Laplace and to Auguste Comte's failed attempt to confirm it mathematically, as discussed in chapter 9—they nevertheless advanced a view of nature as a field of active forces in constant interaction and transformation, undergoing irreversible processes of growth and development.[115] According to this new cosmological orientation, the processes grasped by new sciences and technologies were susceptible in various ways to human modification.

Those modifications could be brought by romantic machines: electromagnetic experiments, geophysical instruments, and the daguerreotype. Attentive to the contributions made by instrumental apparatus and the human senses to the production of phenomena, these researchers did not present knowledge as detached, inhuman, disembodied, or static: instead they showed that human activity played an unavoidable role in the production of knowledge, whether that was in the practices of the workshop and laboratory, as emphasized in the "labor theory of knowledge" or in the action of the will, senses, and muscles, as we saw in Ampère's adaptation of Maine de Biran's philosophy. Humboldt's pregnant definition of instruments as new organs suggested that the activity of science was a means by which humans wove themselves progressively into the field of entities surrounding them.

These scientists' emphasis on the active role of the senses also underlined the aesthetic considerations that underwrote their views of knowledge. They highlighted the importance and variability of individual experience and the moral and intellectual uplift gained through observation of nature; they attempted to represent all fields of science, and nature itself, in harmonious images of unity in which humans, technology, and nature's constituents were joined together. It is also significant that Ampère, Humboldt, and Arago were all strong public supporters of Geoffroy Saint-Hilaire who, as we will see below, was a conduit for German *Naturphilosophie*. At the same time, like Arago and Humboldt, he was an important contributor to the rise of modern popular science, bringing his arguments about the natural world to the attention of a wide audience; he was also an ally of republicans and early socialists. Humboldt's *Cosmos*, like Arago's *Popular Astronomy* and Arago's promotion of the daguerreotype, revealed the harmony between the ambition to popularize science and romantic artists' quest to bring intense sensational and emotional experiences to mass audiences, as well as the reformist ends to which such experiences could be put.

Part 2 moves us yet further out of the enclosed spaces of academies and laboratories into sites of public entertainment and spectacle. Our attention turns to romantic machines in fantastic literature, painting, and music, in the Muséum d'Histoire Naturelle, in the National Expositions, and in popular entertainments from stage magic to illustration. In these settings, we will see a fascination for new machines' power to create new sensations as the correlate to the philosophy of physiospiritualism developed in the works of followers of Maine de Biran; we will also see the parallels between the potentially monstrous modifications that appear in the process of organisms' formation and the demiurgic powers unleashed by engineers. These chapters will also expose the increasing political importance that machines were given in the July Monarchy. The two chapters of part 2 thus prepare for the book's final three chapters in part 3, which lift the curtain on the romantic machines that played leading roles in early socialist projects of political metamorphosis, and show how these visions set the scene for the Revolution of 1848.

PART TWO
SPECTACLES OF CREATION AND METAMORPHOSIS

5

The Devil's Opera

Fantastic Physiospiritualism

TECHNAESTHETICS, OR THE POSITIVE AND THE FANTASTIC

To describe the distinctive mode of knowledge that came of age in this period in the study of chemical forces and invisible fluids, the epistemologist Gaston Bachelard introduced the notion of *phénoménotechniques*. The work of physicists and chemists, he argued, involved the design of apparatus to produce novel, artificial phenomena that could subsequently be brought into theoretical relation. Science was no longer mere observation: it was instead the production of effects—Fresnel's lines of diffraction, the interactions between electric and magnetic wires staged by Oersted and Ampère, "photogenic effects."[1] This view of physics—and of science in general—as the production and correlation of phenomena was folded into a new philosophy of science by Bachelard's distant precursor Auguste Comte, the founder of positivism. "Whether it is a question of the slightest or the most sublime effects," Comte wrote, "we cannot truly know anything but the diverse mutual connections that belong to [the] realization [of observed phenomena], without ever penetrating the mystery of their production."[2] While Comte rejected many of the speculations of his contemporaries, his philosophy was fed by many of the same romantic currents, as we will see in greater detail in chapter 9. Through the coordination of regularly observed or artificially produced phenomena, Comte argued, scientists, guided by philosophers, could assemble a sufficiently reliable "spectacle" of the world. The world was to be built up, section by section, level by level, through the technical production of phenomena and the construction of a harmonious order among them.[3]

Despite his eventual advocacy of a "demonstrated religion," Comte, like other spokesmen for the sciences of this time, sought to establish clear lines demarcating "positive science" from superstition and imagination. This goal seemed particularly pressing in the Restoration, when the Catholic Church reasserted its control over national institutions and when religious feeling, spurred by essays and poems of Chateaubriand and later by Lamartine, led the first wave of romantic literature; for the young Victor Hugo, a feeling for nature and for the divine merged with his support for the restored monarchy. These romantics sought God both in *les voix intérieures* and in the outward pageant of nature.

Yet a darker, stranger note came to dominate the romantic arts from the middle of the 1820s, prompted by the translation of E. T. A. Hoffmann's tales into French. This heralded the rise of a new literature of the fantastic, which pushed effects of the uncanny to a violent irrationality that went well beyond the fairy tales told by Charles Nodier and Walter Scott. Indeed, the term *fantastic literature* was coined in the *Globe* in 1829 by Jean-Jacques Ampère, the son of the physicist André-Marie Ampère and a founder of the field of comparative literature. Ampère the son praised Hoffmann's "natural" and "living" marvels in comparison to the "mechanical" effects of the previous generation's authors, such as Anne Radcliffe and Scott. Hoffmann's texts left the reader wondering whether those episodes that appeared to violate the laws of nature—ghost sightings, lifelike automata, entranced humans in contact with spirits—were merely the imaginings of the protagonist, the result of some material trick, or genuine proofs of the reality of the supernatural.[4] Hoffmann's French imitators—including Théophile Gautier, Alexandre Dumas, Gérard de Nerval, and at times Balzac—likewise featured scenes in which material objects came to life and in which ordinary reality was invaded by worlds of the past or the imagination. In the fantastic arts—whether visual or musical—familiar images or melodies metamorphosed into alien scenes and soundscapes.[5]

Such experiences, if taken seriously, might appear to challenge the stable, predictable world suggested in both late-eighteenth- and late-nineteenth-century views of science. Yet in the early nineteenth century, the fantastic could be seen as the depiction or, more accurately, the re-creation of an experience of the world's plasticity, as demonstrated by science and technology. The fantastic mode in the arts took shape at a moment when the limits of the possible were being stretched; its scenes of animated matter, vibratory communication, lifelike machines, and eerie metamorphoses implicitly and explicitly referenced the scientific and technological transformations of its time. Rather than read the fantastic as a refusal of positive facts, it participated, along with the nineteenth century's confident new sciences, in a dialectic of doubt and certainty. Such was the view expressed in an unsigned review of the 1832 ballet *Le sylphide*:

> The positive sciences have made too much headway today for one to still concern oneself with those that are purely conjectural. If ever one wanted

reality, it is assuredly in our century. We even want too much of it, because this pushes men toward general skepticism after which they do not believe what they see and only put stock in what they can grasp. At bottom, they say, this philosophy is as good as any other! And in spite of this, there is a taste, or rather a vogue for the *fantastic*, which, for a start, has also been timely, thanks to recondite *romantic* writings, where, in the process of looking for the *truth*, one puts everything in question, where in running after *that which is*, one encounters, by a singular piece of bad luck, only *that which is not*.[6]

The critic here defined both positivism and the fantastic by a restless hunger for reality that was bound up with skepticism. Science's craving for certainty had put everything into question; at the same time, romanticism's pursuit of truth in the least likely of places (as in its scenography of remote mountains, exotic lands, imaginary landscapes, or childhood) seemed to lead it into the impossible and frightening scenes of the fantastic. Indeed, the fantastic and the positive traced the same journey in opposite directions. In the fantastic arts, a stable, commonsensical reality was shattered by astonishing, irregular events, while the positive sciences took unexplained phenomena and brought them within the domain of natural, predictable causality.

Another palpable link joined the fantastic and the positive: both were invested in the regular production of novel phenomena. Just as the physical sciences created new devices to control light, heat, and electromagnetism, the fantastic arts were heavily invested in new technical apparatus to produce illusions and to bring about uncanny visual and auditory metamorphoses. To do so, in many cases they used the same technologies as scientists did. Both scientists and artists studied the control of light and color in the camera obscura and the panorama, as well as the sonic properties of musical instruments; scientists were recruited to assist stage designers; and artists, as we saw in the case of the daguerreotype, were recognized for their contributions to the investigation of nature. Just as the sciences were increasingly defined by the technological production of phenomena (not least among them, electromagnetism), the fine arts themselves could now be treated as a branch of the sciences.

One of the new fields invented by André-Marie Ampère in his *Essay on the Philosophy of Sciences* captured this traffic between the sciences and the arts. "La technesthétique" dealt with the "means by which man acts upon the intelligence or the will of his fellows"; its domain was the procedures for "recalling ideas, sentiments, passions, etc., and giving birth to new ones in the spectator of an art object, the hearer either of a piece of music or a speech, or, finally, in the reader."[7] While the Greek root *techne* had long referred to art or craft in general, it was only in this period that "technology" was emerging as a distinct set of objects and production processes, often connected with economic development.[8] In the notion of *technaesthetics*, Ampère captured the calculating, reproducible, and mechanically assisted aspects of the arts: a labor theory of aesthetic effects. He was not alone. Artists of the romantic era reflected on and off the page about the changing technical basis of their craft: the techniques and commerce of printing and the means of making new, striking effects in painting, in music, and in the immersive audiovisual form of the opera. In this chapter, I read the scenes of uncanny transformations and magical powers that defined the fantastic arts as reflections of the metamorphoses and amplified effects made possible by new technologies of communication and spectacle.

The technological dimension of the arts underlined by Ampère the father runs counter to familiar concepts of romantic aesthetics: that artworks form a natural, non-artificial, and organic whole and are thus akin to living things; that art is an expression of freedom; and that true artistic inspiration, the work of the artistic genius, defies any attempt to formulate laws or rules for creation. Under this familiar scheme, whose terms were set in the first half of Kant's *Critique of Judgment*, artists were by their nature hostile to machines and mechanization. In France, such views were advanced by interpreters of German romanticism including Germaine de Staël, Théodore Jouffroy, and Victor Cousin. De Staël's *De l'Allemagne* elevated spirit, imagination, and organisms over matter, reason, and machines: "When man is devoured, or rather reduced to dust by incredulity, [the] spirit of marvels is the only one to return to the soul a power of admiration without which one cannot understand nature.... The universe resembles a poem more than it does a machine."[9] Victor Cousin's manifesto of "eclecticism," *Lectures on the True, the Beautiful, and the Good*, presented "ideal beauty" as a means of leading humans up to the transcendent and the infinite; this "spiritualist" aesthetic was seconded by Théodore Jouffroy, who saw art as the attempt to grasp an infinite domain of ineffable forces.[10] Along these lines, in 1835, the author and critic Théophile Gautier made a slogan of the principle "l'art pour l'art," creating a barrier between the arts and any moral, political, or technical aims. "Pure" art was opposed to the trickery, artifice, and repeatability of mechanism—a critique that later bore fruit in the denunciations of "industrial art" by Flaubert, Baudelaire, and others.[11] These were interpretations of German aesthetics—among them, Kant's and Schiller's—in idealist, otherworldly terms.

Yet, as we will see in this chapter's examples from the literature, visual arts, and music of the early nineteenth century, in spite of this "spiritualist" aesthetics, the use of reason and mechanical devices to produce aesthetic effects was widespread and widely embraced. And just as in chapter 3 we saw that in spite of Kant's emphasis on the "supersensible," and certain idealist tendencies in Schiller's writings, the overall thrust of the *Letters on the Aesthetic Education of Man* was to merge form and sense and to make activity, in common with others, and aided by techniques of all kinds, the basis of the aesthetic state. Aesthetic experience required sense, activity, and objects. Similarly, in France there arose at this time a set of theoretical discourses that addressed the technical and physiological basis of the arts and extraordinary experiences of all kinds. I give this minor tradition — which was never a self-conscious school, but which shared a set of concerns and sources — the paradoxical name of physiospiritualism. What its members had in common, despite their disciplinary and even metaphysical differences, was a serious engagement with the philosophy of A.-M. Ampère's interlocutor Maine de Biran.

Because Maine de Biran was a major influence on Victor Cousin, he is most often read as a founder of the tradition of "spiritualism"; yet his most frequent concern was not spiritual transcendence but rather the dynamic and variable interactions between the mind (*l'esprit*) and the body. His posthumously published journals documented his observation of his changing thoughts, his protean moods, and their dependence on the state of his body and his habits of thought and action, as well as the occasional arrival of unexpected mystical states of "superindividual" well-being. For certain of his readers, including A.-M. Ampère, the science reporter Alexandre Bertrand (whom we met in chapter 2), the philosopher Félix Ravaisson, and the alienist Joseph Moreau de Tours, as well as Bertrand's friend Pierre Leroux, Maine de Biran's preoccupations stimulated reflections on the ways in which physical activity, habits, and external modifications — including, significantly, technical apparatus and drugs — could alter perceptions, judgments, and experiences. These works of physiospiritualism were a philosophical counterpart to the perceptual modifications represented and induced in the same period by the fantastic arts. They emphasized the fragility and the "artificiality" of perception, as well as the role of the perceiver in creating the objects of sense.

Thus their diverse works restated a familiar theme from romantic criticism: that each of us is in a sense a poet or artist, creating our own experience out of raw sensory materials through the activity of our senses and imagination. Coleridge's romantic theory of mind, for instance, spoke of the *primary imagination*, the faculty that synthesizes sense impressions into meaningful experience, and the *secondary imagination*, the faculty that willfully combines fragments of sense in the creation of artistic wholes.[12] The play of the imagination—sometimes identified with, sometimes distinguished from the *fancy*, from the root "fantasia"—was celebrated as the source of the unexpected transitions and chimeras of the fantastic arts.

Paris in the 1830s and 1840s was the scene for a decisive twist on this conception of imagination: the shape-shifting and world-making power of the imagination could be aided and enhanced by technology. New sensory stimuli were created through artistic and scientific experiments and fused together by new techniques. New worlds could be depicted and shared, thanks to new techniques of representation and communication. At the same time, through repetition—often with the help of the regularity provided by external devices—the imagination could be trained, making certain judgments habitual and thus automatic. On one hand, machines helped stabilize a familiar, predictable background of facts and experiences, like Coleridge's primary imagination. On the other, machines could produce the shock of novelty, generating new effects and new combinations of the senses, as in the poetic power of the secondary imagination. In line with our discussion in chapter 4, we might consider this a labor theory of aesthetics.

As an evocative example, consider one of the most famous of the Parisian fantastic arts, Etienne-Gaspard Robertson's *Fantasmagoria*. This spectacle combined the two sides of technologically enhanced fantasia: the reassuring and habitual on one side and the shocking and disruptive on the other.[13] Robertson was a follower of Jacques-Aléxandre-César Charles, an academician who gave performances of popular science near the Palais Royal; Robertson was also a hydrogen-balloon pilot whose journeys inspired Oersted to write *Luftskibet* (airship), a series of poems on the topic.[14] Seen by thousands in the first decades of the nineteenth century, Robertson's *Fantasmagoria* took place in two rooms of an abandoned convent in Paris. Visitors first gathered in a well-lit room where they were invited to look at displays of static electricity machines, Leyden jars, and batteries designed by Robertson's friend Volta.[15] The second room is better known: there, in the dark, to the eerie sound of the glass harmonica—an instrument introduced by Benjamin Franklin as an imitation of the voice, but by the early nineteenth century associated with the realm of spirits—wavering images of ghosts and deceased tyrants appeared and lurched toward and away from the audience. These vague and variable images were produced by a magic lantern on wheels, which made the projection on a translucent sheet grow larger or smaller. The *Fantasmagoria*'s presentation of the predictable, rational effects of electrical machines followed by an optical deception marked as frightening or at least surprising underlined this period's close kinship between scientific realism and fantastic illusion.[16] In literature as well, realist and fantastic techniques were complementary. To produce an effect of uncanniness, the artist first had to master the conventional techniques of realism: only a well-established, predictable world could be effectively thrown into doubt by the sudden eruption of the seemingly impossible. Balzac, a founder of literary realism as well as the author of many visionary texts, enacted this dialectic across his works.[17]

In the phantasmagoric technologies discussed in this chapter, as anticipated in our earlier discussion of the daguerreotype, the fantastic and the "positive," the hallucinatory and the factual, appeared as two sides of a single epistemic coin. *Phénoménotechniques* in the sciences were often the basis for technaesthetics in the arts. Hallucinations and illusions were objects of serious scientific and epistemological reflection. The same constructivist theories of perception, grounded in the active, embodied, and willing self, were used to understand both. Further, the imagery that recurred throughout the fantastic arts—reanimated objects, living machines, and dynamic, protean fluids—captured this period's ambivalent admiration for world-changing machines, as well as its metaphysical and political uncertainties. The fantastic, in both form and content, dramatized the power of technology to remake nature or destroy it, to liberate humanity or enslave it.

THE TRANSFORMATIVE PRESS

French romantic poets did their part to contribute to the cliché that romanticism and machines were undying enemies. Alfred de Vigny wrote poems decrying the railroad;[18] another poet bemoaned

> the somber bunkers
> Where human beings, like automata
> Using up their life and health, mechanically perform
> The same movement from morning until night.[19]

Hugo warned about the "somber machine, hideous monster" that devoured the child laborer.[20] Yet mechanization, especially after 1830, could also be greeted with the opposite response. In his poem cycle *Les voix intérieures*, Hugo urged artists to recognize industry's living, even supernatural power, exclaiming:

> O poets! While you sleep, iron and ardent steam
> erase from the earth the former gravity from all hanging objects
> which had crushed hard cobblestones beneath heavy axles.[21]

An entire subgenre emerged singing of the beauties and demiurgic potentials of mechanization. Victor de Laprade's *The New Age* praised

> the fire which brings metamorphosis.
> It makes all things obey,
> It gives a soul to rude bodies;
> From the mud, at its magic touch,
> Water flows with an energetic thrust
> And moves a forest of steel.[22]

In her novel *Les sept cordes de la lyre* (called "the female *Faust*"), George Sand showed one character's admiration for the "sublime harmonies" created by the "voice of industry, the noise of machines, the whistling of steam, the shock of hammers."[23]

Yet for obvious reasons the most important technological developments for those who worked with the written word were in the domain of printing. A fascination for the impact of the printing press and subsequent technical improvements was inseparable from romantic poetry's oft-noted celebration of the transformative, alchemical, patently physical power of the word—or *la Verbe*—as extolled in Victor Hugo's *Les contemplations*:

> The word is a living being.
> The hand of the dreamer vibrates and trembles in writing it
> . . . . . .
> It comes out of a trumpet, it trembles on a wall,
> And Balthazar stumbles, and Jericho falls.
> It incorporates itself in the people, being itself the crowd;
> It is life, spirit, seed, storm, virtue, fire;
> Because the word is the Verb, and the Verb is God.[24]

According to Hugo, the word vibrates, shakes, destroys, but also creates. Likened to a natural force, a weapon, and a musical instrument, the word has the power to topple empires and forge new societies. Language creates the world.[25] This conception of language drew not only on notions of the *logos*, the biblical "verbe" synonymous with God, but on the emerging discipline of philology. Influenced by the Schlegel brothers and Wilhelm von Humboldt, and spurred by translations and introductions by Eugène Burnouf, Edgar Quinet, and Jean-Jacques Ampère, romantic conceptions of the power of language were informed by what Raymond Schwab called "The Oriental Renaissance." In interpretations of newly translated Sanskrit linguistic texts, language was presented as a form of energy. Romantic poets sought to harness the power attributed to ancient ritual language, understood as both expressing and shaping the essences of things. This conception of signs as a physical force accompanied a conception of nature as fundamentally semiotic as well as subject to growth and alteration. Such potent language was associated with humanity's past; but, in accord with a recurrent romantic temporal structure, it was also seen as a capacity to which humanity might return.[26]

Improvements in printing technology might hasten this return by amplifying the power and reach of words. The notion of rhetoric as an effective, transformative use of signs was reinforced and enhanced by the effects of new technologies of the printed word. Romanticism was inseparable from growth and innovation in the domain of printing. Since the late eighteenth century, authors had no longer been able to turn to aristocratic patrons for support; they were forced instead to piece a living together from the irregular remuneration of the literary marketplace. This was possible thanks to the explosion of printed matter—and the concomitant expansion of the reading public. A massively expanded and increasingly mechanized regime of communication took shape after 1815. Reduction in the price of paper, combined with the rise of lithographic printing and the introduction of cylindrical steam presses—France had one in 1823, and thirty by 1830—as well as changes in copyright law and censorship policies, opened the floodgates of affordable printed works. The daily journal *Le national*, launched in 1830, was followed by the even cheaper dailies *La presse* and *Le siècle* in 1836; 1830 also saw the introduction of the *feuilleton*, or serial novel published in installments along the bottom banner of the newspaper, allowing Balzac, Dumas, Sue, and Sand to publish their engrossing tales incrementally.[27]

It is therefore no accident that the glorification of the press and its quasi-mythical founder, Gutenberg, was a frequent theme of romantic literature. Resonating with Hugo's line from *Notre Dame de Paris*, "Ceci tuera cela"—the printed book will destroy the edifice of Catholicism—many enthused about the epochal effect of the invention of printing. Nerval, who himself designed a new printing machine, co-wrote a play about the print revolution.[28] Balzac was so obsessed with the material format of his writings that he both opened his own publishing house (which like most of his financial schemes was a fiasco) and took printing technology and the effects of words as a frequent topic. His thrilling novel of the world of publishing, *Lost Illusions*, contained lengthy passages on the history of paper and printing. It traced the strategies used by journalists to manipulate public opinion and inspected the *éloges*, polemics, defenses, and apotheoses that encouraged or inhibited the growth of a "great man in embryo." It detailed the mechanics of publicity and the art of the paid applauders, or the *claque*, which could turn a promising obscurity into a larger-than-life public entity, as occurred for Lucien's girlfriend, Coralie.

In *The People*, Michelet bragged of his birth in a church that had been occupied by his family's printing shop: "Occupied but not desecrated, for what is the Press in modern times but the holy ark?"[29] Similarly, in 1829 the poet Ernest Legouvé received a prize at the Académie Française for an epic that retold the history of humanity in terms of communications technology. After centuries in which knowledge was passed down only by song and memory ("souvent infidèle") and papyrus (which preserved sacred texts but restricted their readership), "Gutenberg appeared!" Because the great movements of thought "change their character as they change their instruments," Legouvé detailed the technical improvements, such as movable type and improved ink, that allowed Gutenberg to unleash the press's "puissance magique" and "pouvoir créateur." Thanks to printing, "in a thousand hearts the same soul enters and lives," and a whole people thus "rises up, animated by a single design," bringing revolution and liberty. Gutenberg "made the god of knowledge a popular God."[30] The press was a more-than-human political power balanced between mind and matter, revealing the inadequacies of this world and preparing a new one. According to Lamartine, "Gutenberg spiritualized the world."[31]

Goethe's *Faust*, with its alchemically fabricated homunculus and Mephistophelean dealings, was translated in the early 1820s by Gérard de Nerval; Balzac's *Quest for the Absolute* of 1834 was a modernized version of this story. The notion of the scientist as magician and creator reflected back on the authors of these texts. Literature became a quest for transformative, spirit-altering effects, and literary authors ambivalently likened themselves to mages and sorcerers.[32] Hoffmann, who according to Sainte-Beuve "unleashed and laid bare the power of magnetism in poetry," was exemplary.[33] Like the more familiar "Sandman," his tale "Automata" merged mechanism, mysticism, and reflections on literary power. It began with a group of friends staring "stiff and motionless like so many statues" at a ring oscillating in obedience to an invisible force. The tale depicted an encounter with a speaking machine, the Talking Turk (clearly inspired by the chess-playing automaton that had been introduced by the impresario and metronome-inventor, Maelzel), which might have possessed or been possessed by divinatory powers. Its protagonists met a scientist obsessed with mechanical music and a singer who might have been one of his experiments. In each case the question arose whether characters were moved or were movers, helpless puppets or godlike actors. The tale's unresolved ending linked these uncanny machines to Hoffmann's own philosophy of composition; when Hoffmann's protagonist was asked by his friend how the story-within-the-story finally ended, he replied: ". . . I told you at the beginning that I was only going to read you a fragment, and I consider that the story of the Talking Turk *is* only a fragment. I mean that the imagination of the reader, or listener, should merely receive one or two more or less powerful impulses, and then go on swinging, pendulum-like, of its own accord."[34] Here the pendulum, which could be used to keep musical time, measure the force of gravity, or induce mesmeric sleep, gestured simultaneously to the formal play of aesthetics, mechanistic law, and supernatural forces. Hoffmann presented his obsessed protagonists—and, by implication, his spellbound readers—as autonomous, imaginative agents and at the same time as mechanically determined objects.[35] The text itself entered this chain of associations: as mere paper covered with letters, it was an inert material object. Once read, however, it had a power to entrance and transform the experience of the reader.

The printed word often appeared under such a doubled aspect, as both a potent new technological object and a doorway to supernatural realms. Chapter 6 looks more closely at the fantastic imagery of automata and uncannily animated objects. These could be read as allegories of scientists' and engineers' attempts to control nature and thus challenge God, a concern that was also underlined by the recurrent figure of Satan in romantic literature and the temptations of his "science."[36] Such Mephisthophelean imagery reflected fantastic artists' own efforts to harness the materials of words, paper, and press to produce lifelike effects in the darkened room of the mind.

ROMANTIC AUDIOVISUALITY

The power to create visions and alter perceptions was trumpeted not only in Robertson's *Fantasmagoria* but in the panorama. The craze for panoramas began in the late eighteenth century in London and Paris but was still going strong in the 1830s when Balzac showed the characters of *Père Goriot* playfully "talking '–rama'" by adding the intensifying suffix "–rama" to any word.[37] Balzacorama! The panoramas were cylindrical buildings that visitors entered by climbing darkened stairs into a huge circular room, illuminated by indirect sunlight coming from the center of the roof, surrounded on all sides by a hyperreal landscape that could include distant mountains, cities, ports, or battlefields. The panorama was treated as a solemn, even sacred space, one that critics often compared to the grave; visitors were said to maintain a respectful silence throughout their visit.[38] Like the Egyptian pyramids, it was a machine of reanimation, producing lifelike illusions out of inert matter.

The panorama was of interest to scientists because it stimulated and at the same time made observable the mechanics of natural perception. A report from 1800 to the Institute on the panorama stated: "Let us suppose that the eye, no matter which point on the horizon it is brought to, is constantly struck by a series of images, all in relative proportion, all with the tones of Nature, and that nowhere can it seize the object of comparison which it needs to base its judgment on: in this case it will be fooled; it will believe that it sees Nature, because Nature is no longer there to tell it otherwise."[39] By engaging the same processes of perception as those stimulated by natural objects, the panorama tricked the eye and mind to experience it as the real thing. The neoclassical painter Jean-Louis David told his students: "Truly, sirs, it is here that you must come to study nature." Alexander von Humboldt wrote in *Cosmos* of the pedagogical utility of the panorama's "theatrical illusions," as a means of conducting the aesthetic education brought by popular science on a truly mass scale: "[In] Barker's panorama, by the aid of Prevost and Daguerre, [these images] can be converted into a kind of substitute for wanderings in various climates. More may be effected in this way than by any kind of scene painting; and this partly because in a panorama, the spectator, enclosed as in a magic circle and withdrawn from all disturbing realities, may the more readily imagine himself surrounded on all sides by nature in another clime." He also suggested that entrepreneurs should create "large panoramic buildings containing a succession of such landscapes" and that they should be situated in cities as "a powerful means of rendering the sublime grandeur of the creation more widely known and felt."[40] Panoramas not only instructed the viewer but also transported the uplifting experience of untrammeled nature into the centers of civilization, bringing about that transformation of the ordinary into the magical that was the goal of art.

The panorama spawned a mania for painting on a large scale, including sublime views of the beginnings and ends of worlds. Englishman John Martin displayed his biblically proportioned, prehistoric landscapes and scenes of apocalyptic revelation in London and Paris. Eugène Delacroix, French romanticism's most celebrated painter, who was a friend of George Sand and Franz Liszt, rejected the sharply defined forms and balanced compositions of his predecessors David and Jean-Auguste-Dominique Ingres. Delacroix's salon-shaking works such as *The Death of Sardanapalus* (1827, now held at the Louvre), whose violent subject matter and refusal of classical conventions forced the eye to move restlessly across the entire painting, produced a movement or a shock in the viewer; Michael Marrinan has accordingly described Delacroix's works as introducing an *aesthetics of confrontation*.[41] Delacroix's dynamic, pulsating use of contrasting color and irregular forms also revealed a fascination with energy and its transformations. Much as Michel Serres linked the English painter J. M. W. Turner's amorphous landscapes to Sadi Carnot's writings on the steam engine, we might read Delacroix's works as a visual inscription of the explosive and energetic universe taking shape in the physics of heat and imponderables.[42] Yet Delacroix's quest for effects betrayed further influences. For one, his early *Barque of Dante* was a reworking of Théodore Géricault's celebrated *Raft of the Medusa*, a colossal work shown in London to forty thousand viewers and immediately canonized. Jonathan Crary has argued that the impact of Géricault's *Raft* was due in part to the fact that it depicted a news event that implied a pointed rebuke of government negligence, while its visual impact derived from its use of the size and effects of the panorama.[43] In borrowing Géricault's massive scale and emphasis on movement and effect for his *Barque of Dante*, Delacroix's painting participated in the visual order of the panorama.

The panorama also provided a backdrop for the inventions of Louis-Jacques-Mandé Daguerre. Before attaining worldwide fame for the daguerreotype, Daguerre was already recognized as a theatrical designer and inventor. He had apprenticed in stage painting with Prevost, the builder of the Parisian panorama, and was recognized for his use of gas lighting to produce ethereal effects for the opera *Aladdin's Magic Lamp*. Building on these techniques, Daguerre's diorama, launched in 1822, was an immersive, hallucinatory spectacle. It was housed in a specially made building where audiences gathered in a darkened room (see fig. 5.1). There they watched a lighted screen, showing objects in depth, as it began slowly to transform itself from night to day or from winter to summer, often accompanied by music and other sound effects. Effects of depth and motion heightened the illusion. The diorama was even more explicitly a "machine" than was the panorama: not only was its lighting system mobile; the entire viewing platform rotated to bring visitors face to face with two, sometimes three, distinct views.[44]

FIG. 5.1. The diorama made by Daguerre for the church of Bry-Sur-Marne; the only surviving diorama. Credit: Photograph by Matthieu Lombard, Association Louis Daguerre.

How did it work? In Daguerre's "double-effect diorama," paint was applied to both sides of the canvas, so that a single painting could convey two distinct images, depending on the precise lighting: in green-filtered light, green paint would lose distinctness; in red-filtered light, red paint would lose distinctness. He used this process of "composition and decomposition of form," changing the direction and color of the lights, to change the color and composition of the images seen by viewers. These techniques were used to stunning effect in his *Midnight Mass* of 1834 (see fig. 5.2). Viewers were first shown an empty cathedral in daylight; gradually the sun set, candles were lit, the space was filled with a crowd of worshippers, and an organ burst out into Haydn's "Mass No. 1." One reviewer wrote: "Slowly, dawn broke, the congregation dispersed, the candles were extinguished, the church and the empty chairs appeared as at the beginning. This was magic."[45]

Another mechanically induced marvel was his *View of Mont Blanc Taken from the Valley of Chamonix*, which followed the inspiration of a panorama by Langlois in which the back end of an actual battleship was brought into the hall and joined to a painting of its stern, making it difficult to determine where the three-dimensional object left off and the painting began. In what he called his "hall of miracles," Daguerre built a small mountain chalet, to which he added the sounds of a burbling mountain stream, of alphorns, and of distant singing; there was even a shed in which a braying goat was housed. According to one viewer, "Here is an extraordinary mix-

FIG. 5.2. Poster advertising and describing Daguerre's celebrated *Midnight Mass* diorama. Credit: Musée Adrien Mentienne, Bry-sur-Marne.

ture of art and nature, producing the most astonishing effect, so that one cannot decide where nature ceases and art begins."[46] Even in the front row, "using the best opera-glasses," it was impossible to distinguish between real objects and simulations. One of the most notable visitors to *Mont Blanc* in 1832 was the recently appointed Orléans king, who responded fittingly. "When Louis-Philippe took his family to the Diorama one of the young princes asked, 'Papa, is the goat real?' to which the puzzled king replied, 'I don't know, my boy, you must ask Monsieur Daguerre.'"[47] In the pamphlet publicizing the daguerreotype, Daguerre wrote about his earlier invention: "All of the substances employed by the painter are uncolored; they merely have the property of reflecting such or such ray of light which carries in itself all the colors. The more that these substances are pure, the more they reflect simple colors, but never, however, in an absolute manner, which in any case, is not necessary to bring about the effects of nature."[48] Presented as an example of up-to-date optical science, the diorama produced an "effect of nature" that lay somewhere between the painted screen, the lighting system, and the eye of the viewer.[49]

Yet in what sense were Daguerre's scenes "natural"? As a response to the *Mont Blanc* diorama, the journal *L'artiste* expressed the critical commonplace of idealist aesthetics, as propagated by Cousin and his associates, that true art must avoid the artificial and mechanical: "Should one blame or should one praise M. Daguerre for following the example of M. Langlois by adding to the means which painting gave him, artificial and mechanical means, strangers to art, properly speaking?" Yet others rushed to praise Daguerre as simultaneously an artist, a scientist, and a magician. A poem was written in his honor:

> Knowing the influence of luminous rays
> He knows how to calculate their appearance at will
> Whether in composing or decomposing them
> He shows himself no less an artist than a scientist.[50]

Further, his "creator-like" art was described as a "living painting," as an "enchantment."[51] The diorama and the daguerreotype were seen simultaneously as magical spectacles and as realistic inscriptions of the external world. Their "mechanical" aspect was a source of both trepidation and wonder. These inventions of Daguerre's—boosted by his associations with both romantic playwrights and scientists such as Arago—underline the kinship between the technologies of realist representation and the illusions and metaphysical ambiguities induced by fantastic spectacle.

The production of shared hallucinations also became the stock-in-trade of Parisian orchestral music and grand opera. One of the earliest uses of the term *objective* in English was in Thomas De Quincey's *Confessions of an English Opium Eater*—later translated by Baudelaire—where he described his experience of delusions of grandeur while on opium at the opera. This pregame preparation would have made De Quincey the ideal audience member for Hector Berlioz's *Symphonie Fantastique*. Berlioz's nonstandard harmonies, jarring orchestration, and surprising effects were meant to portray, and perhaps reproduce, the effects of opium. First performed in 1830, it launched the genre of "program music," in which the work's meaning emerged in a dialogue between the music and a written narrative. Berlioz's program told of a jealous lover who witnessed his own decapitation and attended a witches' Sabbath, during a passage in which, musically, the traditional *Deus Irae* of the Catholic Mass was metamorphosed into a freakish death march.[52]

The symphony's novel soundscape was made possible by new instruments. Berlioz planned to include octabasses, oversized cellos designed by the instrument maker Jean-Baptiste Vuillaume, who had assisted the physicist Félix Savart in the latter's pioneering studies of acoustics.[53] Berlioz also collaborated with Adolphe Sax, whose innovations included the saxophone, improvements on the proto-tuba ophicleide, and other brasses. These innovative instruments increased the range of the orchestra and added unprecedented musical "colors": Berlioz's medium was as much sound quality or timbre as it was pitch and rhythm. His *Treatise of Instrumentation* detailed the expressive properties of the instruments of the orchestra: their distinct "personalities," the settings in which their use was appropriate, and the emotions they suggested. It amounted to a conception of the orchestra itself as a composite instrument.[54] In the age of virtuoso performers like Paganini, Chopin, and Liszt—the intimates of Sand, Delacroix, and Etienne Arago—Berlioz presented himself as the virtuoso composer-conductor who played his orchestra like a kind of giant piano, appearing as an orchestra unto himself (see fig. 5.3).

Even further, Berlioz saw music as a heterogeneous assemblage, all of whose aspects had to be controlled: "copyists, players, conductor, the architect who conceived the room, the decorator, the furnisher." More than any of his contemporaries, he also managed the orchestral space, writing, for example: "The place occupied by the musicians, their disposition on the horizontal plane or on the inclined plane, in a three-walled enclosure, or in the very center of a room, with reflectors formed by hard bodies capable of transmitting sound or soft bodies which absorb it and break the vibrations

FIG. 5.3. Berlioz as a one-man band. "L'orchestre de Berlioz, 1850," from *Le dauphiné et les dauphinois dans la charge et la caricature*, in *Collection de portraits dauphinois réunis par Monsieur Maignien*, Collection Bibliothèque Municipale de Grenoble, Cliché BMG.

brought nearer or farther from the players, all have a great importance."[55] To obtain the maximum "acoustic return," his *Requiem* included a giant chorus and twice the standard number of wind instruments. He also increased the strings, added ten tympanists, and set four groups of brass instruments in the corners of the performance space (originally a church), thereby "spatializing" the sound, so that "the fanfare seems to radiate out from the center of the orchestra." Berlioz's interest in innovations in musical technology was a constant. It made him susceptible to the appeal of the industrial religion of the Saint-Simonians, with whom he corresponded in the early 1830s. He composed the "Song of the Railroad" in 1846, for the opening of the first train line in Lille, and at the International Exposition of 1855 (the "Festival of Industry") he staged his *Te Deum* with more than a thousand performers.[56] He also proposed at one point using an electric telegraph to keep players in time. As Alison Winter has suggested in her discussion of depictions of Berlioz and Wagner as dueling mesmerizers, the fantastic phenomena of animal magnetism provided a context for understanding both the power of the conductor over players and audiences and the rise of telegraphy, with its implication of invisible command across vast distances.[57]

Berlioz's uncanny effects aligned him with the optical illusions of the diorama. Heinrich Heine, the German poet, critic, and Saint-Simonian sympathizer who made his home in Paris, compared Berlioz's gargantuan arrangements to the panoramic landscapes of John Martin. Both had an "antediluvian" aspect and possessed "the same bold feeling for the prodigious, for the excessive, for material immensity. With one, the striking effects of shadow and light, with the other, fiery instrumentation; with one little melody, with the other little color, with both of them at times the absence of beauty and not the slightest naïveté."[58] Berlioz's orchestra left mere beauty behind; it opened sublime perspectives upon the violent immensity of prehistory in an auditory idiom. At the same time, his braggadocio about his innovations at the level of musical materiality testified to his faith in the demiurgic powers of the industrial age. As we will see at the close of this chapter, such techniques of both musical and visual hallucination culminated in a decisive form of mass spectacle, the Parisian grand opera, which emerged in the early 1830s.

PHYSIOSPIRITUALISM: A MINOR PHILOSOPHICAL TRADITION

Before we turn to the opera, however, the ties we have seen between fantastic artists and the scientists and philosophers who sought to understand and improve their effects demand further attention. Fantastic audiovisual spectacles were frequently improved with the assistance of scientists who sought to understand and control the properties of light and sound. We have seen how Arago's interest in light—and his support of Fresnel's wave theory—went hand in hand with his development of techniques at the observatory for stabilizing phenomena. Beyond his support of the daguerreotype, his construction of instruments to study polarized light, and his cyanoscope, which measured the blueness of the sky, he also helped design an apparatus with which Léon Foucault and Hippolyte Fizeau measured the speed of light, and he was one of a team of scientists who measured the speed of sound by firing a cannon and noting the time at which an observer at a known distance visually signaled that the explosion had been heard. As we will see in chapter 10, he encouraged Foucault's collaborations with popular presentations of various kinds.[59] Other scientists studied the interactions of sound and light in ways that straddled experiment and performance. Ernst Chladni drew violin bows across the edges of metal plates covered with sand to show how different pitches produced different geometrical figures. These visualizations of sound launched a field of research pursued in Copenhagen by Oersted, in Berlin by Wilhelm Weber, and in Paris by Félix Savart. Savart created experiments to analyze the components of the human voice and constructed an artificial ear to simulate auditory perception. Working with the instrument-maker Vuillaume, he designed new instruments according to his acoustic principles, including a sharp-edged, trapezoidal violin.[60]

Eugène Chevreul, Ampère's friend and the chemist in charge of dyeing operations at the Gobelins tapestry manufactory, oversaw comparable collaborations between craft workers and scientists. His book *The Laws of Contrast of Color* explored the effects of juxtaposing one color with another in space and in time. He explained how certain colors melt together into a third at a distance and how the perceived difference between colors near in tone is heightened when they are seen together, and he provided a rationale of afterimages akin to those Goethe discussed in his *Theory of Colors*. Chevreul devoted much of the book to practical applications: the effect of colored lights on tapestries, the effects of size and distance, the means of producing a third color by weaving two others together. Chevreul also considered painting, including "the difference there is between a colored object and the imitation that the painter makes of it, when the spectator chooses another point of view than his own."[61] Incorporating discussions by Dalton, Young, and Laplace about ocular physiology and the nature of light, the book became a touchstone for the impressionists and the color theorist Charles Henry.[62] Chevreul's central message was that color was a combined effect of technology and human perception. A single color might have a completely different appearance depending on its lighting, the other colors displayed before, after, or next to it, and the location of the spectator. In the eminently practical context of textile manufactures, Chevreul highlighted the artifactual, hallucinatory nature of perception.

Hallucination and illusions have been shown to be keys to the visual epistemology of the early nineteenth century. British natural philosophers detailed personal experiences of disordered senses and nerves in order to understand the basis of knowledge. Thomas Brown asked, "Is it utterly absurd and ridiculous to maintain that all the objects of our thoughts may be 'such stuff as dreams are made of?' Or that the uniformity of Nature gives us some reason to presume, that the perceptions of maniacs and of rational men are manufactured, like their organs, out of the same materials?"[63] Among philosophers, psychologists, and physicians, a "widespread preoccupation with the defects of human vision defined ever more precisely an outline of the normal," leading to the conception that even normal vision was an artifact of the material configuration of our senses.[64] Such a view suggested that the world we experience is an illusion produced by our sense organs and mental faculties—a notion shared with German idealism, recently translated texts of Eastern religious traditions, as well as Western esoteric and illuminist writing familiar to romantic-era thinkers, from Hermes Trismegistus and Jakob Boehme to Saint-Martin to Emmanuel Swedenborg.[65]

As we saw in chapter 2, the notion of perception as an artifact of sensory and psychological processes was also a central theme for Maine de Biran, who was just as interested in hearing as he was in sight and touch. He noted the role of habit and expectation in shaping our perception of melody and harmony: "Habit teaches us to distinguish first the successive terms . . . then to reunite them and to perceive clearly many of them together: it thus creates a harmony for the ear."[66] Cousin's "spiritualism"—which focused on the transcendental will and asserted a radical separation between mind and body—has been seen as the major heir to this philosophy. Yet, as I have argued above, what was most striking about Maine de Biran's writing was his relentless curiosity about the complex relationship between the mind and the body: the fluidity of mental states, their nuances and shifting nature, and the effects of the body—the muscles, the senses, the barely perceptible motions of the nerves and brain—upon one's thoughts and perceptions. This shift of emphasis has consequences for our understanding of the philosophy of the romantic age: instead of valuing transcendence and idealism, it directed philosophical attention to the body and its movements, repetitions, and external appendages as the inescapable ground for the experience of truth and beauty.[67]

Other thinkers, less institutionally entrenched than Cousin, carried these aspects of Maine de Biran's philosophy into new directions.[68] For some, it was the basis for restoring the spirit or soul to a prominent place in the philosophy of the organism. But each of them gave serious consideration to humans' engagement with the world around them—the action of the will on the muscles, the encounter of the senses with external phenomena, and the material objects, or technologies, that gave a repeatable form to those encounters. Further, beyond the question of the multifarious relations between the body and the soul, many of these thinkers also showed an interest in altered states of consciousness. This minor tradition of physiospiritualism was thus the philosophical double to the psychic and metaphysical oscillations of the fantastic.

With respect to Ampère, we saw how Maine de Biran's emphasis on repetition could foster reflection both on scientific methods and on the structuring impact that technical apparatus had on perception and knowledge. Further, Ampère's concept of technaesthetics—of art as a series of rational, repeatable procedures aided by technology—was the perfect complement to the romantic view of the scientist as a kind of artist. Another of Maine de Biran's interlocutors was Alexandre Bertrand, the champion of post-Laplacean physics (including Ampère) and Geoffroy Saint-Hilaire in France's first popular science column.[69] In the *Globe* Bertrand defended Maine de Biran's work against its purely "spiritualist" interpretations in the hands of the eclectic philosophers Damiron and Jouffroy—both of whom downplayed the physiological component of Maine de Biran's thought.[70] For Bertrand, the bidirectional interaction between mind and body was crucial for an understanding of animal magnetism and "somnambulism." Trance states, he held, were not produced by an ambient fluid, as Puységur and other followers of Mesmer assumed, but by a rapid change in the physiological configuration of the patient's organs of thought and perception, brought about in part by the repetitive movements and encouragement of the magnetizer. Certain individuals, because of the way their brains and senses were organized, were more susceptible to these modifications. The change of organization they underwent in the mesmeric séance made new perceptions and experiences possible: this was the basis for the state of "extase," which Bertrand saw as the cause of possessions and trances throughout the ages. It was also the origin of world-changing religious movements.[71] One of Bertrand's oldest friends was Pierre Leroux, a printer and the editor of the *Globe*, who advanced his own form of physiospiritualism in his prophetic religion of humanity, as we will see in chapter 8.

Many of the scenarios that were the obsessions of fantastic art—for example, hallucination, hypnotism, madness—were objects of the nascent field of psychiatry. The alienist Joseph Moreau de Tours, a student of one of the founders of modern psychiatry, Jean-Etienne Esquirol (who restored

Auguste Comte after a nervous breakdown), acknowledged the influence of Maine de Biran in his analysis of madness.[72] As a way to get insight into the altered states experienced by the insane, Moreau de Tours organized experimental simulations of madness. Along with Théophile Gautier, he founded the Club des Hachichins, which met for monthly séances from 1844 to 1849 in sumptuously sybaritic rooms on the Île Saint-Louis, attended by Eugène Delacroix and by the leading lights of romantic poetry: Nerval, Dumas, Balzac (once), and Baudelaire, who described the séances in *Les paradis artificiels*.[73] Moreau de Tours's guiding question was the effect of material causes on perceptions and ideas; conversely, among hashish users he noted the tendency of the mind to make thoughts concrete: "It is remarkable how much, with hashish, the mind is led to transform all its sensations, to dress them in palpable, tangible forms, to materialize them, so to speak!"[74] Like habits, or scientific instruments, or the sympathetic and repetitive passes of the magnetizer, in this view drugs were a technology that structured thought and perception. Hallucinations followed regular patterns: despite the variable content of users' fantasies, all users underwent comparable distortions of time and space, as well as a common sequence in the unfolding of the experience. Moreau de Tours argued that the identification of such regularities in the case of abnormal mental states could provide insight into the normal activity of the senses and the brain.[75]

A similar emphasis on the corporeal foundations of consciousness was found in the works of philosopher Félix Ravaisson, who combined the influence of Maine de Biran with a reading of Leibniz and Schelling, whom he met in Berlin. For Ravaisson, mind and matter were aspects of an underlying force—a version of Schelling's Absolute Ego or World Soul. Through the cultivation of habits (physical and mental) one actualized spirit, drawing nearer to the Absolute, the divine. Ravaisson played an important role in shaping the national philosophical curriculum in the Second Empire. His essay on "le spiritualisme français" gave a prominent role to Maine de Biran, and the close connection he established between material and mental phenomena in the theory of organic life later became tremendously important for Bergson's analysis of matter and memory and "creative evolution."

These followers of Maine de Biran developed distinct metaphysical frameworks, whether the dualism of Ampère, the ecstatic materialism of Bertrand, the organicism of Moreau du Tours, or Ravaisson's biological fusion of Leibniz and Schelling. Despite this plurality, what places these romantic-era thinkers in the same camp is their interest in the ways habits entrain, and thus alter, the organs of perception and thought, creating and eventually stabilizing new perceptions and experiences. Elaborating a conception of the human subject more complex than the sensualists' passive tabula rasa, the spiritualists' unchanging soul guiding the passive body, or Cabanis's and other physiologists' determinist *homme-machine*, this minor tradition suggested that the experience of the world was the product of shared and repeated actions.

Their interest in habit and the mind's relation to matter and physiology also made their thought compatible with reflection on technology.[76] Repetitive experiences or actions, aided in some cases by machines and instruments, reconstituted and maintained the human body and mind. The relentless reformulations to which Maine de Biran subjected his own thought echoed the dynamism of the conception of the self he left to posterity: a self that tried to know itself through a practice of introspection doomed to incompleteness; a self whose thoughts and acts of will demonstrated that it could always be modified. By emphasizing the alterations that habit and repetition brought to both the body and the mind, Maine de Biran and his followers created an image of the human in which will and perception played an ongoing, constitutive role, yet where the flash of self-determination constantly faded—for better or for worse—into the dullness of automatism.[77] This mobile self was susceptible to the metaphysical slippages and perceptual switches that characterized the fantastic arts and was often invoked in the genre's recurrent images. These included such figures as the somnambulist who fell under the determined control of a magnetizer while inhabiting a realm of disembodied spirits; or automata, those machines that appeared to come magically to life; or even, as we will see, such heroes as Robert le diable in Meyerbeer's opera, torn between angelic freedom and satanic enslavement by matter.

THE OPERA OF ATTRACTIONS

The physiospiritualist followers of Maine de Biran offered an alternative to the purely spiritualist view of art as the symbol of the infinite or of art for art's sake. In its place they provided a framework for a reading of the arts—one close to Schiller's—that recognized matter and technology as inescapable constituents of aesthetic experience. To reprise our discussion of the fantastic arts, technology accompanied the artwork at three key moments: first, in the act of creation, when the artist planned effects and suited means to ends; second, in the moment of transmission, when the artist or composer sought out, designed, and employed instruments and techniques to produce specific perceptual effects in sight, sound, or imagination; finally, in the moment of reception, when the observer shaped phenomena into meaningful patterns through the physiological action of the senses and through the faculties of thought and imagination. The first two moments were captured by Ampère's technaesthetics, while the latter was analyzed by Chevreul and by Maine de Biran and his followers.

The fantastic technaesthetics of words, music, and images reached their culmination in Parisian grand opera, a genre that attained its modern form at this time. Through the careful manipulation of light and sound within a strictly controlled environment, participants shared a technically produced, multisensory hallucination: collective experiences that hovered between the spiritual and the material, the imaginary and the concrete, the empirical and the transcendent.[78]

The work that set the bar for the new form was *Robert le diable* of 1831 by Giacomo Meyerbeer. Although his achievements have long been disparaged by music critics, Meyerbeer was the immediate precursor to Wagner and the idea of the total artwork.[79] The plot of *Robert* turned on the threats to the soul of the half-demon Robert. His father, a demon who had passed himself off as a friend, aimed to carry Robert to hell, thwarting his virtuous marriage plans and leading him into corruption and the use of evil magic. The interventions of Robert's mother and his innocent fiancée, Alice, as well as a certain amount of heavenly assistance — signaled by Meyerbeer's unprecedented use of a church organ in a secular performance — combined to defeat the demon's plans. Critical reception focused on Meyerbeer's sonic and visual innovations: reviewers concentrated on his new instruments and harmonic combinations, as well as his sound effects, including a demon chorus placed beneath the stage singing eerily through resonating tubes. Hundreds of lavish costumes were on display, as were several stunning set changes employing illusions of depth and color inspired by other popular visual illusions. "The ruins of the third act," wrote *Le figaro*, "are as perfect in effect and scheme as one of the ingenious Dioramas of M. Daguerre, as the Panorama of M. Langlois; they are of a delicious color; the impression that they produce is completely poetic."[80] For lighting, new gas lamps covered in foil cast unsettling shadows during scenes with Robert's demon father; during a dance of ghosts, exploding fireballs were made by throwing resin dust and moss spores over exposed flames. In the most celebrated scene, a ballet set in the graveyard of a convent, fiendish nuns were resurrected and, in skin-colored costumes on a moving platform, performed a lascivious dance of seduction to the accompaniment of chilling bassoons (see fig. 5.4).

FIG. 5.4. Meyerbeer's Robert le diable at the Paris Opera, Salle Le Peletier, first night, 1831. Credit: Opera Garnier, Paris, France; Erich Lessing; Art Resource, New York, NY.

To understand the appeal of the genre of grand opera, we need not only to look at its score and librettos but also to consider it as part of what we might call an "opera of attractions." Discussing the earliest motion pictures of the late nineteenth and early twentieth centuries, film scholar Tom Gunning wrote of a "cinema of attractions," a genre of very short films that were focused not on narrative and characterization, but on illusions of motion, visual novelties, and exhibitionism. He did not see these short early movies as abortive steps toward the conventions of narrative cinema but argued that they aimed to show something previously unseen and thereby produce an effect of shock or pleasure: "a unique event, whether fictional or documentary, that is of interest in itself."[81] In adapting Gunning's term, I do not mean to deny that Parisian opera involved intricate plots and rich characterizations; yet neither these traditional dramatic concerns nor the music alone was sufficient to spark the explosive success of the genre. Much like the audiences for the panorama and the diorama, people flocked to the opera for its opulent, abundant, and shockingly mobile sights and sounds; they sought surprise, thrills, and gratuitous pleasure. As much as the plot and the music of an opera, these perceptual thrills were the object of its creators' obsessions and the focus of its critical reception.[82]

Robert was frequently staged, imitated, and discussed over the next fifteen years. Balzac's short story "Gambara" contains extended reflections on its innovations. It told of a musician who composed a work so sublime that it could be played only on the gigantic contraption he had built, a *panharmonicon* that reproduced all the sounds of an orchestra along with human voices. The machine emitted pure cacophony, except when the composer, Gambara, had drunk enough alcohol to lift him to a rare state of exaltation. In an attempt to cure him of his deluded notions about music, another character took him to see *Robert*. The tale concluded with Gambara's ten-page-long exegesis of Meyerbeer's work. Although he described the performance as a cosmic struggle between good and evil, light and dark, order and chaos, at the same time Gambara—who had conducted experiments on the connections between sound, light, the ether, life, and thought, much like the protagonist of Balzac's *Quest for the Absolute*—made it clear that the only way to grasp the meaning of the plot's spiritual battle was to focus on the material detail, expense, and technical care that went into its sets, lighting, instrumental textures, and orchestration. In this sense, the composer, Gambara, argued against taking sides in the critical debate between German idealism (Beethoven) and Italian sensualism (Rossini). Instead he defined music simultaneously by its physical and spiritual aspects; his experiments on instruments and on the ether—a substance he and Balzac saw as joining mind and matter—provided a scientific and metaphysical basis for these views.[83]

Although at one point Gambara attributed an inspired performance of an opera he composed to "The Spirit!"[84] Balzac's description of the performance revealed the author's awareness that much more was required to produce these effects: just as Gambara had described Meyerbeer's sets as "the material representation of thought," Balzac made it clear that the ceremonial arrangements of room, audience, instrument, and wine were all required to summon the otherworldly presence that animated Gambara's music. Likewise, after his exalted disquisition on *Robert*, Gambara "fell into a sort of musical ecstasy," and improvised "a divine song" that forced his audience out of its materialistic illusions. Within the listener's mind, "the clouds dispersed, the blue of the sky reappeared and raised the veils which hid the sanctuary, [and] the light of heaven fell in floods"; yet this spiritual epiphany was at the same time a precise transcription of the action of the stage machinery at the conclusion of Meyerbeer's staging of *Robert*. Gambara punctuated the ensuing silence with a single word: "*God!*"[85] The precise arrangement of charismatic technologies—both in Gambara's retelling and in the opera hall—was what called forth "The Spirit" and opened the heavens for the audience. Balzac presented Gambara's performance as an artistic Eucharist.[86]

Both Meyerbeer and Balzac were staking out a metaphysics in which materiality and machines were not the opposite of spirit and organisms but were instead their necessary concomitants and preparations. Under the impulse for new forms and new experiences, in the 1820s and 1830s there appeared a range of artistic, philosophical, scientific, and political experiments that posed a direct challenge to the dualism of matter and spirit. They proposed ecstatic materialisms and concretizing spiritualisms that converged on the contact points between the inner and the outer worlds: instruments, fetishes, symbols, and machines. Through the physical and metaphysical quests of his characters, Balzac's literary philosophy and philosophical literature both described and embodied this impulse.

Meyerbeer's operas were experiments along similar lines. Not only did they assemble unprecedented technologies of sound and vision, hallucination and ecstasy, but they made charismatic technologies themselves characters in the drama. As Balzac's commentary made clear, in the opera the struggle between good and evil was expressed through the score and the orchestration, through the opposed colors and melodies associated with Bertram and Alice, and through the temptation of the bassoons and the redemptive chords of the organ. We might read the ambiguous cypress branch that Robert was tempted to seize at the opera's climax as a figure of the desirable but morally dubious powers of technology. In the drama, this power was renounced, when Robert was recalled to morality thanks to the heavenly organ; yet this power was heartily embraced in Meyerbeer's own enchanting stage design and orchestration. We might read the fantastic character of Robert himself—half human, and therefore drawn to salvation and spirit, and half demon, hence the victim and agent of enslavement to matter—as one more charismatic technology: as the iconic intersection of this period's eclectic currents of spiritualism and mechanics. Robert's oscillations between matter and spirit resonated with physiospiritualism's philosophy of the self as the automatic and willful producer of hallucinations; they were echoed by the fantastic metamorphoses on display in other popular spectacles of this time.

6

Monsters, Machine-Men, Magicians

The Automaton in the Garden

FABLES OF MODIFICATION

Whether critics celebrated the "magic" and "enchantment" achieved in Daguerre's diorama, Berlioz's symphonies, or Meyerbeer's operas, or fretted over their use of "means foreign to art," a large part of the appeal of such performances came from the fact that their unprecedented effects were the result of technical novelties. What's more, their fantastic imagery—animated instruments, sudden metamorphoses, protean fluids, the beginnings and ends of worlds, inert objects or machines that spring to life—commented reflexively on the arts' own enchanting techniques and, more generally, on the world-altering changes promised by the dawning industrial revolution: uncertainty over the increasing presence of new technologies in everyday life was reflected by the fantastic's alternations between wonder and dread. We saw that fantastic works also paralleled physiospiritual theories of perception that saw not only artists but their audiences as creators. Just as artists' "creativity" mirrored the magical powers depicted in fantastic works, perceivers' internal organic movements, will, desire, and habits all actively inserted themselves into experience.

In this way, the fantastic presented allegories of cosmic order: about what is in the world, how we know it, and how it can be altered. Other spectacles of the romantic period conveyed the order of the cosmos more didactically. As we have seen, this was a time in which the "vulgarization" of science was on the rise, and changing its nature from polite diversion to mass entertainment.[1] The volume of scientific publication rapidly expanded, and scientific journals and textbooks were featured in new public reading rooms and lending libraries. Alexandre Bertrand's writings for the *Globe* encouraged physician Alfred Donné to write a weekly science column, which occupied the bottom of the first page of the *Journal des débats*, the newspaper of record; in the 1840s Donné transferred the writing of the "feuilleton scientifique" to the experimentalist Léon Foucault. In the lecture halls of the Athénée, Gall and Spurzheim spoke on phrenology, and Auguste Comte preached the social mission of the sciences.[2] Going beyond publications and speeches, popular science of the 1830s and 1840s often aimed, as music and the opera did, to involve audiences in complete, fully embodied experiences. I have mentioned Arago's dramatic and colorful lectures on popular astronomy in the expanded lecture hall next to the Observatory, and his efforts to bring the public into the process of science, as well as the musically inclined Humboldt's dream of instructive panoramas. Sharing romantic spectacles' goal of appealing to multiple senses, scientists and their representatives aimed to bring truths—and disputes—about the order of nature to a growing public.[3]

Two other important Parisian locales in which the public came into contact with innovations in science and technology on an expanded scale were the Muséum d'Histoire Naturelle and the recurring Exposition Nationale des Produits de l'Industrie Française. In both places, face-to-face encounters with nonhumans—exotic animals and new machines—prompted reflections about human nature, its powers of transformation, and its relation to the rest of the universe. These displays, whether of natural or of man-made creatures, were focal points for discussions in which understandings of the ultimate nature of reality wavered between materialism and spiritualism, pantheism and vitalism, and static and historical views of nature. The correlations between different organisms and their relations to their environments raised the question of the origins and transformations of species; the ensemble of industrial products and machines testified to the power of humans, aided by the technical arts, to remake the world. In both cases, the modifiability of the natural order by artificial means was presented in dramatic form. The engineer-scientists' vision of the transformative power of industry was made plain in the Expositions Nationales; just as their dynamic view of nature was put forth as a supersession of the Laplaceans' emphasis on equilibrium and detachment, Geoffroy Saint-Hilaire's impassioned claims about the modifiability of species at the Muséum challenged the static natural order defended by Cuvier.

The heady possibilities the Muséum and the expositions evoked of humans' ability to act as creators—indeed, to become like gods—were brought to a head in two popular works: the *Soirées fantastiques* of Eugène Robert-Houdin, the pioneer of modern stage magic, and *Un autre monde*, J. J. Grandville's madly heterogeneous storybook. Robert-Houdin and Grandville made serious play out of the idea of a technologically modifiable nature, showing the susceptibility of organisms, society, and the natural milieu to human modification as a possibility at once disorienting, hopeful, and dangerously close to blasphemy. This chapter traces routes through which the mechanical metamorphoses of the sciences were brought to an eager public in romantic-era Paris.

ANIMALS GOOD TO THINK

The Muséum d'Histoire Naturelle, with its adjoining Jardin des Plantes and Ménagerie, was a site for research in which new species were classified, old classifications were challenged, and new techniques of cultivation were tested. Its beautiful plants and strange animals on display for the pleasure and curiosity of visitors also made it an idyllic setting for walks and rendez-vous. These sites had been woven into French science for nearly a century, thanks to their expansion at the hands of Buffon. In the revolutionary period, the nature collected and arrayed there was a model of and a model for the new political order. Experiments in acclimatizing exotic plants and animals to French environmental conditions suggested the power of republican institutions to shape a new human species.[4]

In the Restoration and the July Monarchy, the Jardin des Plantes continued to foster such speculations on animals as models for human society. By looking at the animals collected at the Ménagerie, visitors were provoked to reflect on the special kind of animal that is the human.[5] The Muséum provided ideas for Balzac's "Natural History of Society" and the literary genre of the "physiologies," which satirized Parisian mores from a quasi-scientific point of view. J. J. Grandville depicted this mirror play in illustrations for the tales collected in his *Scenes from the Private and Public Life of Animals* (1839–40); the stories themselves, by Balzac, George Sand, and P. J. Stahl (the pseudonym of Balzac's editor, Hetzel), riffed on *Les Français peints par eux-mêmes*, a celebrated collection of physiologies: they restaged stock situations found in the capitol and satirical morality plays with animals in the place of humans. In the frontispiece, Grandville depicted himself seated in the Jardin, caricaturing the literary caricaturists collected within as though they were animals on display (see fig. 6.1). This mirroring relationship was likewise visible in *Un autre monde*'s image of a bourgeois family with beaks, wings and fins paying a visit to the Jardin, peering at a humanoid family of birds-fish-insects that distinctly resembled their observers.[6]

The animals at the Muséum and the Jardin were good to think with: they were focal points for reflection about the distinguishing characteristics of humans, the differences among species, the relationship between thought and instinct, and the possibility, and possible causes, for changes in species. Frédéric Cuvier, brother of the great anatomist and a member

FIG. 6.1. *Left*, Grandville sketching Balzac, Sand, and others at the Ménagerie of the Jardin des Plantes. Frontispiece of Grandville, *Scènes de la vie privée*. Courtesy of Rare Book and Manuscript Library, University of Pennsylvania. *Right*, "The Perch," from Grandville, *Un autre monde*, 114.

of the circle of Maine de Biran, was in charge of the collection of living specimens in the Ménagerie. His work, along with that of Isidore Geoffroy Saint-Hilaire (Etienne's son) on the study of animal behavior helped launch the field of ethology, whose observations of animal behavior contributed to debates about the relation between habit and instinct.[7] Of particular note were exotic species, which served as novelties, as tokens of French colonial power, and as exemplars or anomalies with regard to theories in zoology and comparative anatomy (see fig. 6.2).

For example, in 1824 the pacha of Egypt offered France a giraffe. Etienne Geoffroy Saint-Hilaire, the professor of invertebrates who had been part of Napoleon's Egyptian expedition, personally accompanied the creature from Marseille to Paris. The giraffe's unusual gait and odd elegance brought vast crowds to the Jardin and sparked a trade in souvenirs, from plates and drawings to songs and wallpaper. The species had also been used as an example in Lamarck's *Zoological Philosophy* to illustrate the central planks of his theory: that changes in the environment produced new behaviors in animals, that the use and disuse of organs led to their growth or atrophy, and that these changes were passed on to following generations. The giraffe's ancestor was assumed to be a horselike creature that had responded to a lack of low-lying vegetation by progressively stretching its neck to eat from trees. Such a modification, to satisfy needs in response to a changing milieu, contradicted Cuvier's view of both the stability of species and God's providential harmony between organisms and their conditions of existence. In the 1820s, with Cuvier ensconced as both theoretical and institutional authority, Geoffroy revived the transmutationist theories of Lamarck. Associating himself with the giraffe was part of this controversy.

The Cuvier-Geoffroy debate built on long-standing disagreements among students of living things. Following eighteenth-century materialist arguments of Diderot, Helvetius, and d'Holbach, several thinkers at the turn of the century had argued that life and thought depended on the specific organization of matter, explaining the moral or mental life by the physical one. In zoology, Lamarck had argued for two distinct moments in the formation of living things: nature and its laws may have been cre-

FIG. 6.2. The Bear-Pit at the Menagerie of the Jardin des Plantes, in Acaire-Baron, *Album des Jardins des Plantes*. Lithograph, 1838. Credit: Réunion des Musées Nationaux; Art Resource, New York, NY.

ated by God, yet a force of *production*, inherent to nature, was constantly making changes in living things. Life arose spontaneously from a propitious combination of matter, light, heat, and electricity. Thereafter, further modifications of organisms ensued, both in the gradual perfection of species according to their place in the series of animals, and as a result of the accidents they encountered in their milieu.[8] Around the turn of the century, the ideologues, chief among them the philosopher Destutt de Tracy and the physician Jean-Pierre Cabanis, launched a project for a science of man that would synthesize medical, historical, physiological, and zoological considerations.[9] Cabanis assumed a continuity between humans and other animals, with a difference only of degree between thought and the sensitivity and irritability inherent to organic tissue. This view was supported in the phrenology promoted by Gall. In the Restoration, physician François-Joseph-Victor Broussais used the notion of matter's intrinsic irritability as the basis for a militant assertion of the metaphysical continuity between humans and animals and between thought and matter.

Cabanis' and Broussais's materialism clashed with assertions of the existence and independence of the soul or spirit. Catholic apologist Joseph de Bonald depicted the denial of a separate principle of life or thought as leading inevitably to the denial of the eternal soul, and thus to even graver ends. De Bonald's definition of man as "a soul served by organs" paralleled his theocratic justification of the Restoration. Anyone who refused to acknowledge the soul's dominion over the body would also deny God's dominion over nature; such a person would also refuse to recognize the king's sovereignty over his subjects, defined as the instruments of the will of the divinely legitimated monarch. Both this traditionalist spiritualism and the materialism it opposed had physiological, political, and theological ramifications.

Historians have tended to reduce these metaphysical debates to a contest between two cosmologies, with mechanistic and atheistic republicans squaring off against spiritualist and devout monarchists. Yet the antagonists' actual positions were more mobile than such a dichotomy suggests.[10] Those who held that life must depend on a substance or principle distinct from ordinary matter might well support the idea that a transcendent God directed or enforced the laws of nature just as the immaterial soul used the body's organs as its instruments. But another option was a materialist (and potentially atheistic) vitalism, in which the distinct principle that caused life was simply different from ordinary matter—as, for instance, in Bichat and Broussais's notion that life resided in "irritability." Similarly, one might be a monist, claiming that thought and life were due only to the configuration of matter. Yet such a view did not necessarily entail atheism. Another stripe of monism was pantheism, the belief that the entire world was saturated with or indeed identical to the substance of God. Catholic authorities and defenders of tradition used "pantheist" as a term of abuse for Victor Cousin; yet the Saint-Simonians and Pierre Leroux, inspired by Goethe and Schelling's rediscovery of Spinoza, willfully embraced it.[11] Materialist monists tended toward a determinist reductionism, while many pantheists emphasized the ongoing process of divine creation.

As shown in our discussion of physiospiritualism in chapter 5, even authors who shared a set of questions and methods — such as the correlation and reciprocal causality between mental states and physical activity — might hold quite distinct metaphysical views. The organisms displayed and discussed at the Muséum prompted the same multiplicity of metaphysical positions among anatomists, physicians, and natural historians. Without claiming completeness, the following chart at least suggests the range and distinguishing features of the available perspectives:

	ATHEIST	THEIST
one substance (monist)	physicalist/materialist reductions of life and consciousness	pantheism, Spinozism
two substances (dualist)	vitalism	Catholicism, Cartesianism

Of course, even these categories were slippery. Although vitalism was often compatible with materialism, as in the case of Bichat, it could just as easily be presented as a rejection of monistic materialism, as it was by Bérard, a Montpellier physician who was rewarded by the Restoration government for his attacks on materialist and mechanistic treatments of life.[12]

Furthermore, the political correlates of these metaphysical views were also far from stable. For one, as we will see in later chapters, the religious dimension of early socialism suggested that the political dividing line was not so sharp as one between godless republicans and Catholic monarchists. And there was a defense of the spirit that came not from Catholic defenders of the faith but from the liberal opposition. The eclecticism of Victor Cousin and his friends Jouffroy and Damiron sought to reconcile eighteenth-century materialism with skepticism, Cartesian dualism, and mysticism.[13] Cousin saw each of these views as one step in the progress of the mind, although the deepest foundation of the process was introspective knowledge of the self. Cousin's eclecticism was the philosophical equivalent of the July Monarchy's "juste milieu" between traditional monarchy and liberalism.[14]

Yet it would be unfortunate to reserve the term "eclectic" as the name for just one school of philosophy in this period: the adjective applied equally well to the entire landscape of philosophical debate concerning the nature and relationships among mind, matter, and life.

A BATTLE OF TWO NATURES

During the Restoration of the 1820s, however, there was a clear imperative to ban heterodoxies of all kinds. The Medical Faculty of Paris, a hotbed of republicanism and materialism, was purged in 1822; by government decree, faculty members who denied a distinct spiritual principle as the basis of life were relieved of their functions.[15] Among the members of the committee behind the purge was one closely identified with the Muséum d'Histoire Naturelle, Georges Cuvier. The antagonisms of that episode—between defenders of the traditional religious order and anti-establishment challengers—flared up again in different forms around 1830 in Cuvier's debate with Geoffroy Saint-Hilaire. While this was an argument about classification and about the variability of species, it was also about the order of nature and the sciences' relationship to the public; and as we will see below, Geoffroy and his ideas—topics that have been mainly studied only by historians of biology—turned out to be pivotal for the social and industrial turn in romanticism.

Cuvier was a founder and leading light of comparative anatomy. His grasp of the functional unity of organisms was suggested by a famous anecdote in which he was able to reconstruct an entire mammoth on the basis of a single bone. He received early training from the German physiologist Kielmeyer, and he shared his mentor's view of the teleological adequacy of all aspects of animal life, especially the harmony between organisms and their "conditions of existence."[16] He introduced a new principle to bring order to the animals collected at the Muséum: this was a classification based on the functional systems of circulation, digestion, and sensation. He divided the animal kingdom into four branches (*embranchements*): *vertebrata* (animals with spines), *articulata* (insects, crustaceans, and segmented worms), *mollusca* (clams, squids, and slugs), and *radiata* (jellyfish, corals, and starfish). Within each branch there was a shared plan or "prototype," the same functional systems with directly comparable forms. Cuvier disparaged the eighteenth century's dominant view that life could be organized according to a continuous chain of being or animal series, in which every link was occupied and the missing links simply had yet to be discovered (or, as Charles Bonnet's ingenious solution had it, would appear in the future according to a divinely preestablished plan). Instead, for Cuvier, these branches were divided by an abyss, a "hiatus": the vertebrates and the insects were different beasts entirely. Cuvier's focus on functional systems and their adaptation to conditions of existence was also read as an argument on behalf of natural theology and against Lamarckian transmutation. Cuvier argued that the forms of these types had been set at the moment of their creation: despite some superficially different varieties, it was unthinkable that genuinely new species could emerge out of old species.

Yet the view that organisms developed over time and were united by a single plan or scale of being was gaining traction in the Restoration, in the generalizing turn taken by Cuvier's colleague Geoffroy. Geoffroy was one of the first professors at the Muséum during its reform in the 1790s, working alongside Lamarck and helping secure a position for Cuvier. He was sympathetic to Lamarck's materialist explanations of the origins of life via the "ambient milieu" of light, heat, electricity, and magnetism.[17] While in Egypt between 1798 and 1802 as part of Napoleon's expedition, under the command of physicist Joseph Fourier, he sketched a materialist philosophy of nature: "In accepting the general principles I have established ... one can explain all galvanic, electric, and magnetic phenomena, the nervous fluid, germination, development, nutrition, generation ... I say also that one can explain the intellectual function by physics [*la physique*]."[18] On his return to France, he collaborated with Cuvier to place comparative anatomy on solid ground.

Over time, however, as Cuvier's influence grew, Geoffroy carved out an approach to classification that was due to clash with Cuvier's. While Cuvier's classification focused on functions, Geoffroy looked to form. Morphological analogies between the parts of different animals led him to a general, "philosophical" approach to comparative anatomy. His "principle of connections" led him to identify commonalities between species not on the basis of specific organs, but on a relation between parts that held through "its different metamorphoses" across different species.[19] For example:

> The foot in the bear uses the whole sole or the totality of its bony parts to form the basis of the column that serves to support the trunk. It uses only the metacarpals and the digits in the martens, the digits only in the dog; two in the three of the digital phalanges in the lions and the cats; the last of these phalanges in the wild boar; finally, it touches the ground only at one point in the ruminants and the solipeds, dedicating to this not even one part of that last phalanx, but only the nail that encases the extremity.[20]

Thus, different bones were in contact with the ground in different animals; yet despite being developed and used in different ways, the underlying connections between the bones themselves remained the same. It was not the "foot" that was the unit of comparison, but the formal connections of parts, many of which were not directly used for locomotion or standing in several creatures. Soon he moved beyond comparisons among mammals to considering the homologies between features of vertebrates and what Cuvier considered distinct branches: he argued that the carapace of insects, for example, was simply a skeleton turned inside out.

From such comparisons emerged Geoffroy's signature concept, the unity of composition or unity of type. He argued that a single underlying type, pattern, or plan was realized in different proportions and configurations in all species. This was an explicit revival of aspects of Lamarck's neglected "zoological philosophy," as changes in the milieu were seen as causes for the appearance of morphological differences. Unlike Lamarck, however, who emphasized the mediating role played by habits, Geoffroy argued that surrounding conditions could directly accelerate or slow, interrupt or prolong, a universal process of development. Thus Geoffroy was arguing that species changed due to material causes in their environments. Yet his claims about differentiation were less important to him than the profound unity underlying the animal world:

> There are no longer any different animals. One fact alone governs them; it is as one single being that it appears. This fact is Animality: an abstract being that is perceptible by our senses under different shapes. These forms do indeed vary, as is ordained by the conditions of special affinity of the ambient molecules that are incorporated with them. The infinity of these influences, ceaselessly and profoundly modifying the animal's contours as well as points on its surface, corresponds to an infinity of distinct arrangements from which arise the varied and innumerable forms spread throughout the universe. Thus all these diversities are limited to certain structures, according to the way the elements are displaced and reengaged.[21]

There is a single, abstract being, or "Animality"; it unfolds to various degrees and with different contortions, exaggerations, compensations, expansions, contractions, and inversions. An infinite number of "distinct arrangements" are possible, depending on the "ambient molecules" with which an organism engages. Quoting Leibniz, Geoffroy defined nature as "unity in variety." His "universal plan" or "animal type" was a field of possibilities—what he called, again using Leibnizian language, a set of "virtual conditions"—

whose actualizations depended on the circumstances in which the animal was found.[22]

Geoffroy's doctrine led to the creation of a new field, teratology, the study of monsters. While other classifiers sought "typical" specimens and avoided extreme variants, Geoffroy boasted, "I confront directly the most shocking anomalies"; surveying collections of monsters, he argued that "all these very diverse types of organization converge on the same trunk, and are only more or less differing branches of it."[23] His son Isidore also contributed to teratology, as did his devoted assistant, the physician-turned-anatomist Etienne Serres. As we will see again in chapter 8, Serres fused Geoffroy's philosophy with the concept of the single animal series—the cosmological principle that was Cuvier's bête noire. Serres also extended Geoffroy's thought by developing the concept of organogenesis, advanced by Schelling's follower, Lorenz Oken. This was the study of "the rules that nature follows in the successive formation of organs." In his *Researches on Transcendental Anatomy* of 1832, a work that he dedicated to Geoffroy, Serres presented his theory of the formation of embryos through a discussion of a monstrous birth: the two-headed Rita-Cristina (see fig. 6.3).[24] He wrote: "An organ that is normal or abnormal, regular or irregular, is for [nature] the same organ; it elaborates one like the other, one or the other advances towards its goal according to the same rules; and whether they reach it or

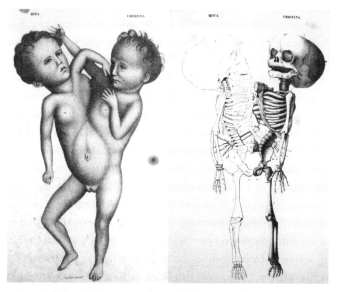

FIG. 6.3. Monstrous symmetry demonstrates principles of organogenesis: the two-headed girl, Rita-Cristina, in Étienne Serres, *Anatomie transcendentale*.

not, or even if they should surpass it, they remain circumscribed within the circle of laws that preside over organic formations."[25] The laws Serres identified concerned the formation of organs out of a single cluster of matter, the "ex-centric development of animals," following patterns of double symmetry and later fusion.[26] Serres attributed the different degrees to which organs were formed and divided and the differences between normal and abnormal specimens, and between normal specimens of different species, to their variable degrees of "formative force."[27] For Serres, abnormality illuminated the norm; monsters became a "subject for meditation" on the perpetuation of species and the appearance of new ones.[28]

Geoffroy's interest in environmentally provoked modifications of organisms took an experimental turn in his teratological experiments of the 1820s. By exposing chicken eggs to heat and cold, piercing their shells, and removing and adding oxygen to their atmosphere, he was able to produce blind, contorted, small-headed, and multi-clawed baby chickens.[29] Monstrous as these results may have been, the experiments confirmed Geoffroy's conviction that the processes of transformism could be technologically induced.[30] According to Serres, in Geoffroy's "sublime" thought, "the earth becomes a vast laboratory where a succession of beings continually develop, following a progressive and ascending path," and "the whole animal kingdom appear[s] as but a single being that, during its formation, stops in its development, here sooner, there later."[31] Geoffroy's embryological experiments suggested that the "vast laboratory" of nature could be simulated in the scientist's laboratory. They demonstrated that what Lamarck had called nature's power of production could be redirected and extended by humans' technical interventions.[32]

Tensions between Cuvier and Geoffroy mounted throughout the 1820s. Geoffroy published research and philosophical reflections on classification that drew on German *Naturphilosophie* and challenged the four branches and the fixed view of nature put forth by his younger, more successful colleague.[33] In response to the specter of a living, creative nature, and a blurring of his careful delineations between groups of species, Cuvier threw down the gauntlet in a dictionary entry titled "Nature" in 1824.[34] The concept of nature, he argued, was both physical and moral; in other words, it was both descriptive and normative. Although expressions like "nature abhors a vacuum" employed personification, it was "puerile" to take such language seriously by imagining the real existence of, for instance, a vital force within matter: "It was only in attributing a reality to the phantoms of abstractions that [physiologists] could deceive themselves and others concerning the profound ignorance in which they stand where vital movements are concerned."[35] Nature, he proclaimed, depended entirely upon the laws, properties, and forms with which God endowed it. Notions such as the "scale of nature" and the "unity of composition of organized beings" likewise implied a kind of necessity within nature that would be independent of God and that would set limits on his omnipotence. Instead, "the beings that exist in the world are coordinated in such a way as to maintain a permanent order."[36] Most scandalously, the ideas of a scale of being and a unified animal type lead to "a new form of the metaphysical system of pantheism."[37] On the contrary, nature must be seen as a "production of the all-powerful," ruled by his wisdom and by laws that conserve the harmony of the whole; the idea of a "successive appearance of diverse forms" would disrupt that order. In the midst of the Restoration, Cuvier was deploying some of the biggest rhetorical guns at his disposal: not only do the philosophers of nature err empirically by denying the real differences between branches of being, but they commit religious heterodoxy by constraining the power of the creator, introducing rival principles into nature, and advocating pantheism.

Yet in the later years of the rule of Charles X, the Restoration's authoritarian pieties were met with increasingly vocal challenges. In an 1829 dictionary entry of his own on "nature," Geoffroy responded to Cuvier. He first defended himself as a Christian; he then argued that theological views should not place restrictions on research, and defended the wide scope of his views. The opposition perceived between the scientific schools of France and Germany—between observation and speculation, between sensualism and spiritualism—had been greatly exaggerated.[38] French scientists employed abstractions as much as the Germans, and the Germans placed as high a value on observation as the French. And although *Naturphilosophie* may have occasionally been guilty of "precipitation" and "indiscreet ardor," even its speculations contained many fruitful ideas.[39] The "principle of unity" had led to the discovery of new laws, the reduction of phenomena to a more limited number of elements, and the discovery of relations among all the parts and levels of the universe. The notion of polarity, developed by Oken, had been shown to be at work throughout nature; Schelling's identification of magnetism as a basic principle of all "material formation" as well as "the germ of all animation" had led to the demonstration of very promising relationships between life, light, heat, and electricity.[40] Geoffroy defended these theories and the "a priori method" in general as providing an "active and subtle means of feeling one's way around" (*un tâtonnement actif, subtil*) that led to important principles.[41] Such principles included the unity of animal organization, the evolutionary claim that that "the ambient *milieux*" of animals transform their corporeal structure and thus their actual composition; and the underlying principles of combustion and electrification that rule over the decomposition and composition of both matter and life.[42] Natural philosophy did not court disaster by refusing to set limits on thought, but instead contributed to the expansion of the human domain. "Whatever man's industrious activity may be," he declared, "the scene surrounding him expands even as he remains nothing but a point in a field without a horizon"; each new discovery is a further step in man's "indefinite perfectibility." Knowledge brings mankind into an infinitely open, aesthetically uplifting theater, one he transforms and possesses with thought and activity, with the mind but also with the hands. "For such is the aptitude of his genius, that the more he sees, the more he learns and finds worth seeing. Entering into the most ravishing spectacle, the one provided by the relations among things, mankind holds them, he grasps them in his powerful hand; and in effect, *if* he knows them, and *as soon as* he knows them, they are his. Has he not already touched and penetrated them by the principle of his thought?"[43] The opposition between Cuvier and Geoffroy can be read not only as a conflict between a commitment to fixed nature and to a progressive nature, but between two views of the role of humans and technology within that nature. Like his ally Laplace, Cuvier advanced a conception of a fixed and stable order of nature, known through detached observation and reflection; Geoffroy, like his supporters Arago, Ampère, and Humboldt, put forth a view in which nature was transformed over time both through its own internal processes and by the individuals who were part of it. In the latter view, technological modification—this *tâtonnement*, this grasping, touching, and penetrating of things by thought and hand—was seen as part of the universal tendency toward transformation, remaking the spectacle in which humans and all other organisms are players.

Thus the highly publicized debate between Cuvier and Geoffroy that raged in the Parisian and international press in 1830, when Geoffroy attempted to show that the invertebrates were further manifestations of the unity of type, had been long prepared. It went beyond the thinkers' disagreement over the number of animal types (four versus one). Like his ally Laplace, Cuvier was cast as the man of power, exploiting the patronage system to secure posts for himself and his supporters, avid to keep science in the hands of the elite and shut others out from its workings and its rewards. He was closely allied to the government, and his emphasis on the fixity of species from the time of creation harmonized with the official Catholic and royalist ideology of the Restoration. Geoffroy was portrayed as the man of the people, the outsider who challenged the establishment;

his emphasis on "unity" resonated with criticisms of the July Monarchy's fragmentation and egotism. Geoffroy's support came from republicans and anti-Laplaceans within the scientific community and from liberal journals including the *Globe*; as we will see in chapter 8, after 1830 he was also allied with socialist reformers. In fact, one of the major issues in the debate was the kind of arena in which science should make its mark. While Cuvier sought support primarily in the Academy, Geoffroy (with the assistance of Arago) decided to "enter the domain of publicity (*publicité*)" and had his entries to the debate of 1830 printed as newspaper articles and pamphlets, in order, he claimed, to prevent excessive passion from affecting the exchange of ideas. Cuvier—by far the more agile orator—scorned this appeal to the press.[44] The much-used and abused term "publicité" captured the ambivalent aspect of the new media regime emerging in the 1820s and 1830s.[45] Geoffroy turned to this new resource in support of his theories, inviting curious Parisians into the same Academy whose doors Arago, as perpetual secretary, opened further. In reports on the Cuvier-Geoffroy debate, the daily paper, like the Ménagerie, became another politically charged place in which to contemplate the similarities and the differences between humans and other animals.

Goethe, who saw this conflict as more significant than the Revolution of 1830, characterized it as an opposition between two kinds of thinkers: Cuvier was the facile, easily appreciated man of facts, while Geoffroy was the more obscure visionary who followed a single idea to its limits.[46] It was also an opposition between two natures: one stable, unchangeable, created long ago, and divided; and one that was in motion, constantly producing novelty, and characterized by an underlying unity. Yet the debate should not be seen as simply opposing mechanical fixity and romantic growth. One of the ways in which Geoffroy's living, unified nature expressed itself was through organogenesis, the creation of new organs. Among these were human tools and machines, which could be applied back to nature itself to direct the course of its development. As we will discuss at greater length in chapter 8, for the mechanical romantics who drew inspiration from Geoffroy's theories, machines were part of nature's growth.[47]

PROGRESS EXPOSED

A visit to the Jardin resembled a visit to another much-publicized display in Paris: the Expositions of the Products of National Industry, which became major international events in the Restoration and the July Monarchy.[48] Here, reflections on the nature and power of human beings arose in response to a confrontation not with animals but with machines. Charles Dupin of the CNAM, one of the expositions' major boosters, and a frequent member of its Jury, wrote in a summary of the 1827 exposition aimed at workers, "For fifty years, each class of the arts . . . has undergone immense changes to satisfy the metamorphosed needs of a society itself transformed by a complete and profound revolution. I wanted to present the *tableau* of this industrial revolution, commanded by the social revolution."[49] In the 1830s, the term *industrial revolution* had only recently come into use.[50] The goal of the exposition, for Dupin, was to make this revolution immediately apparent to visitors. Further, the exhibit implied a relation of both analogy and causality between industrial transformation and the "social revolution" launched in 1789. In quasi-Lamarckian terms, Dupin describes the social revolution as changing the needs of the society, resulting in new adaptations to meet them: the instruments and machines that Humboldt called "new organs." Just as the garden of acclimatization at the Muséum demonstrated the modifications that species underwent to adapt to their milieus, so the expositions made visible the technological adaptations that society generated in response to changed conditions.

The expositions were a legacy of the Revolution. After the success of the "Festival of the Supreme Being" in 1794—with its marches, fireworks, frenetic music, and artificial mountain crowned with a Tree of Liberty and a Phrygian hat—Robespierre had proposed a "Festival of Industry," and in 1798 a "striking and novel spectacle" was organized to celebrate the mechanical arts. Robespierre was guillotined before he could see it. The festivities took place on the Champ de Mars, the traditional location for drills and victory celebrations, and featured a "Temple of Industry" that contained a statue of "Commerce" and the products named best by the jury. Trumpet-players, mounted cavalry, artists, and inventors all marched in a ceremonial procession. The organizer, the minister of the interior, François de Neufchâteau, declared that the time when "the arts helped industry to become at the same time the instruments and the victims of despotism" was in the past; instead, "now the torch of liberty belongs to industry."[51] The display of goods from 110 exhibitors was declared a victory in military terms: the exposition was "a campaign disastrous to the interests of English industry"; the French manufactures were "arsenals most fatal to the power of the British."[52] The elevation of the mechanical arts was joined with a celebration of military strength, commerce, republican ideals, and the glory of the nation.

Napoleon renewed festivities with the support of the chemist and minister Chaptal, who organized industrial fairs in 1801, 1802, and 1806. The first two were held in the courtyard of the Louvre, making a symbolic association between the fine arts and mechanical industry—between the work of *artistes* and of the less-instructed *artisans*.[53] The automatic loom of Jacquard, which promised to replace the skilled silk weaver, was displayed, but it was destroyed by a workers' association when it was taken back to Lyon. For the 1806 exposition, whose jury was headed by Gaspard Monge, there were more than 1,400 exhibitors, including many from nations conquered by Napoleon. The emperor attended the opening ceremonies, toured the exposition, and spoke with participants about their inventions. In 1806 the military aspect of the display was emphasized not only by the creation of a new prize category for "fabrication of armaments" but by the exposition's relocation to the Champ de Mars; the awards for the best products and processes were printed in *Le moniteur* alongside reports of recent military victories.

Thirteen years passed before expositions resumed in the Restoration, with fairs held in 1819, 1823, and 1827, again in the courtyard of the Louvre. These displays were notable for their textiles, including cashmeres said to rival British imports from India and colorfast white silks. Displays of metal were also steadily increasing. The catalog of 1827 stated, "We have seen new branches of metallurgy born and develop as if by enchantment," primarily in purifying and reinforcing iron.[54] A commentator praised the appropriateness of the Louvre as the exhibition site, transformed from a sanctuary of fine art to "the sanctuary of national industry and the depot of this host of marvels."[55] The expositions' increasingly wondrous impact was accompanied by their growth in size: the 1827 exposition had 1,693 exhibitors and six hundred thousand visitors—many of whom, however, may have been in town primarily to see Geoffroy's celebrated giraffe, the other great tourist attraction that year.

Under the July Monarchy, the expositions were promoted and planned by engineer-savants at the CNAM, as well as by members of the Saint-Simonian movement, whose religious glorification of industry was echoed in their displays. The official report of the Jury for the 1834 exposition was written by Charles Dupin, and a popular book summarizing it was written by the Saint-Simonian polytechnician Stéphane Flachat, who argued that what once had been a display of an "arsenal" was now a contribution to international peace, to the goal of industrial development.[56] The emphasis shifted away from curiosities and luxury goods of limited reach, such as ostrich-feather carpets or cat skins, to products cheap enough to reach a mass market of consumers. Industrial machinery was represented on a much greater scale, and it included steam-powered techniques for produc-

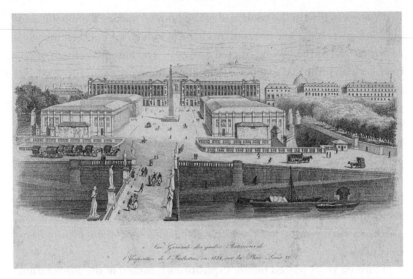

FIG. 6.4. Buildings constructed for the Exposition Nationale des Produits de l'Industrie Française, 1834, Place de la Concorde. Cover of the catalog by Saint-Simonian scientist-engineer Stéphane Flachat, *L'industrie: Exposition de 1834* (Paris, 1834). Credit: Avery Architectural and Fine Arts Library, Columbia University.

ing paper. To house these displays, new buildings were constructed on the Place de la Concorde, featuring the obelisk of Luxor donated in 1829 by Mehmet Ali, viceroy of Egypt (see fig. 6.4); a new national ordinance mandated staging an exposition every five years. In 1839 the exposition moved to buildings on the Champs Elysées.[57]

The expositions' successes spawned imitators: the exposition in Berlin in 1844 was a rival, as was the 1851 Great Exhibition at the Crystal Palace in London—an event often presented by historians, erroneously, as the first of its kind. In Paris in 1844, the scale of operations had increased to 3,696 exhibitors; daguerreotypes were on display, along with more and more powerful steam engines. The expositions' galleries of luxury goods were impressive, but, said a report by the jury in 1839, the truth of national glory was pressed in metal: "As soon as we leave the dazzling spectacles of magnificence and richness and enter the vast circle of machines, which shows us iron, and more iron, and yet more iron, the illusion vanishes, truth emerges, and the soul is at once seized by the grandeur of these silent devices, so productive when active. Iron is the agent of force. The power of nations can in a sense be measured by the amount of iron they use."[58] Like Napoleon before him, King Louis-Philippe took a strong interest in these events, giving an inaugural address reprinted in the official rapports (in which exclamations of "Vive le Roi!" are duly noted) and visiting regularly. He handed out medals at an awards ceremony, which for one observer was proof that "the industrialists were the new nobility of France."[59] The products and processes on display at the expositions showed the increase of humans' power to alter their environment; they also revealed, as Dupin suggested, a transformation in the organization of industry that had political dimensions: a reorientation of the national interest toward inventors, entrepreneurs, and profits.

But whether this metamorphosis was to the advantage of workers or to those who invented or who owned the machines was a poignant question.[60] Despite the harsh greeting that met Jacquard's loom in Lyon in 1801, by 1844 tens of thousands of these mechanized weavers were at work in France and were recognized as crucial for French textiles' international competitiveness.[61] Charles Babbage, Dupin's English correspondent, advocated lowering costs of production by introducing machinery wherever possible to reduce the amount of skilled labor and even, with his calculating engine, to perform intellectual tasks. One of the greatest successes of the exposition of 1844 was a machine that claimed to take this process of mechanical transmutation yet farther: an android that could write, draw, and answer questions. Its creator was a clockmaker from Blois, whose name looked backward to *Robert le diable* and forward to Harry Houdini: Eugène Robert-Houdin.

THE FANTASTIC AUTOMATON

On his tour of the exposition's mechanical products, King Louis-Philippe lingered at Robert-Houdin's stand. His attention was held by a man made of metal, attired like an eighteenth-century nobleman, sitting at a writing table. The king asked him questions and received clever replies written on paper in a fine hand. The "bourgeois monarch" and the writing automaton formed an intriguing pair: the king, a merely human holder of an office only recently demoted from divine status, met a machine suddenly elevated into "esprit"—wit or mind.[62] Thanks to the publicity surrounding this bagatelle, Robert-Houdin was lauded as the heir of Vaucanson and Maelzel. He moved his business from Blois to Paris and acquired a theater near the Palais-Royal, where, starting in 1845, he launched a new kind of performance: the *Soirées fantastiques* (see fig. 6.5).

The ambiguities we have seen in earlier chapters surrounding animated machines, ether physics, and animal magnetism were central to Robert-Houdin's performances. In his "experiments," as he called the acts of his show, he shared the stage with a harlequin gymnast, an android who

FIG. 6.5. Publicity poster for Robert-Houdin's *Soirées fantastiques*. Note the automata around its border ("the baker of the Palais Royal," acrobats, magic orange tree, floating clock) and his self-attribution as "Physicist and Mechanic." Courtesy of Maison de la Magie, Blois.

emerged from an egg, a baker who delivered miniature *pâtisseries* to order, and a tree that rapidly flowered with orange blossoms. His performances flirted with suggestions of divine or diabolical mastery over the powers of creation, as in his *Memoirs*' description of the moment when he breathed life into his writing automaton: "'Who is the author of your being?' I pressed the spring, and the clockwork began acting. I dared hardly breathe through fear of disturbing the operations. The automaton bowed to me, and I could not refrain from smiling on it as on my own son. But when I saw the eyes fix an attentive glance on the paper—when the arm, a few seconds before numb and lifeless, began to move and trace my signature in a firm handwriting—the tears started to my eyes."[63] Heightening his role of father-creator, Robert-Houdin also rotated his flesh-and-blood sons through his act as fellow performers. In the experiment he called "the Etherian Suspension," his son drank "ether"—a substance recently introduced as a surgical anesthetic but whose name evoked both physical theory and animal magnetism—whereupon the young man began to levitate, an effect that one viewer assumed was due to the fact that the boy was actually a machine.[64] Robert-Houdin modeled another experiment on animal magnetism and somnambulism, a state often associated with the power of seeing at a distance. The advertising poster promised: "The son of M. Robert-Houdin, gifted with a penetrating second sight, will offer with his father an entirely new experiment." Twelve-year-old Emile's abilities were supposedly confirmed by the fact that his "pale, intellectual, and ever-thoughtful face represented the very type of a boy gifted with some supernatural power." After blindfolding him, Robert-Houdin rang a "mysterious little bell," prompting the boy to describe objects submitted to him by audience members.[65] With these scenes of somnambulism, magnetism, ethers, and uncanny creations, as well as his animated machine-men, Robert-Houdin lodged his spectacles within the imagery of the fantastic (see fig. 6.6).

We saw in chapter 5 how Hoffmann and his French imitators made automata a central feature of fantastic literature. Images of inert objects coming to life often involved a superimposition of modern technology and ancient beliefs: an animated painting in Théophile Gautier's tale "Omphale" and a haunted coffeepot in his "La cafetière" both channeled the souls of departed aristocrats, invoking the lost individuality and passions of the ancien régime.[66] Other works evoked Renaissance-era alchemy and hermeticism. At this time, Egyptian occult science and the Tarot had been "rediscovered" by Fabre d'Olivet and Abbé Constant; Emmanuel Swedenborg's doctrines of correspondences between nature and the spirit world became popular sources for salon conversation and literary creation, and they played a central role in works of Balzac, Sand, and eventually Baudelaire.[67] In the Renaissance, automata, like amulets and idols, had been understood as magnets and mouthpieces for magical powers; in the early nineteenth century's patchwork of revived illuminism, automata could be seen as allegories for the infusion of spirit in the material world.[68]

FIG. 6.6. Two vignettes from a brochure distributed at Robert-Houdin's *Soirées fantastiques*: "Ethereal Suspension" and "Second Sight," both of which use the magician's son as a prop. Maison de la Magie, Blois.

As symbols of both forward-looking technological rationality and ancient supernatural belief, fantastic imagery of animate machines raised the question of the proper limits of human modification of nature. Poet Laurent-Pichat's "Ce siècle est en travail" reinvested his century's innovations with ancient myth:

> At the risk of being an atheist
> I love Pygmalion and I love Prometheus.[69]

Le magicien, a novel of 1838 by Alphonse Esquiros, a promoter of animal magnetism, a socialist critic, and a member of the Club des Hachichins, was set in sixteenth century Paris; it featured a brass automaton built by an Alexandrian magician rumored to have "the gift of giving life to stones or to metal," like "Pygmalion himself." The magician was overheard murmuring: "*Homo animale rationale*: here is the definition of the great philosopher, Aristotle.... If I could prove that this definition holds for the man made by my own hands, I would be truly a creator and a God." In the Renaissance's dreams of ensouling a mechanical creature, Esquiros found an anticipation of the Faustian ambitions of modern industry.[70] Gérard de Nerval's novel *Aurélia; or, Dreams and Life* featured visionary hallucinations, eerie imitations of life, and mergers of art and nature. In it, a mechanical bird introduced the delirious narrator to his dead ancestors; the narrator read announcements whose "typographic setting was made of garlands of flowers so well represented and colored that they seemed natural." In one vision, he saw workers molding in clay an enormous llama; as limbs grew from its body, he rephrased Esquiros's question: "Can we not also create men?" *Aurélia*'s metaphysical slippages echoed a fundamental uncertainty of Nerval's time: What would be the rewards or punishments for reproducing and modifying the natural order?[71]

Robert-Houdin's *Soirées fantastiques* played on these ambiguities in his images of lifelike machines and uncanny humans. He has been recognized as the creator of modern stage magic—not least by Harry Houdini, whose name was an homage. Robert-Houdin himself always spoke in a low-key patter, wore unadorned evening dress, and gave his proceedings a scientific air by calling his tricks "experiments." With his animated machines and mechanized humans, Robert-Houdin played a complicated epistemological game with his audience. Current interpreters suggest that viewers were surprised by curious phenomena while knowing that they were all merely tricks. According to this view, audiences' pleasure lay in the puzzle of the deceptions' hidden mechanisms: the only "magic" at work, they knew, was an exceptional mastery of stagecraft and technical ability.

While such analyses capture the bluff and misdirection in Robert-Houdin's presentations, they take for granted that a single, well-defined understanding of nature was shared between the performer and his public.[72] Yet it was precisely the absence of such a firm common ground—among audiences composed of kings and workers, atheists and believers of various stripes—that made these evenings, like the presentations of somnambulism to which they alluded, "fantastic."[73] Did the child see at a distance, or was there a trick?[74] Was the machine's writing and speaking accomplished by ordinary engineering, or some new wondrous method? For some, scenes of animal magnetism confirmed the existence of a soul distinct from the body, communicating beyond the limits of matter; others read them as proof that even the phenomena of thought were subject to control, once the definition of matter was expanded to include the medium of magnetic influence; others assumed they were all bunk. Similarly, displays of automata could suggest an uncanny mastery over matter; they could imply that thought, reflection, and memory could be performed by machines; or, on the contrary, they could be read as playful illusions, as clever tricks that mocked such beliefs.[75] Robert-Houdin's performances capitalized on the uncertain status of ether, electricity, and animal magnetism, along with the constantly expanding possibilities of technology. Experiments like "The Second Sight" encouraged a hesitation between multiple conceptions of the natural order, a movement around the chessboard sketched on page 159. Similar ambivalence appeared in the devilish Robert-Houdin's claims of mechanical prowess. By debunking the illusions of other magicians with mere "mechanical" explanations, he strengthened a mechanistic reading of his own life-imitating inventions. For instance, in the passage quoted above, on the quickening of his filial writing automaton, he never let on to his readers that what he was describing was impossible by the technical standards of the day. *We* might know that such a lifelike, quasi-intelligent machine could not have been built, and some audience members might have known it too; but that is no reason to attribute an identical threshold of belief to all of his viewers.[76] Robert-Houdin's *Soirées* could be taken in without succumbing to either complete belief or utter refusal. Like fantastic literature, the performances made audiences hesitate, oscillating between mutually contradictory interpretations of the scenes before them.[77]

In this period, scientific controversies like those between the magnetizers and their critics, between the Laplaceans and their dissenters, or between Cuvier and Geoffroy were occasions to shift, and frequently to dissolve, lines of demarcation between rational science and mystical speculation. The theories of ether that opposed notions of action-at-a-distance, for example, drifted easily into inquiries about vital fluids and spirit communication; further, as we have seen, and will see even more in part 3, mystical and religious threads were woven through many discourses ostensibly concerned with rational science and industrial progress. Performances of lifelike machines in fantastic literature and stage magic made the most of such mixtures and oscillations to generate fear and delight, profits and fame. They enacted the central concerns and metaphysical uncertainties of a society undergoing a thorough technological metamorphosis.

APOTHEOSES AND OTHER THINGS

The transformations implied by the animals of the Muséum, the machines of the expositions, and the combinations of the two in the fantastic arts were summarized and worked over in the book *Another World* of 1843, illustrated by J. J. Grandville. Grandville explored the plasticity of all levels of the world—of matter, minds, organisms, and society—and the techniques available to take advantage of it. His book catalogued the dizzying possibilities raised by the new role in which humanity found itself: that of a second creator.

In *Un autre monde*, every domain of reality (animal, vegetable, mineral, technological, sociological) was presented, in every possible style (philosophical, religious, portentous, satirical, mock-heroic, taxonomic) and with every possible political attitude (revolutionary, reactionary, liberal, socialist). Its subtitle indicated a hysterical portfolio of the period's encyclopedic, cosmological, and messianic impulses: *Transformations, Visions, Incarnations, Ascensions, Locomotions, Explorations, Peregrinations, Excursions, Stations, Cosmogonies, Phantasmagorias, Reveries, Foolishness, Facetiousness, Whimsies, Metamorphoses, Zoomorphoses, Lithomorphoses, Metempsychoses, Apotheoses, and Other Things*. The rhyme of the last four words—"*apothéoses et autres choses*"— placed divine ascension alongside derisive miscellany, and was just one figure of the book's disarming bathos and disorienting mixtures. Grandville was a major influence on John Tenniel, the illustrator of *Alice in Wonderland*. Like the *Alice* books, *Un autre monde* showed a world turned upside down—in fact, many different worlds turned in many directions—in which animals, vegetables, minerals, and machines exchanged roles with humans and each other.

The book was published in thirty-six installments that were mailed to subscribers one by one, following a wide-scale publicity campaign; as with many "books" of this period, subscribers would have the installments bound once they had all been received. The text was unsigned but later revealed to be by P. J. Stahl, the pseudonym for Pierre-Jules Hetzel, the fiercely republican publisher of Balzac and Jules Verne. The preface narrated this non-standard form of authorial attribution as a quarrel between a quill and a pencil: although the usual order of composition had images created to match the text, here it was the writer, represented by the quill, who would follow the lead of the illustrator, or the pencil. After this allegory of the book's material construction came scenes that pushed the logic of the fantastic to the hilt, making the familiar strange and the strange familiar. Yet for all of its kaleidoscopic arrays of distortions of the present, the book's very interest in the process of reconfiguration and transformation was entirely true to its time. As an inventory of ways of world making—a cosmogram of cosmograms—it showed how acutely this age felt the interdependence between the natural order, the social order, and the technical order. It revealed the close ties between romanticism's organic mutations, aesthetic ecstasies, voyages into infinity, and passionate dramas; technologies of production and knowledge; the media of steam, light, electricity, and human labor; and the instruments these powers brought to life.

The opening pages declared: "For a long time the earth has not been enough." Responding to the "need for another world," Dr. Puff, a down-on-his-luck inventor, publicist, and poet, "transforms himself." "Would it not be better," he asked, "following the celebrated myths of the Saint-Simonians, the Fourierists and other socialists, to found a new religion?" It was easy: adding "neo-" to any term was enough to launch a "fresh new theogony." Thus, he announced, "I will reunite the joyful fictions of Greek mythology to the no less laughable incarnations of Indian religion, and I'll serve it all up under the name of Neo-Paganism." To research and promulgate his doctrine through the water, earth, and air, he recruited two other "Neo-Gods," Kracq and Hahblle. Puff (whose name echoed both literary and critical hype and the image of the cosmic bellows depicted in the book's pages) physically resembled the anarchist reformer Proudhon; Hahblle (whose name suggested manual dexterity and chicanery) had the wild hair and wide-eyed enthusiasm of Pierre Leroux, a socialist I will discuss below; and the older, fussy Kracq, a former captain adorned in military clothing, seemed a visual merger of Charles Fourier and Saint-Simon, recalling their serial, military organizations. Both Hahblle and Kracq also found themselves in need of a "total transformation," not least at the level of linens and boots. They happily accepted Puff's invitation to be elevated into "supreme potency."[78] The book's chapters traced their journeys, observations, and creations, revealing the building blocks of the cosmos and the different scales and consequences of creation.

The "Zoomorphoses" of the subtitle suggested the transformation of humans, inert objects, and machines into animals and vice versa. In *Un autre monde*, the natural-historical gaze that determined the arrangement of species in the Jardin des Plantes and the Muséum d'Histoire Naturelle became sociological.[79] *Un autre monde* reflected the classificatory impulse in the literary genre of the "physiologies," Geoffroy's philosophical anatomy, Balzac's natural history of society, and the nascent social sciences, with their studies of "typical" specimens illustrating the classes of workers, bourgeois families, vagabonds, divas, judges, salon musicians, affected painters, platitudinous professors, confidence artists, duped crowds, washerwomen, nursemaids, babies, and landowners. In dressing up animals in these disguises, Grandville transposed the animal kingdom and the human world. This ontological transvestism went even farther in his imagery of underwater creatures growing into the form of manmade objects, whether medals, obelisks, cathedrals, dominos, mirrors, wigs, fans, or hairbrushes.[80] With several references to Laplace's famous *Mécanique céleste*, the book also anthropomorphized superlunary beings. It introduced the complicated domestic arrangement between the moon and the wayward sun; readers watched, along with animated telescopes, their breakups and periodic reconciliations—which astronomers mistakenly called eclipses. They followed, likewise, the social intrigues of the celestial *beau monde* in its sumptuous soirées, including one hosted by Uranus, a relatively new arrival to the social firmament (Uranus was discovered by Leverrier under the guidance of Arago), with a dance performed by the members of the Zodiac.

One implication of these systematic disorientations was that the same kinds of logic were at work in understandings of society and of the different domains of nature. Another was that nature's order itself could be rearranged, that hierarchies of both society and the comsos could be reversed and upset, shuffled and recombined like the letters of a compositor's tray: not just at the level of representation, but at the level of things themselves. Such unchecked demiurgical play could result in the monstrous, the hybrid, and the chimerical. An episode set in the Jardin des Plantes showed creatures that illustrated the grotesque consequences of Geoffroy's "unity of plan": if there was only one animal, then mixed creatures like Grandville's snake-headed dog and the two-mouthed "doublivores" must, in potential, be part of the natural order too. Grandville's imagery sprung from an accurate understanding of the implications of Geoffroy's transcendental anatomy: no combination, however monstrous, was forbidden.

The book also ran through new ways of seeing and experiencing. The first vignette discussed Puff's new symphony, "Of the Self and the Non-Self," whose title recalled Fichte, Schelling, and Cousin's emphasis on the creative activity of the "I" in constructing the world, while the automaton orchestra it depicted applied the phrase to the uncanny mirror play between humans and machines. The book repeatedly depicted the activity of the eye, in imagery of disembodied orbs and animated ocular aids including telescopes, opera glasses, microscopes, and an ambulatory magic lantern.[81] Scenes were presented from disorienting angles and scales. The "bird's-eye view" enjoyed by the airborne Hahblle revealed the relative "flatness" of humans: from the sky, ordinary mortals were barely shorter than the heroes who occupied the heights of commemorative columns. Another episode discussed the gap between appearance and reality, the "war between physiognomy and character"[82] giving rise to a "philosophy of disguise": "The more disguised the man is, the more he makes himself known."[83] At a carnival, Puff wondered: "Did I see men disguised as beasts, or beasts disguised as men?"[84] These distortions correlated closely with the mutations featured in the arts, psychology, and physiology of the time. The chapter entitled "The Metamorphoses of Sleep"—with its serial depiction of the gradual substitutions accomplished by the imagination (see fig. 6.7)—began with a sonnet in which only the last word of each line was provided, laying out a basic lexicon of romantic hallucination: "*cieux/dieux/extase/gaze/coeur/Bonheur.*" Next came the one-word poem of an opium-eater: "aaahhhhh." The key term of Bertrand's theory of animal magnetism summarized the chapter, "which is just one long ecstasy."[85]

These internal voyages paralleled other chapters' voyages in space and time. A balloon in his pocket enabled Hahblle to travel the skies, and his journey landed him in a strange capital where marionettes performed ballets and submitted paintings to juried exhibits. Kracq got to know the undersea community, and took a journey through the topology of Western myth, including Dante's inferno and the Elysian Fields. Research into new technologies of transport aided Puff on his journeys: he launched himself through the air with spring-, steam-, and lightning-based aerial locomotives. He also discovered the change of morals that accompanies changes of latitude on a visit to the Îles Marquises, to "Young China," and to an island where the rulers were very tall and the people very short, teaching him "the first cause of all the social distinctions." Popular travel accounts and the beginnings of a comparative science of mankind—in which temporal distance

FIG. 6.7. "Metamorphoses of Sleep," from Grandville, *Un autre monde*, 243.

was superimposed on geographical distance, much as Grandville comically mapped social standing on literal stature—were echoed here. These journeys replayed the effect of defamiliarization brought about by the encounter with unknown civilizations, the sudden copresence of multiple times and places in the crowded capital, and the strain imposed by growing class divisions.

Much like Balzac's *Lost Illusions*, the book diagnosed the conditions of its own production: the new regimes of mass marketing, publicity, criticism, and popular science. It lampooned the vast growth of print, publicity, and mass-produced literature, and the fashions of the salons and boulevards. It showed a worker chopping texts like sausage, a wheel of fashion determining the fate of authors and artists, the desperate dog race of fame and success, a pump designed to flood the city with prospectuses and reviews, and the apparatus required to place a new "star" in the heaven of fame. Characters bemoaned the thoughtless rapidity of literary production, the predictable (mechanical) conventions of literature (as in the fill-in-the-blanks dream poem), and the appearance of literature done "by automaton." Throughout, readers saw the mechanics of success for Parisian theaters, ballets, and operas, with their well-coordinated campaigns of publicity, advance notices, and reviews. Yet the book's own prospectus and front and back writings were full of the "puffery" and inflated critical currency it mocked; it had been advertised widely; each installment advertised the next. The book's awareness of this ambiguity—its participation in what it ridiculed—was highlighted by placing the words of its laudatory introduction in the mouth of Robert Macaire, the legendary grifter and practitioner of the subtler forms of the Parisian scam.[86]

Grandville's critique of publicity overlapped with his relentless political satire. In the penultimate chapter, Kracq reviewed all of the forms of government, pointing out each of their flaws, and concluded that only an "omniarchy," the simultaneous existence of all forms of government, could last. Grandville also depicted aspects of Fourier's system— and did so quite faithfully, as if in Fourier's case, literal transcription was satire enough. The struggle between forces of tradition and those of reform figured throughout the book, as in the vegetable uprising, where a hitherto oppressed class—chopped, grated, stewed without recourse—rose up: "The plants had long been organized in a secret society, subdivided into centuries, decuries, ventricles and sub-ventricles; this classification, as little anticipated by Linnaeus as by the police, gave me a high opinion of the forces of the conspirators. Nothing is more dangerous than an eloquent proletarian."[87] For vegetables and proletarians, self-legislated organization into divisions, ranks, and series was the precondition for revising the imposed order that skinned, ground, and devoured them (see fig. 6.8). Another image captured the political stalemate of the "juste milieu" in a constitutional clock kept inert by the counteracting effects of a withered, aristocratic lobster dragging the government toward the past and a strapping, mustachioed steamboat man pulling it toward the future.

Along with its rethinking of the kingdoms of nature and the foundations of government, the book was obsessed with new powers of industrial technology. It abounded in imagery of steam, gears, and sprockets. Accord-

FIG. 6.8. The vegetable uprising, from Grandville, *Un autre monde*, 59.

ing to an epic poem to which it alluded, *Du vapeur*, "The piston of Papin is the scepter of the world." Grandville repeatedly showed humans replaced with machines. Puff's first scheme, we saw, was a "humano-mechanical concert" demonstrating that "in this century of progress, the machine is a perfected man."[88] The "Steam Concert's" frenetic musicians, with appendages resembling the pistons of steam engines, were a grotesque mockery of the charming appearance and harmonious demeanor of Enlightenment-era automata such as Jacquet-Droz's harpsichord player or Vaucanson's drummer and piccolo player (see fig. 6.9). Yet even if steam introduced an element of violence to automaton imagery, other pictures, like that of the automaton nursemaid/harp who rocked the cradle of the world's youngest musician, prevented a purely negative or monstrous reading. Rather, mechanism was one more element in Grandville's all-encompassing game of ontological transpositions and hybrids. New machines took the place of animals and humans, even as they extended humans' capacities and shifted the boundaries of the world. Grandville made machines both symbols and agents—both repulsive and endearing—of an open-ended, monstrous, sublime, and gleeful social metamorphosis.

In its broadest reach, the book took as its subject the birth and death of worlds. Traveling into space, Hahblle encountered "the mysteries of infinity" and "the origin of all things." The proprietor of the Celestial Mechanics was a wizened magician blowing bubbles that inflated to create the elements of the world—into which a demon added confusion and trouble. He also saw a cosmic jester, "more skilled than any Indian juggler," throwing and catching universes and displaying the hidden "great law of the equilibrium of worlds."⁸⁹ Hahblle also observed a giant bellows that caused the winds, but he decided not to record its location for fear that some well-meaning human would destroy it, to the detriment of makers of wigs, hats, and umbrellas, who depend on strong winds to keep themselves in business. He also floated by a set of sidereal iron bridges, the height of contemporary industrial design, bridging planets.⁹⁰ In such imagery Grandville aligned cosmogony with play, the aleatory whimsy of children and fools, and the combination of beauty and utility in modern engineering. The book also ran through ancient myths of origins and ends: a visit to the Champs-Elysées mixed the diversions found on that famous boulevard with the assembly of

FIG. 6.9. "The Steam Concert," from Grandville, *Un autre monde*, 17. Note the symphony's title, "Le Moi et le Non Moi," in the upper left, and the human hand in the lower left.

heroes dwelling in its afterworldly namesake. While such imagery might be suited to "Neo-Paganism," the book also referenced the genesis and Apocalypse of the Judeo-Christian tradition. In the final pages, the "neo-Gods" gave up their lives after hearing a prophecy that their world would end when they saw "matter animated." A flood arrived, along with the steam ark to carry away animals dressed as for a railway journey, two by two. Technological transformations were reimagined as theater and as myth. In the book's most poignant indication of the notion of nature as theatrical artifice, in the image at the start of this section, Grandville showed stagehands lifting the curtain of night, lighting the gas lamp of the sun.

THIS WORLD REMADE

Grandville and his writer and publisher, Hetzel, had strong republican and reformist sympathies. Although *Un autre monde* exaggerated aspects of socialist and reformist doctrines, the images and text never settled into obvious mockery of fixed targets. In the "Steam Concert" image, for instance, what held the symphony together appeared to be the baton of a pressure indicator at the center; yet at the lower left there was a human hand adjusting a valve. The suggestion appeared to be that human interests still ran this otherwise automated show and, perhaps, that "mechanization" was really just a new means of human domination. If such an image was intended as a satire of socialist critics, it was a strange one, because at this level, the message—that mechanization and spectacle can be tools of oppression—would have been wholeheartedly endorsed by most social critics.

Un autre monde was neither a condemnation of mechanization nor a straightforward satire of the new and vocal social reformers attempting to remedy the ills of industrial society. Instead, it affirmed the polymorphous, world-making potentials of new machines—their ability to serve as sources of oppression, liberation, creativity, inhibition, prophecy, and absurdity.[91] Rather than undermine contemporary reformers and revolutionaries, the book intensified and repeated their central moves, by showing the inseparable relationship between exaggeration, critique, and utopian construction.[92] The new, fantastic world that Grandville presented was distorted; yet, in its distortions, it was essentially a faithful mirror of the uncanny modifications displayed and discussed at this time in sites including the Muséum d'Histoire Naturelle and the Expositions of the Products of National Industry. Like Robert-Houdin in his *Soirées fantastiques*, Grandville began from the central, giddy discovery of his era—the plasticity and modifiability of nature, due both to nature's inherent processes of change and invention and to the technical interventions of humans—and pushed that discovery to its limits, both feeding and mocking audiences' credulity and optimism about the world's perfectibility. *Un autre monde* was a landmark in a shift in emphasis within the currents of romanticism that began in the 1830s: a move away from concern with the nostalgic, fantasizing subject and toward a still-fantastic concern for the plight of the newly visible working class.

Grandville's imaginings of *Another World*—of multiple other worlds—are a fitting transition into this book's part 3, which focuses on utopian thinkers and activists of the late Restoration and the July Monarchy. Building on the techniques of discovery, creation, and spectacle examined in parts 1 and 2, we now turn more deliberately to social technologies—techniques of conversion, communication, and temporal coordination—in several of the works in which modern social science and socialism were born. To be complete, our discussion would need to grant further attention to republican reformers and conspirators, the politicization of popular Christianity, the plans and failures of the Fourierists, the sources and impact of the feminist writings of Flora Tristan and Suzanne Volquin, the Icarian communism of Cabet, and the specific impact of these ideas on Marx.

Yet the closely related cases on which we concentrate—Saint-Simon and those who called themselves his followers, and the two Saint-Simonian apostates Pierre Leroux and Auguste Comte—led three of the most influential movements of social reform in their time and in the long term. More important for the concerns of this book, they were also the "romantic socialists" who were the most deeply engaged with the creative power of technology and the social necessity of developing and reorganizing the sciences. The Saint-Simonians' ranks were filled with polytechnicians, their visions of conversion were informed by engineering practices, and they went on to build the technical and administrative infrastructure of modern France. Pierre Leroux compiled and edited writings from all the fields of science and was one of the age's major interpreters of both Geoffroy Saint-Hilaire and German *Naturphilosophie*; furthermore, his ideas were inseparable from his extensive work as a compositor and printer. Finally, ending part 3 with a new look at Auguste Comte brings the book full circle in two respects. First, chapter 9 returns to the Academy of Science and the Ecole Polytechnique and the issues surrounding "the second scientific revolution" that were the focus of part 1. More significantly, however, approaching Comte after passing through the preceding chapters gives us a new context for understanding the origins of the disciplines he founded, sociology, sociology of science, and French "epistemology"—the very fields from which the present book draws many of its questions. That the founder of positivism was in many respects an exemplary mechanical romantic forces us to pause before taking for granted the oft-remarked soullessness and reductionism of positivism, and before identifying "modernity" by a split between the positive sciences and the subjective arts, between murderous machines and soulful humans.

By showing the connections between these authors' visions and the sciences, as well as their obsessions with new technologies, I do not mean to deny their often outlandish, fantastic aspects. On the contrary, my aim is to show how the speculative, cosmological, and utopian themes for which these political works are both remembered and dismissed drew upon the same currents of inspiration and practice that lay behind the era's developments in electromagnetism, thermodynamics, geophysics, and natural history; they also were immersed in the emerging, sensational mass market of romantic arts and spectacle; in addition, they drew upon romanticism's reflection on the protean powers of nature and mind. The chapters of part 3 concentrate again, as in part 1, on the central role these authors granted to transformative technologies or "magical instruments" (as Leroux described Volta's battery): their romantic machines.

Like the scientists in part 1, these authors insisted on the human intentions, needs, and actions that went into making natural knowledge. Yet even if they shared views on how to know the world, they offered three compellingly different responses to the question of what the world is like. The Saint-Simonians resolved the opposition between spiritualism and materialism by embracing pantheism. This solution appealed to Leroux, although his own ontology, which relied on a distinction between the virtual and the actual, was more complex and in many ways more intriguing. Comte's positivism bypassed such metaphysical questions by presenting science as simply the establishment of relations among phenomena, without inquiring about underlying essences. Yet his revision of natural philosophy relied on a logic of embryonic development; and he later evangelized about the social necessity of speculative fictions. Most importantly, these authors arrived at quite different views about the nature of the social bond and the ideal distribution of power. The Saint-Simonians' centralized, hierarchical autocracy can be contrasted with both Leroux's ground-up, commune-based democracy and Comte's religious federation of positive states.

Strikingly, these projects of social reform also announced a theme that has only been touched on until now—although it was the organizing conceit of Grandville's *Un autre monde*, with its neo-Gods and retellings of myth. While to modern eyes it might seem that mechanistic and positivist reform must be antithetical to religion, or that a mixture of science and technology with emotion or faith is taboo, these actors claimed that to renew society, the religious spirit of the past had to be revived. For this revival, the character of dogma and rites and the objects of worship had to evolve: both science and religion would be changed in the passage to a more harmonious future. After 1830, as romantic literature took a social turn, so too were the sciences and industry drafted, with evangelical force, into projects for social metamorphosis. As we will see in the conclusion, this turn provided oxygen both for the political explosion of 1848 and for the short-lived republican conflagration that ensued.[93]

PART THREE
ENGINEERS OF ARTIFICIAL PARADISES

SYSTÈME DE FOURIER.

PART THREE

ENGINEERS OF ARTIFICIAL PARADISES

7

Saint-Simonian Engines

Love and Conversions

A PARABLE

In 1830 Thomas Carlyle—the Scottish author of "Signs of the Times," a celebrated 1829 article warning of the dangers of "The Mechanical Age"—received a package from Paris.[1] Its sender, Gustave d'Eichtahl, was the apostle of a new religion, Saint-Simonianism. After reading the pamphlets and journals, Carlyle wrote back. He found much to admire in the new movement's diagnosis of the modern age, especially its observations about the crippling effect of excessive individualism and competition and its recognition of humanity's religious needs. But he had one serious reservation. "[In] all religions recognized up till now, an indispensable and even essential element" is *"a symbol or a symbolic representation in which the divinity perceptibly manifests itself.* . . . There must be a symbol that offers itself up to the believers, as it is only in this way that the imagination, the true organ of the infinite in man, can harmonize with the intelligence, which is the organ of the finite. Up to this point, I find nowhere in your doctrine such a symbol."[2] This view of the power of symbols to capture the divine and infinite—bringing the infinite imagination into harmony with the finite understanding—was drawn from Carlyle's readings of German romanticism: he had translated and introduced Goethe, Schiller, and others. Yet whether or not Carlyle's assessment of the Saint-Simonians in 1830 was accurate, by 1832 members of the movement were engaging potential converts with a barrage of symbols, targeting both the intelligence and the imagination. They forged a new iconography with saints and statues; songs, poems, and temples; a new Bible and a new creation story; a sacred college; prophecies of a new creation; and a new theology that identified the universe with God.

By plunging headlong into symbolism and a metaphysics based on nature's infinite powers, the Saint-Simonians might be seen as endorsing Carlyle's "romantic" preference for "the primary, unmodified forces and energies of man" against the stultifying forces of "The Mechanical Age." Yet perhaps one of the reasons for Carlyle's gradual cooling toward the Saint-Simonians was their refusal to take seriously his central opposition between the dynamical—the "mysterious springs of Love, and Fear, and Wonder, of Enthusiasm, Poetry, Religion"—and the mechanical. For among the most portentous symbols for the Saint-Simonian religion were *machines themselves*: "The railroad is the most perfect symbol of *universal association*," wrote an apostle in an influential work.[3] This group of educated and often wealthy malcontents combined a fantastic outpouring of emotion, speculation, and aesthetics with a fastidiously practical interest in technical efficiency and the increase of productive force. Unlike Carlyle, for whom the "romantic" opposition between the mechanical and the living, dynamic, or spiritual was crucial, the Saint-Simonians saw machines as mediators between the spirit and the world.

Their gospel was conveyed in an 1832 essay by Charles Duveyrier, the "Poet of God," titled "The New City of the Saint-Simonians," which was published in one of the major collections of physiologies. It began by depicting contemporary Paris as "a great satanic dance," replete with jarring, unseemly contrasts: orphans in the children's hospital next to astronomers at the Observatory, the home for wounded soldiers located between the laundries and the Chamber of Deputies, the scientists of the Latin Quarter face-to-face with the "howling animals" of the Jardin des Plantes. The city's infernal chaos reflected the dissolution and moral anarchy of its denizens.[4]

In its place, Duveyrier—speaking in the voice of God—poetically detailed a prophetic vision of "the new creation which I want to draw from the heart of man and the entrails of the earth." The rebuilt city of Paris he foresaw was "the engagement ring of man and the world." The new city took the form of a man, whose head was on the Île de la Cité and whose feet pointed northwest toward England ("the bazaar of the world"); his left leg was bent as if about to take a step, a sign of his great dynamism. The man-city's left side was devoted to science, with a giant pyramid, the central building of the university, as his breast, surrounded by steep, shining buildings that "climb[ed] to the sky in crystals of light." Cooled by regularly planted trees, this area enjoyed the reign of "silence and mystery." The man's right flank, devoted to industry, was noisier. His right arm reached up to the Port de Saint-Ouen, filled with small businesses and shops displaying "the marvels of human labor"; his right thigh contained heavy industry, "where copper and iron" were "kneaded and molded like dough"; at the horizon could be seen "the parabolic curves of the foundries and forges, the blackened cones of the furnaces, the cylindrical chimneys opening jaws filled with flames." Sparks and clouds of steam filled the air, to "the rhythm of hammer blows and axes." His right breast was a purple sphere, the great central bank that regulated relations among production, distribution, and consumption.

Between the man's legs sat a menagerie, a hippodrome, and, at the Place de l'Etoile, a half circle of theaters: "all the buildings consecrated to the ecstasies of the mind and the delirium of the senses." Outward from these buildings radiated "the operas and all the theaters with their apparatus of instruments, costumes and décors, their impassioned troupes." At the head of this "new colossus," overshadowing the bank and the university, was the focal point of the new city and the new religion: a temple in the form of a giant woman. Passengers took to balloons and crowds gathered on surrounding hills to contemplate the city, this "giant man of fire," admiring "the new creation in all of its glory."

Duveyrier's spectacular vision went beyond anthropomorphism to animism. Where the parts of old Paris were "monstrous, unformed, inanimate, dead," Duveyrier's God announced, "[With a] jolt from my will, I will draw up this inconceivable and frightening mass into a harmonious and living being," in order to "bring the members and the organs of my city out of their hideous chaos." All the city's parts, now enlivened and "endowed with movement," would "take their places": the beds of the old and sick would "lift up into the air"; "the laboratories, the Observatories with their machines and lenses, the Ecole Polytechnique, and the Ecole des Arts et Métiers," would join the parade, all marching to the new university. At the same time, warehouses, factories, workshops, furnaces, hammers, bellows, lathes, carpenters, and smiths would all rise up, along with clockmakers, tailors, and jewelers, with each member of "this whole active, noisy, animated army, . . . making the dust of the earth fly around them like a cloud of incense," and arriving at its proper place. Yet the universal force of God's will would not contradict the will of any member of the scene: "For none of these does my will create either scandal or servitude, because there will be not one nail, not one hair of these men and women, these old men and children, these buildings, shops, and work-sites, that moves otherwise than by its own will." The inclinations of the city's human members were to be soldered to the demands of God and nature; individual functions would become identical to the overall order.

This new city was lying in wait beneath the pavement of the present. Already one heard a sound like the "monotonous groaning of water compressed under these paving stones, or like a hidden fire" that would burst through the stones, the sound of the earth "swollen with the desire to live from the life of man." Duveyrier's language deliberately recalled the words of an earlier apostle, Paul: "the whole creation groans and suffers the pains of childbirth together until now."[5] This new creation—a new gospel, a new Genesis—would harmonize the "petit monde" of humanity with the "grand monde" of the cosmos, by placing human functions in their proper order and unleashing the forces hidden in matter with industry, science, and art.

The Saint-Simonians' religious beliefs have often been seen as a colorful but inessential aspect of an ultimately technocratic movement. Yet the group did not dream of a society ruled solely by reason and facts: the arts and the emotions were indispensable to their vision. They openly espoused a form of pantheism, the belief that God and nature are the same thing and that thought and matter are simply two modes of this single substance. As far-fetched or scandalous as this philosophy has been seen—whether by atheists or those who believe in a God distinct from nature—the Saint-Simonians' pantheism was directly tied to their technological and scientific interests. It was one aspect of their obsession with conversion: their simultaneous concern with releasing the hidden powers of physical nature and transforming human hearts. Their new religion was a radiant, consequential point of fusion between romantic-era aesthetics and the engineering sciences of the mechanical age, giving a mythical depth to the practical task of industrialization.

NEWTON, VICQ D'AZYR, BEAVERS, AND JESUS

Henri de Rouvroy, Count Saint-Simon (1760–1835) lived a life worthy of myth. A distant relative of the chronicler of the court of Louis XIV, he fought alongside Lafayette in the American Revolution; in France he made and lost a fortune speculating on confiscated lands; he made proposals to build and provide workers and troops for a dam in Spain and a canal in Panama. He struck contemporaries as both far-seeing and ridiculous, capable of detailed visions and persuasive rhetoric, while inclined to emotional plunges and outlandish proclamations. Fundamentally, his claim was that society needed to be reorganized: the rule of aristocrats and priests had to be replaced by one of scientists and industrialists. The chaos of existing government functions should be streamlined in a central administration to allot the means and rewards of industry according to individuals' capacities. Scientific knowledge and research should also be centrally reorganized to form a new "spiritual power."

Convinced that a society, like an organism, required a unity among its functions, Saint-Simon focused many of his writings on unifying knowledge beneath a single principle. Newtonian gravitation was the leading candidate in his writings of the first years of the nineteenth century. In his "Letter from a Citizen of Geneva" he appealed for donations to fund scientific research in the name of Newton, who had earned a place in heaven at the right hand of God for having brought all of nature under the reign of a single explanatory principle, universal gravitation. In another crucial work written during the empire, the "Mémoire sur la science de l'homme" of 1813, Saint-Simon argued for Newton's social importance. The reorganization of science and its applications "according to the conception of a unique law" was the only way to end Europe's self-destructive wars.[6]

Yet despite his embrace of Newtonian universality, in the "Mémoire sur la science de l'homme"—composed while Laplace held considerable academic power—Saint-Simon was already expressing hostility toward mathematical physicists for their arrogant and unjust domination of the sciences. He dismissed the students of *corps bruts* as "brutists." Drawing on lessons learned at the Ecole de Médecine, he now argued that physiology had to take its rightful place alongside physics at the head of the sciences. He believed in an underlying material and mechanistic basis for physiological phenomena, in line with Cabanis and Lamarck, holding that the interactions of particulate, imponderable fluids—light, heat, electricity—were the causes of life. In inanimate bodies, or *corps bruts*, solids predominated over fluids, while in *corps organiques* there were more fluids than solids. This difference meant that physiology had to develop methods and concepts distinct from those of physics—an argument that had been decisively made, from a vitalist perspective, by the great physiologist Xavier Bichat.

A reformed physiology would be the basis for a new science of mankind—a dream shared by Cabanis, Destutt de Tracy, Degerando, and others associated with the philosophy of Ideology.[7] Although Saint-Simon's physiological thought is usually attributed to the influence of Bichat, Cabanis, and the organic metaphors used by the Catholic reactionaries de Maistre and de Bonald, the source he cited most in the "Mémoire sur la science de l'homme" was the anatomist Vicq d'Azyr. In addition to attempting to reform the life sciences under the rulership of anatomy and creating institutions to monitor public health, Vicq d'Azyr had strongly affirmed the concept of the animal series, a chain of organisms ranked in order of increasing perfection, with humans as the highest point.[8] An organism's degree of perfection was indicated by the complexity of its structure, the number and diversity of organs and tissues, channels, tubes, and fluids.[9] Saint-Simon accepted this view and pushed Vicq d'Azyr's interest in organisms' relationships to their milieus in a new direction. According to Saint-Simon's view of the animal series, the most complex animals had the greatest control over their environments: "The more varied are the tubes that an organized body contains, in the dimension of length and diameter, the more they form distinct viscera and senses, the more the body is elevated on the scale of beings, which is to say, *the more action this phenomenon has on that which is external to it.*"[10] As a result, Saint-Simon suggested that the animal series be redrawn: in the slot next to humans, monkeys would be replaced by beavers, thanks to their manipulation of their environment by building dams and lodges.

Yet the decisive innovation of the "Memoire" was to extend Vicq d'Azyr's physiological principles—analysis in terms of organic unity, complexity, and the graduated scale of perfection—to human society. Like an animal, a society is "a true *organized machine* of which all the parts contribute in different ways to the functioning of the ensemble"; as with an organism, its health depends on how well its organs complete the "functions which are entrusted to them."[11] Following his new formulation of the animal series, the increasing perfection of human societies would also imply an increasing mastery over the environment. Thus the history of human civilization would be a history of both ideas and technologies. To chart this new series, he sketched a "physiological tableau" of "the life of the human species, that is, the physiology of its different ages." This twelve-step program began with humanity in a state hardly distinguishable from other animals, much like the Wild Boy of Aveyron. Language, chiefs, architecture, and religion emerged by stages. The seventh step, Egypt, was a turning point because of the massive public works projects along the Nile and the emergence of a distinct social class of scientists, or natural magicians. Further steps were claimed by Greek polytheism, Roman-era monotheism, and Islam ("the Sarrasids") with its attempts to bring all explanations back to God, a single "animated cause." In the next phase, the Middle Ages, appeared a central principle for Saint-Simon, as it had been for de Maistre: the distinction and harmonious balance between the spiritual power of the Church and the temporal power of kings. This unity was shattered in the penultimate stage, which began in 1500 and had persisted until the present: an age of "critical strife" brought about by the Protestant reformers and antireligious philosophers.[12]

The twelfth and final term of the series lay in the future: this would be a society reorganized, with the spiritual and temporal powers once more made distinct and brought into balance. The "spiritual power" would be taken from the Church and placed in the hand of the sciences; at the same time, the temporal power would direct its attention to developing industry. Saint-Simon's physiological series was aligned both historically (although societies could be observed still living in "earlier" stages) and in terms of increasing perfection and complexity. In organisms, a higher rank in the series meant more intricate and specialized organs—tubes, canals, and fluids. Analogously, in societies, a higher rank meant a greater division of labor, as well as more, more varied, and more harmoniously disposed "vessels" for the circulation and communication among its parts. Thus the highest stage would be a society that constantly developed its channels and networks for the movement of goods, people, ideas, and credit. As in the animal series, this would also be the society with the greatest power over its environment.[13]

Physiology thus implied a social mission. In another famous essay, his "Parable," he invited readers to imagine France on the day when the king and the royal family, the nobles, the ministers, and the clergy all disappeared. What would change? Well, "we would be sad," he said, because we love humanity. Overall, however, life would continue as before. At present, those with the most power and prestige were those who made the smallest contributions to society; instead, it was the "industrials"—entrepreneurs, professionals, and laborers—who kept society moving forward. In place of an unjust hierarchy determined by birth and tradition, then, society had to adopt a natural, just hierarchy of tasks and rewards established according to members' abilities and accomplishments: "To each, according to his capacities; to each capacity according to his works." His *Du système industriel* and *Catéchisme des industriels* (1823–24) advocated a meritocratic, industrial society, placing decisions in the hands of the most able and rewarding individuals for their labor instead of for the rank into which they were born. In the Restoration, such arguments were a threat to the status quo. Saint-Simon was tried in 1820 as an instigator of the assassination of the Duc de Berry.

These stances also earned him a place within the liberal opposition. Yet the third and final phase of Saint-Simon's writings, encapsulated in his pamphlet, *Le nouveau christianisme* of 1825, laid the groundwork for his followers' clash with the liberals.[14] At this time he returned to the question of the single unifying principle that every society required—the function of "spiritual power" that had been provided, in the medieval period, by the Catholic Church. Because of the Church's increasing worldliness, dating from around 1500, and its continuing failure to accommodate new scientific discoveries in its cosmological teachings, it had lost intellectual authority, and a new spiritual power was now needed. Saint-Simon first looked to Newton's gravity for the unifying principle; he subsequently built up a notion of social unity modeled on physiology and the anatomy of Vicq d'Azyr. Now he turned to the teachings of Jesus Christ. As in the Middle Ages, the coming society would be held together by religion. But dogmatic teachings about the order of nature would now be provided by the sciences. Even more importantly, the moral code of Christianity—to love one's neighbor as oneself—would be given an industrial slant: the goal for each individual and for society was "to work with all one's forces for the improvement of the poorest and most numerous class." In his New Christianity, Saint-Simon presented a faith in the universal brotherhood of humanity as the intellectual and moral engine and end for the industrial exploitation of the globe.[15]

Saint-Simon's ideas attracted many of the brightest young minds of the Restoration. The historian Augustin Thierry and Auguste Comte both served as his secretaries. After his death, in 1825 a new movement began in his name, including a small coterie of his followers and a growing number of recruits, many of whom had never actually met him. Their leaders were educated and wealthy; several came from Jewish families.[16] Many had been students at the Ecole Polytechnique, which he targeted as the "canal" for his teaching. The similarities between the school's technocratic and meritocratic ideals and Saint-Simon's industrial vision have often been noted. Its disciplinary regime, ethos of republican citizenship, and distinctive curriculum—including the key sciences of roads and work we discussed in chapter 4—made it a constant point of reference in Saint-Simonians' projects for social and technological metamorphosis.[17]

But another resource for the new movement should be noted. In printed lectures, an apostle proclaimed that "the doctrine of Saint-Simon does not want to bring about an upheaval or a revolution. It comes to predict and to complete a transformation, an evolution."[18] This use of the term *evolution*—relatively unusual for the time—was inspired by Ampère's old Lyon friend Pierre-Simon Ballanche, whose theory of "Social Palingenesis" offered a mystically flavored and prophetic Christianity that appealed to romantic circles. Ballanche himself had taken the concept of palingenesis from the Leibnizian naturalist Charles Bonnet and the latter's discussions of the unfolding, or evolution, of embryos. Bonnet's "temporalization of the great chain of being"—in which each link in the chain would perfect itself simultaneously in preestablished harmony with changes on the earth's surface—anticipated Saint-Simonian concepts of a providential and progressive unfolding of society.[19] The better-established use of *evolution*, however, was military: it referred to the deployment of formations of troops or ships.[20]

The Saint-Simonians fused these two senses of this pregnant term, "evolution": the natural, chrysalis-like development of society would be guided by careful, quasi-military strategy. The attitudes and the skills required to bring about this conversion, at once mechanical and moral, had a common source in the Ecole Polytechnique.

AN ARMY OF INDUSTRIAL SALVATION

Yet the Saint-Simonians redirected the nationalistic, military orientation of the Ecole Polytechnique into international and explicitly pacifist directions. Like Lazare Carnot, the "organizer of victories" who had created a patriotic corps of citizen-soldiers through the *levée en masse*, they made themselves into both an army and an extended family. Saint-Simonian members were ranked according to "degrees" and functions, with corresponding uniforms—from the two Supreme Fathers (Bazard and Enfantin), to the "sacred college" of sixteen "fathers" and two "mothers," with a hierarchy reaching down through several more degrees, including workers (see fig. 7.1).[21] Several members of "the family," including couples, lived together on the Rue Taranne near the Palais Royal, holding lively evenings of discussion and music that were open to the public.

The movement spread through a vast propaganda effort coordinated by the Directors of the Teaching, led by Lazare Carnot's oldest son, Hippolyte. Like his father, Hippolyte saw that the most efficient army (even one orga-

FIG. 7.1. "Le Père," a.k.a. Prosper Enfantin in Saint-Simonian uniform. Bibliothèque Nationale de France.

nized for peace) was the most enthusiastic one. The aim was to "change profoundly, radically, the system of sentiments, of ideas, of interests" and bring "a new education, a definitive regeneration to the world," in terms recalling Ballanche's palingenesis. All classes and nations were targeted: "It is humanity as a whole that we have come to teach and to convert."[22]

The movement also took on the obsessions of the engineer-scientists of the Ecole Polytechnique who filled its ranks. The Saint-Simonians were among the first to define the challenge of the nineteenth century as the struggle of workers against the idle owners, inheritors, and hypocritical priests of the past. They propounded the transformation of existing armies into a "pacific army of laborers" that would undertake large public works; social life would be organized as a family, a monastery, or a barracks, with goods held in common. Against the vagaries of the market and the cruel law of supply and demand, they advocated a new regime of property: land and all instruments of labor would be allocated according to abilities, and the rewards of labor would be redistributed according to the importance, difficulty, and quality of work.[23] They would optimize the *organized machine* of society by use of the strategic and organizational techniques they had learned in the engineering corps. Their three slogans evoked these aims, in the terms of Saint-Simon's New Christianity: "All social institutions must have for their goal the improvement of the moral, physical, and intellectual conditions of the most numerous and poorest class; all privileges of birth, without exception, will be abolished; and to each according to his capacity, to each capacity according to his works."[24] The polytechnical corps was metamorphosed into a pacific army fighting for social reform.

Two divergent tendencies pulled at the movement. Saint-Amand Bazard, a former republican conspirator, or Carbonaro, emphasized the continuity of Saint-Simon's ideas with the ideals of the Revolution. Among those who shared Bazard's republican and rationalist orientation were Philippe-Joseph Buchez, a physician who left the movement to develop his own form of Christian socialism (one respectfully dramatized in Balzac's *Lost Illusions*),[25] and Hippolyte Carnot. The other tendency took its inspiration from Saint-Simon's late religious writings and came to be identified with Barthélémy-Prosper Enfantin. Enfantin's mathematics teacher Olinde Rodrigues, a former banker who had been close to Saint-Simon in his last years, urged Enfantin to read *Le nouveau christianisme* closely. Enfantin had studied at the Ecole Polytechnique but had been forced to leave early when his family's bank collapsed; he turned to a number of jobs, including wine sales, yet maintained close contact with his polytechnician comrades on missions as far from Paris as Saint Petersburg.

The speeches or "predications" that the Saint-Simonians gave in the years 1828 to 1830 were edited and collected as the *Doctrine de Saint-Simon*. Its first chapters presented a historical vision that expanded and complicated Saint-Simon's "physiological tableau." In place of his simple linear progression, they emphasized a new law of history, the alternation between harmonious, *organic* periods, and *critical* periods in which shared beliefs were criticized and destroyed and when egotism and conflict were the rule.[26] Progress thus appeared as a constant movement upward, though with regular, necessary setbacks—a sinusoid constantly ascending. The unity of ancient Greece with its polytheistic Pantheon was dissolved under the criticism of the first Greek philosophers; likewise, the harmony of medieval Europe was disrupted by the Protestant reformers, whose modern successors were the destructive *philosophes* of the Enlightenment. Europe had been in crisis for three hundred years, as shown in both the social and the intellectual order. The *Doctrine* lambasted the egotism and hedonism of the wealthy, the dry specialization and fragmentation of the sciences, and the futile despair and self-involvement of the poets and artists. It decried the waste and competition in the contemporary economy; the oppression of women, seeing marriage as a hypocritical prostitution; and the exploitation of the poor, with workers and owners as the historical successors to serfs and lords and slaves and masters. The *Doctrine*'s stirring analysis of economic injustice, expressed with "incredible ardor," provided "the phraseology of future socialist parties"—indeed, it is in many ways the template for *The Manifesto of the Communist Party*.[27]

In presenting their solutions to this crisis, the Saint-Simonians again deepened and extended Saint-Simon's suggestions. The task ahead was to organize—to make organic—the social and intellectual orders. Saint-Simon saw human life as a synthesis of intellect, action, and emotion, which for his followers led to the tripartite classification of social functions: thought, represented by science, and action, represented by industry, would be reconciled through the emotions, the love of humanity that was embodied and strengthened by a new class of priests.[28] Priests were to be chosen and ranked according to their love for humanity; their key role was to administer "human resources," determining which individuals were suited for which tasks. A centralized administration would replace the current chaos of government functions and would be placed in the hands of a "sacred college" or hierarchy of priests. Some would look after the needs of industry, others would be concerned with science, and a third group would coordinate the relations between the other two. A central bank would keep production, distribution, and consumption in balance, pooling resources and ending wasteful competition. The result would be a global society organized no longer for conquest and defense, but for the peaceful exploitation of the earth for the benefit of mankind as a whole.

Another major development was that where Saint-Simon himself had been a staunch materialist in physical matters—even when stressing the importance of the emotions and morals, in both his "Mémoire" and *New Christianity*, thereby continuing the eighteenth century's "sensible empiricism"—*The Doctrine of Saint-Simon* instead embraced pantheism.[29] The Saint-Simonians spoke often of the "rehabilitation of matter." All religions before Christianity, they claimed, had emphasized the material and external world, with punishments and rewards delivered on earth; Christianity, on the contrary, gave exclusive attention to the spiritual and interior world, demanding the renunciation of the flesh and placing a paradisiacal Kingdom of God in the afterlife. The Saint-Simonians synthesized these poles in their new doctrine, marrying the West (seen as male and spiritual) to the East (seen as materialist and female, in contrast to another set of Orientalist myths in which the East was inherently spiritual).[30] Matter and spirit needed to be preserved and yet unified. Pantheism was the logical solution. They also employed the language of Fichte, then being disseminated by Victor Cousin and merged with a reading of Schelling, in speaking of the reconciliation of the self (or *moi*) and the world (the *non-moi*), by means of a third term that contained both: the *infinite*, or God. The mind, and the material world it confronted, were but finite modes of the infinite divine.[31]

These beliefs, defended in Year Two of *The Doctrine*, appeared in the "credo of Saint-Simon":

> God is all that is
> All is in him, all is through him
> Nothing of us is outside of him
> But none of us is him
> Each of us lives from his life
> And all of us are in communion in him,
> Because he is all that is.[32]

There is a debt here to Spinoza's notion of "God or nature," mind and matter as two parallel modalities of a single underlying substance. Yet it was common at the time, thanks to Cousin, to read Spinoza's philosophy as having dissolved the distinct parts of nature into a single abstract unity, in which human agency disappeared in a vision of universal determinism.[33] The Saint-Simonians rejected this "pantheism of the past," because their

"*God-all* [would] be living and loving, and individualism, instead of annihilating and immobilizing itself in God, [would] progressively develop itself in him."[34] The individuality of each part would be enhanced by its participation in this vital, loving whole—in something like a cosmic division of labor, a freedom-in-connection much like that suggested by Schiller's notion of autonomy and Humboldt's polity of instruments, as discussed in chapter 3. Their sense of nature as an organism undergoing constant, progressive development echoed Schelling, although a more direct source was Ballanche's embryological view of society. And just as the embryologists of this moment were arguing that the prenatal forms of the higher animals developed through the entire series of animals preceding them in the animal series, the Saint-Simonian religion also sought to actualize all earlier forms of life in the final state they foresaw: "The religion of the future will be greater, more powerful than any of the religions of the past, as it will *recapitulate* all of them; its dogma will be the synthesis of all the conceptions, of the means of being man."[35] They would unite humanity's earlier, embryonic stages of fetishism, polytheism, and monotheism into an adult, pantheistic synthesis.

And how does one spread a religion? By evangelism, of course. In the 1820s the Saint-Simonians started the journal *Le producteur*, where they published their views alongside those of liberal publicists and theorists of industrialization, including Auguste Comte, who had broken with Saint-Simon several years earlier. Yet after the Revolution of 1830, their teachings became overtly hostile to liberalism. Freedom of religion was good, but only because it left a ruin that could be replaced by a single faith; likewise, liberal economics was useful, but only because its opposition to monopolies prepared a new organization of industry. The Restoration's central mouthpiece of liberalism and romanticism, the *Globe*, lost its raison d'être after the Revolution of 1830 as its key authors entered the government's ranks. At this point, Pierre Leroux, its founder, printer, and editor, converted to Saint-Simonianism and transferred the journal to the leaders of the new faith. Michel Chevalier became its editor.

Yet the preferred method of evangelism was through direct contact and preaching. According to a report of August 1831 from Hippolyte Carnot, sermons previously presented on the Rue Taranne to small audiences were replaced by one weekly "central" teaching in the much larger Salle Taitbout, right next to the Paris Opera, at four o'clock each Thursday afternoon, on the present and future organization of science, industry, education, and the fine arts. An audience of four to five hundred at the Athénée heard a lecture each Wednesday on the moral and scientific progress of society; discussions "according to the needs of the room" were held there on Sundays. Further, lectures were given specifically to students of the Ecole Polytechnique, and still others were given weekly to two hundred men "involved in the study of science." Artists also received their own teaching. In each case—including the lectures to women given by Claire Bazard—the style of speech and the thrust of the argument were adjusted to the interests of those in attendance.[36]

Considerable reflection and effort were also devoted to reaching the "poorest and most numerous class" directly. Regular meetings in workers' quarters sought to "reunite as often as possible the workers who are already converted, to operate upon them a more and more complete transformation."[37] In addition to teachings, material aid was provided. Each *arrondissement* (district) was assigned a medical doctor and a male and a female director: medicine was dispensed and workers were given materials and equipment to repair and make clothing. Cards were handed out as marks of different degrees of membership—*fidèles, catéchumènes*—and "partial chiefs" were named to spread the doctrine on their own. The ultimate goal was to create a harmonic although hierarchical union among classes, the *association* that was at the heart of the Saint-Simonian social vision. "Houses of association" were formed, each with more than 330 *fidèles*, 100 of them women, 100 children. Diplomas were given to fifteen hundred sympathetic listeners labeled as *aspirants* and "functionaries."

Meanwhile, the gospel reached well beyond Paris. The Eglise du Midi was established in Montpellier, with missions to surrounding cities, including one to Lyon, led by Reynaud and Leroux, to preach gradual reform to *canuts*, textile workers who had recently rioted. Other missions went to Dijon, Besançon, and the industrializing cities of the north—Mulhouse, Strasbourg, and Metz. In February 1831, Hippolyte Carnot and Pierre Leroux preached together in Brussels, and fifteen hundred listeners received the teachings in Liège; Duveyrier and d'Eichtahl toured workers' meetings in England, where they met Carlyle and J. S. Mill in person.

Although their sermons were later published in the *Globe*, the preferred instrument of conversion was direct, sympathetic contact between Saint-Simonian priests and their audiences. The sermons of Barrault, the most persuasive of the preachers, were marked by "tumultuous passion, rhetoric that is resplendent and penetrating by turns, but always elevated, a call to feelings, to ideas, to sense—all is put to work to convince, or even more to seduce and to lead. The Apostle suffers, he hopes, he cries, he is joyful with the crowd. The force, the audacity, the tenderness that the philosophers and philanthropists never had—these he draws upon, in a constant communion with God."³⁸ These emotional appeals went along with an intensified interest in the arts. In an essay, *Le savant, l'industriel, l'artiste*, written in the last year of his life, Saint-Simon argued that artists must form an "*avant-garde*": "Let us unite. To achieve our one single goal, a separate task will fall to each of us. We, the artists, will serve as the *avant-garde*: for amongst all the arms at our disposal, the power of the Arts is the swiftest and most expeditious. When we wish to spread new ideas amongst men, we use in turn the lyre, ode or song, story or novel; we inscribe those ideas on marble or canvas.... We aim for the heart and imagination, and hence our effect is the most vivid and the most decisive."³⁹ The view that an *avant-garde* of artists should anticipate and guide social progress has cast a very long shadow on modernism: one influential echo was Baudelaire's *Salon of 1846*, which closely followed Saint-Simon's notion of an alliance of artists, industrialists, and scientists in its appeal to "the natural friends of art... on one hand the rich, on the other the *savants*."⁴⁰ The gospel was proclaimed in Barrault's sermon "Aux artistes," which laid particular stress on the social function of music. Saint-Simon had commissioned a song, "To the Industrials," from the composer of "La Marseillaise" in 1813.⁴¹ Hector Berlioz expressed great sympathy for the movement in 1830, and his *Chant des chemins de fer* shows its influence; yet the Saint-Simonians' greatest musical catch was the young Félicien David, who, even after leaving the movement, composed celebrated symphonies—*Le désert, Moïse*—inspired by his life as a Saint-Simonian.

INSTRUMENTS OF ASSOCIATION

The propaganda of the Saint-Simonians was launched as a peaceful form of total war, with the aim of converting the hearts and minds of all levels of society. The training many of them received at the Ecole Polytechnique shaped more than their recruitment strategies. In their careers as engineers, they made important contributions to the sciences of roads, work, and engines discussed in chapter 4. These pursuits, in turn, provided them with concepts for their plans for social and industrial regeneration as well as their metaphysics, which was based on equivalences and conversions.

As discussed above, Saint-Simon's physiological tableau in the "Mémoire sur la science de l'homme" can be read as a series of states of civilization or *organized machines* ranked according to their complexity, much as organisms were ranked according to the variety and complexity of tubes, channels, and vessels making them up. To rank a stage of civilization in this way, one considered the extent and harmony of the division of labor, but also, quite literally, the quantity and efficiency of the channels of circulation that allowed the society's parts to communicate. The multiplication of roads, canals, railroads, journals, and lectures made for a more complex organism.[42] Furthermore, Saint-Simon had argued that an organism's greater complexity was accompanied by a greater power over its environment. Thus the very projects with which the scientist-engineers were most deeply involved could be seen as ways of increasing their society's state of "organization," its complexity, and its control over its milieu. The goal was to progressively incorporate that which surrounded society into society itself.

Several of the most prominent applications of "optimization theory" to railroads were made by the engineers Gabriel Lamé and Emile Clapeyron. These polytechnicians had spent the 1820s in Saint Petersburg, detailed on a mission for the state corps of engineers, teaching math, mechanics, and "communication" at the Russian Institute of Roads. There they met Enfantin, who was visiting the region on business. When they returned to Paris in 1831, they joined the Saint-Simonians and combined the practical, reformist ambitions of that group with their theoretical sophistication in the study of engines and efficiency. Along with their coreligionists, the brothers Stéphane and Eugène Flachat and the Conservatoire National des Arts et Métiers lecturer Perdonnet, they published analyses showing the comparative benefits, in increased speed and reduced costs, of transporting goods by road, canal, and rail. They also made contributions to the design of engines.[43]

The optimization of technical systems formed the background for Michel Chevalier's pamphlet *Système de la Méditerranée*, first published serially in the *Globe*, in the wake of the 1830 Revolution.[44] He argued against those who held out hope for a Europe-wide revolution or war as the means to bring about a definitive end to aristocracy. There were obstacles in each country to such a strategy, but more importantly, the growth of industry throughout Europe in the previous fifty years, and with it the spread of financial credit, now made war impossible. "Industry is eminently pacific. Instinctively it repels war. That which creates cannot be reconciled with that which kills." The goals of a liberal or republican revolution had fallen out of step with the times: "The freedom of the liberals has lost its prestige, because it is not the freedom of the future. *Freedom without association, that is, without hierarchy, is isolation.*"[45] Freedom was only possible with connection. The future lay with the carefully planned implementation of technologies of transport and communication.

As a way to reach this future, he sketched an "industrial plan" centered on the Mediterranean. For centuries, this sea had been the "arena" for the constantly renewed struggle between Europe and the East. After developing transport and communication among its previously divided peoples, said Chevalier, "the Mediterranean will become the marriage bed of the Occident and the Orient."[46] The large gulfs that open on that sea would develop ports; each of these would be the terminus for a major railway line running through its host nation, connected by secondary networks crossing the territory, accompanied by telegraphs. "In the eyes of those who have faith that humanity marches towards *universal association*, and who devote themselves to leading it there, the railroads appear under a completely different light. The railroads, which along with men and products can move with a speed which twenty years ago we would have judged as a tall tale, will singularly multiply the relations of people and cities. In the material order the railroad is the most perfect symbol of *universal association*. The railroads will change the conditions of human existence." Steam-powered trains were symbols and instruments of organization and regeneration.[47]

Chevalier's vision of freedom through association also implied hierarchy and a division of functions: every nation would have its place, those of Europe as well as the East. Railroads would line the coasts of Africa; Spain would awaken from its lethargy; Russia, whose inhabitants had long been "mere instruments," would happily choose this future; France, of course, would take the lead, with the assistance of the industrial and commercial power of England. Railway communications were new blood, awakening all these peoples to an "intoxicating activity." There would be a place for all capable men ("hommes de capacités"), "whether their preferred chimera had been republicanism or liberalism or the *juste milieu*"; there would be a place for scientists, for men of art and engineers, for "the industrials into whose hands nature pours its products, and who metamorphose them in a thousand ways for the beautification of humanity and of the globe which it inhabits," a place for traders and sellers, a larger and larger place for "the poor people of workshops and the fields," and a preeminent place for bankers, "dispensers of credit, depositories of the wealth of individuals and states."[48]

As mentioned above, for the Saint-Simonians, the West, the place of action and spirit, was male; the East, the place of emotion and matter, was female. The reconciliation of these two directions would be "the political consecration of the agreement which must exist between *matter* and *spirit*, which until now have been at war." The same fusion grounded the project as a whole. At the close of his pamphlet, Chevalier wrote: "This is our political plan. Combined with the moral work conceived by our SUPREME FATHER, of which it is the material translation, it must guarantee one day the triumph of our faith." A political plan—a highly technical project of railroads, waterways, and credit—presented moral metamorphosis in material terms: the physical transformation and the spirit realized within it were two sides of the same reality. The science of roads of the Ecole Polytechnique and CNAM, along with a pantheist metaphysics, underwrote Chevalier's vision of the Mediterranean as a unified system, the matrix of reborn humanity.[49]

ENGINES OF CONVERSION

The science of work and early thermodynamics offered another scientific correlate to the Saint-Simonians' metaphysics. Chevalier's view that "the railroad [was] the most perfect symbol of *universal association*" described a paradise of communication that would be achieved among peoples previously in competition or at war. Yet the railroad—or rather, the steam engine that made it possible—symbolized in another way the universal association implied by the Saint-Simonians' pantheism, their goal of the "rehabilitation of matter" and the material extension of spirit. Hippolyte Carnot organized the efforts of other engineers, many of them trained at the school his father founded, to spread the doctrine that spirit and matter are ultimately the same and can be converted into one another: both were manifestations of God. Hippolyte's brother, Sadi Carnot, wrote the landmark text of thermodynamics, *Reflexions on the Motive Power of Heat*, suggesting that that heat and productive force are ultimately the same and can be converted into each other: both were manifestations of energy. The degree of Sadi's exposure to Saint-Simonianism is unknown; but Hippolyte, the leader of Saint-Simonians' campaigns of spiritual conversion, read his brother's work before its publication.

The *Reflexions*' first pages depict a universe propelled at all of its levels by heat. As Robert Fox has shown, Carnot's analysis derived from his interest in the new, high-pressure engines designed by Arthur Woolf being introduced in France at that time by his partner Humphrey Edwards; their hulking frames, along with models of their action, can still be visited in the museum of technology that the CNAM now houses in Paris (alongside Vaucanson's automata, and reconstructions of experiments by Lavoisier and Foucault).[50] These were "compound" or "double-expansion" engines; they channeled the exhaust expelled from the first cylinder of the classic Watt engine into a second cylinder, using the steam's expansion as an additional

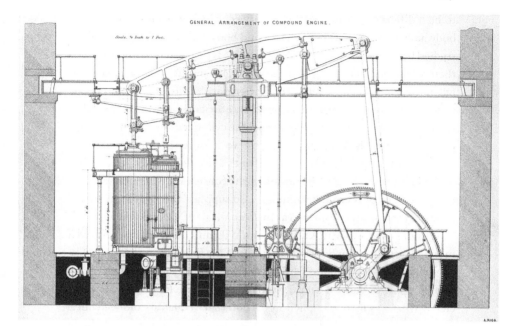

FIG. 7.2. The Woolf compound steam engine, from Arthur Rigg, *A Practical Treatise on the Steam Engine* (London, 1878).

source of power (the two cylinders are visible at the left in fig. 7.2). This innovation allowed the engine to operate at higher pressure and with less loss of force. Carnot's short book analyzed the heat cycle of an idealized engine of this type, tracing the phases through which the substance inside a cylinder passed as it heated up, expanded, and pushed a piston out, then cooled, letting the piston fall. The cycle had four distinct phases. First, the cylinder was brought into contact with a heat source, at temperature T_1; this heated the substance in the cylinder, causing it to expand (this was called the *isothermic expansion*). The cylinder was then removed from the heat source but continued to expand (known as *adiabatic expansion*). It was then brought into contact with a cold source, at temperature T_2; this was the equivalent of injecting cold spray into the cylinder (known as *isothermic compression*). As it cooled, the steam condensed and its volume contracted, bringing the piston back down. Finally, it was removed from the cold source while it continued to compress, returning to its starting temperature (*adiabatic compression*). Then the cycle could begin again.

Carnot demonstrated that increasing an engine's efficiency required maximizing the difference between the temperature of the hot source (T_1) and that of the cold source (T_2). To extract work from an engine, there had to be a difference, however slight, between the temperature of the first body and the temperature of the second body. Just as in a waterwheel, as studied by his father, where there had to be a fall of water to produce work, there had to be a fall in heat to make the engine productive; the greater the fall in the heat, the more "efficient" the engine.[51] Thus, among the conclusions that Carnot drew was that a perfectly efficient engine was impossible: some heat had to be lost in order to cool the cylinder down. The underlying notion of his study was that productive work, or motive power, could be studied as the equivalent of heat. Carnot's analysis has been seen as one of the first expressions of the notion of the conservation of energy, the so-called first law of thermodynamics. It also supplied the logic for the principle of entropy, later enshrined as the second law of thermodynamics.[52]

Carnot's essay was little known before William Thomson rediscovered it in an 1832 article in the *Bulletin des élèves de l'Ecole Polytechnique* by Emile Clapeyron, while Thomson was working at the laboratory of the physicist and photographic enthusiast Victor Regnault. In his article, Clapeyron stated clearly what was merely implicit in Carnot: that "a quantity of mechanical action, and a quantity of heat able to pass from a hot body to a cold body, are quantities of the same nature, and that it is possible to replace one by the other."[53] Just as importantly, he presented Carnot's ideal cycle in what became an iconic diagram. While in Saint Petersburg, Clapeyron and Lamé would have had the chance to learn a secret method for monitoring the efficiency of steam engines from English engineers also stationed in Russia: the Watt indicator diagram. This device involved a pencil attached to a piston that moved up or down with changes in the engine's pressure, and left or right with changes in volume; the result was a tracing of an oblong quadrangle, with each side corresponding to one phase of the heat cycle.[54] Clapeyron correlated the parallelogram traced by the indicator diagram—and the relation it made manifest between pressure and volume over the course of an engine's cycle—to Carnot's four stages of isothermic and adiabatic expansion and compression (see fig. 7.3).[55] Thus he made it possible to visualize the conversion produced by an "ideal engine." At the same time he made the diagram a mathematical tool, since by calculating the area of the parallelogram it was possible to quantify the work performed by the engine. The famous "governor" added by Watt to his steam engine as a means of avoiding extremes of pressure has been seen as one of the first instances of a cybernetic feedback machine. The indicator diagram, as a tool for optimizing the machines, lifts the engine's self-regulation to a higher order; it allowed the engine to trace its own process, creating a diagram that could be mathematically analyzed to guide subsequent improvements.

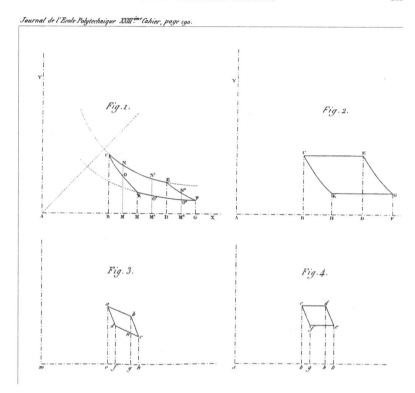

FIG. 7.3. The first diagram of the Carnot cycle, in Emile Clapeyron, "Mémoire sur la puissance motrice de la chaleur," *Journal de l'Ecole Polytechnique* 14. no. 23, 1832.

This device did not simply regulate the machine to maintain it in its state; it showed how it might be perfected.

The fact that the indicator diagram worked without direct human intervention has also led historians of science to align it with the graphic method of the later nineteenth century and with mechanical objectivity.[56] Yet while the engineers clearly aimed at precision, the overall context of this research in France was not that of impersonal, automatic, ascetic machinery. Instead, Clapeyron and Carnot's work had its roots in the labor theory of knowledge, with its valorization of the epistemological contribution of laborers and a strong sense of the continuity between human work and the work of engines. Further, Clapeyron's decisive reworking of Carnot followed his immersion in Saint-Simonian metaphysics and politics. It was Clapeyron, Lamé, the Flachats, and the Pereire brothers who arranged the preliminary studies, engine improvements, route analysis, and financial credit needed to build France's first passenger train line, which opened between Paris and Saint-Germain in 1834. Emile Pereire undertook this project, he said, in order to "write his idea on the earth, to give it body and consistency"—the language used repeatedly by the Saint-Simonians of making the spirit flesh, often by means of technology.[57] Clapeyron's essay united Monge's descriptive geometry and cartography, studies of optimization and efficiency, the study of work, and Saint-Simonian metaphysics. His graphic and practical reappropriation of Carnot's cycle of the conversion between heat and work was itself part of a cycle of conversion between objects and ideas—a cycle in which human laborers guided machines to humanitarian ends. The subsequent scientific importance of Clapeyron's iconization of Carnot's essay underlines the fact that far from being delirious dreamers, the Saint-Simonians—founders of socialism at its most romantic—were preeminent conduits for the theories, practices, and machines that built the industrial world.

"DIVINE MOTORS, SUBLIME INSTRUMENTS"

Despite this strong practical vocation, their ideas often went beyond the bounds of traditional logic and rational science. At a critical moment in the movement's history, Enfantin spoke of certain men who "incessantly see connections between acts or facts or ideas that most people do not see as connected. Pushed to a certain degree of abstraction, these men are what we call mad[;] they live in the fantastic." Chevalier replied that such people "animate everything: stones, metals, engines, plants." In these polymorphous sympathies "there is a form of animal magnetism" that he imagines must have been present at the invention of all the arts, an ability to intuit the principles of agriculture or iron-making through an identification with objects. The apostles believed there was wisdom in such madness. As Enfantin answered, "We must follow the lead of the imaginary; there is a lesson to be derived from the fantastic."[58] In 1832, pursued by the government, weakened by the departure of key allies, and threatened with imprisonment, Enfantin and the core of his followers entered boldly into this domain to create *Le livre nouveau*, a new Bible founded on a reworking of the sciences and the sympathetic identification with nonhumans.

Enfantin's teachings around 1830 showed intense reflection on the metaphysics of sympathy. He saw a spiritual connection between himself and his followers; this was the model for the enthusiastic obedience required for the new political order. Writing to Charles Duveyrier, Enfantin criticized the proponents of animal magnetism for condemning man "to being no more than a passive tool (*organe*) of the milieu that surrounds him.... Whereas I, I see you *awakened*, and if my saddened eye meets yours, I see yours grow moist. . . . You do not *sleep*, you are awake, and you and I are *doubly* happy about it. I say more—we will each of us have our distinct consciousness, each reacting upon the other, not as *master* and *slave*, not as *agent* and *patient*, but as powers of *authority* and *obedience* by *reciprocal* LOVE."[59] If this new gospel included dependence, hierarchy, and obedience, that was because these relations were enacted by an "awakened" choice and transmitted through the medium of love, in such a way that both individuals remained free. Enfantin rejected a soulless materialism in favor of pantheism ("God is everything," he wrote), arguing, "What does a well-made machine with perfectly meshing gears mean to me? I wouldn't lose a hair for it. Let the machine come alive, let it speak a human language to my heart, let it ask me to lend it the aid of my arms, my intelligence, let it implore me as does humanity to destroy all that is opposed to its free movement. [Then] I admire no more, I love, I give myself to it. Could the world be a machine without life? Impossible!" The New Christianity and the action of the priest would awaken the dormant powers of machines, nature, and human relationships; individuals—human and otherwise—became autonomous through their participation in a vast organization that was the size of the world.[60]

Enfantin also turned increasingly to the question of female oppression and its domestic forms. As has been pointed out by Claire Goldberg Moses, Naomi Andrews, and others, feminism and socialism were closely linked in the 1830s and 1840s; Saint-Simonian "predications" denounced the hypocrisy of bourgeois marriage and the subordinate position of women, seeing feminine liberation as a crucial aspect of the society that was to come; women furthermore played an important and active role in the movement.[61] Enfantin pursued these views to the point of challenging the sacredness of marriage and argued for the inevitability of serial sexual relationships; this crusade alone was enough to chase the Catholic Buchez away. Much more corrosive to the unity of the movement, however, was Enfantin's increasingly close and often uncomfortable involvement in the personal lives of many of the followers and his public revelations of their confidences, among them an affair that Claire Bazard, the wife of Saint-Amand Bazard, had undertaken with another follower. This revelation set off a schism that divided allegiances between the two "Pères Suprèmes." Among the Saint-Simonians who departed with Bazard were those associated with the republican wing, including Leroux, Reynaud, and Hippolyte Carnot.

After the Bazards' departure, Enfantin instituted a new leadership: himself and a chair, held empty at his side at public meetings, reserved for "La Femme-Prêtre," the female half of the spiritual parents of the church. This couple would be completed once "La Femme" was discovered and brought to sit by his side. Enfantin assumed that she would be found somewhere in the East and that this couple would embody the prophesied marriage of the two directions. A further schism followed on the heels of Enfantin's description of the role of the future priest and priestess. He held that they would at times personally and corporeally intervene to awaken or quiet the physical ardor of emotionally frozen or over-stimulated followers—a proposal that led even Enfantin's most imposing ally, his former math tutor Olinde Rodrigues, to break away.

This rupture coincided with more serious external threats from the government. In early 1831, workers' revolts took place in Paris and Lyon. Although the Saint-Simonians advocated peaceful solutions to the problem of social order, the government (quite correctly) saw them as a source of instability. They not only promoted views that threatened the established order, calling for state appropriation and distribution of labor and its instruments and the dismantling of the power of inherited wealth. Their pantheism was an affront to the Trinity, and the priesthood was a repudiation of the authority of the Catholic Church. The writings of Enfantin and Chevalier, which denied the sacredness of marriage and advocated free love, were also taken as affronts to public morality. The police shut down their gathering place at the Salle Taitbout; their accounts were frozen and their papers seized, and a trial was prepared.

Enfantin adopted a radical strategy to respond to this crisis. He pared the movement down to its core members and led these forty males on a retreat to the estate his family owned in northeastern Paris, in the working-class district of Ménilmontant. The retreatants took a vow of celibacy and lived their lives in common, enacting in microcosm the fraternal society they prophesied. This cell of true believers embodied the cult of labor and industry they taught, by performing physical tasks outdoors, an activity they widely trumpeted and which was widely ridiculed. To see these sons of wealthy families and members of the liberal professions getting dirt and callouses on their hands was in fact remarkable. Their point was didactic: to demonstrate the community of all men as laborers and to celebrate the centrality of industry.

Enfantin threw open the gates of the Ménilmontant garden to visitors. Two thousand reportedly turned out for the ceremony of the "taking of the *habit*," in which the members one by one donned their new uniform: a white undershirt with a red border beneath a blue-violet overcoat, signifying that Saint-Simonianism "was applied in love [white], was fortified in labor [red], and was enveloped in faith [blue-violet]."[62] The overcoats could be laced up only in the back, requiring assistance, to remind each individual every day of his dependence on others. The retreatants staged other elaborate religious rituals, with music and speeches. Upon the death of Enfantin's mother, they held a large parade through the streets around the estate. A much larger parade was held for the follower Talabot, a victim of the cholera epidemic that raged in Paris that year. Thousands lined the streets and joined the progress of the parade to the Père Lachaise cemetery. Enfantin spoke in this period of the "play" that forms social rites as means of experimentation and instruction: "Think of the Catholic mass, and the play involved there! How much everyone learns from it!" The Saint-Simonians' rituals, according to Antoine Picon, were spectacles that realized in the present the future order of the world.[63]

Félicien David's musical contributions were particularly important. He composed rousing songs in praise of labor and laborers and joyful calls to leave the old injustices behind, including a hymn devoted to Ménilmontant itself (see fig. 7.4). Others were expressions of longing for *La Femme* or triumphant choruses in praise of *Le Père*. The painter Machereau also recorded scenes of this period, filling a sketchbook (now held at the Bibliothèque de l'Arsenal) with images of new costumes and projected architectural works. Enfantin and his chief disciples also reflected on new forms of architecture that included more dynamic bricks in the shape of interlocking but slightly yielding teardrops—drawing on Monge's paradigmatic course of applied descriptive geometry, stereotomy, for cutting stones. Preparations for building a temple, digging out holes in the ground in which to sink the foundation, were the object of another ritual.

In Ménilmontant, Enfantin launched a project to systematize the metaphysical, grammatical, and scientific bases of the doctrine. This was *Le livre nouveau*, expounded in four late-night sessions.[64] Plunging into the fantastic, the apostles fused their symbolic, metaphysical, and religious preoccupations with the sciences of the engineer. They first outlined new scientific principles. Enfantin, with Chevalier and the young polytechnician Charles Lambert at his side, envisioned a religious metamorphosis of the axioms of polytechnical training: "ALGEBRA and GEOMETRY will have been redressed in a LIVING, POETIC, RELIGIOUS character." Leibniz, because of his use of the concept of the infinite, was the prophet of this new science: "May algebra at last take its place in MORAL life, so that the veritable INFINITESIMAL era of the human mind, indicated by LEIBNIZ, will have begun."[65] Calculus and descriptive geometry would be the "points of departure for the new sciences, divine motors, sublime instruments which place the INTELLIGENCE of man into communion with time and space."[66] Monge's descriptive ge-

FIG. 7.4. Cover for sheet music, "Ménilmontant: Chant réligieux," by Felicien David, apostle. Image by Machereau, showing a Saint-Simonian ceremony with onlookers. Bibliothèque Nationale de France.

ometry already offered a "living IMAGE of the PROGRESSIVE HARMONY of *humanity* and of the *world*"; religiously transformed, it would be "the base for the form of ceremonies of the cult and in the construction of religious monuments." Natural logarithms were praised, as their convergence toward asymptotes presented "the singular and entirely living, entirely human notion of a progress both indefinite and continuous." Building on the trinity of intellect, action, and emotion, Enfantin proposed a new set of meanings for geometry, with the *line* as thought and the *surface* as action; merging them and literally adding a new dimension was the priest, who appealed to the emotions, as *volume*. In mechanics, movement was the "LIVING expression" of its components, "rotation" and "translation," their "harmonic and ungraspable combination."[67]

Chevalier had already anticipated a new science of steam in *La Méditerranée*: in addition to the "marvels that steam already births under the fingers of man," he predicted "a new scientific inspiration" that would unify theories concerning "calorific force" and illuminate the present "darkness and the heart of chaos." In a later conversation in the garden, Chevalier suggested to *Le Père* that the unifying term he sought for a truly living mechanics would come from the engineers, in the concept of *forces vives*, which combined speed and mass. This was a concept directly related to *work*, a direct and generative importation of engineering science into Saint-Simonian metaphysics.[68]

The text also acknowledged the "war in the sciences" whose battle lines we have traced in previous chapters: Cuvier versus Geoffroy Saint-Hilaire, (Laplacean) emissionists versus wave theorists of light, and materialist physiologists of Paris versus the vitalists of Montpellier. Their own earlier writings, the Saint-Simonians claimed, had "virtually" pacified the political oppositions of the period; what was now needed was a virtual peace in the sciences, a reconciliation between all the oppositions that had structured thought: "SPIRIT and the FLESH, TIME and SPACE, NUMBER and EXTENSION, FORMULA and FORM, THOUGHT and ACT, UNITY and MULTIPLICTY, IDENTITY and DIFFERENCE, OBSERVATION and EXPERIMENTATION, the PAST and the FUTURE, AUTHORITY and LIBERTY, the SELF and the NOT-SELF, MAN and WOMAN, HUMANITY and the WORLD."[69] It was up to the "men of love"—those with a feeling for theory and practice, science and industry, reality and appearance—to convey to the rest of humanity that these oppositions were actually just moments in a process of constant, progressive harmony. Each pair was a "double hypothesis" that would be "sympathetically subordinated" to a single concept that defied definition—like "life" or "progress," just as the "generative couple" of the male and female priest formed a sublime and harmonizing principle for humanity as a whole. The goal was a "universal dualism whose two terms [were] linked by a vital term."[70] The next séance applied this logic to words themselves: the noun was thought, the adjective was action, but the crucial link was provided by the priest, *La verbe*, the principle of movement and unity.[71] This ramifying dialectic in *Le livre nouveau* wrote into the order of ideas the sympathetic subordination expected of individuals under the loving guidance of the priest.

A further séance discussed a new "Genesis" that synthesized the Saint-

Simonians' preferred scientific myths and industrial prophecies. They retold the story of creation. The earth longingly awaits her beloved; God incessantly beautifies her to prepare for the arrival of mankind: to water, trees, and "monstrous beasts and unformed mollusks" are added more perfect beings; the atmosphere grows temperate; new metals, stone, and mountains cover the earth. These new tidings bring joy to the earth, yet she suffers still from a "hot fever" in the molten metals in her veins. Mankind at last arrives. Centuries still must pass before he reaches the intellectual and technical stage needed for the "marriage of humanity and the world" and the birth of a new creation. This will be engendered by three "harmoniously placed couples" of the globe: new countries formed by the merger of Europe and Africa, North and South America, and Asia and Oceania. "The three husbands who inhabit the North will go to call the three brides who inhabit the south, and will draw them to the nuptial bed, which will be, for one, the Mediterranean; for the second, the Archipelago of the Antilles; for the third, the great bays of China and India."[72] The arousal of humanity into a procreative association in the Mediterranean became a global affair, whose offspring would be the opulent harmonies of the industrial world.

Duveyrier's *Ville nouvelle*, with its glorious Temple-Woman in symbolic union with the city-man, was drafted in the midst of these visionary séances. The temple became an object of intense longing for these celibates: she was the Great Mother of the East, bride of the masculine West; she was, in God's words as spoken by Duveyrier, "[my] living love, the joy of my heart, the beauty of my face, my hand of caressing and charity," nothing less than "the hope of the world."[73] In Duveyrier's description, one of the Temple-Woman's hands rested on a crystal sphere containing a sacred theater painted with panoramas showing scenes from around the world; the other hand held a blue and silver scepter, visibly blended at the horizon with the steeples of the university, representing the sciences. Staircases spiraled up her sides, past shops and glass galleries, like jeweled belts and garlands; music poured from an organ at her front; behind her, the train of her dress was draped over an amphitheater (see fig. 7.5).

An ecstatic tribute by Chevalier offered this temple up as a representation of the universe as a whole, as the Saint-Simonian cosmogram. He described it as

> A temple that is a Voltaic battery; a temple built of colossal magnets, a temple of melody and harmony, a temple through whose mechanism enormous lenses throw torrents of heat and light; a temple which vomits forth light and fire by gases; the life of the earth manifested by magnetism and

FIG. 7.5. The Saint-Simonian "Woman-Temple." Album Machereau. Courtesy of Bibliothèque de l'Arsenal, Bibliothèque Nationale de France.

electricity, its pageantry by the vividness of metals and fabrics, by marvelous waterfalls, by splendid vegetation appearing across the windows of the temple; the life of the sun manifested by heat and light; the life of men manifested by music, by all the arts, by the profusion of paintings and sculptures, by panoramas and dioramas that will reunite in a single point all space and all time.[74]

Awed by his own creation, he concluded: "What an immense communion! What a gigantic moralization of a whole people! What a glorification of God, of his messiah, and of humanity!" All parts of heaven and earth—including the vital fluids of electricity, light, magnetism, and music, all metals and fabrics, water, fire and vegetation, not to mention the human arts, music, painting, sculpture, dioramas, and panoramas from around the world—went into this artificial woman. Like the Tabernacle whose construction is spelled out in *Exodus*, she contained both God's creation and all of the technologies of mankind, telescoping the stages of human history: "The building recapitulates all the religions of the past, the ancient temple of the Jews, the ruins of Thebes and Palmyra, the Parthenon, Alhambra and Saint Peter's, the Kremlin, mosques of the Arabs, pagodas of India and Japan."[75] The temple was a scientific wonder and technical feat containing architectural vestiges of all earlier systems of intellectual and social unity. All the romantic arts were present, especially in their Parisian, theatrical modes; the temple itself was both church and spectacle. The adulation of nature's bounty and beauty, the longing to grasp the totality of existence, the ecstatic dance of past and future, and the merging of all the peoples of the world were incarnated in this living sacrament of a worldwide communion. At the center of the man-city, at the center of the globe, the woman-temple-machine was the living symbol that held together the Saint-Simonian cosmos.

* * *

This was the doctrine and iconography drafted by engineers who had received the best scientific training available in the world. Their goal was not simply to build bridges, but to remake society and humanity. Machines articulated this vision and made it possible to imagine a truly living, connected "organic machine," in which all individuals, classes, and nations were in their proper place. This was not a naïve daydream. This very combination of elements later drove the Saint-Simonians into Egypt, where they laid the foundation for the Suez Canal and provided humanitarian justifications for imperial expansion; it contributed to the formation of large banks for savings and credit; it helped build the French national railroad network. Further, the Saint-Simonians' plans for a logically ordered, aesthetically unified city, with common spaces, open air, spectacles, healthy sanitation, and wide roads for the passage of people and goods (and, as it turned out, armies), directly inspired Haussmann when he rebuilt Paris in the 1850s and 1860s.[76]

The Saint-Simonians imagined a continuity between human-built technologies and the living order of nature. As jarring as it may sound to modern ears, this interpretation of technology was in harmony with their doctrine that matter and spirit were two modes of a single divine substance. The Saint-Simonians' pantheism was also expressed in a theory of "association": each individual element expressed its freedom through its participation in a much wider whole, through its submission to "the Will of God" (as determined by self-appointed priests). Their arguments laid the groundwork for French socialism, while their actions built the infrastructure for French industrialization. Thanks to labor, technology, and love—fraternal and procreative—their technical networks were fruitful and multiplied.

In the hands of the Saint-Simonians, political and scientific terms such as *labor* and *industry* were joined together in ways that reconciled activities and even social groups that might otherwise appear to be at odds. They insisted on the equivalence between the labor of the skilled artisan and that of the unskilled proletarians—at a time, as William Sewell has made clear, that the nascent workers' movement's imagination drew upon the tradition of skilled *compagnons*, rather than on that of unskilled factory workers. The Saint-Simonians also equated workers' labor with that of managers, entrepreneurs, and speculators. They further merged it with the transformative force of steam engines, theorized and optimized by Sadi Carnot and Emile Clapeyron; ultimately this force was synonymous with the will of God.

To be sure, the Saint-Simonian vision of social harmony through industry was never realized according to the plans of the poets, artists, and evangelists who proclaimed it; the few sad holes dug in the garden at Ménilmontant in preparation for the pillars of a first temple never supported a bridge across the great division opening at this time: that between the offspring of the bourgeoisie and those who whose birth destined them to labor either in the workshop or in the new Satanic mills being plotted by technocrats and speculators. Jacques Rancière has written the tragic-comedy of the crossed communication between the Saint-Simonians and the workers for whom they claimed to speak, showing how workers cherished a dream rather different from the "majestic" ambition of being defined by their work. Instead, Rancière showed, the utopia for many workers was to cease to be defined by work, to have the time to write poetry; they wanted not to be fitted into a vast international hierarchy but to organize their own local lifeways that might operate on a smaller scale.[77]

These quite different rhythms and ambitions informed the projects of the worker-philosopher Pierre Leroux, whose machines and ideas are the heart of the next chapter. A founder of the *Globe* who helped direct the turn of romanticism toward the cause of workers, he later railed against the Saint-Simonian hierarchy in favor of a society based fundamentally on equality. The practical side of his schemes focused on technologies of communication and subsistence, operating on a scale that could support a small colony of workers, like the one he led in Boussac. Leroux's alternative vision—less monolithic, less directive, but equally attuned to aesthetics, religion, science, technology, and the living powers of nature—grew out of Saint-Simonian soil.

8

Leroux's Pianotype

The Organogenesis of Humanity

AN INSTRUMENT IN SERVICE TO THE IDEAL

In 1842 a new device to compose printed words from iron letters was presented at the Academy of Sciences. No longer was it necessary to line letters up into a composing stick by hand and drop them into place; the new device composed automatically. Its operator pressed on keys like those of a piano to select individual letters, which would then slide, by gravity, into alignment first in lines and then onto the page (see fig. 8.1). The technique was introduced by a Frenchman, Gaubert, and resembled a similar machine by the inventors Young and Delcambre that had made inroads in the printing industry.[1] The introduction of the composition machine, like the introduction of mechanization to other trades controlled by powerful guilds, led to anxiety, complaints, and even physical attacks on machines by workers.

The response of one printer, Pierre Leroux, was different. In 1822, while apprenticing for Didot printers, Leroux had invented a similar device. Rather than simply a mechanization of the work of the composer, Leroux's invention would transform the entire process of printing: it included a composing machine which, like Young and Delcambre's, and Gaubert's, made use of an inclined plane to distribute and arrange letters into serial arrangement within a frame. Yet instead of printing from a composing stick, Leroux's invention immediately cast a new metal mold of several lines of type and produced a fixed, stereotyped page which could be stored and reused according to the printer's needs. He later referred to this "synthetic" transformation of printing, "the industry that holds the most closely to thought," as his pianotype.[2] Unable to get financial backing to realize his vision of the 500 such systems that would be needed to "liberate the press by mechanization," he abandoned the invention for twenty years, until he read about Gaubert's invention, and once more sought to realize his idea.[3] He wrote to François Arago, the permanent secretary of the Academy of Sciences—who just three years earlier had secured a pension for the inven-

Fig. 1.

FIG. 8.1. Leroux was cagey about the details of the composition machine that was a key element of his new system of typography; neither the working models he built nor diagrams of the device survive. Yet his description made it clear that it followed the same basic principles as similar devices that appeared in the 1840s, including Young and Delcambre's, Gaubert's, and the one depicted here by Captain Rosenberg: a pianolike keyboard with letters falling into lines under the force of gravity. From *Mechanics Magazine, Museum Register, Journal, and Gazette*, no. 1003 (October 29, 1842): 1.

tors Daguerre and Niépce, and who traveled in the same literary and political circles as Leroux—to claim priority for his invention. Republishing his letter in *La révue indépendante*, the journal he ran with George Sand, Leroux conjectured that one day writers would all have their own printing presses: "Then liberty of the press will truly exist." True freedom required independent ownership of the means of communication. He continued:

> *May these conjectures be prophecies!* It is to make these conjectures truly prophecies, and to make these prophecies real, that we have chosen to expose once again, in a forum not usually devoted to technological subjects, our old ideas. . . . It is still a question of perfectibility in the aim of an ideal; it is still a question of liberty, fraternity, equality. All kinds of progress hold together; all discoveries form a chain. We will set the human spirit free

and we will reorganize human society by dogma, by experimental science, by art, by industry—and not by any one of these things in isolation. A machine is an instrument in service to the ideal.[4]

A machine, no less than art, philosophy, or science, can work for the highest principles; and all must work together to fulfill the Revolution's goals. While a "labor-saving" machine in the hands of an unjust employer could pose a threat to those skilled workers it replaced, such a device, if collectively owned and deployed as part of a thoughtfully organized and democratically administered system of labor and consumption, could be a prophetic tool of liberation.[5]

As the founder, editor, and printer of the *Globe*, Leroux played a pivotal role in the early romantic movement in France; he helped guide its transition to social concerns after 1830, when he joined the Saint-Simonians and transferred the journal to them. After leaving the movement, he went on to proclaim his own religious philosophy of social progress. Leroux was one of the coiners of the term *socialism*, which he intended as the opposite of *individualism*: his own philosophy would overcome the limitations of both. Despite the variety of Leroux's arguments and interests—some curious, some ominous—the core of his project was the effort to transcend the polarization between liberalism, which affirmed the reality of individuals alone, and the "absolute socialism" of the Saint-Simonians, in which society appeared as an organism to be served by its members, a machine served by its cogs.

In place of this polarity, he presented one of the first and most persuasive articulations of a democratic socialism.[6] Leroux became a leading advocate of *association* and was at the forefront of the religious turn taken by reformists in the 1840s. His philosophy—informed by German romanticism, Saint-Simonianism, the philosophy of the eighteenth century, and the history of religions—concentrated on the nature of society: "Society is not a being, in the same sense as we are beings. Society is a milieu, which we organize from generation to generation to live there.... Life is a multitude of relations between man and the different beings which co-exist with him in the world."[7] This definition of society as milieu expresses Leroux's attempt to avoid the extremes of reductive individualism and a reified collectivity. Society must be understood as a halfway point, a "*milieu*" between these extremes. Scholars have linked Leroux's use of the term milieu to his romantic-tinged theory of the symbol as a concrete instance, and to his participation in debates over pantheism in the period between Kant and Marx.[8] As suggested by his reference to "the different beings that co-exist" with humans, another set of connections tied Leroux's theory of the milieu to its historical setting. From the founding of the *Globe* through to his final writings, Leroux was enthralled by the natural sciences and the need to communicate and unify their findings. The term milieu had for him the combined biological and physical resonances we have seen in earlier chapters: it was both the nutritive envelope surrounding organisms, whose modifications could incite the formation of new organs, and the ether or medium through which light, heat, and other invisible fluids traveled. Further, Leroux was one of the most vocal champions and interpreters of the transformist advocate of the unity of composition, Geoffroy Saint-Hilaire. I argue that Leroux's conceptualization of "humanity," as an eternal and progressive virtuality that manifests itself in individuals, was a transposition of Geoffroy's concept of an abstract "animality." Many have noted the impact of organic metaphors on the formation of sociology; yet Leroux's unique organicism—a democratic socialism rooted in the transcendental anatomy and organogenesis of Geoffroy—has not had its due.[9]

Like the Saint-Simonians, Leroux entwined an influential utopian vision with a fascination for science, technology, and the emotional effects of the arts. Yet his arguments emphasized equality, participation, and bottom-up processes directly opposed to their "absolute socialism." Leroux's ecologically tinged social philosophy and his experiments in communal living were a strikingly distinct condensation of the themes of mechanical romanticism. The practical and artisanal orientation of his philosophy reveals that along with the arts and the sciences, machines—like his pianotype—were inseparable from the incessant processes of natural development that shaped the human milieu. In his words, industry and art are "no longer nature left to itself," but instead, "nature continued by man"; and mankind was "placed on the face of the earth in order to achieve the work that God charged him with finishing."[10]

ROMANTIC PRINTER AND ENCYCLOPEDIST

The son of a Parisian drink-seller, Leroux won a scholarship to a private college, where studies in philosophy led to a loss of his Christian faith. Like other ambitious young men under Napoleon's reign, he planned to enter the Ecole Polytechnique, but the death of his father forced him to become a typographer. After an apprenticeship in England, he returned to Paris and joined the revolutionary Carbonari; he and two co-conspirators, the eclectic philosopher Théodore Jouffroy and the historian Paul Dubois, professor of history at the Ecole Normale Supérieure, cofounded the *Globe*. Leroux was a typesetter, an editor, and a frequent contributor. After his liaison with the Saint-Simonians from 1830 to 1832, he developed his own doctrine, elaborated in journals he owned and edited, often with the assistance of George Sand and his Saint-Simonian fellow-traveler Jean Reynaud. After shuttering the *Revue encyclopédique*, which he edited alongside Hippolyte Carnot, he began editing, with Reynaud, the multivolume *Encyclopédie nouvelle*, which they intended as a compendium of contemporary learning in letters, history, comparative religions, science, and industry. In the late 1840s, his relations with Reynaud grew strained as Leroux promoted his new "religion of humanity"; he then took on the *Revue indépendante* alongside George Sand. Throughout this life in journalism and philosophy, Leroux retained his identification with the working class and his typesetting *compagnons*: he was frequently caricatured wearing a dirty blouse and the long, unkempt hair that earned him the binomial *Philosophicus hirsutus* (see fig. 8.2).

Leroux's writings in the *Globe* laid out themes that he pursued in his

FIG. 8.2. Pierre Leroux around 1854. Daguerreotype by Auguste Vacquerie. Musée d'Orsay, Paris, France; Réunion des Musées Nationaux; Art Resource, New York, NY.

later philosophy. Most historical analyses of the journal have focused on its contributions to romantic criticism and the liberal politics of the Restoration. It was, for example, the locus of an important debate between representatives of classicism and of romanticism, as well as debates about political economy. Yet the natural sciences were also central to its mission. According to Leroux, the *Globe*'s aim was to "keep its readers abreast of all the discoveries made in the sciences and in all branches of activity in the principal nations."[11] We noted in chapters 2 and 5 the science reporting of Alexandre Bertrand and his enthusiastic reports on the progress of post-Laplacean physics, his clarifications of Maine de Biran's physiospiritualist philosophy, and his popular writings on animal magnetism. Another of Bertrand's fascinations was comparative anatomy: the Cuvier-Geoffroy debate was stoked in part by his articles on the ideas of Geoffroy and those of his student Etienne Serres.[12] Bertrand's theory that the organic state of "ecstasy" induced by magnetism was at the origin of religions also resonated with the recurrent interest expressed by the *Globe* in the essence and function of religion. This included the publication of selections from Benjamin Constant's *Histoire des religions*, which saw religious sentiment as both necessary and essentially individual, and Jouffroy's "How Dogmas End," a much-discussed analysis of the decline of Catholicism and the need for a revival of religious enthusiasm.

In the Restoration, the *Globe* was an important channel through which German philosophy and criticism entered France.[13] Leroux contributed to this dissemination. He reworked a translation of Goethe's *Sorrows of Young Werther* and published it with a lengthy appreciation of the work. Leroux saw Werther's transcendent longing and its tragic consequences as the diversion of a religious impulse that no longer had a place in the self-involved and materialistic world opened by the French Revolution. Chateaubriand's *René*, de Staël's *Corinne*, Vigny's *Chatterton*, and the life of Lord Byron, Leroux argued, were so many variants of the species *Werther*.[14]

The central focus of Leroux's literary criticism was a theory of the symbol.[15] The symbol reconciled antinomies. Both concrete and spiritual, specific and vague, it realized the infinite in the finite while at the same time revealing the distance between these poles:

> Poetry is this mysterious wing which floats at will in the entire world of the soul, in this infinite sphere of which one part is colors, another sounds, another movements, another judgments, etc., but which all vibrate at the same time following certain laws, such that a vibration in one region com-

municates with another region, and the privilege of art is to feel and to express these relations, profoundly hidden in the very unity of life. For from these harmonic vibrations of the diverse regions of the soul there results a *chord*, and this chord is life; and when this chord is expressed, it is art; now, this chord, expressed, is the symbol; and the form of its expression is rhythm, which itself participates in the symbol: this is why art is the expression of life, the repercussion of life, and life itself.[16]

This was more than an intellectual, idealist, or even emotionalist conception of symbolism. If the symbol is the expression of a discrete vital chord—as well as the instrument that strikes this chord in others ("repercussion")—the artistic work participates in vibratory patterns of action and reaction in a medium that is simultaneously physical, mental, and spiritual. Leroux wrote that the poetry of Lamartine, Mérimée, and Hugo tended toward pantheism, a divinization of nature that presupposed a unity of mind and matter. He read these poets' worship of nature and advocacy of poetry's religious calling as a series of tentative responses to the religious vacuum of the post-Revolutionary age.[17]

Leroux's emphasis on the pantheist undercurrents of romantic symbolism as well as his sense of the historical impasse faced by his contemporaries prepared him for conversion to Saint-Simonianism immediately after the Revolution of 1830. Yet his adherence to the eighteenth-century ideals of equality and individual freedom also portended his rapid desertion of the cause. Leroux was one of the most prominent defectors from the movement after Enfantin's proclamations about free love and his reckless involvement in members' personal affairs, departing along with Reynaud and Hippolyte Carnot. These three went on to take over the *Revue encyclopédique*, spreading a message open to republicanism and democracy, while placing a heavy emphasis on the unity of the sciences.

The essay with which Leroux introduced this new journal, the *Revue encyclopédique*, spelled out the implications of the journal's slogan, "Tradition, Progress, and Continuity." This seemingly anodyne formula made clear his differences with the Saint-Simonians, who had dismissed the philosophy of the eighteenth century as destructive and merely symptomatic of the chaos of a "critical" age.[18] In contrast, Leroux emphasized the continuity between the Revolution and the present by showing the philosophical ideals of the encyclopedists and the revolutionaries as the most recent instantiations of a universal tradition that had been developing at least since the Renaissance. At the heart of this tradition was the doctrine of progress, which seemed a new basis for unifying all contemporary sciences:

> Take nature or society, contemplate the formation of worlds or the formation of civilizations, dive into cosmogonic sciences or into the depths of history, be physicists or historians, consider the animal type (*le type animal*) in the series of its developments or in any animal whatsoever in its particular life from the fetal state up to death, the earth in the order of its successive constructions or the matter of stars, inasmuch as our weakness is allowed to pierce the secrets of the heavens—always you will see life developing itself by an incessant creation and a continual series of progress.... To transform the formula of Leibniz: *The present, born of the past, is pregnant with the future.*[19]

Leroux saw this tradition flowering in the most striking new approaches in contemporary geology, astronomy, human history, and zoology: "By an admirable synchronicity, all contemporary discoveries reveal to us the continuous change and incessant creation of the universe." Discoveries regarding the vast stretch of human history, the history of the solar system (such as the nebular hypothesis), the earth, and animal species, all revealed this progress.[20] "Life" and its incessant development connected embryology, anatomy, geology, astronomy, and cosmogony.

Yet much like the Saint-Simonians' retelling of Genesis, these cosmogonic discoveries pointed to a higher destiny: "The current labors of geologists, anatomists, historians, the labors that we call science, are on the road to religion."[21] Science, once it recognized that its sole object was the process of life in all of its forms, would prepare a new faith and institutions of worship. Philosophy's task was therefore to provide the ends toward which to strive: "To aspire to the universal tradition is above all to live, to live with a life of hope and desire. It is to have already a faith, a belief, a goal, an ideal." Leroux called himself an "idealist"—as did Sand—but not just any ideal would do.[22] In his introduction to the *Revue encyclopédique*, Leroux argued that we are born with a "historical innateness," defined by the particular moment in which we are born. For those living in Leroux's time, the central article of faith of a true religion had to be progress. An examination of the philosophical and religious traditions of humanity would thus follow the same impetus as the newest historical discoveries of the sciences. Both would furnish elements for a new religion that would also incorporate the "historically innate" ideals of the French Revolution.[23]

Leroux developed his philosophy throughout the 1830s and into the

1840s, in part through a selective dialogue with ideas coming from Germany. In the early 1840s, Leroux seized in particular upon the philosophy of Schelling—much to the consternation of German social reformers, who saw Schelling's late philosophy as a reactionary retreat toward the pieties of church, state, and individualism.[24] In a prologue to a review of Schelling's inaugural Berlin lecture, Leroux reflected on the unreliable French reception of German philosophy. He objected in particular to Victor Cousin's statement that "Descartes produced Kant without any intermediary," and pointed out that by skipping over Spinoza and Leibniz, Cousin kept hidden the real sources of the recent progress in contemporary German philosophy. Schelling, according to Leroux, was "Spinoza after Kant"; rebuilding what Kant had destroyed with the *Critique*, Schelling's philosophy allowed a return to metaphysics, the affirmation of the unity of being, and a resurgence of a theology grounded in the concrete. The great guiding theme of German philosophy, as Leroux saw it, was *"the question of God and his intervention upon creatures,"* a topic he expounded on with references from chemistry, physics, linguistics, and natural history.[25] Leroux took inspiration from Schelling's notion of the "absolute identity" between mind and matter and his central idea that through the unfolding of the process of life in individuals and in society, especially in arts and the sciences, this broken whole, these two fissured halves, might once again unite. As we saw in chapter 2, both the artistic creation of symbols and *Naturphilosophie*'s experiments with interactions among physical, chemical, and physiological phenomena were routes toward the realization of this underlying unity and identity among the seemingly diverse elements of the world.[26]

Yet just as important for Leroux and those in his circle was an intellectual resource much closer to home, one deeply embedded in the methods, debates, and alliances of Parisian science: the philosophical anatomy of Geoffroy Saint-Hilaire, whom Leroux presented as a homegrown *Naturphilosophe*: "What is the *unity of composition* of M. Geoffroy, if not the *absolute identity* of Schelling? . . . M. Geoffroy has never presented any metaphysics; he has not dissertated on the finite and the infinite, on the ideal and the real, and yet the philosophy of nature invented by him is precisely the analogue which can best make the metaphysics of Schelling understood. Only M. Geoffroy has made the greatest discoveries and posed more certain principles about *absolute identity*, as the Germans say, than the entire school of naturalists who followed Schelling."[27] Although he has often been treated as an also-ran in histories of nineteenth-century biology— less clear than fellow transformists Lamarck or Darwin, less systematic and institutionally successful than Cuvier—Geoffroy's importance has begun to be recognized, both by historians of evolution and by readers of Gilles Deleuze, who transformed Geoffroy's writings on the composition of organs into a biological nomadology.[28] Inheritors aside, we have already seen the importance of his work as a rallying point for the anti-Laplaceans: Geoffroy was a major figure of the French romantic era, and central for the aspect of it being traced in this book. The list of those who attended his funeral in 1844 reads as the social register for the mechanical romantics: Charles Dupin was a pallbearer, orations were given by Serres, the chemist Dumas, and the translator of Herder, Edgar Quinet; also named as in attendance were Arago, Blainville, Poncelet, the geologist Elie de Beaumont, Ballanche, Reynaud, and Victor Hugo.[29] Leroux's use of his work further adds to our understanding of why this was the case. Many of his contemporaries on literary and philosophical quests were drawn to Geoffroy's monistic metaphysics and to his openness to the notion of a progressive series of life forms that were different yet ultimately the same. Pierre Leroux found this natural philosophy ripe for adaptation into a philosophy of social progress, one in which technological development and the encyclopedic reform of the sciences would help guide the unfolding of humanity.

ECSTATIC MATERIALISM AND PROGRESSIVE SERIES

As we have seen, Geoffroy Saint-Hilaire had long made a point of addressing an audience outside of the Academy. He first became a celebrity as the chaperone of the Egyptian giraffe, and his fame only grew during his debate with Cuvier, at which point he reached out to authors of the generation of his son, Isidore—the children of the century. In the early 1830s, he hosted a salon at his home near the Muséum on Sunday nights where he mingled with various authors and social thinkers of the younger generation, many of whom had proved themselves skillful in publicity and the spread of ideas. Leroux attended, along with other ex-Saint-Simonians Hippolyte Carnot, Jean Reynaud, and Gustave d'Eichtahl, as well as George Sand and prominent romantic artists including Lizst, Hugo, and Balzac.[30]

Balzac's contribution to the Grandville-illustrated *Scenes from the Private and Public Life of Animals*, titled "Guide-âne à l'usage des animaux qui veulent parvenir aux honneurs," dramatized the Cuvier-Geoffroy debate. The narrator was a donkey whose master, the schoolteacher Adam Marmus, wondered why donkeys learn all they need to know instinctively, while humans require years of training. The pair went to seek knowledge and research funding in Paris; Marmus dreamt of a professorial chair, while the donkey dreamt of the comforts of the Jardin des Plantes. They met a journalist who persuaded Marmus to launch a new field, "instinctologie," which would be "a new science against the baron Cerceau [Cuvier], in favor of the great philosophical naturalists who advocate zoological Unity [Geoffroy]." They treated the donkey's dark skin with a corrosive chemical that left him covered in yellow stripes, and with a press release they announced Marmus's discovery of a new zebra, a species whose unusual appearance and behaviors challenged the fixity of species in Cuvier's system.[31]

A disciple of the character based on Geoffroy saw the hapless donkey as proof of the modifying power of the environment: "If the instinct of animals changes according to climates, according to their milieu, our conceptions of Animality will be revolutionized. The great man who dared to claim that the principle of *life* would accommodate itself to anything would be definitively proven right, against the ingenious baron who held that each class is an organization apart." Yet such arguments were risky, as they challenged the existing social and religious order: "They will spread calumny about us as much as they did about your great philosopher. Look what happened to Jesus Christ, who proclaimed the equality of souls, as you would like to proclaim zoological unity." The Cuvier character, however, held the keys to prizes, chairs, and publications; Marmus thus agreed to have his theory presented by the Baron's promoter—a parakeet who faithfully repeated what he was told at the Athénée. The new zebra had become merely a monstrous anomaly that confirmed the stability of the existing classification, and the turncoat journalist denounced the philosopher as "a dreamer, as the enemy of scientists, as a dangerous pantheist." Marmus was fêted; he entered the Legion of Honor and was awarded with a professorship. For a small fortune he sold the donkey-zebra to an English naturalist, and our donkey narrator happily ended his days as a distinguished specimen in a prominent collection, as "Museums are the Pantheon of animals."[32]

The fawning reference that Balzac made to Cuvier's feats of skeletal reconstruction in the *avant-propos* to the *Human Comedy* has led many to associate Balzac's philosophy of nature with that of Cuvier; yet this story, published by Hetzel (author of the text for *Un autre monde*) in a collection illustrated by Grandville, suggests that he held Geoffroy in far greater esteem. Balzac's *avant-propos* in fact clearly showed Geoffroy's as the guiding philosophy of his multivolume work. He wrote: "There is only one animal. The creator made use of only a single pattern for all organized beings. The animal is a principle which takes on its external form, or to speak more exactly, the differences of its form, in the milieus in which it is called to develop itself. Zoological Species result from these differences. . . . In this respect, Society resembles Nature. Does not Society make of mankind, according to the milieus in which its action is deployed, as many different men as there are varieties in zoology?"[33] Indeed, Balzac's recurrent narrative of the *ingénu* who deploys a range of social, technical, and media-based machinations to transform himself and rise to wealth and power in Paris replicated the metamorphoses followed by Geoffroy. In what amounts to a cosmic *arrivisme*, Balzac showed any number of techniques—printing presses, tailored clothing, musical and artistic or scientific instruments, plus, of course, publicity—as tools, or organs, that allowed these creatures to adjust themselves to their milieu and rise in the chain of beings.[34]

The influence went in both directions. Geoffroy cited the slogan of Balzac's novel about a young visionary, *Louis Lambert*, in his *Notions synthétiques*: "SCIENCE IS ONE." He joined this illuminist doctrine to his own: science is "placed under the authority of one and the same principle and is marked in its diversity by the characteristic of *Unity of Composition*."[35] Geoffroy also favorably reviewed and wrote the introduction for a book by the dissident Saint-Simonian reformer P. J. Buchez, a Catholic physician who saw the law of cosmic progress at work in the formation of planets, continents, animals, and human history.[36]

Another enthusiastic appraisal of Geoffroy's philosophy came from George Sand. After shocking readers with frank depictions of female desire in her novel *Indiana* of 1830, this cross-dressing and cigar-smoking baroness became the star of Paris's romantic salons; passionate liaisons with Alfred de Musset, Frédéric Chopin, and Liszt heightened her reputation for scandal. Yet her intellectual friendship with Leroux was the inspiration for her philosophical and social novels, from the mystical *Lélia* of 1832 to *Le compagnon du tour de France* of 1840, which celebrated the traditions of workers' guilds (see fig. 8.3). *Lélia* exposed the pantheist ideas of Leroux and, in an early draft, included her religious interpretation of Geoffroy's work.[37] Sand saw Geoffroy's ideas as aids in the battle "against the atmosphere of an atheistic and materialistic century," in which science seemed to rule out religion and hope:

> Threatened by science either with being forced to deny the God of our hearts, or with returning to the fetishism of our fathers, we tremble before the scientists, like our fathers before the alchemists and the sorcerers. Led to the discovery of the truth by an invincible attraction, by the destiny of our century, we stopped at each instant to mourn the naïve faith of our childhood, and to ask again for our fabulous empyrean, our apocalyptic heavens.... But this prophet [Geoffroy] has come to console us, to reconcile us with the painful labor of intellectual regeneration, to guide us to-

FIG. 8.3. George Sand, *Compagnon de la Tour de France*, promotional poster. Bibliothèque Nationale de France.

ward still-veiled sanctuaries, but from which the *wisdom* of God begins to shine behind the cloud. Creation brings itself into order, order completes itself, the unity of the creative principle is physically demonstrated and the first principle of the instinctive idea of Christianity, the law of Moses, no longer leaves any doubt.[38]

Sand presented Geoffroy's science as a solution to her generation's crisis of faith, a means of reconciling religious aspiration with the discoveries of the sciences, and a tool of "regeneration." She extended his doctrine of the unity of animal types to the notion of a divine plan guiding an emerging order through the action of an underlying creative principle. The notion expressed in Geoffroy's later philosophy of a universal motor, in "the attraction of *soi pour soi*," of like for like, was the cosmic generalization of the Golden Rule.

Leroux likewise granted Geoffroy the status of prophet. Two aspects of his thought in particular were appealing for Leroux: his metaphysics and the notion (developed by E. R. A. Serres) of the animal series. Geoffroy's materialist monism—which at times veered into a pantheism—tended to efface the line between the animate and the inanimate.[39] At the same time, it made the concept of a radically distinct, incommensurable domain of "ideas" or "archetypes" unnecessary. Borrowing from Leibniz, Geoffroy's writings of the 1830s frequently used the language of "virtuality." For Leibniz and the mathematicians he inspired, the "virtual" was a way of analyzing a given point as the intersection of infinite series of mathematical functions, which themselves extended into infinity. It was also applied to describe the dynamic processes that were at work in apparently stable situations. In the physics of D'Alembert and Lagrange, for example, "virtual velocities" was a way of describing the balance of forces in a dynamic system in equilibrium; later authors have seen this as a precursor to the nineteenth century's concept of potential energy.[40] Geoffroy—inspired perhaps by Charles Bonnet, who explicitly announced a Leibnizian natural history in his *Palingénésie Philosophique*— applied the notion in his philosophical anatomy, using it to refer to the latent conditions and organizing patterns that shaped the emergence of living things. In a letter to George Sand, he wrote: "God created matters predisposed to organization, by attributing the virtual conditions required to pass through all possible transformations according to the prescriptions of the incessantly variable atmospheres that surround them (*milieux ambians*)."[41] We may hear in the quote an echo of Lamarck's Deism, according to which God created a material world that was independently capable of producing novel forms over time, along with the emphasis on nature's dynamism and diversity ("incessantly variable atmospheres") that we saw, for example, in Humboldt's *Cosmos*. Yet Geoffroy's use of the Leibnizian language of "virtual conditions" took him further into metaphysics. His conception of the virtual made it possible to conceive of creatures' development as guided by an underlying form—a set of quasi-mathematical potentials—without recourse to a Platonic otherworld of unchanging, disembodied ideas. It also left an organism's process of becoming open to modifications according to the molecules, forces, and other entities in the milieu in which it unfolded, thus preserving the possibility that new species, monsters, and other unforeseen evolutions might emerge—notions that led Deleuze to claim Geoffroy as an ally.[42] In Geoffroy's philosophical anatomy, each animal was simultaneously itself and the abstract field of potentialities acting within it; his metaphysics was an *ecstatic materialism*.[43] As we will see, this metaphysics informed Leroux's discussions of both God and humanity.

Another reason for the appeal of Geoffroy's work was its association with the concept of progressive series. The "animal series" was the segment of the great chain of being of interest to anatomists, physiologists, and natural historians.⁴⁴ This notion underwrote the classificatory obsessions of the eighteenth century, including those of Vicq d'Azyr. It was allegedly shattered by Cuvier, who replaced the single, ascending series with four independent branches. In Cuvier's work, "micro-series" might be made according to the gradations of a single function or organ as it appeared in different species, although these never cohered to form a single scale or ladder. Despite his refusal to consider transmutation, evolution, or the gradual emergence of new species, Cuvier has been celebrated by Michel Foucault for initiating a modern conception of *life* based on functional systems and animals' adaptation to their conditions of existence.⁴⁵ But despite Foucault's now-familiar narrative, the concept of the animal series continued to thrive in France in the July Monarchy, and allied itself to the new concepts concerning the living; it was especially prominent among opponents of Cuvier. It also served as a crucial link tying natural history to social theory. A static, Catholic version of the animal series underwrote the work of anatomist Henri de Blainville and, through him, the biological and social thought of Auguste Comte. Further, we have already discussed Ballanche's theories of social regeneration as an application of the Leibnizian naturalist Charles Bonnet's temporalized chain of being to human history.⁴⁶ Ballanche's *Palingénésie sociale* was a key source for the social prophecies of Leroux and his close associate Jean Reynaud.

Beyond the chain of animals, the term *series* had a wide resonance among social reformers: according to the editors of the *Doctrine of Saint-Simon*, "in the first half of the nineteenth century the expression *series* seemed destined to a great philosophical future."⁴⁷ The Ecole Polytechnique had trained many future socialists in the analysis of divergent and convergent mathematical series. The "serial law" of the Saint-Simonians, with its alternation of organic and critical periods, was the template for the historical science announced by Buchez. For the visionary social theorist Charles Fourier (and his successor, the polytechnician Victor Considerant), the laws of social harmony, and of the universe in general, were serial: his ideal community, the Phalanstery, was organized according to "Progressive Series," the range of tasks suited to the diverse passions of all members of the society. Rotating individuals through activities and work groups intensified affections and rivalries, spurring production: "The more the passions, struggles, and leagues between the Series of a canton can be aroused, the more they will compete in their enthusiasm for work, and the more they will perfect the branch of industry their passions incline them towards."⁴⁸ Serial organization of labor provided the "gears" and "springs" (*engrenages* and *ressorts*) of the "social mechanism," and following its dictates promised luxury and riches. It would also turn a page in the history of the earth and the cosmos, ushering in the copulation of the planets and the birth of new suns. In one of Fourier's most famous prophecies, the aurora borealis would form a "Northern Crown," emitting heat and light, raising the temperature at the extremities, making the sea taste like lemonade, and producing "a host of amphibious servants to pull ships and help in fisheries."[49]

In one of his many polemics against the phalansterian school, Leroux accused Fourier of lifting the central concept of "series" from "Saint-Simon's serial law of history."[50] Yet it was from Fourier, not from Saint-Simon, that the anarcho-syndicalist philosopher Pierre-Joseph Proudhon took the inspiration for his extensive use of "series" in his *On the Creation of Order in Humanity*; his life was changed when, working as a young printer, he set the type for one of Fourier's works. For Proudhon, series were the universal principle of knowledge and organization: "Everything that can be thought by the mind or perceived by the senses is necessarily a series."[51] Another reason for the appeal of "series" to social reformers may be the fact that most of them, like Proudhon, were deeply involved in the production of serial publications: Leroux and Proudhon, as well as Ballanche and the prophetic historian Michelet, had all worked as printers or typesetters. Proudhon saw the process of composition as a model for the combinatory possibilities of all creation: "The typographic case is nothing but a series whose movable units can serve indistinguishably to reproduce all imaginable words."[52] The reformers' immersion in the press's logic of monthly, weekly, or daily installments may well have helped form their serial imaginings. "Series" was thus a foundational principle of historical sequence, epistemological order, social organization, and the temporality of mass communication.[53]

The concept of series was also associated with Geoffroy Saint-Hilaire's philosophical anatomy. Although Geoffroy did not himself argue for a chain of being (and, according to Toby Appel, had actually argued against it in his youth),[54] German transcendental anatomists such as Oken, who shared his idea of "the unity of animal type," did assume such a hierarchy of animals, as did Blainville. The animal series was assimilated to Geoffroy's philosophy thanks to the work of his champion, the physician turned anatomist, E. R. A. Serres.[55] For Serres, Geoffroy's unity of type and the animal series were mutually reinforcing concepts: graduated series were the basis for classifying and comparing embryos of different species as well as teratological inquiries into monstrous deviations in the course of development.[56] The concept of unity of type also justified comparisons between the developmental stages of embryos of diverse species; on the basis of such comparisons, Serres claimed that the embryos of "higher animals" in the animal series passed through the stages corresponding to the adult phases of each of the "lower animals." This was one of the earliest expressions of the theory of recapitulation—also known as the Meckel-Serres Law—whereby the adults of the "lower species" were understood as "frozen embryos" of the higher species.[57] Among Serres's most successful acts of popularizing (and modifying) Geoffroy's philosophy of life was the entry he published in Leroux and Reynaud's *Encyclopédie nouvelle* on "organogenesis," the impact of which was confirmed by George Sand and Jules Michelet. Geoffroy's metaphysics and his doctrine of a single animal type was thus merged with the doubled concept of series: an embryological sequence of developmental stages for each individual animal ("evolution" in the sense first used by Bonnet) as well as a hierarchy of species ranked according to their complexity—and, potentially, according to a historical sequence (a view more akin to evolution in today's sense).

Many commentators have noted the "organicism" that permeated nineteenth-century social science. But a quite specific and dynamic organic analogy was at work in Leroux's appropriation of Geoffroy's philosophy. Geoffroy's notion of an ideal, universal "animality" that was the "virtual" set of possibilities for every real animal, actualized according to the material composition of the surrounding milieu, was directly analogous to Leroux's notion of a virtual "humanity" that unfolded itself to varying degrees in every historical moment. Further, the idea of an underlying series, one of Serres's decisive additions to Geoffroy's ideas, provided a structure for Leroux's historically based prophecies.[58]

VIRTUAL HUMANITY AND FREEDOM THROUGH CONNECTION

Leroux's most famous work, *L'humanité*, of 1840, was a speculative philological study of the idea of humanity that investigated the traditions and historical settings in which the concept had developed. The first half of the work expounded the general concept, defining it as "an ideal being composed of a multitude of real beings who are themselves humanity, *humanity in the virtual state*."[59] Our finite existence depends on this larger entity, which exists eternally, manifesting itself both in individuals and in distinct historical arrangements and societies. To explain, he offered the example of a mirror, an object that gives us knowledge of our own bodies by producing an image. But this image would not exist without the body before it, just as our understanding of ourselves depends on that external image. The two entities are distinct but indisputably dependent on each other. The same is true of individuals and the collective entity, society. "Human life is the knowledge, the sentiment, and the sensation that result from the co-existence of man and society. Suppress one or the other and life stops and disappears, like the image.... Man and society, even so, are just as distinct, just as independent as are our body and the mirror in which we look at ourselves. But it is the case that between the man and society, between society and the man, there is a mutual penetration by which they merge without ceasing to be distinct."[60] Society is a mirror that reflects the individual back to himself, creating the image in which he exists in a state of "mutual penetration." At a higher order, humanity was also a historically developing, collective entity realized in, but distinct from, individuals; a society was the form this vast collective entity took in a given place and time. The aim of Leroux's book was frankly evangelical: Leroux believed that once we realize that we are part of this current flowing through the centuries and learn to hear "the actual life within us," we will be inclined to serve humanity in the concrete forms it now assumes.

Leroux discovered hints of his key ideas, progress and humanity, in both the distant and the more recent past. Anticipations of "humanity" were found in Confucianism, Buddhism, Hinduism, and Islam, as well as the Western classical tradition; he saved particular praise for Pythagoras's and Plato's allusions to reincarnation and metempsychosis, and like his admirer Michelet, he paid homage to the Renaissance for having retrieved the humanistic tradition.[61] Leroux traced "a universal and truly catholic tradition" that carried the germs of progress and humanity over time and across space. The second half of *L'humanité* focused on the Jewish and Christian tradition, unveiling the social meaning hidden in its symbols and rites. Moses was the first "people-man" [*homme-peuple*] who combined all of humanity in his individuality; the ascetic Jewish sect of the Essenes presaged the emergence of a radically egalitarian material and spiritual community in Christianity. The rite of Communion enacted the idea that the sharing of material goods was inseparable from humans' shared essence: the Eucharist's hidden meaning was simply "We want to share the cup!"; "We will all eat at the same banquet!"[62] In his own period, Leroux saw the idea of humanity expressed in romantic poetry: following from his criticism in the *Globe*, he saw the sorrows of modern poets as testaments to the incompleteness of the present, while their oscillation between despair and hope prepared humanity for its future. In the 1840s, the ideal of humanity took the form of freedom, brotherhood, and equality, which meant participating in the movement of workers' liberation.

Leroux sought a course between the twin dangers of liberal individualism and the "absolute socialism" of the Saint-Simonians, showing that individuality was both real and relative. The individual alone was a partial being, attaining fulfillment only through physical, emotional, and intellectual interaction with others (what Leroux called the "Triad" of sense, emotion, and knowledge). Failure to recognize our interdependence was the root of all harm: "Evil, in principle, is egotism; because it is *separation*, the destruction of *unity*, it is the contrary of being. Being is not only the *self*, it is the *self* united to a *non-self*. Being is not the *individualized* life, it is the individual life united to the Universal Life. Man began by *separating* himself, *individualizing* himself in an absolute manner: hence evil."[63] At the same time, society should not be seen as an entity that erases individuals by making them passive instruments of a general will. Leroux thought that this danger was as acute in Saint-Simonianism and Cabet's communism as it was in the "organic" vision of neo-Catholic theocrats. Instead, he insisted on both the interdependence and the difference that existed between individual and society. As he argued, "Society is a milieu, which we organize from generation to generation to live there."[64] Society was a fluid entity halfway between a discrete individual and an independent collective body.

Philosopher Pierre Macherey has related this political conception of inbetweenness to Leroux's metaphysics, and in particular to Leroux's participation in the controversy over pantheism in the 1830s and 1840s. The issue had been brought to a head in the polemical book, *Du panthéisme dans la philosophie moderne*, by the Abbé Maret, a Jesuit and ally of Philippe Buchez. Maret lumped the Saint-Simonians, Victor Cousin, Hegel, and Leroux together as proponents of the heretical view that God is identical with the world, that soul and matter are but two modes of a single substance.[65] Although Leroux did in fact identify himself at times as a pantheist, he made a point of distinguishing his philosophy from the pantheism he attributed to Hegel and Spinoza. In this form, pantheism implied a form of abstraction and fatalism in which particulars disappeared into the rational concept of a deterministic and encompassing totality. The pantheism Leroux preferred (which at times he denied was a pantheism at all) insisted on both the specificity of each individual part of the universe and its connection to the others. Every stone, animal, or human being was "of" God, "came from God," was "in" God, but was not equivalent to God. God was the totality, both manifest and latent, actual and virtual, the sum of entities that exist, have existed, and will exist; yet no single person or object or idea could be that totality.[66] This position indicated a middle point, a milieu between, on one hand, a theological spiritualism that would give ontological precedence to God and the soul, and, on the other, a materialism that would deny the existence of both.

The idea that society is a milieu also formed an analogy with Leroux's theory of the symbol. Catholic and medieval symbolism had mirrored a stable and knowable (if abstract) theological order; in German romantic criticism, the symbol became a sensuous and emotional gesture toward a reality more complex, mysterious, and elusive than any discrete sign might capture.[67] Warren Breckman has linked Leroux's romantically infused recognition of symbols' incomplete signifying power to his dissatisfaction with talk of "incarnation," whether as theology or as politics. While Moses may have been the first "people-man," he was distinct from those for whom he stood; Christ was the son of God, Leroux argued, but was different from God. The fallibility of aesthetic representation and the incompleteness of divine incarnation also converged with Leroux's view of the fallibility of representation in the political sphere. Any government or administration could only be a partial and temporary approximation of the individuals for whom it stood. As a result, governmental structures could never be complete, and the law could never be finished; instead, representation always had to be renewed and reorganized.

The notion of milieu thus accomplished a crucial work of synthesis, or sublation, in relation to *politics*, because society was a milieu in the sense that it was halfway between being a unified entity standing over and dictating to its members and being nothing but an aggregate of autonomous individuals; to *aesthetics* because the symbol was a finite element that participated in but did not capture an infinite reality; and to *religion* because the universal divine was both separate from and contained in all of its parts. Leroux's use of the concept of milieu was part of what should now be a familiar theme among the mechanical romantics, that of freedom-in-connection—of participation without submersion—that wove through his writings.

Leroux expressed this metaphysics of the in-between by using the Leibnizian language of virtuality. He wrote, for instance, that each individual is "a real being in whom lives, in the virtual state, the ideal being called humanity."[68] Geoffroy Saint-Hilaire likewise described the animal plan as a "virtuality" that unfolded itself at different speeds, angles, degrees, and proportions according to external circumstances, or its "ambient milieu." Leroux's "humanity" was an abstract being directly analogous to Geoffroy's "animality": it was an ideal entity, a matrix of potentialities that realized itself to varying degrees in specific social arrangements and in concrete individuals. In this analogy, a "society," halfway between universal humanity and an individual, compares to a species, which likewise mediates between the universal animal and a discrete organism. Like Geoffroy, Leroux recognized that that the physical environment and the earth were determining factors for the shapes humanity has assumed in any given moment. Similarly, he argued, the historically grounded institutions, language, and ideas that determined a person's "historical innateness" also shaped humanity. Both of these milieus—the physical environment and the social and historical order—set the terms for the physical, emotional, and intellectual unfolding of virtualities into actual forms.

Beyond its uses in the life sciences, we have also noted the importance of the term milieu in physics and chemistry, where it referred to the medium of transmission for imponderable fluids; both Oersted and Ampère identified the composing and decomposing luminiferous ether with electricity. The milieus of physics and chemistry offered Leroux further food for thought. In the same essay in which he likened Geoffroy to Schelling, entitled "On God," Leroux advanced a theology of the milieu. His question was that of the relation between the universal Being, or God, and particular beings. He considered examples from natural history and linguistics before turning to optics. Just as light was the background or milieu for perception, so should we consider God the milieu for all that exists; and just as light was distinct from the objects it made visible, God contained entities and made them possible, and yet remained distinct from them. The science of chemistry also illustrated the "interpenetration" between universal Being and individual beings. Speaking of chemical decomposition, Leroux argued that the entire universe was at work in maintaining the existence of any discrete being: "Because you consider this body as a permanent compound until the moment when you put fire to it, it is thus necessary that the whole entire Universe contributes to this permanence."[69] Things maintain their state through the collusion of all elements of the whole; a modification of one element, such as exposing a solid to fire, disrupts the pattern of activity that maintains other entities.

In this exegesis, Leroux gave particular importance to scientific instruments. Chemical compounds had to be understood in relation to their circumstances as well as to the specific means through which they were known: "Every compound appears to chemists as existing under the specific conditions, as they say, of the general milieu of the Universe, which is to say at such a temperature, at such a barometric and hygrometric degree, under such electric and magnetic intensity, etc. [The chemists'] art consists in large part of the multitude of instruments they have invented to characterize with assurance the influence of all the general milieus on the portion of the universe they are considering."[70] As Humboldt so vibrantly demonstrated, instruments revealed the individual components of a given ecological niche or milieu and made it possible to trace their harmonic interactions; they articulated the relation of part to whole. For this reason, for Leroux, scientific instruments and experimental devices could be cosmic symbols, truly romantic machines: "At its current point of development, chemistry finds itself, so to speak, theoretically summarized by Volta's battery. This magical instrument, which is to chemistry what the steam engine is to industry, is not only an instrument, it is a symbol. . . . What a strange phenomenon, what a power, what magic! Does electricity not appear to you as sovereign?"[71] Leroux denied that chemistry was merely a science for classifying atoms and analyzing compounds; indeed, in terms Balzac's hero in *The Quest for the Absolute* would have approved of, Leroux argued that such a conception denied the "life" that characterizes even the lowliest chemical phenomena. Instead, "in an ordinary battery, you cannot separate the very production of electricity from the decomposition of the water and the metals." The universal milieu, manifested in electricity, was inseparable from the individuals it united and dissolved; chemistry was now charged with theology. For Leroux, the participation of the material milieu in all phenomena suggested a potent analogy—and indeed was continuous—with God's simultaneous participation in and difference from all beings. "Life exists only in the union of two beings, the universal Being and the particular being, God and its atoms."[72] For Leroux, the instruments of geophysics and electrochemistry contributed to theology. In the same way, the anatomical arguments of Geoffroy Saint-Hilaire about the universal animal's participation in each animal nourished Leroux's prophecies about the divine progress of humanity.

Leroux's statement "Society is a milieu that must be organized" was therefore more than an expression of his modified pantheism, his aesthetics, or his reconciliation of liberalism with socialism. The milieu nurtured an organism, standing between it and other beings; it was the medium of transmission of light, sound, and communication. To organize this "space between" required the interface of organs: Geoffroy described the hand, for example, as "the part most especially allocated to the communications of a being with all that surrounds it."[73] With the notion of organogenesis, Geoffroy and Serres attended to the progressive formation of organs in embryos and the interruptions or deviations such evolution could undergo. Leroux's comparative anatomy of human societies noted the morphological and functional changes, over time, in the distribution of property and power, as well as the development of new concepts, arts, and machines. These were the external organs, emerging out of virtual conditions, that served to organize the milieu and extend the progress of humanity.

ORGANIZING THE MILIEU

We now have a stronger grasp of the meaning of Leroux's slogan "We must organize the social milieu for the free development of individualities."[74] Organizing the social milieu meant creating the external conditions for the dynamic process by which individuals unpredictably actualized themselves. Symbols could play a role: like the Saint-Simonians, Leroux emphasized the religious mission of the arts in formulating ideals and training our emotions to direct us toward their realization. The progressive reorganization of humanity would also arrive through the invention, use, and reformed ownership of technology. He argued that society should "place at the disposition of each, according to his faculties, the instruments of labor and the diverse means of development which form the common heritage of humanity. Thus, each individual freely comes to draw from this source, through his labor, a certain fruit which, then, becomes his own domain, where he is king and as free as society is over its own domain."[75] Because printing was "the one among all the industries which has the most to do with thought, and which is its most direct messenger," Leroux believed his pianotype had a unique social destiny. He originally refused to develop and patent the device, as he saw it achieving its ends only when it was in the hands of hundreds of printers and editors; owning a monopoly over its use would run counter to this goal. Around 1820 he secretly presented the invention to Lafayette as a means of printing encoded documents for the Carbonari, and as a tool of "pacific conspiracy" that would make the censors' job impossible by vastly increasing the number of publishers.[76] In the 1840s, he also hoped it would lift him from his perpetual financial difficulties. In the late 1840s he retreated to a commune in the town of Boussac to develop the invention, surviving by means of donations from George Sand and the profits of a traditional press he and his fellows operated. In this socialist "colony," surrounded by his followers and his children, he put many of his social ideas into action.

To explain his views of the division of labor and the progressive role of machinery, in an essay on "social science" published in 1848 Leroux applied his "Triad" of sensation-feeling-knowledge to the analysis of the occupation he knew best, printing. He correlated knowledge with the activity of the corrector, feeling with the compositor, and sensation with the press operator. Even as these tasks became mechanized, the overall process of printing would continue to require a division of labor to adjust to the industry's new functions: "Machines can thus add themselves to the synthesis of sensation-feeling-knowledge, which alone, in industry, is the true power, without being able to replace or render this synthesis useless."[77] The fraternal traditions of *compagnonnage* and the principles of association would persist in spite of changes in the means of production, fluidly incorporating machines into harmonic, triadic arrangements of ownership and use.[78]

While he did not oppose privately owned property, he argued that it should be allotted according to its social function. In place of free-floating finance capital, which allowed some individuals to accumulate wealth without active labor, he proposed that all wealth should be put to direct use in production. For this lifelong printer and typesetter, collective administration of "the instruments of labor" was of central importance. He proposed that governmental structures should also be understood as collective resources. Democratic elections were essential: "Representative government is . . . the permanent and necessary instrument of progress and the improvable but indestructible form of the society of the future."[79] At the same time, social adjustments were subject to revision: democratic procedure would decide which structures should be preserved and which should be abandoned. Even religion could undergo democratic modifications.[80] He argued for state-funded public works projects, universal public education, and workers' savings banks—all apparently modest reforms that aimed at a balance between equality and individuality. His philosophy has been recognized by his successors in socialist thought as an important source for the later ideas of anarchist cooperatives and mutualism.

Further, his philosophy had implications not just for the relations between humans and the means of production, but between humans and nature; many of Leroux's ideas anticipate contemporary notions in deep ecology. Humanity's constitutive milieu was not only social but natural. A man's life, he believed, "does not belong to him entirely, and is not in him only; it is in him and outside of him; it lives . . . undivided, in his fellows and *in the world that surrounds him*."[81] He came up with a program for recycling human waste as a means of perpetually renewing the fertility of the earth. The oft-ridiculed idea (discussed below) of the "*circulus* between production and consumption" was for Leroux just one of the means by which human discoveries could strengthen the ties that linked humans and nature via technology.[82] New inventions, including agricultural schemes, were means of realizing the "virtual" plan and vital forces of nature and humanity in new collective actualizations. Yet Leroux—like Comte, whose concept of "biocracy" is discussed in chapter 9—differed from the Saint-Simonians in insisting that human life was inseparable from the nonhumans that surrounded it. Where Saint-Simonians saw the earth as passive matter (gendered as female) waiting to be fertilized by (male) activity, Leroux and Comte avoided these Baconian tropes of dominion; they also insisted on the limits set by nature to human modifications. Leroux suggested that social improvement was possible only so long as humans maintained a delicate balance with their milieu.[83]

The ecological resonances of his view of humans and their activities as "organs of the earth" were set forth in his discussion of the *circulus*, in his 1846 essay "L'humanité et le capital." Once more, Leroux drew upon the logic of virtuality, but no longer only in relation to animality or humanity, but to nature as a whole. The article was framed as a refutation of Malthus's so-called law of nature, the prediction that because human population increases geometrically, the number of people on earth will necessarily outstrip the food supply, which can supposedly increase no faster than arithmetically. Exhibiting sarcasm as astringent as that deployed by Marx in his readings of English political economy, Leroux called capital a new "God on Earth" and compared Malthus's "law of nature" to "the law of capital," which was simultaneously "the law of inequality." In a statistical table, he juxtaposed the geometrical expansion of population with that of a sum increased by compound interest over one hundred years, showing that capital grows even faster than human population. While it accumulates, however, capital actively "kills humanity" by creating unproductive wealth for those who do not need it while denying others the means of survival. Thus, capital was "one of the most odious forms of the internecine war of Humanity." Political economy employed a false notion of property, as "the right to draw a profit, or to use the consecrated term, an interest, from the single fact of accumulated wealth, *without participating in the slightest in the useful employment of that wealth.*" Yet, unlike Marx or Cabet, Leroux did not advocate the abolition of property. A true understanding of property, he said, would recognize it as a social function, when it was employed in the direct and personal creation and use of wealth.[84]

Malthus's "law of nature" and its corollary—the inevitability of natural "checks" on population growth in the form of famine—contradicted the biblical injunction to "be fruitful and multiply." Further, it was only "because capital rules over production and consumption, that human multiplication always exceeds the means of subsistence"; in other words, the current regime of property created artificial scarcity.[85] Thus Leroux based his refutation of Malthus on two claims, the infinite potential of nature, and the artificial limits set by human injustice: "Here is a proposition as certain as any mathematical truth: 'Human subsistence was created by God to be infinite; thus was it created, thus it is virtually. It is thus by the very essence of things, by virtue of the infinite fecundity of all species, and by the gift given to man to be able to profit from all of nature.' And here is another which is but the consequence of the first: 'Human subsistence, being by essence infinite, is rare only through the fault of the human species.'"[86] He framed his critique of Malthus in terms of a virtuality, one accompanied by a moral and historical obligation (and Germanic Capitalization): "What all true science must be able to show to man is the means of returning to this Eden that Nature contains virtually, and which the Human Species left, through imperfect knowledge and egotistical love, only in order to return to it one day through complete knowledge and through enlightened love."[87] At this point he rolled out the *circulus*: a material and practical means of actualizing the "infinite fecundity of Nature" and returning to this virtual Eden.

Leroux next delivered one of the best lines in the history of economic thought: "All that is needed to answer Malthus is human excrement." The nitty-gritty of his proposal was that France would simultaneously save on imports of guano and increase the fertility of its land by saving human waste and applying it directly to fields. He cited the physiologists and chemists Thaër, Wogth, Payen, and Boussingault, who had demonstrated that human waste contains as much nitrogen and ammonia as that of cows. He also cited Justus Liebig as an authority for the claim that every kilogram of urine that is disposed of leads to the loss of a kilogram of wheat. For Leroux, the equation between consumption and production was secured by God: "All is complete in the divine synthesis; all is fragmentary in the pretended science of our savants."[88] The recycling of wastes would close the (endlessly expanding) circle formed by human needs, technological innovation, and nature's limitless fecundity.

Although Leroux was mocked for making this proposal in the National Assembly after the revolution of 1848 (see fig. 8.4), in his biographer's words, "as much of an idealist as he was, Leroux was not above a preoccupation with material things."[89] The *circulus* was a concrete means of articulating and intensifying the relationship between individuals and the larger whole of which they are a part. Likewise, his cherished invention in the realm of communication, the pianotype, was conceived as another contribution to humanity's ongoing organogenesis—a new tool drawing people closer together and weaving them more tightly into their surroundings while making them increasingly free.[90]

FIG. 8.4. Leroux in the National Assembly in the Second Republic, caricatured by Honoré Daumier in *Physionomie de l'Assemblée*. "Pierre Leroux, having exposed to the Tribune his social doctrines, no less tangled than his hair, shakes the hands of his friends so that it looks as if they have understood him." Bibliothèque Nationale de France.

EARTH'S ORGANOGENESIS

After the revolution of 1848, Leroux was elected mayor of Boussac; he returned to Paris as a deputy, where he was seen as one of the most impassioned, if at times bewildering, advocates of the workers' cause. He argued that the liberation of all workers—like his own corporate brothers, the typesetters—would be achieved only when laborers controlled the instruments of labor and enjoyed their fruits. In a speech in September 1851—just before the coup d'état—at a banquet of the corporation of typographic composers, he prophesied: "Soon it will be known in all of Europe that it is in ASSOCIATION around the instruments of labor according to the diverse functions of science, of art, and of industry that the true human society is found, that which makes all men solidary while rendering them free.... The profession, thus understood, is a religion.... Yes, you have made a great invention that will count for as much in centuries to come as Gutenberg's invention.... You wish to proclaim THE TYPOGRAPHICAL REPUBLIC! [thunder of applause]."[91] We have already seen (in chapter 5) the religious, transformative role that was granted to the printing press in romantic and fantastic literature around 1830. Twenty years later, Leroux again framed this romantic machine as performing a sacramental function, but one whose efficacy required a reorganization of ownership and labor: it had to be owned by those who actually employed it to perform a "social function." Labor or work—celebrated in epistemology, in aesthetics, and in the productive force of engines and of the earth itself—had become a religion. Organized around the tools of industry, it would be the foundation for an enthusiastic republic in which members' freedom would be directly proportionate to their solidarity.

Leroux saw symbols, governmental institutions, scientific instruments, and other machines as ways to actualize the virtual powers of nature, or God, within the human community. The discrete stages of this development traced a process of cosmic growth.[92] This evolution was expressed plainly in yet another retelling of Genesis, an essay by his follower Ange Guépin, called "The Earth and Its Organs" in the journal Leroux printed (and composed) in Boussac. Much like Buchez's *Philosophy of History*, it described the progressive formation of the earth, beginning with the nebular hypothesis, in which the sun and all the planets had condensed out of a primal cloud of fiery gases. As the earth cooled, rocks formed, water covered it, and continents appeared. Guépin cited both Lamarck and the "animal series" to describe the progressive emergence of animals with increasingly complex organs; mountain ranges and seas prepared the terrain for the dinosaurs and eventually the mammals that came to cover them. Culminating this millennia-long efflorescence was "man, in this great history of eras anterior to his, finding the secret of his mission, [to] draw up the courage needed to accomplish his destinies."[93] Humanity, and human inventions, the external organs of humanity, were the most recent outgrowths of the earth's ongoing organogenesis.

Leroux's fragmentary philosophy—combining political interventions, literary criticism, polemics, and commentaries on the state of the sciences—flowered into a cosmic narrative in which all human creations had a place. Leroux's pianotype, like other instruments of labor and industry, modified and perfected the relationship among humans. It was, as Leroux argued, an organ of perfectibility and a machine dedicated to serve the ideal. Like the symbols in poetry and religion, these "magical instruments" coordinated the particular with the universal, the finite with the eternal; they enabled individuals to be preserved in their independence even as they became one with their milieu. Denied an opportunity to train as an engineer, Leroux reached this conception of the technical interdependence of humans and their supports after a journey that began in the workshops of romantic-era printing; Auguste Comte, whose intellectual trajectory was first set in the halls of the Ecole Polytechnique, nevertheless arrived at a similar point. And while he rejected the *naturphilosophisch* metaphysics that fed Leroux's schemes, Comte's plan for extending the progress of science, industry, and nature also culminated in a religion of humanity.

9

Comte's Calendar

From Infinite Universe to Closed World

> Strictly speaking, there is no phenomenon within our cognisance which is not in the truest sense human, and that not merely because it is man who takes cognisance of it, but also from the purely objective point of view, as man summarizes in himself all the laws of the world.
>
> COMTE, *System*

TIME MACHINES

The preceding chapters have examined early-nineteenth-century devices that contemporaries relied on to learn about and make use of elusive, frequently invisible phenomena. Ampère's experiments harnessed and quantified electromagnetic attraction and repulsion; Humboldt's instruments wove the fluids and gases of local milieus into an aesthetic unity in which connection was the basis for freedom; Arago's optical devices and the daguerreotype put light to work in creating images of the visible and the invisible worlds; theatrical techniques produced stunning new effects of sound and vision; steam engines depended on the explosive power of heat; and Leroux's pianotype, like other innovations in communication, aimed to channel and diffuse thought more rapidly.

In this chapter, the romantic machines under inspection are *paper* technologies. Auguste Comte's theories took the material form of serial timelines, charts, and above all, the Positivist Calendar. This device worked upon an even more elusive phenomenon than electricity, light, heat, or even thought: Comte's calendar harnessed, reorganized, and coordinated the flow of time.[1] Introduced as mnemonic devices and liturgical guidelines for the Religion of Humanity which grew out of his *Course of Positive Philosophy*, his calendars inscribed the rhythms of daily prayer and weekly and monthly collective rites into an annual, repeating order, which itself retraced, day to day and month to month, the historical steps undertaken in the gradual maturation of the organism of society (see fig. 9.1). Adopting the traditional

CALENDRIER POSITIVISTE

POUR UNE ANNÉE QUELCONQUE

OU

TABLEAU CONCRET DE LA PRÉPARATION HUMAINE, dessiné surtout à la transition finale de la république occidentale formée, depuis Charlemagne, par la libre connexité des cinq populations avancées, française, italienne, espagnole, britannique et germanique.

		PREMIER MOIS. **MOÏSE.** LA THÉOCRATIE INITIALE.	DEUXIÈME MOIS. **HOMÈRE.** LA POÉSIE ANCIENNE.	TROISIÈME MOIS. **ARISTOTE.** LA PHILOSOPHIE ANCIENNE.	QUATRIÈME MOIS. **ARCHIMÈDE.** LA SCIENCE ANCIENNE.	CINQUIÈME MOIS. **CÉSAR.** LA CIVILISATION MILITAIRE	SIXIÈME MOIS. **SAINT-PAUL.** LE CATHOLICISME.	SEPTIÈME MOIS. **CHARLEMAGNE.** LA CIVILISATION FÉODALE
Lundi.....	1	Prométhée ... *Cadmus.*	Hésiode.	Anaximandre.	Théophraste.	Miltiade.	Saint-Luc ... *Saint-Jacques.*	Théodoric-le-Grand.
Mardi.....	2	Hercule ... *Thésée.*	Tyrtée ... *Sapho.*	Anaximène.	Hérophile.	Léonidas.	Saint-Cyprien.	Pélage.
Mercredi...	3	Orphée ... *Tirésias.*	Anacréon.	Héraclite.	Érasistrate.	Aristide.	Saint-Athanase.	Othon-le-Grand .. *Henri-l'Oiseleur*
Jeudi	4	Ulysse.	Pindare.	Anaxagore.	Celse.	Cimon.	Saint Jérôme.	Saint-Henri.
Vendredi...	5	Lycurgue.	Sophocle ... *Euripide.*	Démocrite ... *Leucippe.*	Galien.	Xénophon.	Saint Ambroise.	Villiers ... *La Valette.*
Samedi....	6	Romulus.	Théocrite ... *Longus.*	Hérodote.	Avicenne ... *Averrhoès.*	Phocion ... *Epaminondas.*	Sainte-Monique.	Don Juan de Lépante. *Jean Sobieski.*
DIMANCHE.	7	**NUMA.**	**ESCHYLE.**	**THALÈS.**	**HIPPOCRATE.**	**THÉMISTOCLE.**	**SAINT-AUGUSTIN.**	**ALFRED.**
	8	Bélus ... *Sémiramis.*	Scopas.	Solon.	Euclide.	Périclès.	Constantin.	Charles-Martel.
	9	Sésostris.	Zeuxis.	Xénophane.	Aristée.	Philippe.	Théodose.	Le Cid ... *Lancrède.*
	10	Manou.	Ictinus.	Empédocle.	Théodose-de-Bythinie.	Démosthènes.	Saint-Chrysostôme ... *Saint-Basile.*	Richard ... *Saladin.*
	11	Cyrus.	Praxitèle.	Thucydide.	Héron ... *Ctésibius.*	Ptolémée Lagus.	Sainte-Pulchérie ... *Marcien.*	Jeanne-d'Arc.
	12	Zoroastre.	Lysippe.	Archytas ... *Philolaüs.*	Pappus.	Philopœmen.	Sainte-Geneviève-de-Paris.	Albuquerque ... *Walter Raleigh.*
	13	Les Druides ... *Ossian.*	Apelles.	Apollonius de Tyane.	Diophante.	Polybe.	Saint-Grégoire-le-Grand.	Bayard.
	14	**BOUDDHA.**	**PHIDIAS.**	**PYTHAGORE.**	**APOLLONIUS.**	**ALEXANDRE.**	**HILDEBRAND.**	**GODEFROI.**
	15	Fo-Hi.	Ésope ... *Pilpai.*	Aristippe.	Eudoxe ... *Aratus.*	Junius-Brutus.	Saint-Benoît ... *Saint-Antoine.*	Saint-Léon-le-Grand .... *Léon IV.*
	16	Lao-Tseu.	Plaute.	Antisthènes.	Pythéas ... *Néarque.*	Camille ... *Cincinnatus.*	Saint-Boniface ... *Saint-Austin.*	Gerbert ... *Pierre Damien.*
	17	Meng-Tseu.	Térence ... *Ménandre.*	Zénon.	Aristarque ... *Bérose.*	Fabricius ... *Régulus.*	Saint-Isidore-de-Séville. *St-Bruno.*	Pierre-l'Ermite.
	18	Les théocrates de Tibet.	Phèdre.	Cicéron ... *Pline-le-Jeune.*	Eratosthène ... *Sosigène.*	Annibal.	Lanfranc ... *Saint-Anselme.*	Suger ... *Saint-Eloi.*
	19	Les théocrates du Japon.	Juvénal.	Epictète ... *Arrien.*	Ptolémée.	Paul-Emile.	Héloïse ... *Béatrice.*	Alexandre III ... *Thomas Becket.*
	20	Manco-Capac ... *Taméhaméa.*	Lucien.	Tacite.	Albategnius ... *Nassir-Eddin.*	Marius ... *Les Gracques.*	Les archit. du moyen âge. *S.-Benezet.*	St-François-d'Ass... *St-Dominique.*
	21	**CONFUCIUS.**	**ARISTOPHANE.**	**SOCRATE.**	**HIPPARQUE.**	**SCIPION.**	**SAINT-BERNARD.**	**INNOCENT III.**
	22	Abraham ... *Joseph.*	Ennius.	Xénocrate.	Varron.	Auguste ... *Mécène.*	St-François-Xav. *Ignace de Loyola.*	Sainte-Clotilde.
	23	Samuel.	Lucrèce.	Philon d'Alexandrie.	Columelle.	Vespasien ... *Titus.*	St-Charles-Borrom.. *Fréd.-Borrom.*	Ste Bathilde. *Ste-Math.-de-Toscane.*
	24	Salomon ... *David.*	Horace.	Saint-Jean-l'Evangéliste.	Vitruve.	Adrien ... *Nerva.*	Ste-Thérèse. *Ste-Cather.-de-Sienne.*	St-Etienne-de-Hong.. *Mat. Corvin.*
	25	Isaïe.	Tibulle.	Saint-Justin ... *Saint-Irénée.*	Strabon.	Antonin ... *Marc-Aurèle.*	St-Vinc.-de-Paule *l'abbé de l'Épée.*	Sainte-Elisabeth de Hongrie.
	26	Saint-Jean-Baptiste.	Ovide.	Saint-Clément-d'Alexandrie.	Frontin.	Papinien ... *Ulpien.*	Bourdaloue ... *Claude Fleury.*	Blanche de Castille.
	27	Haroun-al-Raschid. *Abdérame III.*	Lucain.	Origène ... *Tertullien.*	Plutarque.	Alexandre-Sévère.	W. Penn ... *G. Fox.*	Saint-Ferdinand III... *Alphonse X.*
	28	**MAHOMET.**	**VIRGILE.**	**PLATON.**	**PLINE-l'Ancien.**	**TRAJAN.**	**BOSSUET.**	**SAINT-LOUIS.**

		HUITIÈME MOIS **DANTE.** L'ÉPOPÉE MODERNE	NEUVIÈME MOIS. **GUTTEMBERG.** L'INDUSTRIE MODERNE	DIXIÈME MOIS. **SHAKESPEARE.** LE DRAME MODERNE	ONZIÈME MOIS. **DESCARTES.** LA PHILOSOPHIE MODERNE.	DOUZIÈME MOIS. **FREDERIC.** LA POLITIQUE MODERNE.	TREIZIÈME MOIS. **BICHAT.** LA SCIENCE MODERNE.
Lundi.....	1	Les Troubadours.	Marco Polo ... *Chardin.*	Lope de Vega ... *Montalvan.*	Albert-le-Grand. *Jean de Salisbury.*	Marie de Molina.	Copernic ... *Tycho-Brahé.*
Mardi.....	2	Boccace ... *Chaucer.*	Jacques Cœur ... *Gresham.*	Moreto ... *Guillen de Castro.*	Roger Bacon ... *Raimond Lulle.*	Côme de Médicis l'Ancien.	Kepler ... *Halley.*
Mercredi...	3	Rabelais.	Gama ... *Magellan.*	Rojas ... *Guevara.*	Saint-Bonaventure ... *Joachim.*	Philippe de Comines. *Guicciardini.*	Huyghens ... *Varignon.*
Jeudi	4	Cervantes.	Neper ... *Briggs.*	Otway.	Ramus ... *Le cardinal de Cusa.*	Isabelle de Castille.	Jacques Bernoulli. *Jean Bernoulli.*
Vendredi ..	5	La Fontaine ... *Robert Burns.*	Lacaille ... *Delambre.*	Lessing.	Montaigne ... *Érasme.*	Charles-Quint ... *Sixte-Quint.*	Bradley ... *Rœmer.*
Samedi....	6	Foë ... *Goldsmith.*	Cook ... *Tasman.*	Gœthe.	Campanella ... *Morus.*	Henri IV.	Volta ... *Sauveur.*
DIMANCHE.	7	**ARIOSTE.**	**COLOMB.**	**CALDERON.**	**SAINT-THOMAS-d'Aquin.**	**LOUIS XI.**	**GALILÉE.**
	8	Léonard de Vinci ... *Le Titien.*	Benvenuto Cellini.	Tirso.	Hobbes ... *Spinosa.*	Coligny ... *L'Hôpital.*	Viète ... *Harriott.*
	9	Michel-Ange ... *Paul Véronèse.*	Amontons ... *Wheatstone.*	Vondel.	Pascal ... *Giordano Bruno.*	Barneveldt.	Wallis ... *Fermat.*
	10	Holbein ... *Rembrandt.*	Harrison ... *Pierre Leroy.*	Racine.	Locke ... *Malebranche.*	Gustave-Adolphe.	Clairaut ... *Poinsot.*
	11	Poussin ... *Lesueur.*	Dollond ... *Graham.*	Voltaire.	Vauvenargues ... *Mme de Lambert.*	De Witt.	Euler ... *Monge.*
	12	Velasquez ... *Murillo.*	Arkwright ... *Jacquart.*	Métastase ... Alfieri.	Diderot ... *Duclos.*	Ruyter.	D'Alembert ... *Daniel Bernoulli.*
	13	Téniers ... *Rubens.*	Conté.	Schiller.	Cabanis ... *Georges Leroy.*	Guillaume III.	Lagrange ... *Joseph Fourier.*
	14	**RAPHAEL.**	**VAUCANSON.**	**CORNEILLE.**	**Le Chancelier BACON.**	**GUILLAUME-le-Taciturne.**	**NEWTON.**
	15	Froissart ... *Joinville.*	Stévin ... *Torricelli.*	Alarcon.	Grotius ... *Cujas.*	Ximenès.	Bergmann ... *Scheele.*
	16	Camoens ... *Spenser.*	Mariotte ... *Boyle.*	Mme de Motteville ... *Mme Roland.*	Fontenelle ... *Maupertuis.*	Sully ... *Oxenstiern.*	Priestley ... *Davy.*
	17	Les Romancistes espagnols.	Papin ... *Worcester.*	Mme de Sévigné .. *Lady Montague.*	Vico ... *Herder.*	Colbert ... *Louis XIV.*	Cavendish.
	18	Chateaubriand.	Black.	Lesage ... *Sterne.*	Fréret ... *Winckelmann.*	Walpole ... *Mazarin.*	Guyton-Morveau ... *Geoffroy.*
	19	Walter Scott ... *Cooper.*	Jouffroy ... *Fulton.*	Mme de Staël ... *Miss Edgeworth.*	Montesquieu ... *d'Aguesseau.*	D'Aranda ... *Pombal.*	Berthollet.
	20	Manzoni.	Dalton ... *Thilorier.*	Fielding ... *Richardson.*	Buffon ... *Oken.*	Turgot ... *Campomanes.*	Berzélius ... *Ritter.*
	21	**TASSE.**	**WATT.**	**MOLIERE.**	**LEIBNITZ.**	**RICHELIEU.**	**LAVOISIER.**
	22	Pétrarque.	Bernard de Palissy.	Pergolèse ... *Palestrina.*	Robertson ... *Gibbon.*	Sidney ... *Lambert.*	Harvey ... *Ch. Bell.*
	23	Thomas A'Kempis. *Louis de Grenade, et Bunyan.*	Guglielmini ... *Riquet.*	Sacchini ... *Grétry.*	Adam Smith ... *Dunoyer.*	Franklin ... *Hampden.*	Boërhaave ... *Stahl.*
	24	Mme de Lafayette ... *Mme de Staël.*	Duhamel (du Monceau). *Bourgelat.*	Gluck ... *Lully.*	Kant ... *Fichte.*	Washington ... *Kosciusko.*	Linné ... *Bernard de Jussieu.*
	25	Fénelon ... *Saint-François-de-Sales.*	Saussure ... *Bouguer.*	Beethoven ... *Handel.*	Condorcet ... *Ferguson.*	Jefferson ... *Madison.*	Haller ... *Vicq-d'Azyr.*
	26	Klopstock ... *Gessner.*	Coulomb ... *Borda.*	Rossini ... *Weber.*	Joseph de Maistre ... *Bonald.*	Bolivar ... *Toussaint-Louverture.*	Lamarck ... *Blainville.*
	27	Byron ... *Elisa Mercœur.*	Carnot ... *Vauban.*	Bellini ... *Donizetti.*	Hegel ... *Sophie Germain.*	Francia.	Broussais ... *Morgagni.*
	28	**MILTON.**	**MONTGOLFIER.**	**MOZART.**	**HUME.**	**CROMWELL.**	**GALL.**

Jour complémentaire .... Fête universelle des **MORTS.**

Jour bissextile .... Fête générale des **SAINTES FEMMES.**

FIG. 9.1. Comte's Positivist Calendar, from Système de philosophie positive.

Catholic notion of saints' days, he named each day for a "hero" and each month for a "God" who had contributed to one of the decisive dimensions of the positive state, from Moses ("the initial theocracy") to Gutenberg ("modern industry") and Bichat ("modern science").

Comte used the history of science and its relation to social formations as a vantage from which to organize the sciences after the unified "metaphysical" systems of natural philosophy had been replaced by highly specialized fields and subfields.[2] Many of Comte's recurrent terms for describing the sciences and their objects—including *regularity*, *aberration*, *organization*, *consensus*, *coordination*, and *series*—allude to processes in time and to the identification and control of temporal sequences. His philosophy tied several such sequences together into a mobile, harmonious system, one that retained its internal relations even as it expanded and progressed. Comte saw this intellectual organization as a necessary step toward a perfected social organization, which in turn was the condition for a harmonious organization of humanity's technologically aided modifications—for humanity's domestication of itself and the entities that surrounded it. He rejected the term *nature* in favor of *world* because he believed that phenomena had meaning only to the extent that they contributed to the processes by which humanity makes itself at home on the earth.

Comte's *Course* and his *System* summarized, refined, and added complexity to themes we have traced in previous chapters. He squarely confronted the rivalry perceived by many between mechanics—especially Laplacean reductionism—and the life sciences for the status of model for all other sciences. Although he disdained the use of notions like ethers, weightless fluids, vital principles, spirit, and soul, these concepts returned in his later works as deliberate fictions that would saturate "objective analysis" with humane values and "altruism"— a term Comte coined as an antonym for the "egotism" he saw dominating his present. More than any of his contemporaries, he emphasized the artificiality of any scientific system, its relativity to human needs and capacities, and above all its multiplicity and heterogeneity. Finally, Comte was a decisive contributor to the understanding of the human as the technological animal—the being who intentionally remakes its own environment, and itself, by means of machines.[3] In contrast to the Saint-Simonians, however, Comte did not celebrate the absolute dominion over nature that the industrial age seemed to promise. Instead, in his work he spoke again and again of limits to human beings' power to modify nature. Recognition of human dependence on its milieu was the moral cornerstone of the new religion he developed.[4]

By the end of the nineteenth century, Comte was commemorated by a statue in the Place de la Sorbonne; he has subsequently been acknowledged as a founder of France's traditions of sociology and *épistémologie*. Yet today his works are seldom read by philosophers or historians of science, especially outside of France. In the twentieth century, the Viennese logical positivists took inspiration from Comte's refusal of metaphysics for their projects to reduce the sciences to empirical observations and logical relations. Recent studies have shown the complex routes by which the Vienna Circle searched for a single scientific language and have revealed the tight links that tied their attempts at scientific reform to their projects to transform society.[5] After the catastrophe of the Second World War and the pressures of the Cold War, however, a new object appeared: a notion of positivism as divorced from politics and naïvely devoted to the project of a unified, disembodied, and ahistorical scientific language.[6] This straw-man version of neopositivism has been an easy target for historians and philosophers of science interested in the diverse forms science has taken and its social contexts, from Thomas Kuhn onward. By exploring the richness of Comte's views of knowledge and society—and by showing them as responses to the volatile historical situation of Paris between the Napoleons—this chapter adds to the revisitation of the multiple strands of thought subsumed under the word *positivism*. In a way quite different from that of the Viennese positivists, Comte arrived at a new conception of the unity of the sciences and the unity of nature—or rather, of the "world." The striking coherence and radical stance of his views appears most clearly when we focus on the connections he forged between technology, the biological notion of milieu, and the Religion of Humanity.[7]

Comte's religion aimed at instilling new, socially oriented habits of thought and action.[8] He practiced what he preached, aiming to "live in the light of day" and speaking publicly of his own regular habits and "hygiene." In the 1840s, he would walk every Wednesday from his home to the Père Lachaise cemetery; in the morning and at night he would kneel before a green chair in his rooms in the Rue Monsieur-le-Prince, saying prayers and visualizing his departed muse, Clotilde de Vaux, while flooding his brain with feelings of love and veneration.[9] These habits were grounded in a view of philosophy as a way of life, a means of cultivating specific virtues. But this was more than an individual "care of the self."[10] Comte's practices of the self were prescribed as technical applications of his science of society. He inscribed daily individual rites, and collective rites unfolding over days, weeks, and months, within a much vaster temporal order: that of the progress of humanity. Fusing subjectivity and objectivity in explicit ways, these practices of collective asceticism were in direct continuity with procedures for the technical exploitation of the planet.

The overview this chapter provides of Comte's thought and its decisive institutional and personal contexts must be partial. The central object that I want to make visible is a chain of connections that are essential keys to his work as a whole but that are rarely shown in their close interdependence. In short: The concept of the animal hierarchy or animal series was the indispensable skeleton of Comte's presentation of biology; this series provided both a model and an organizing logic for the central innovation of his system, the hierarchy of the sciences. This hierarchy, in turn, was not merely a classification of abstract knowledge but was a guideline for humanity's technological modifications of its milieu, for entering into the temporal sequences and processes of each level of reality. In the case of the most modifiable and most complex domain, society, technological modification meant the introduction of new individual and collective practices to create a unity of thought, feeling, and action.

The chapter thus reveals the final form of Comte's positivism, the Religion of Humanity, as a technological innovation that coordinated previous and future technological innovations; it was an outgrowth of the specific organization of the human species.[11] To provide the background for this chain of ideas, and to flesh it out, we will need to return—as Comte did throughout his life—to that hothouse of mechanical romanticism, the Ecole Polytechnique.

THE ECOLE POLYTECHNIQUE AS SPIRITUAL POWER

The Ecole Polytechnique, the homeland of engineer-scientists, had a powerful effect on the young Comte. After entering in 1815, he wrote to a friend in his native Montpellier: "I would be much happier if you had been admitted with me, because the two of us would be in paradise here. You can't imagine what a fine spirit rules among the Ecole's students; the most perfect unity exists between us. . . . In many rooms, altars have been set up to friendship; one of them bears these words: *To friendship*, and on the pediment you read: *Unity and Strength*. These ceremonies are powerfully moving."[12] Born in 1798, Comte was another of the alternately optimistic and despairing "children of the century."[13] Although raised in a strongly Catholic family, Comte lost his faith at age twelve. The unity and equality he found in the Ecole's secular rituals provided an image of nontheological solidarity and faith that echoed through his later writings.[14]

He later wrote to John Stuart Mill that the Ecole Polytechnique was "almost a prototype of the spiritual power."[15] It was a *total institution*.[16] Classed and ranked, from their entrance exams to their daily mathematics drills and military exercises, cadets were part of a functionally and hierarchically ordered "corps," one that was devoted to finding technological solutions to social problems. The daily schedule controlled students' time, enforced by strict sanctions. The sense of a regulated temporal order was also built into the curriculum, which prepared students to "react with speed and intelligence when faced with any kind of situation." It was assumed that a polytechnician could "learn much more, much more quickly than anyone else." Rapid instruction was aided by blackboards—introduced here on a wide scale for the first time.[17] This pedagogical technology allowed students and examiners to watch a mathematical demonstration unfold as a series of steps. Speed helped determine the ranking among students. The institution's first instructors followed *la méthode révolutionnaire*, aiming at compression of time: Laplace, for example, would deliver rapid-fire lectures of bold new proofs of fundamental principles of analysis before they were even published.[18] Despite this pedagogical uniformity, the disciplines taught there—descriptive geometry, stereotomy, mechanics, calculus— were not joined in a single axiomatic system but were presented as distinct, irreducible approaches to be deployed according to the demands of a situation. The Ecole's temporal coordination of heterogeneous sciences was a template for Comte's later writings.

Even though he was expelled from the school in his second year—as the ringleader of a group who protested an incompetent, royalist-leaning professor—the Ecole Polytechnique was a constant point of reference in Comte's life. He subsequently attached himself to Saint-Simon as his secretary and assistant, contributing to and editing *L'industrie* and *Le producteur*. In the 1820s, Comte therefore had extensive contact with many of the future leaders of the Saint-Simonian movement, notably Olinde Rodrigues. Gustave d'Eichtahl had been Comte's student, and Gabriel Lamé had been his friend and comrade at the Ecole Polytechnique. At this time, Comte laid out many of his key themes in a series of "Opuscules"; Saint-Simon's failure to credit Comte's contributions was one of the major causes of Comte's departure.

In his "Fundamental Opuscule" of 1822, the "Plan of Intellectual Works Necessary for the Re-organization of Society," he wrote: "Any system of society whatsoever, whether it is made for a handful of men or for several million, has for its definitive object the aim of directing all the particular forces towards a general goal of activity, because there is no *society* but where a general and combined action is exercised. In any other hypothesis, there is merely an agglomeration of a certain number of individuals on the same soil. This is what distinguishes human society from that of other herd animals."[19] He defined society—and the human species itself—by united activity toward a shared purpose. Like Saint-Simon, Comte's pre-eminent goal was the establishment of a new, independent spiritual power—the institution or social group with the authority to direct beliefs and goals. In medieval Europe this power, held by the church, was separate from but in harmony with the "temporal power," the physical rulers of men and things.[20] Due to the breakdown of this harmony and the decline of Catholic authority, society faced a choice between "conquest, that is, violent action on the rest of the human race, and industry or production, that is, action on nature in order to improve it for the advantage of mankind." To ensure that humanity chose the latter, intellectual and moral authority would have to be assumed by scientists as Saint-Simon also argued. A new science of society would provide cardinal points.[21]

Following from these concerns—systematizing the sciences and setting the direction of the progress of civilization—Comte had arrived, by the end of his apprenticeship with Saint-Simon, at the foundational "laws" of his philosophy: the law of three states and the hierarchy of the sciences. The entire outline of his *Course of Positive Philosophy* came to him after "a continuous meditation of eighty hours."[22] He then went to spread the good news to those he imagined were prepared to hear it. The audience for his introductory lectures of the course, given in 1826 in his rooms in the Faubourg Saint-Antoine, could not have been more auspicious. It included the leading lights of the Parisian scientific scene, many of whom we have already encountered: Joseph Fourier, François Arago, Alexandre von Humboldt, Claude-Louis Navier, Hippolyte Carnot, and several other polytechnicians and Saint-Simonians (Ampère was, to Comte's dismay, a no-show); from the life sciences, the physiologists Broussais and Blainville attended. These were among the most prominent and reform-minded scientists of the two institutions that were the best candidates to house a new spiritual power, the Ecole Polytechnique and the Academy of Sciences.

According to Adolphe d'Eichtahl, the evening was a success. Yet after his first three lectures, Comte suffered an attack of insanity, marked by delusions of grandeur, incapacity to work, and a suicide attempt, followed by treatment by Esquirol. Remarkably, he saw his madness as a retreat to the earliest state of knowledge, "a vague pantheism"; returning to health, he retraced the stages of mankind, ascending to a state of healthy positivism.[23] He published the first volume of the *Course* in 1830 and the last in 1842, earning his living by preparing students for the entrance exam and as a teaching assistant at the Ecole Polytechnique.

A GREAT CHAIN OF DISCIPLINES AND TIMES

What was this first version of the positivist gospel? Comte began the *Course* like a good military cartographer and strategist: he delimited his territory with "a general circumscription of the field," a "general overview of the progressive march of the human mind, envisaged in its ensemble."[24] What stood out in this landscape was the law of three states: "Each branch of our knowledge passes successively through three different theoretical states: the theological or fictive; the metaphysical or abstract; the scientific or positive."[25] Every science, like society as whole, and like every individual, first saw the objects of the world as animated by supernatural beings — whether in the direct "waking hallucination" of fetishism, the multiple personified causes of polytheism, or the single, omnipotent cause of theology. Doing away with these fictitious personifications, each science next arrived at the equally fictitious notion of absolute causes or unified "nature," exemplified by the all-encompassing systems of natural philosophy of the eighteenth century. Finally, in the positive age, reason had the upper hand on imagination, and limits to knowledge were recognized:

> [In] the positive state, the human mind, recognizing the impossibility of obtaining absolute notions, renounces searching for the origin and destination of the universe and knowing the intimate causes of phenomena, in order to attach itself uniquely to discovering their effective laws, that is, their invariable relations of succession and resemblance, by the use of reasoning and observation well combined. The explanation of facts, now reduced to its real terms, is no longer anything more than the link established between the diverse particular phenomena and a few general facts, of which the progress of science tends to diminish the number.[26]

We see here the most familiar elements of "positivism." Absolute notions and hidden causes must be renounced. Instead, knowledge must content itself with studying the circumstances of phenomena and establishing the lawful relations among them.

Many have seen such passages as advocating a monolithic "scientific method" or logic of discovery: the arrangement of observed facts in series according to similarities or temporal succession.[27] Yet Comte in fact rejected the notion of a single method, offering instead a differential and pluralist theory of science.[28] As shown in his second "great law," the hierarchy of the sciences, each science had its distinct objects, concepts, and methods. There was a progressive logic to this heterogeneity. Each science's objects, methods, and history corresponded to its place in a hierarchy (sometimes called a series, chain, or ladder), running from mathematics (in a class of its own) through astronomy, physics, chemistry, and biology, up to the final science, Comte's coinage, sociology.[29] Each science represented an increase in complexity from the one preceding it, according to gradients I will discuss below. Further, each science developed at its own rate; the later ones were still on their way to becoming "positive."[30]

Comte's view that each science possessed distinct, autonomous concepts and methods was a decisive break with earlier traditions of natural philosophy, which sought to subsume all domains of nature under a single set of laws or causes.[31] In introducing this differentiated series, he distinguished between a "historical" presentation of this order, in which historical contingencies were recognized, and a "dogmatic" presentation, according to a logical progression. The "dogmatic" history presented the sciences in their idealized form, according to the final state Comte assumed they would eventually reach. Yet the two modes of presentation agreed on the crucial turning points and distinguishing characteristics of the sciences. Astronomy was the first positive science; the others were added in time, in response to concrete experience, establishing regularities that combated the terror arising from the unexpected.[32] In each of the sciences in turn, the use of metaphysical explanations and the search for causes are replaced by the discovery of relative laws; yet in the historical sequence, certain theological and metaphysical notions continued to mar contemporary sciences—as with the "fantastic fluids and ethers" of physics, the vital principles still used by some physiologists, and the notion of "soul" in the sciences of mankind. Crucially, although later-emerging fields were less "general" than those that preceded them and on which they depended, their phenomena were not simply magnifications or extensions of those beneath them. Comte's scheme was explicitly opposed to reductionism. The application of purely mechanical or mathematical explanations in biology failed to account for the specificity of biological phenomena, just as human physiology could not be used to explain the specific properties of social life.

Comte modeled his series of the sciences on the "animal series," which was a central notion in his presentation of biology—as it had been for two of his major influences, Lamarck and Blainville.[33] The series of the sciences was ranked according to four axes: generality, proximity to humans, complexity, and modifiability. In the gradual progression from astronomy up to sociology, each science dealt with phenomena that were less general, nearer to humans, more complex, and more modifiable.[34] For example, astronomy's phenomena were the most general, in that they affected all entities on the earth; they were extremely distant from humans (even though Comte thought we should restrict astronomy to only the five nearest planets, since only these could possibly have any effect on the earth). Further, celestial motions relied on the least complex laws, involving the smallest number of intervening and complicating factors. Finally, because astronomical phenomena were beyond human reach, they were the least susceptible to human modification. At the opposite end of the series was social physics, which Comte eventually called sociology. This science dealt with phenomena that were the most specific, concerning only one species; its phenomena were very near to humans (indeed, its object was humanity itself); these were also the most complex phenomena. Yet this complexity meant that there was considerable leeway for humans to modify these phenomena. This progressive scheme recognized the specialization that increasingly divided the sciences, while maintaining the unity and completeness that had characterized natural philosophy.

Although Comte himself did not highlight it, one other important gradient defined the hierarchy of sciences. The place of each science was also determined by the way in which it aligned specific phenomena in temporal series. Comte analyzed every science according to both static and dynamic aspects—as in biology's division into (static) anatomy and (dynamic) physiology. Yet going further, each science's set of phenomena had its own rates, rhythms, and periodicities. The *Course* showed the harmony among the methods, instruments, and concepts of each set of phenomena as well as the actual modes of regularity found in each domain. The rates of these phenomena were also relevant factors for the ranking of the fields.

Astronomy, again, was exemplary. In addition to instructing humans to focus only on those aspects of the universe that were relevant to humans he pointed out that astronomy taught the fundamental scientific lesson of regularity. Science, Comte believed, was the study of phenomena that repeat themselves. Proof of the regularity of phenomena was always a blow to theology, because it destroyed the idea that nature could be affected by the caprice of supernatural beings. He discussed the importance of devices for measuring time in the field's emergence: although the sky itself, with its repeated cycles, was "the first chronometer," non-celestial clocks were needed to measure its more precise movements. Astronomy also gave us the first example of science as definitively aimed at prediction. Astronomical phenomena provided the very model of measurement, regularity, and prediction, with the identification of diurnal rotation, the phases of the moon, and seasonal revolutions. As it developed, it also provided the impetus for more refined chronometry to capture transits and to calculate latitude and longitude. Laplace's refinements to celestial mechanics took these tendencies still further: they laid open the past and future states of the heavens: "It is to celestial dynamics that we owe our power of ranging up and down the centuries, to fix the precise moments of various celestial events, such as eclipses, with certainty."[35] Astronomical tables granted power over time, which was the mark of positive science. Astronomy made plain the fundamental relationship between a science's particular concepts, methods, and objects and the specific temporalities and kinds of events it encountered.

Another temporal dimension structuring the hierarchy of sciences appeared in Comte's arguments about the distinct "modes of molecular activity" studied by each science. Physics, chemistry, and biology, he said, "may be considered as having for their object the molecular activity of matter, in all the different modes of which it is susceptible. From this point of view, each corresponds to one of three successive degrees of activity, which are profoundly and naturally distinguished from each other. The chemical obviously presents something more than the physical action, and something less than the vital action."[36] The molecular activity proper to physics modified "the arrangement of particles in bodies; and these modifications [were] usually slight and transient, and never alter[ed] the substance." In chemical activity, "alterations in the structure and the state of aggregation" were the object of study, along with the changes "in the very composition of the particles" brought about by chemical interactions. In physiology, molecular activity took on "a much higher degree of energy" because, as Comte discussed at length in later sections, "the vital state [was] characterized, beyond all physical and chemical effects that it constantly determines, by a double continuous motion of composition and decomposition, more or less rapid, but always necessarily continuous," a process that sought to maintain "within certain limits of variation, over a greater or lesser time, the organization of the body by incessantly renewing its substance."[37] This "gradation" of the "modes of activity" proper to physics, chemistry, and the sciences of life was as much a defining criterion of the series of the sciences as were the degrees of generality, distance, simplicity, and modifiabilty.

While it was not molecular activity proper that was in question in the final science, sociology, it was in sociology that the complexity of temporal scales was the greatest. This temporal complexity defined the field. The law of three states and the hierarchy of the sciences were sociology's first and most fundamental discoveries; thus, in addition to reckoning the history of humanity and the rhythms of social life, this field was tasked with reconstituting the time scales of each of the other sciences. The recursive nature of this project was striking. When Comte's exposition reached this final science, he revealed the historical grounds for the journey his readers had just taken through the previous ones. In other words, the history of the progress of human knowledge—the temporal orders established by each of these sciences, as well as the order in which they emerged—could be told properly only once sociology had been put on a secure base. Sociology thus

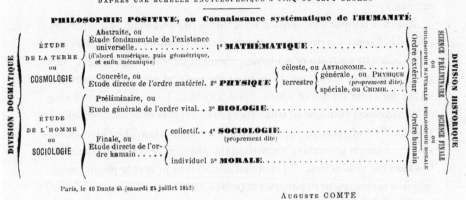

FIG. 9.2. Diagram of sciences, from Comte, *Catéchisme positiviste*.

coordinated the progress of all of the sciences. In addition, it established the laws and temporal series of society itself; in later formulations, sociology was followed by another science, morals. The sciences of humanity thus set the template for the reorganizing of the time series of individual and collective life; they provided the theory for what, in the Religion of Humanity, became practice.[38]

Comte thus ranked the sciences in temporal terms in at least two senses: first, according to the moment when they left behind theology and metaphysics, and second, in terms of the modes of temporality entertained by each of their objects. From the slow revolutions of astronomy, to the increasing molecular activity of physics, chemistry, and physiology, to the historical series and complex social metabolism studied in sociology and the science of morals, each science had its own relationship to time (see fig. 9.2).

Comte's analysis of the sciences showed not only the differences between the sciences but the joints at which they could be coordinated. For instance, he suggested links between the periodicities of astronomy and terrestrial life.[39] The influence of the earth's static factors was significant for living things (its shape, mass, dimensions, distance from the sun), but as Comte noted, dynamic factors were decisive: "Bichat pointed out that the intermittence of the animal's own life is subordinate in its periods to the diurnal rotation of our planet; and we may extend the observation to all the periodical phenomena of any organism, in both the normal and pathological states.... Moreover, there is every reason to believe that, in every organism, the total duration of life and of its chief natural phases depends on the angular velocity proper to the rotation of our planet."[40] This linkage of the cycles of sleep and waking, as well as the phases and length of life, to the movements of the earth and sun was one form of "coordination" between these two orders of facts, highlighting both the dependence between them (terrestrial life is shaped by astronomical phenomena) and their conceptual autonomy (distinct concepts are used to understand organisms and to understand planetary movements). Likewise, he suggested that life on earth would be transformed "if the earth were to rotate much faster," "if the duration of the year were changed," or "if the ellipse became as eccentric as a comet."[41] Comparable speculations about the correlation between the changing heat of the earth and occurrences on its surface were at the heart of discussions of the nebular hypothesis in the 1830s: these aimed to link the temporal scales of the solar system, geology, natural history, and in some cases (such as Ange Guépin's "The Earth and Its Organs," discussed last chapter), the history of mankind.[42]

For Comte, each science was a set of concepts and tools for marking time, for registering the repetitions and regularities that appeared at each level of reality. Positive sociology, with its series of the sciences, showed the points at which these series intersected. Comte's slogan, "Order and Progress," captured this structure: his philosophy wove multiple rates and scales of progress into a single, developing order. If each discipline was a train line moving at its own rate, the *Course* was the station master who synchronized their arrivals. Furthermore, the order laid out in the *Course* was not a closed and static inventory but rather a framework to guide further research, to make progress in each field "regular," to harmonize it better with other fields, and to secure "correspondences" between the different lines. Comte extended the template laid down at the Ecole Polytechnique—the home base, as we saw in chapter 4, for other scientist-engineers working to optimize the efficiency of roads, networks, and steam engines—to the cosmos as a whole. He presented his philosophy as a coordination of diverse time scales, a convergence of distinct but intersecting developmental series within a single expanding framework, at the center of which was humanity.[43]

A HYPOTHESIS COLLAPSES

By 1830 Comte had published the *Course*'s first two volumes, containing his presentation of mathematics and astronomy. The latter was the focus of his efforts in the early 1830s: he delivered courses on popular astronomy at the Athénée, and every Sunday morning at the Mairie of the Second Arrondisement, located between the Palais Royal and the Opera.[44] It was also to astronomy that he turned in his doomed attempt to secure a firm institutional base. Although Comte was renowned as a teacher, the proposal he wrote in 1832 to Louis-Philippe's prime minister, Guizot, asking for a chair at the College of France in the "General history of the positive sciences," came to nothing. As he often complained, the sciences of his time were becoming increasingly specialized and isolated; a general overview like that of the *Course of Positive Philosophy* could not stand on its own as a claim to scientific expertise. Ambitious polytechnicians were encouraged to concentrate on the specialized research programs of their teachers, and Comte did in fact undertake a handful of researches on the pressure of gases and published articles on canal building.[45] Despite these efforts, and the friendship and patronage of the engineer-scientists Hachette, Navier, Poisson, and Lamé, as well as the anatomist Blainville, Comte's refusal to carve out a niche within the established scientific specialties excluded him from the prestige and financial security his classmates were beginning to enjoy in the early 1830s.

Professional advancement in the sciences now required the publication of specialized research.[46] To meet this demand, in 1835 Comte made his first and only appearance in the Academy of Sciences—the body with the power to legitimate scientific authority both in France and worldwide.[47] This work was a mathematical justification of Laplace's nebular hypothesis, entitled "Cosmogonie positive."[48] The nebular hypothesis held that planets and suns were formed out of spinning clouds of very hot gas; as these "nebulae" condensed, they left behind them planets that continued to revolve around the final kernel of condensation, the sun. Joseph Fourier employed the theory in his work on terrestrial heat, as a possible explanation of the increase in heat found as one went deeper into the earth's core; further, the notion of a gradually cooling and condensing earth was widely discussed as a concomitant to theories of gradual transformation in both geology and natural history.[49] In Britain, the nebular hypothesis's vision of a progressive universe was embraced by reformers and oppositional thinkers: if planets had a history that was caused by materially observable forces and progressed by stages, so might organisms; likewise, new social orders might also progressively emerge.

Intriguingly, however, neither Humboldt nor Arago endorsed the hypothesis. This hesitation may appear counter-intuitive, given their embrace of a dynamic conception of nature, support of transcendental anatomy (Serres and Geoffroy), and their advocacy of "progress" in political as well as scientific forums. Equally puzzling may be the fact that it was

Laplace—the theorist of balanced equilibrium and cosmic stability—who put forward this narrative of cosmic development. Yet Laplace presented this hypothesis not as an argument about constant change, but merely as an explanatory preface for the stable, unchanging order of the solar system. Further, one suspects that if the nebular hypothesis had been persuasively advanced by anyone other than Laplace, Arago and his allies would have gladly embraced it, as it was precisely the sort of unifying narrative, with the potential to join together multiple distinct approaches, that both sought in their attempts to popularize the sciences. It is from this perspective that Auguste Comte's presentation of his "Cosmogonie positive" in 1835 at the Academy of Sciences, while Arago was permanent secretary, makes sense—as does its fallout. Arago had previously offered Comte assistance at several turns, attending the *Course*, supporting his appointments at the Ecole Polytechnique, and at one point putting him in touch with a potential English patron; further, in the 1820s and early 1830s Comte was very much part of the scientist-engineer milieu of the Ecole Polytechnique and the CNAM, as well as a proponent of anti-Cuvierian anatomy through his alliance with Blainville. Comte's choice of subject for his debut performance in the Academy drew upon his sense of the foundational status of astronomy, his expertise as a mathematics tutor, and the notion of "progress" as a fundamental principle; for Arago, who set the Academy's agenda, it may also have appeared as a chance to free the progressive and unifying implications of the nebular hypothesis from being associated with Laplace, the empire, and the Restoration.

In his "Cosmogonie positive," Comte calculated the period of rotation of the supposed primary nebular mass at the moment at which its circumference was that of the orbit of each of the planets. He found, for example, that when the surface of the sun would have been where the earth's orbit is now located, the sun would have had a periodic rotation of about 357 days, "which really only differs by about 8 days from our sidereal year."[50] The same relation appeared between this primitive mass and the rotation of each planet at each stage of its condensation. From this coincidence he concluded (in italics in the text): "*supposing the mathematical limit of the solar atmosphere successively extended as far as the diverse planets, the time of the rotation of the sun was, at each of these epochs, noticeably equal to the present sidereal revolution of the corresponding planet.*"[51] He also concluded that this inward condensation was now complete: in the final state of the solar system, the sun was clearly distinguished from its satellites, with which it had been previously merged. Limitation and separation, as in his *Course*, signaled an orderly progress; the argument has also been seen as a version of Comte's insistence on the harmony between the static and the dynamic, because the current order of the planets, their static series, was directly correlated to the series of their formation in time.[52]

Soon after the publication of an account of his report, however, a French mathematician named Person published a refutation, pointing out that Comte's "discovery" was in fact nothing but a restatement of assumptions built into his question. To arrive at his figures for the successive revolutions of the nebular mass at the moment at which it would have left behind each planet, Comte had used equations that had themselves been derived from the periods of the current planets' revolutions. Comte's logic was viciously circular. In the first draft of the paper, Comte announced a second report; on the copy of this report which is found in the archives of the Academy of Sciences is a note in the hand of François Arago: "The report did not take place."[53] It is likely that Person's exposure of his argument's fallacy is what made Comte's follow-up dispensable.

While French commentary on Comte's paper ended there, in Britain, where Person's response was unknown, Comte's "Positive Cosmogony" was a smash hit. Many British scientists, especially those with Jacobin sympathies, looked with envy at French science and mathematics. The numerical confirmation proposed by Comte, a former student of the illustrious Ecole Polytechnique, added considerable authority to the hypothesis.[54] Its confirmation was also sought through visual observation of nebulae, with increasingly powerful telescopes like that of Lord Rosse in Ireland, either to resolve them into fields of stars or to show that they could not be so resolved.[55] Comte's essay was translated into English; the *Course* received a lengthy commentary in the *Edinburgh Review* by David Brewster, and John Stuart Mill began an admiring correspondence with its author. The ideas of Comte (though spelled "Compte") were even featured in Robert Chambers's anonymously published *Vestiges of Natural History*, the hugely successful work of popular science that proved crucial in setting the scene for acceptance of Darwin's evolutionary theory.[56] Comte's article may well have influenced Darwin more directly: Darwin was exposed in Edinburgh in the 1830s not only to the transcendental anatomy of Geoffroy Saint-Hilaire, but to Comte's cosmogony; his eventual theory of evolution by natural selection had strong commonalities with both.

In 1844, however, John Herschel virulently attacked Comte and exposed the error in his calculations, both in public at a meeting of the British Association for the Advancement of Science and in private. Herschel walked Mill through his demonstration, after which Mill's relations with Comte cooled.[57] Comte later renounced his own paper as a "vicious concession to the last habits of metaphysical atheism that pursue in their manner questions that the healthy (*saine*) philosophy must finally leave to the side."[58]

Following this "fiasco," Comte found himself progressively excluded from the spaces of established science, in part through his own knack for provoking a "decisive crisis" with powerful opponents, and above all Arago. The cruelest blow came in 1840, when his candidacy for the position at the Ecole Polytechnique vacated by Poisson, for which he had the support of several professors, was rejected in favor of Arago's favorite, the geometer Sturm. Comte began to target Arago as the chief of the "pedantocracy," the mathematicians and mechanists with undue influence and who directed research toward sterile, specialized topics. As perpetual secretary of the Academy, Arago held considerable power over elections and thus over research directions; his *éloges* of departed scientists also made him France's most influential historian of science, a position coveted by Comte; finally, the fact that Arago collected his hugely celebrated lectures at the Observatory under the title "Popular Astronomy"—the name of the much more modestly attended course that Comte gave on Sunday mornings at the Mairie of the second *arrondissement* from 1827 to 1844—may also have spurred Comte's rivalrous rage. With his penchant for burning bridges, Comte published a "Personal Preface" at the beginning of the sixth volume of the *Course*, in which he insulted Arago and accused him of leading a conspiracy to interfere with Comte's publications and block his career. In the margins of a copy of this text in his library, Arago scrawled, "Unbelievable buffoonery!"[59]

MILIEU, MODIFICATION, AND THE TECHNOLOGICAL ANIMAL

This professional disaster preceded Comte's intensified interest in biology and sociology; it was in its wake that he composed the lengthy volumes dealing with these topics in the *Course*. This shift in emphasis also corresponded with a strategic shift in his communications: from his failure to reformulate the mechanics of the heavens for the educated elite, he increasingly put his energies into transforming the organism of society by preaching to untutored plebeians. The justification for these interventions can be found in the strong connection his own theory established between biology, social science, religion, and his technological orientation. If the hierarchy of the sciences was a ranking of increasingly rapid and complex temporalities, Comte's theory of life was a guide for the modification of the external milieu at each of these time scales.

According to Comte, the sciences that dealt with living things represented an enormous increase in complexity over the sciences of the nonliving. As suggested above, this complexity could be understood in part in temporal terms. Biology incorporated multiple temporalities: metabolism, "intermittent" cycles of rest and activity, and the unfolding of developmental phases undergone from conception to death. This increase of complexity in the domain of the living was also reflected in its methods, as another aspect of the ladder of the sciences was its gradual inclusion of new methods of investigation. Astronomy depended on observation; physics added experiment; in chemistry both were present, although expanded—since chemistry made use of observations of smell and taste and a wider variety of experiments—along with a limited use of comparison, the basis of the classification of compounds and elements. Biology multiplied comparisons: between the parts of the same organism, between the sexes, between the different races or varieties of each species, between all of the organisms of the "biological hierarchy," and between the different phases of the animal's development.[60] Sociology added the method of historical comparison.

In his discussions of the classification of animals, Comte integrated Cuvier's notion of "conditions of existence" and its focus on the organism's "life"—the elusive depth of which Foucault wrote as an "empirical transcendence"—with the animal series, as put forward by his ally Henri de Blainville.[61] Although Comte celebrated the works of Xavier Bichat for arguing that biological phenomena required their own methods and concepts (for this, Comte named the month in the Positive Calendar devoted to "Modern Science" after him), he disagreed with Bichat's famous definition of life as "the ensemble of forces which resist death."[62] Instead, Comte radically displaced life from the confines of the animal's body: in his view, life existed in the interaction between an organism and its surrounding environment, or milieu. "Life," Comte said, supposes "not only a being so organized as to support the vital state, but such an arrangement of external influences as will also make it possible. The harmony between the living being and the corresponding milieu obviously characterizes the fundamental condition of life; whereas, on Bichat's supposition, the whole environment of living beings tends to destroy them."[63] Rather than define life as a tragically fated war between an organism and its environment, Comte located life in this very nexus, in the interactions between an organism and its surroundings.

Similarly, while Comte approved of Blainville's definition of life as a "double internal motion, general and continuous, of composition and decomposition," he felt that Blainville had failed to give sufficient attention to the external half of the equation: "It is from the reciprocal action of these two elements [organism and milieu] that all the vital phenomena proceed."[64] Comte did not ignore internal functions; in fact he elaborated a quasi-political concept of organic consensus to describe the internal harmony among them. Yet the overarching research program Comte set for biology was to inscribe the study of organs and their functions, which were so central to Cuvier's natural history, within this double movement between inside and outside: "It immediately follows that the great problem of positive Biology consists in establishing, in the most general and simple manner, a scientific harmony between these two inseparable powers of the vital conflict, and the act which constitutes that conflict: in a word, in connecting, in both a general and special manner, the double idea of organ and milieu with that of function." This dynamic interaction—both continuous and intermittent—between an organism and its environment was a guideline for his construction of the animal series; it was also pivotal for his conception of sociology.[65]

Following Blainville, Comte ranked animals (anatomically) according to the increasing complexity, diversity, and specialization of their organs. This gradient corresponded (physiologically) to "a life more complex and more active, composed of functions more numerous, more varied, and better defined." This life is also more susceptible to modification, just as the sciences were ranked according to the increasing modifiability of their objects. This modifiability, however, corresponds to a greater power of modification over the environment. A simple organism had little ability to alter its environment, yet its needs were satisfied with comparative ease and it was resilient to external changes. At the other extreme of the scale, the complexity of the higher animals (for instance, the more elaborate system of circulation of warm-blooded animals) made them dependent on a wider range of factors in their environment and, accordingly, more vulnerable to small modifications in their milieu. This greater vulnerability was compensated, however, by a greater power to modify their environment—by building nests, dams, or burrows. Climbing the animal series, "the living being becomes, as a necessary consequence, more and more susceptible of modification, at the same time that it exercises a continually more profound and more extensive action on the external world."[66] Both this vulnerability and this power were greatest with humanity. "Man . . . can live only with the help of the most complex ensemble of exterior conditions, atmospherical and terrestrial, under various physical and chemical aspects; but, by an indispensable compensation, he can endure, in all these conditions, much wider differences than inferior organisms could stand, because he has a greater power of reacting on the surrounding system." Like Saint-Simon in the

"Mémoire sur la science de l'homme," Comte placed mankind at the summit of the animal hierarchy because of his preeminent ability to alter his environment.[67]

The supreme rank of humans in the animal series had significant consequences for Comte's conception of sociology. As early as his *Plan of Scientific Works*, he had argued that a true social science "conceives of the goal of this social state as determined by the rank that man occupies in the natural system. . . . It sees, in effect, resulting from this fundamental relation, the constant tendency of man to act on nature, to modify it to his advantage. It considers subsequently the social order as having for its final object the collective development of this natural tendency, to regularize it and to concentrate (*concerter*) it, so that the useful action may be the greatest possible."[68] Social science coordinated human action upon nature, which required a greater integration among individuals, producing a new, collective organism. In mankind Comte observed a tendency "increasingly to transform the species artificially into a single individual, immense and eternal, endowed with a constantly progressive action upon external nature."[69] Just as there was a "consensus" or "harmony" within the individual animal—a coordination of its diverse systems into unified activity—sociology considered humanity as a whole as a collective organism, divided into multiple functions acting in concert upon their shared milieu.

In the science of sociology, a new and temporalizing tool was added to those of observation, experiment, classification, and comparison: the historical method. Human history offered the spectacle of successive "political combinations" and the progressive emergence of each science according to the order sketched above. As a result, history traced humans' increasing knowledge of their milieu at its variable and particular rates and speeds, making it possible for them to intervene and modify it. Our relation to the environment is not just the basis of the organic life of humanity; it is the basis of knowledge. "In a word, every phenomenon supposes a spectator: since the word phenomenon implies a definite relation between an object and a subject." Yet this relation between subject and object went well beyond mere spectatorship. Life depended on exhalation and inhalation, composition and decomposition, of a milieu; in the case of humans, this dependence meant alteration. "The true conception of life is indistinguishable from that of the world. For life incessantly requires the maintenance of a constant harmony, both active and passive, between any organism and its suitable milieu. . . . The most eminent of all beings, Humanity, is that which is most dependent on the World, but also that which modifies it the most. Thus are united at their elementary source the healthy ideas of submission and power, because activity always increases with dependence."[70] The *Course* systematized the different scales through which humans interacted with and modified their milieu. Thus the history of humanity was not just the rationalized narrative of its social forms and ideas; it also recounted the means by which this collective organism came to know its surroundings (in turn astronomically, physically, chemically, and physiologically), grew progressively united, and thereby learned to act with increasing force and concentration upon this environment.

More explicitly and more coherently than Saint-Simon or Leroux, Comte presented a theory of the human—both individually and collectively—as the technological animal. As mentioned above, the "great problem of positive biology" was that of correlating the vital interaction between organism and milieu with the functions of distinct organs and systems. In the case of the collective organism of humanity, this problem became that of coordinating distinct social organs—classes and individuals—with those aspects of the exterior world with which they dealt. Technologies were extensions of humanity's organs, improving and rendering more exact humanity's knowledge of the limits set by its milieu. Every innovative individual and every successfully deployed technology represented an adaptation of the "social organism" to the external world.[71]

This technological conception of the social organism—a direct consequence of Comte's theory of life as the composition and decomposition between organism and milieu—prepared for the decisive turn in Comte's writing in the late 1840s. The *Discourse on the Positive Spirit*, appearing in 1848, the first volume of *System of Positive Polity*, dealt explicitly with the functions of each of humanity's four major organs: the proletariat, the "patriciate" (holders of the temporal power, bankers and industrialists), women, and priests, and the role each played in maintaining "consensus" within the social organism. To direct these organs, Comte introduced a new set of technologies: he called these the Religion of Humanity. His religion was a necessary outgrowth of his biologically grounded history of science and technology.

TECHNOLOGIES OF SELF, SOCIETY, WORLD

Another mental breakdown struck Comte after the publication of the last volume of the *Course*. Thanks to the restorative influence of Clotilde de Vaux, after this point his "second career" began. De Vaux was a young widow, an early feminist, and a novelist of ideas who saw herself as a rival of George Sand; she and Comte shared an intense and unconsummated relationship in the year before her death.[72] Transformed by their exchanges and the tragedy of her disappearance, Comte later deified her as an icon of humanity. The Saint-Simonians had attacked Comte publicly for a lack of feeling, but thanks to Clotilde de Vaux, Comte at last took up the religious and emotional legacy of his former teacher's *New Christianity*.[73]

Comte's exchanges with Clotilde awoke him to two fundamental oversights in the *Course*: its excessive emphasis on the intellect at the expense of emotions and activity and its failure to consider properly or make direct overtures to anyone other than scientists and industrialists—in particular, to women and workers, who were society's heart and muscles. His *Positive Catechism* of 1848 was a conversation between a priest and a woman whom his readers were to imagine as Clotilde herself, presenting the chief outlines and articles of faith of the *System*.[74]

Comte redefined religion not as the worship of supernatural beings but as a linking (*lier*) or "tying together." "Religion," he wrote, "expresses the state of perfect *unity*, which is distinctive of our existence, both individual and social, when all its parts, moral as well as physical, habitually converge towards a common purpose. . . . Religion, then, consists in *regulating* each individual nature, and in *rallying* all the separate individuals; which are but two distinct cases of one problem. For every man, in the successive periods of his life, differs from himself not less than at any one time he differs from others."[75] The new Religion of Humanity was an ongoing "linkage" among individuals, as well as the integration of the momentary states and atomized faculties of any single individual into a unified self, by training its parts to "habitually converge." In his later writings, Comte reproached more violently than ever the "pedantocrats" and "algebrists" of the scientific establishment; he even bemoaned his own education's failure to address human emotions. While we might see his "second career" as a repudiation of the rigid discipline and mathematical emphasis of the Ecole Polytechnique, we can also see this training as a preparation for the religion's continued imperative to systematize and coordinate diverse temporal series: Comte had simply expanded the scope of the *Course* to include social life and the economy of individual minds. Into the diverse but intersecting time series of the natural world, the *System* proposed to integrate the multiplicities of human time: from the passing of weeks, months, and years to the everyday, habitual rhythms of deliberately cultivated habits.

Comte's *System* expanded the slogan of positivism from "Order and

Progress" to "Love for Base, Order for Principle, Progress for Goal." The goal was "the perfect co-ordination of the three essential modes of human existence, collectively or individually regarded": speculative life, affective life, and active life."[76] The doctrine (sociology) established a true order; the worship (sociolatry) used art and ritual to cultivate love; and the regime (sociocracy) guided the ethical and just progress of the collective. Its aim was to bring the fundamental aspects of human life into convergence, forming habits that would draw distinct elements into a shared focus or unity.

Doctrine

Positivism was not a revealed religion but a "demonstrated religion." Whereas the *Course* presented the sciences from the "objective viewpoint," in terms of the logical relations among phenomena, in the *System*, Comte moved to a synthesis according to what he called the "subjective method." This meant putting the intellect in service of the heart and the "social sympathies": "Subjective inspiration must persistently bring [the intelligence] back to its true vocation ... The universe must be studied, not for itself, but for man, or rather for humanity."[77] Scientific knowledge would be ordered according to the needs of humanity. Materials provided by the *Course*'s "objective analysis" would be crafted by means of "subjective analysis" into a new, systematic presentation, no longer proceeding from world to man (from astronomy to sociology), but from man to the world. The priest of humanity would idealize and systematize the doctrine in terms of the object of each science's relationship to human life on the planet. "Laws are necessarily plural," he wrote.[78] "In the midst of this growing diversity, the dogma of Humanity gives to the whole of our real conceptions the only unity they admit, and the only bond we need."[79] The doctrine taught worshippers their place within the universal order.

The "subjective method" also implied a reevaluation of the law of three states. Comte's reflections had led him to realize that fetishism, the earliest intellectual and religious order of humanity, in fact had numerous advantages as the basis of social unity. Belief in the intelligence and loving support of nonhuman nature had powerful moral and social effects. Although the detached relationship to nature implied by the "objective method" of the *Course* was still needed to provide the materials for Comte's later synthesis, fetishism was now seen as the mode of thought best suited to the popular diffusion of the sciences, to showing ordinary people the harmonious relationships among the different phenomena of the world and revealing humans' place within it. Comte's return to the primitive notion of an animated, conscious nature was a strategy of science popularization and moral uplift.

Worship (culte)

If fetishism offered a model for conceiving of the environment in personalized, humanized terms, it also provided means for cultivating socially oriented feelings of love, adoration, and submission. The doctrine was therefore complemented by the worship, or "the cult." While earlier humans had worshipped nature, then multiple gods, then a single god, in the positive religion, humans were to worship—and thereby strengthen—a new divinity, humanity itself, the new "Supreme Being." Yet the Religion of Humanity confronted a problem common to all societies: the tendency toward fracture and separation. The doctrine—the summary of the underlying laws and principles of the sciences—contributed to social unity by teaching the inevitability (or "fatality," a term Comte used increasingly) that all humans shared by being subject to the same laws.

A more direct step toward unity was to work directly upon individuals' minds, to make social virtues predominate over antisocial impulses, to supplant egotism with "altruism," the now-familiar term that Comte coined. The goal of the worship was to modify the relationships among the diverse parts of the human brain, and thereby to reorient the self toward humanitarian ends. Revising Gall's phrenology, Comte identified ten affective motors in the brain: seven personal motors (a new version of the seven deadly sins) and three social motors, corresponding to the three distinct kinds of love: veneration toward superiors, attachment toward equals, and benevolence toward inferiors. The goal was to make the stronger and more numerous selfish motors submit to those that, though worthier, were less numerous and far weaker. The only means to strengthen the altruistic motors was through practice: positive prayers were *psychotechnics*. This mental training took only two hours per day, the amount typically wasted in idleness or "bad reading." One hour in the morning was devoted to invocation of guardian angels; in a brief session during the day, one emptied the mind of all thought; and in a final session before bedtime, the worshipper "saturated" the mind with feelings of gratitude and love.

To strengthen these feelings, Comte also suggested visualizations, in an echo of the spiritual exercises of Ignatius Loyola. Imagining the face, the attitudes, and the physical presence of departed loved ones and embellishing these "subjective beings" until their image took on a living vivacity in the mind, worshippers stimulated an "effusion" or an "outpouring" of altruistic feeling. Depending on the loved one imagined, it might be veneration or benevolence that grew strongest, love for someone in particular or for the Great Being of Humanity, as personified (he suggested) by a child held by a thirty-year-old woman with the face of Clotilde de Vaux. Through these absent intercessors—whose "subjective" existence, Comte argued, was nevertheless very real and powerful—one could train altruistic impulses to dominate egotistical ones. This tendency was suggested by the positivist secret handshake, or "universal sign of recognition": worshippers moved the hand from the back of the head to the front, signifying the dominance of the altruistic impulses (at the back of the brain) over the egotistical ones (at the front).

Comte described the external aspects of the worship in hallucinatory detail. There would be green banners—the color of the earth and of hope—with an image of the goddess on one side and positivism's "noble slogan" on the other, banners that would be placed at the head of "our solemn processions" during seasonal festivals. Positivist temples would also be built. His diagrams detailed their construction, growing larger in each draft. They were placed in the midst of a wood that served as a cemetery. Those who worshipped or studied in the "positivist school" attached to the temple would therefore be constantly surrounded by the segment of society to whom the most was owed: the dead. Of the sacraments he created to mark life transitions, one of the most important was that of "incorporation," a special funeral for those who had helped society the most and were deemed worthy of joining the "living dead" of humanity.

The dependence of the living on the dead was also reinforced by the Positivist Calendar, with its assignment of days, weeks and months to remembered greats who had contributed to the intellectual and social progress of humanity (see fig. 9.1). Comte broke the year into thirteen months of 28 days each; the 365th day was the "Universal Festival of the Dead," and a leap year was celebrated as the "Festival of Women." To this "concrete calendar" he added a second "abstract calendar," in which the months were named for humanity's essential relationships, functions (the classes of priests, workers, women, and the "patriciate"), and stages (fetishism, polytheism, and monotheism) (see fig. 9.3). In the temple, these elements were represented on signs around the central altar. Each god (those for whom a month was named) had its own small shrine, and at the altar stood a representation of the Great Being of Humanity itself, personified by Clotilde holding a small child (see fig. 9.4). The positivist temple incarnated the *System*. It was a machine for traveling in time. After passing through the

CULTE CONCRET DE L'HUMANITÉ, POUR PRÉPARER L'OCCIDENT AU CULTE ABSTRAIT, SEUL DÉFINITIF.			CULTE ABSTRAIT DE L'HUMANITÉ, OU CÉLÉBRATION SYSTÉMATIQUE DE LA SOCIABILITÉ FINALE.			
	TYPES MENSUELS.	TYPES HEBDOMADAIRES.				
Janvier......	MOÏSE .... (la théocratie initiale.)	Numa, Bouddha, Confucius, Mahomet.	Janvier.....	L'HUMANITÉ....	Fêtes hebdomadaires de l'Union......	Occidentale. Nationale. Provinciale. Communale.
Février.......	HOMÈRE ... (la poésie ancienne.)	Eschyle, Phidias, Plaute, Virgile.	Février.....	Le MARIAGE.		
Mars.........	ARISTOTE .. (la philosophie ancienne.)	Thalès, Pythagore, Socrate, Platon.	Mars......	La PATERNITÉ.		
Avril.........	ARCHIMÈDE .. (la science ancienne.)	Hippocrate, Apollonius, Hipparque, Pline l'Ancien.	Avril......	La FILIATION.		
Mai..........	CÉSAR .... (la civilisation militaire.)	Thémistocle, Alexandre, Scipion, Trajan.	Mai.......	La FRATERNITÉ.		
			Juin.......	La DOMESTICITÉ.		
Juin........	SAINT-PAUL ... (le catholicisme.)	Saint-Augustin, Hildebrand, Saint-Bernard, Bossuet.	Juillet......	Le FÉTICHISME.		
Juillet.......	CHARLEMAGNE . (la civilisation féodale.)	Alfred, Godefroi, Innocent III, Saint-Louis.	Août.......	Le POLYTHÉISME.		
			Septembre...	Le MONOTHÉISME.		
Août........	DANTE .... (l'épopée moderne.)	Arioste, Raphaël, Tasse, Milton.	Octobre....	LA FEMME, ou la vie affective.		
Septembre...	GUTTEMBERG .. (l'industrie moderne.)	Colomb, Vaucanson, Watt, Montgolfier.	Novembre...	Le SACERDOCE, ou la vie contemplative.		
Octobre.....	SHAKESPEARE .. (le drame moderne.)	Calderon, Corneille, Molière, Mozart.	Décembre...	Le PROLÉTARIAT, ou la vie active.		
Novembre....	DESCARTES ... (la philosophie moderne.)	Saint-Thomas-d'Aquin, le chancelier Bacon, Leibnitz, Hume.	Final......	L'I.DUSTRIE, ou le pouvoir pratique.	Fêtes hebdomadaires.	Banque. Commerce. Fabrication. Agriculture.
Décembre....	FRÉDÉRIC ... (la politique moderne.)	Louis XI, Guillaume-le-Taciturne, Richelieu, Cromwell.				
Final........	BICHAT .... (la science moderne.)	Galilée, Newton, Lavoisier, Gall.	Jour complémentaire...........		Fête générale des MORTS.	
Jour complémentaire..........		Fête générale des MORTS.	Jour additionnel des années bissextiles.		Fête générale des SAINTES FEMMES.	

FIG. 9.3. Concrete and abstract cults of humanity, from Comte, *Calendrier Positiviste*. The "concrete cult" (shown in detail in fig. 9.1) recaps the positive calendar in its full form and prepares the "abstract cult." The "abstract cult" celebrates sociability. Month by month, each year recapitulates the basic stages of the development of human society, its fundamental relationships, and the four main organs or "functions" of the healthy society: women, priests, workers, and industrialists.

memorial "fields of incorporation" to arrive at its door, one contemplated the major contributors, divisions, and stages of humanity, arriving at last in the holy of holies, at the final stage of humanity and its object of worship: the Supreme Being, humanity itself (see fig. 9.5).

Regime

The Religion of Humanity was an instrument to build a unified body or corps, not unlike the republican *corps d'état* formed at the Ecole Polytechnique. Comte's political vision was less centralized than that of the Saint-Simonians. In the positive state, the governmental apparatus of the nation-state would be no more; instead, social life would be organized around independently administered cities that would form the core of "intendancies"—seventeen of which would cover the present territory of France. Comte also foresaw a much sharper division between the spiritual and the temporal powers: while the Saint-Simonian priests were closely involved in technical questions concerning industry, commerce, and the everyday functioning of large nations, in Comte's plan the only centralized authority joining his medium-scale intendancies was the spiritual power, in the hands of the priests. The priests would advise but not control the temporal power, by means of a new "patriciate" that would control the "nutritive reservoir" of each intendancy; the patriciate was responsible for banking, trade, manufactures, and agriculture. The priests also administered justice, "like ancient theocratic judges," arbitrating disputes and determining punishments, from "moral rebuke," to "ostracism" and, in extreme cases, execution. They also instituted a series of festivals to commemorate both the necessary functions of society and its greatest representatives over time.

FIG. 9.4. Humanity as represented by Clotilde de Vaux and child. Statue in the Positivist Temple, Rio de Janeiro. Photo by Margarete Vöhringer, with thanks to Danton Voltaire Pereira de Souza.

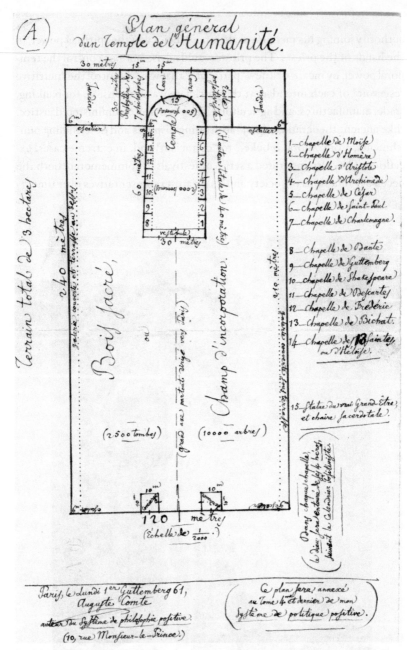

FIG. 9.5. Comte's "Plan A" for the Positivist Temple. The vertical parenthesis at lower right reads: "In each shrine, the god will be surrounded by the four heroes, following the Positivist Calendar." A similar "Plan C" is also preserved in the Maison Auguste Comte, with the main difference one of scale: it is nearly twice as large as "Plan A." The Templo Positivo in Rio de Janeiro faithfully follows Comte's blueprint for the temple, down to the order of statues, the structure of the altar, and the ubiquity of the color green. Archives de la Maison Auguste Comte; thanks to Aurélia Giusti.

* * *

In coordinating these three divisions—dogma, worship, and regime—the Religion of Humanity, spelled out in the *System*, was the technical and practical application of the abstract laws of society discovered in the *Course*. After teaching the regularities and limits of modification of each level of external reality—as guides to further research and industrial development—Comte's attention pivoted inward, to consider the most intimate, complex, particular, and modifiable phenomena known to man, humanity, as both individuals and a collective. In the *System*, sociology took on the responsibility of directing human history, regulating the future progress of this organism. The Religion of Humanity did so by systematizing the temporal unfolding of individual daily life and the collective rites of society, just as the *Course* had coordinated the time series of the chain of sciences and the various temporal sequences each of them studied. Comte's calendars—like his charts of the series of the sciences and the affective motors of the brain—offered constant reminders of that order and the steps through which it would progress.

In his last work, *The Subjective Synthesis*, which appeared in 1856, Comte theorized the emotional foundation of the positive state.[80] The book was a reconceptualization of mathematics in aesthetic and altruistic terms.[81] In its introduction, Comte expanded the reevaluation of fetishism he had presented in the *System*. As a means of strengthening humans' affective attachments, two new objects of worship were introduced to the cult. The "Great Being" of Humanity was now joined by the "Great Fetish," the earth—the ensemble of nonhuman entities, now conceived of as possessing an intelligence and a benevolent intention toward humans—and by the "Great Milieu," the space, "universal fluid," or ether that was the empty but necessary foundation for all phenomena: "the theater, as passive as it is blind, but always benevolent . . . whose sympathetic suppleness facilitates the abstract appreciation of both our hearts and our minds." Following the "subjective method," these fictions—in which humanity, the external world, and the ultimate ground of phenomena and thought were symbolically united—would create a real unity in human thought and activity. By performing external rites for and internal meditation upon the Great Being, the Great Fetish, and the Great Milieu, worshippers developed love, gratitude, and recognition of their inescapable dependence upon all that exceeds mere individuality. Their actions were directed outward, toward others and toward the world.[82]

TOWARD BIOCRACY

Comte's lifelong project of organizing and coordinating the sciences led to the construction of a hybrid being that erased the "metaphysical" opposition between the natural and the artificial. His conception of a unified politics of nature may sound wildly ambitious, with regard to both his own capacities and those of humans in general. Yet repeatedly he warned that "the enormous preponderance of inorganic nature" imposed limits upon humans' control over other organisms and over the earth. "The familiarity and appreciation of these invincible limits will be of great intellectual and even moral importance, for directing our efforts and restraining our pride. In freeing us from chimerical scruples and oppressive terrors, the final regime would make us liable to extravagant projects and mad presumption, were not such tendencies easily checked, as here indicated, by systematic education. But this necessary discipline should never hinder the natural outpouring of wise hopes."[83] Comte never argued for a return to "nature" as a pristine wilderness to be preserved; yet neither did he see nature as a passive material for humans to shape according to their will. A recurrent lesson of positivism was to recognize the firm bounds set by the external world upon human plans.

Comte's sense of nature's limitations, as well as his boundless inventiveness, are conveyed by another curious coinage in the *System*. He gave the name "Biocracy" to the planetary alliance formed between humanity and those other creatures that help the "Great Being" survive and improve its conditions:

> Under the positive system, a lawful cooperation and a just fraternity will establish a solidarity among all the biocratic organs that is appropriate to their places in the shared service of the true Great Being. In a word, Biocracy and Sociocracy [the regime] will be equally ruled by Altruism; whereas during the long period of theological and military training Egotism prevailed. Thus it is that biology, now presented in a systematic form, places us at last at the best viewpoint in which human politics, or rather animal politics, can be considered, as it makes the social regeneration of our species, henceforth destined to govern with dignity all other species, of interest to all other species. . . . In allowing humanity to succeed animality, just as animality succeeds vegetality, the true hierarchy of life is synthetically instituted.[84]

With the concept of biocracy, Comte applied a modified notion of the great chain of being to the practical question of tending to humanity's garden:

humans (ascendant over animal life, just as animal life dominates plant life) are responsible for the care of all life forms on the planet, just as these creatures may be brought into the service of humans. His religion was founded on the twin foundation of each individual's dependence on humanity, and humanity's dependence on its environment, articles of faith that resulted in an explicit biopolitics and an ecopolitcs.[85]

In 1957, the historian of science Alexandre Koyré defined the break between medieval and modern cosmologies in a book entitled *From the Closed World to the Infinite Universe*. Just as Comte sought to end the fragmentation and chaotic oscillations of the modern world by a return (and improvement upon) the medieval social order, so too did his systematic replacement of natural philosophy close the universe again.[86] He presented the order of the sciences as a means of coordinating and technologically aligning the diverse temporalities of the "external spectacle" of the world. This was accomplished by bringing all conceptions back to humanity: by saturating facts with altruistic emotion and ordering them according to social ends.

Comte's positivism—which included an historically oriented epistemology and a formulation of sociology that were two of the most influential products of the age of romantic machines—was both more grandiose and more humble than has been recognized. It presented "the world" as an artfully composed and necessary fiction, one produced in an ongoing way by orchestrating series of phenomena and their specific temporalities—first intellectually (or spiritually), and then practically (or temporally). This philosophy, like that of his fellow Parisian, the young Karl Marx, was not just the description of the world but a bid to change it. But unlike Marx, Comte was extraordinarily specific about the transformations that were required.

Building on the plural, non-reductive sciences he acquired at the Ecole Polytechnique, Comte provided guidelines for the engineering of improvements in the most complex organism of all, humanity. Social science became a technical art. The new priesthood would advise, guide, and regulate the functions of the temporal powers directly engaged in industry, agriculture, banking, and everyday life. This new spiritual power was a power of temporalization, a basis on which to make and coordinate time. Further, Comte made clear the interdependencies that placed limits on technical development. A sane, reasonable, and healthy "progress" would emerge only through submission to the natural order. The demands imposed by that order could be fully recognized and met only upon a base of love—altruistic love for others and for the Great Fetish, the earth. In its final form, the moment of its literal apotheosis, positivism was a user's manual for remodeling the human habitat. Comte's closed world prefigured the more recent ecological vision of *spaceship earth*.

* * *

Comte's was one of the many new universes that sprang from the meeting between the machines of the early industrial age and the diverse cultural currents of romanticism. Comte sought to reset a time that was "out of joint," as did other reformers and visionaries.[87] In ways both troubling and inspiring, they aimed to make the forms of life of the modern world adequate to the demands of history and biology: they sought to be the logical sequitur predicted by the historical and scientific series whose laws they proclaimed.[88]

The last three chapters have largely focused on the common ground shared by these three influential utopian philosophies. The Saint-Simonians, Leroux, and Comte all embraced the romantic arts, urged the reorganization of the sciences, and put forth a conception of technology informed by developmental theories of organs, the milieu, and the animal species. Further, they each argued for the psychological, social, and intellectual necessity of a new religion. The Saint-Simonians advocated pantheism and the providential marriage of humanity and the earth. Leroux nuanced their monism with the conception of a divine and historical virtuality actualizing itself in particulars. And despite Comte's unwavering refusal of metaphysics, the complete form of his philosophy introduced a new pantheon of deified ancestors and fetishistic fictions. All three held that the findings of the natural sciences needed to be organized into a unified dogma that would teach the order of the cosmos and humanity's place within it; they also proposed new rites to implant a common goal and habits of altruism in the minds and bodies of society's members. Thanks both to a reimagination of France's Catholic past and to encounters with "primitive" systems of belief and knowledge, the age of positivism, it turns out, was a religious age.

This embrace of religion was also a response to the central political question inherited from the Enlightenment: how to create a society that preserved cohesion while maintaining individual freedom. Kant had phrased this as the challenge of lifting humanity from its "self-imposed tutelage"; subjects who freely submitted to the rational laws they gave themselves would form a kingdom of ends and would each be "more than a machine." For Schiller, "the perfect symbol of one's own individually asserted freedom as well as of one's respect for the freedom of the other" was an orderly but spontaneous English dance.[89] External forms might harmonize the two halves of what Foucault called the "empirico-transcendental doublet," the internal polarity of matter and transcendence that made human subjects themselves romantic machines; social planning could organize the romantic machine they collectively brought into being at the level of a society.

The social reformers we have discussed built on these late Enlightenment arguments to envision structures that likewise aimed at a situated autonomy, a freedom-in-connection, but each with distinct political and ethical emphases. As Naomi Andrews has argued, romantic socialism wrestled to find a way out of the excessive individualism that was the reigning ideology of the July Monarchy.[90] The Saint-Simonians followed their founder's insistence on reconstituting a new intellectual or "spiritual power" as a means to unite hearts and minds. Yet Enfantin's conception of the priest as a "living law" possessed of absolute (if "loving") authority over a rigorous hierarchy gave unfortunate support to those who have viewed the movement as a forerunner of totalitarianism. Leroux himself broke away from "the family" in protest against its tendency to bend individual will, initiative, and spontaneity to an impersonal whole. His plans for the reconstitution of society were radically egalitarian and democratic, with a flexible, constantly renegotiated local base of production, along with provisions (however sketchily drawn) for coordinating these small-scale communes into a national administration and a "national religion."[91] Despite the fierce debates that divided reformers in the Second Republic, Leroux's views had much in common with both Proudhon's syndicalism and the Fourierists' "association" based on attractive labor. These movements also yielded at times to nativist xenophobia.[92]

Comte may be seen as splitting the difference between the centralized techno-theocracy of the Saint-Simonians and Leroux's devolved "communionism." He divided the spiritual and temporal powers more strictly: his intendancies (each approximately the size of Massachusetts) that would take over the functions of nation-states would administer their temporal affairs independently; international centralization was the province of the spiritual power. The priesthood was responsible for the moral and intellectual development of humanity—the dogma and cult—while the details of its physical well-being would be under the control of local temporal powers. Yet this system, not unlike the Saint-Simonians', relied everywhere upon subordination (defined by Comte as "voluntary"). If this society's members could be considered free, it was not because they possessed self-determination, which Comte dismissed as a metaphysical illusion. Rather, Comte incorporated individuals into the "Great Organism" of humanity in much the same way that he arranged the individual sciences within the series of the sciences: each retained a certain autonomy and irreducibility, even as they were fixed in an inevitable and naturalized hierarchy.

Romantic philosophy is well known for its obsession with the relations between fragments and totalities, the means of incorporating or sublimating parts within wholes, and the effort to balance subjectivity with objectivity; yet commentators have almost exclusively associated this set of issues—at once metaphysical, epistemological, ethical, and political—with reflection on organisms. However, in Paris in the Restoration and the July Monarchy, we have seen these same philosophical obsessions staged again and again in the context of a serious and sustained engagement with machines. The dilemma of "freedom-in-connection," or how to conceive of autonomy for beings necessarily dependent on one another, penetrated every domain of life.[93] It was a tension that was manifest in the mergers we have discussed (some seamless, some fraught) between the determinate order of mechanism and the freedom associated with spirit, mind, or organism. Examples have included Humboldt's cosmic republic of disciplined but spontaneous instruments, Arago and the Ecole Polytechnique's ideal of the citizen-soldier-engineer, physiospiritualism's subtle interplay between body and soul, the fantastic arts' sudden animations and sublimations, transcendental anatomy's dialectic of monstrosity and normality, and even the struggle between diabolical enslavement and divine inspiration waged by the romantic antihero, Robert le diable.

These themes also underwrote the utopian visions that served as the philosophical background for the Revolution of 1848. The Second Republic, which ensued, brought these tensions to the fore. As we will see, its attempts to establish a harmony between classes that had grown increasingly polarized and between demands for equality and the temptations of autocracy were embodied in collective spectacles that featured increasingly volatile images of machines.

10

Conclusion

Afterlives of the Romantic Machine

THE ROAD TO 1848

Throughout this book we have seen how wondrous new devices—electrical apparatus, geophysical instruments, daguerreotypes, musical instruments, stage sets, printing technologies, calendars, and the mother of them all, the steam engine—helped remake Paris in the first half of the nineteenth century. We have seen these machines merging with and extending the capacities of humans—now understood as a species whose perceptions, actions, and technical interventions transformed its milieu and itself. We have seen the vital contributions of romantic machines to the coalescence of a new image of knowledge as the product of active and embodied engagement with the world and to the emergence of a concept of nature as constantly developing and susceptible to technological modifications. And finally, we have examined the roles these devices played in redefining and guiding the progress of society—a collective entity considered as a growing organism whose functions needed to be harmonized to increase both the connection and the autonomy of the individuals that constitute it.

As we saw in part 1, physicists' efforts to identify the relationships among invisible, weightless fluids, using new and newly celebrated machines, were frequently spurred by *Naturphilosophie*-related interests in conversions and identities. The quest for unity or an underlying "absolute" also led scientists, philosophers, and authors to pursue the connections between matter, life, and thought; this quest underwrote epistemologies in which scientific instruments were seen as active, mobile, and at times nearly human mediators between minds and external nature. Scientists' projects of popular science aimed to provide growing audiences with an aesthetic education that would also be a liberation: ideologies of enlightenment and republicanism underwrote the development and diffusion of new machines.

In part 2, we saw how musings on the role of mental faculties, bodily activity, and sense organs in shaping experience—especially when aided by machines—opened a common ground between scientific attempts to see things as they are and the fantastic arts' drive to produce unsettling collective hallucinations. Techniques like the daguerreotype and the diorama embodied this doubling between realism and phantasmagoria; physiospiritualist philosophies explored the dynamic relations between the body and mind and the creative activity that went into even ordinary perception. The creative power of humans was made even more apparent in the National Expositions honoring the technological inventions that were transforming the landscape, the economy, and social relations. These metamorphoses resonated with the fascination in the life sciences for transmutations and monstrous variations, demonstrating both the infinite variability and the underlying unity of animal forms and of nature in general. The automaton and the monster—whether naturally or artificially produced—embodied this power of modification at the level of the individual.

At the level of society, as part 3 demonstrated, new projects of reform tied together the recognition of the transformative power of industry, the discovery of the progressive tendency of nature and history, and the call for a new science of society. Utopian socialists wove an emphasis on labor and technology into projects to reorganize knowledge and rebuild the social order on a more just, more truthful basis. These collectivist projects themselves came in diverse guises, including the large-scale systems engineered by the Saint-Simonians, the egalitarian, craft-based communities urged by Leroux, and the detailed religious technologies of Comte. The visions of these "romantic socialists" of a perfected, resolutely artificial order of humanity and nature were not merely starry-eyed fantasies; they were grounded in deep knowledge and familiarity with cutting-edge science and technology. Reciprocally, all the chapters have shown how themes that have been erroneously associated solely with a nostalgic opposition to machines—aesthetic and organic holism, protean fluids, creative perception, and active imagination—in fact permeated the sciences and the technological ethos and practices of this period. In all these ways, romantic machines were at the heart of projects to know and to remake society and its surrounding milieu.

This constellation of ideas, practices, and techniques provided kindling for the Revolution of 1848. The uprising was spurred in part by injustices and miseries brought about by the emerging industrial regime. Yet the reformers we have studied did not argue—as "hopeless romantics" might have—that what was needed was an end to machines, or a return to a pure and unmodified nature. Rather, these thinkers and activists—and among them we could include the Fourierist Victor Considerant, the "organizer of labor" Louis Blanc, feminists including Flora Tristan, Claire Démar, and Jeanne Deroin, the communist Etienne Cabet, the anarchist Pierre-Joseph Proudhon, and yes, Karl Marx—argued instead for continued inventiveness in the administration of machines, a rethinking of their use and ownership, and careful forethought about their consequences for society and its milieu. Nor were they "soulless mechanists," advocating cold reason and detached calculation as the means to direct nature and society. Instead, the mechanical romantics sought to forge a unity by means of art, science, and technology between human consciousness and the nature from which it emerged. They aimed at the recreation of an organic society and cosmos, a fully human form of life, through mechanical artifice.

What remains is to sketch the unfolding of that revolution and the three-year republic that followed it: a series of events and a process of development in which many of our protagonists, both human and nonhuman, played important roles. Romantic machines were very much on display in those uncertain years; their public representations oscillated between liberatory promise and repressive threat. The coup d'état of 1851 in many ways closed the era of romantic machines, although the questions they raised have remained—and press upon us now with even greater insistence.

FROM ACCIDENTAL REVOLUTION TO TEETERING REPUBLIC

The Revolution of 1848 was a haphazard series of events that seemed to accelerate despite both official opposition and attempts of even republican politicians to impede them. As in 1830, the spark of revolution was struck by the government's attempt to suppress the exchange of ideas. For the previous two decades, the national economy had alternated between growth and crisis with discombobulating rapidity. Strikes and uprisings in Lyon and Paris in response to unemployment and the mechanization of industry came with violent regularity. The worst economic downturn yet had begun in 1846, when a financial crisis was coupled with bad harvests. By 1847 the country was in a depression, with one-third of Parisians unemployed. Popular protests about ostensibly economic matters gave renewed urgency to long-standing demands for electoral reform. The Prime Minister, François Guizot, when responding to those who demanded that the requirement to become an elector be lowered, had glibly replied, "Enrichissez-vous!" These words came back to haunt him in the years of crisis: protestors sought to force the government to hear the voices of those who suffered the most. Yet rather than listening and compromising, the government responded in a brittle manner: in 1847 all political gatherings were deemed illegal. To circumvent this repression, workers' groups and republicans began a campaign of "banquets," which featured food, music, and political speeches among associations of workers.[1]

The key agitators in these movements were speakers whose names had become closely associated with the workers' movement in the 1840s. Ledru-Rollin was a lawyer and an author of books on jurisprudence, a legal defender of arrested participants in workers' uprisings in the 1830s and 1840s. This incendiary speaker had joined with François Arago to oppose a government plan to build fortifications around Paris, which they saw as instruments not of defense but of repression; when even the republican journal *Le national* found Ledru-Rollin's arguments on behalf of labor too extreme, he founded his own journal, *La réforme*, in the early 1840s. As for the reformer Louis Blanc, he had written a text, *History of Ten Years, 1830–1840*, denouncing "bourgeois" interests and their domination of the July Monarchy; in 1839 in the *Revue du progrès*, he published an essay, "The Organization of Labor," that expressed the key ideas that he developed for the next ten years. He argued that the evils faced by workers—and by society in general—came from excessive competition and a lack of coordination between production and consumption (a point that had already been clearly made by the Saint-Simonians in the *Globe* in 1831). To ensure that work was constant and attuned to actual needs, Blanc advocated the establishment of "Social Workshops," combining the features of a cooperative and a union, which would allocate jobs and wages according to individual abilities and productive needs. Ledru-Rollin and Blanc were joined in the banquets of 1847 by the orator Odilon Barrot.

In February 1848, Guizot outlawed the banquets—aiming in particular at a large gathering planned for February 22 in the working-class twelfth *arrondissement*, where Barrot and Blanc were scheduled to speak. Blanc decided to cancel the event; crowds assembled nevertheless and began to march toward government buildings in the seventh *arrondissement*. A huge mass formed in front of the Ministry of Foreign Affairs and was met by armed troops. Soldiers were ordered to attach bayonets, in order to avoid gunfire; but one gun went off, and many more followed: forty-two protestors were killed. Crowds filled the streets and some tipped over omnibuses to form barricades. But rather than break up the barricades, members of the National Guard either did nothing or, in many cases, joined the insurgents. Further, liberal deputies, including Thiers, who had been losing patience with the constitutional monarch Louis-Philippe, rapidly formed an alliance with the republicans and prevented the Parliament from intervening. Guizot submitted his resignation on February 23, but it was too little, too late. The Palace of the Tuileries was stormed; the king abdicated and fled the country. On February 26 the republic was declared and a provisional government announced, consisting of members of the republican and liberal opposition. Among its members were Blanc, Barrot, and Ledru-Rollin, as well as the romantic poet Lamartine and the physicist and astronomer François Arago.

This change of regime did not, of course, bring an end to the social agitation and experimentation. New political clubs started, and more than four hundred new newspapers were launched.[2] In the months after the Revolution, according to Baudelaire, "utopias sprung up like Spanish castles."[3] Radical authors who had been long excluded from official institutions now entered the government: Leroux, Considerant, Blanc, and Proudhon all obtained seats in the Assembly. Outside the realm of formal politics, an intellectual insurgency was brewing; the list of names Jules Michelet put forward for membership in the revitalized Academy of Political and Moral Sciences was a Who's Who of left-leaning intellectuals, many of whom were in the circle of the *Encyclopédie nouvelle*, including Louis Blanc, Edgar Quinet, Jean Reynaud, and Pierre Leroux.[4] Working together, and sometimes at odds, republican and socialist reformers pushed for a more complete realization of the utopian promises that had fed the Revolution. As suggested by Flaubert's satirical descriptions of these events in *The Sentimental Education*, and by the caricature "The Ideas Fair" from 1848, the marketplace for utopian projects was laughably crowded (see fig. 10.1). This prompted an anxious response from property owners and those who had benefited under the previous regime. As utopian proposals multiplied, growing numbers of the urban middle class, provincial elites, and the conservative countryside sought a return to a more manageable, traditional, stable order.

The immediate goals of the provisional government were universal suffrage and relief for the unemployed. The first ambition had a bitter outcome: the election in April 1848 of a National Assembly dominated by conservatives who had campaigned for "order." As for the second goal, the government established the Luxembourg Commission to oversee the organization of labor—in this way a Saint-Simonian slogan became a mainstream political platform. Blanc was put in charge of the National Workshops, coordinated from the Luxembourg Palace, but he was given at best half-hearted support by the state. Logistical and financial difficulties undermined the workshop's guarantee of work and minimum financial assistance. To cope with the difficulties faced by the unwieldy structure of the provisional government, the executive branch was consolidated on May 6 around a five-person Executive Power Commission, consisting of

FIG. 10.1. Cartoon of "The Ideas Fair" by Bertall. The hawker in the foreground of the image says, "Take advantage of it, these farces won't last long . . ." Shown here are (*from right to left*) Considerant, Proudhon, someone advertising a dental utopia, Louis Blanc, Leroux (in wild hair and printer's blouse, accompanied by George Sand holding a lyre), and the archetypal con artist Robert Macaire; at far left is the "Icarian" communist Cabet with his "Cabêtisorama." Bibliothèque Nationale de France, with thanks to the University of Pennsylvania Image Collection.

Arago, Lamartine, Garnier-Pages, Pierre Marie, and Ledru-Rollin. Arago also briefly served as prime minister, becoming the most powerful man in the government. Through the advocacy of Victor Schoechler and the support of Arago, slavery was banned in the colonies, an event still celebrated in overseas departments.

The implementation of domestic policies was less successful, however. The chaos and financial instability of the National Workshops led to their official closure on June 24. Thousands of unemployed and hungry workers who had flocked to Paris after being promised a job or financial support were suddenly left without options, and riots ensued. The Executive Power Commission granted emergency powers to the republican general Cavaignac, who had tested his mettle on battlefields in Algeria. Unlike the feeble resistance offered by the army and the National Guard in February, the suppression led by Cavaignac in June was swift and brutal. Thousands were killed outright, and still more were imprisoned under horrendous conditions below the ground near the Hotel de Ville. The uprising was suppressed in three inglorious days, with some of the most violent exchanges taking place outside the Pantheon (see fig. 10.2). The tenuous bond between the government's liberal bourgeoisie and the workers on whose behalf they claimed to lead had been broken.

Over the course of 1848, nationwide attitudes grew ever more conservative, especially outside of Paris, where the issue of protection for industrial workers seemed a very remote concern. Another election was held in the fall, this time for a national president, with General Cavaignac running against Louis-Napoleon Bonaparte, the nephew of the former emperor. This new Bonaparte's brilliant publicity and seemingly providential name allowed him to win in a landslide. The three remaining years of the republic were contentious and jagged. Although all sides invoked the ideals of harmony and order, these years were anything but harmonious or orderly.

During the Second Republic, which lasted from 1848 until Bonaparte's coup d'état in December 1851, the clashing and often contradictory tendencies we have seen in previous chapters were amplified. Throughout the Restoration and the July Monarchy, we saw machines and mechanism

FIG. 10.2. *Taking of the Pantheon, June 24, 1848*, by Nicolas Gabe. The battle, portrayed as relatively placid, is presented from the point of view of the (ultimately victorious) forces of order; no revolutionaries are shown. Musée de la Ville de Paris, Musée Carnavalet, Paris; Bridgeman-Giraudon; Art Resource, New York, NY.

playing a double role: in many cases they represented (and enacted) familiar schemes of restriction and repression, but at other times, as I have emphasized, they offered hope for freedom, new social forms, and a fuller expression of natural and human capacities. In the Second Republic, the meanings of machines were even more polarized: they could offer testimony for the inevitable progress of republican ideals of liberty, equality, and brotherhood. But they also appeared as instruments of dictatorial control.

In what follows, we briefly consider the paradoxes of three new romantic machines of the Second Republic: a lighting device used in an opera by Meyerbeer that opened in 1849; the collection of machines in the last of the National Expositions; and the triumph of popular science, Foucault's pendulum. My goal here is to show how the romantic machines that contributed to the re-imaginings of nature and society in the previous decades continued, in the Second Republic, to be invested with liberatory potential; yet at the same time, spectacles of world-changing machines prompted darker imaginings, anticipating the ways in which technology might be used—under subsequent regimes—as a tool of repression. Frequent oscillations in the way machines were viewed were a defining trait of the Second Republic, a political order that was a fraught intermission. In some ways it fed the hopes of the reformers of previous decades, while at the same time it prepared the way for the Second Empire.

ELECTRIFYING THE SUN

The Second Empire was a regime characterized by theatricality, as suggested by Marx's famous description of Louis-Napoleon Bonaparte's repetition, as farce, of his uncle's tragic coup d'état.[5] The techniques of stagecraft and illusion developed in this period were undeniably precursors for what was later called by Guy Debord and others "the society of the spectacle"—a visual variation on Adorno and Horkheimer's "culture industry." For both the situationists and the Frankfurt school, mass entertainment, particularly in the form of the cinema, was a means of manipulating audiences, offering ideological reinforcement of an exploitative status quo in the guise of harmless diversion. Yet, as I have argued, quite different ends were attributed to the new devices for producing collective sensations and immersive environments that emerged in the romantic era. Balzac read Meyerbeer's *Robert le diable* as a machine of collective transcendence—a technological Eucharist. The Saint-Simonians celebrated the "play" involved in ritual as a means of educating both the intellect and the emotions. Humboldt and

Arago championed innovations such as the panorama and early photography in their groundbreaking campaigns of popular science, with the ultimate aim of using aesthetic experience as a way to educate the public and increase citizens' participation in government. What's more, Comte's view that science sought to create an artificial spectacle of the world through the careful assembly of phenomena undermined the very distinction between truth and "phantasmagoria" upon which many critics of spectacle have relied. The political and epistemological ambivalence of romantic-era "techn-aesthetics," to return to the term coined by Ampère, was heightened by the close connections between innovators in the sciences and the arts.

The roles of scientist, popularizer, and showman were combined by Arago's protégé Léon Foucault, who first made his name as the author of the popular "scientific *feuilleton*" in the *Journal des débats*. Foucault's predecessor was the physician and science writer Alfred Donné, whom he had assisted in an earlier collaboration, the first atlas of microscopic daguerreotypes, which catalogued hidden organic structures and the composition of bodily fluids.[6] One of Foucault's contributions was to create a more reliable light source than the sun. To this end, he designed an arc lamp, in which electrified filaments of carbon were briefly touched together, generating a spark, then pulled apart, to produce a glowing arc. The problem with this construction was that as the carbon rods burned down, the space between them grew, diminishing the intensity of the lamp. To maintain this distance, and the brightness of the lamp, Foucault created a self-regulating mechanism: as the rods grew shorter, a system of triggers and springs drew them closer together (see fig. 10.3).

The steady, brilliant light of Foucault's arc lamp produced images remarkable for their sharpness and detail. The atlas caught the attention of Arago, whose failing eyesight made him keen to find an assistant. He encouraged Foucault, along with his classmate from medical school, Hippolyte Fizeau, to take the first solar daguerreotype in 1845. He then put them to work on his plan to measure the speed of light, an experimental setup that depended on the unprecedented precision of contemporary instrument design. It made use of a mirror spinning eight hundred times per second, powered by a small steam engine; a beam was passed through the teeth of small gears and reflected off the mirror many times over a small space.[7] In this way Foucault and Fizeau arrived at the most reliable measure yet of the speed of light. They were also able to show that light traveled faster through air than through water, a finding that lent support to the wave theory of light. The extremely delicate and regular mechanism of

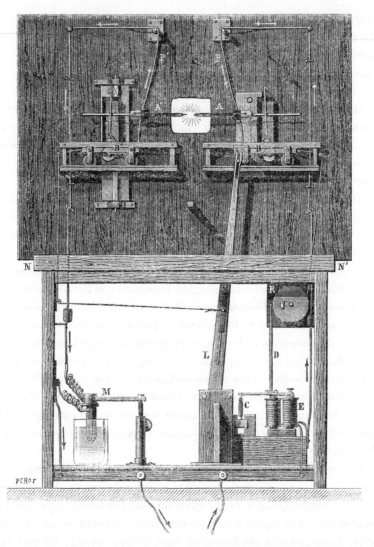

FIG. 10.3. Foucault's self-regulating electric arc lamp as used for micro-daguerreotypes and stage lighting. In Léon Foucault, *Recueil des travaux scientifiques publié par Mme. veuve Foucault sa mère, mis en ordre pas C.M. Gabriel, et précédé d'une notice sur les oeuvres de L. Foucault, par J. Bertrand* (Paris: Gauthier, 1878). Courtesy of Rare Book and Manuscript Library, University of Pennsylvania.

their device also caught the attention of the precision-mad composer Berlioz, who wrote to Foucault for technical assistance in his performances.[8]

Foucault was enlisted in another musical adventure—this time in the immediate wake of the Revolution—when Meyerbeer's long-awaited opera *Le prophète* debuted in 1849 (see fig. 10.4). The opera dramatized the takeover of the city of Munster in 1534 by the Anabaptist Jan van Leyden (called Jean in the opera), who set up a theocracy that enforced collective ownership of wealth; van Leyden was executed eighteen months later. In the Parisian opera's ever-escalating arms race of spectacular effects, Meyerbeer had once again outdone all rivals. *Le prophète*'s orchestration was as rich, dense, and full of unstable contrasts and reworked allusions as that of his previous works. Just as remarkable were its visual attractions. The flames of hell were "a sublime horror": "spectators shuddered and all looked behind them to see if the doors were open, if they could escape in time." According to the fantastic author and critic Théophile Gautier, "Perhaps never has the art of decoration been taken farther: it is no longer painting; it is reality itself."[9] Critics also admired the fourth act's ice-skating ballet, made possible with roller skates.

The most brilliant effect, however, came at end of the third act. It consisted of a sunrise simulated by the self-regulating electric lamp invented by Léon Foucault and, for the first weeks of the performance, personally operated by him. Critics raved that it was "a dazzling dawn with a sun, a veritable sun at which no one can look directly," one that "inundat[ed] the theater with a light so bright that the actors [were] reduced to shadows."[10] Having created the first portrait of the sun drawn with its own light using the

FIG. 10.4. Meyerbeer's *Le prophète*. Sketch of the decor for act 4, scene 2, from *Le prophète*. Bibliothèque Nationale de France.

daguerreotype, and having measured the speed of light, Foucault had now created, by means of electricity, the sun's artificial replacement: a machine that could repurpose electricity and light toward an enchanting effect.

Many critics saw the events of 1534 staged by Meyerbeer as an allegory of the religiously informed uprising of 1848; indeed, one of the opera's sources may have been an essay on the Munster rebellion by the prophetically inclined historian Jules Michelet.[11] Yet the staging and the script conveyed considerable ambiguity. The visionary protagonist originally rose to power as the dupe of a trio of conspirators who took advantage of his charisma and prophetic dreams to transform him into a demagogic figurehead. Yet as Jean came to believe their propaganda, he began to rule despotically. Only after tense confrontations with his devoted mother and his soprano girlfriend was he led to an act of contrition. In the end, he perished in a fire along with the Anabaptists.[12]

Meyerbeer's earlier music had been received, in Heine's phrase, as the "sound of the masses."[13] In 1849, by contrast, some critics saw *Le prophète* as an indictment of the despotic inclinations of the revolutionaries of 1848. Was Meyerbeer's work a reactionary mockery of visions of social metamorphosis, or an earnest attempt to defend laudable ideals against manipulation? This uncertainty paralleled the ambivalent status of the science and technology put to work by Foucault—following the lead of Arago and Humboldt—in entertaining the public. Was Foucault's electric sun, the direct precursor of today's electrified mass spectacles, an enlightening affirmation of rational mastery of the hidden powers of nature, or was it merely an instrument of illusion, as deceptive as it was dazzling?

LABOR HONORED AND REPLACED

We have seen how the Expositions of the Products of National Industry, held in Paris regularly from 1794, were moments when a large public was both informed and entertained by the spectacle of technological progress. For the statistician and engineer Charles Dupin, their lead organizer, the expositions showed the close connection between industrial progress and social metamorphosis. They were also celebrations of patriotism and entries in the rivalry between nations, provoking an answer in London's Crystal Palace of 1851.[14] Paradoxically, the exposition of 1849, the last of the national expositions, coming directly on the heels of the Revolution, was perhaps also the last in which France's revolutionary ideals allowed it to stand in for universal humanity. In the subsequent "International" and "Universal" exhibitions, France was one nation among others.[15]

In the exposition of 1849, productive forces of the nation were again put on display, with church, science, and state represented at the prize ceremony by the participation of the archbishop of Paris; the president of the jury, Charles Dupin; and the president of France, Louis-Napoleon Bonaparte. Their speeches all echoed the giant banners at either end of the hall that proclaimed "HONOR TO LABOR." The archbishop's opening sermon strained to link the church to the resurrected republic, reminding the audience of Christianity's original charity, long before the Saint-Simonians, toward "the poorest and most numerous class." Just as Arago, another member of the exposition's jury, had emphasized the artisanal origins of great inventors, the archbishop pointed out that Jesus and the apostles had been laborers.[16]

Dupin's speech celebrated what he saw as five years of unprecedented progress in the arts and sciences since 1844, thereby glossing over any rupture that might have been caused by the Revolution of 1848. As part of his insistence on continuity, he linked the origins of religion and the domestication of animals to the industrial employment of active powers. "In the first ages of the world, grateful mortals erected altars to the inventors of the means of adding living forces to human labor, by taming animals. Today, we are content to honor the memory of the men who teach us how to tame, and I would almost dare to say, to grant intelligence and life" to electricity, heat, steam, and gas.[17] This link between animal power and the power of the quasi-vital and intelligent fluids driving modern industry was in line with one of this exposition's novelties. For the first time, the products of art and industry were accompanied by the products of agriculture, implying a common interest between Paris and the rest of the still largely agricultural nation.[18] Putting animals and plants under the same roof as machines and products of mechanical industry highlighted the common denominator of productive force underlying all these activities, and it may well have been an attempt to bridge the growing divide between Paris and rural France—a divide that had at first benefited Bonaparte. While the National Workshops of 1848 had failed, Dupin presented the exposition as a microcosmic tableau of the nation as a workshop, with both inorganic and organic forces contributing to an endless saga of unity and growth.[19]

In contrast to Dupin's utopian vision of permanent growth supported by the state, President Bonaparte offered a stern reality check: "Perhaps the greatest danger of modern times comes from this false opinion . . . that it is of the essence of any government to respond to all demands, to remedy all evils. Improvements cannot be improvised; they are born from those that precede them; like the human species, they have a descent that allows us to measure the extent of possible progress and to separate it from utopias. Thus do not let us give birth to false hopes, but hold ourselves to accomplishing those that it is reasonable to accept."[20] The president set limits on populist hopes for the Second Republic: utopianism, he suggested, led to dangerous and false opinions about what government can do. His reign would embody reasonable compromise after the idealism of the preceding year.

Yet the expositions still played on hope and wonder. One of the great successes of 1849 was the "Arithmaurel," a calculating engine built by two young inventors from the provinces who claimed to know nothing about the works of Charles Babbage. This small box with handles and gears could add, subtract, multiply, and divide "with an incredible rapidity and a magical precision" (see fig. 10.5).[21] A pamphlet about the device quoted Arago's praise of the invention and reported an overheard conversation:

> A grand lady exclaimed: "Those emanations of human intelligence designated by the prosaic names of *watch* and *clock* are beautiful things. But in a watch and in a clock, it is as if human intelligence is condemned to forced labor for eternity; it is always the same movement, executed always in the same manner. In the arithmetic machine, on the contrary, the mind of man remains what it is essentially: a sublime Proteus which escapes all limits, and surprises us endlessly with new and powerful feats."
>
> "Well then, my dear," replied her witty neighbor (*voisine spirituelle*), "a watch is *instinct*, the calculating machine is *spirit* or rather *genius*."[22]

Much as Babbage's calculating engine had been received—by David Brewster and others—as a machine of a new order, one whose incarnation of thought left other machines in the past, these great ladies saw the "Arithmaurel" as more than an ordinary or classical machine. Instead of the instinctual mindlessness of repetitive labor, the calculating machine imitated—and thus somehow preserved—the essentially human or spiritual, the genius that defies rules.[23]

A device capable of thought can be read as the culmination of attributing living powers to material. This romantic machine, like Ampère's science of "La cybernétique," was a harbinger of things to come. Yet did this device, like its precursors, including the Jacquard loom, actualize and enhance the capacities that define humans, or did it automate them, discipline them, and take their place? The exposition celebrated ingenuity and labor, insisting on the harmony between church and government, science and workers,

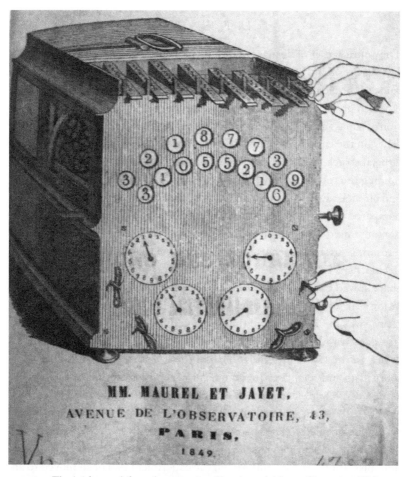

FIG. 10.5. The Arithmaurel, from the promotional brochure *Arithmaurel inventé par MM. Maurel et Jayet, rapport à l'Académie et opinions des journaux sur l'arithmaurel* (Lille: L. Lefort, 1849).

urban centers and the countryside. Yet the current president (and future emperor) circumscribed the utopian aspirations that these transformative machines had previously encouraged. Despite the ceremonial honors given to work and workers, did the spectacular display of these industrial inventions increase workers' power and status, or instead merely offer a reassuring and distracting display? The calls made by reformers of the previous decades for unity and machine-assisted liberation were hardening into a set of propositions about "honor to labor" and "progress through machines" that would be effective as a cover for newly leveraged inequalities.

A PENDULUM'S SWINGS

The political uses of displays of machinery and of popular science were not lost on President Bonaparte. In the last year of the Second Republic, he came to the aid of Léon Foucault's plan, once again, to organize a spectacle in which precision technology would harness cosmic forces for the edification of a mass audience. While working on the question of the effect of the ether on the earth's movement, Foucault had observed that a bar held in a lathe, if struck, would vibrate in the same plane whether or not the "chuck" holding the rod was spinning—an effect previously noted by the musician and physicist Ernst Chladni.[24] That the inertia of the vibrating rod was not hampered by any force suggested to Foucault that the same effect could be demonstrated by a pendulum swinging in a fixed plane suspended above a rotating surface—such as the earth. In January 1851, in the basement of his mother's house, he suspended a two-meter pendulum from a bracket that allowed the wire to pivot in all directions. Watching its plane of oscillation as it seemed to shift slowly in a clockwise direction, Foucault argued that it was in fact the plane that remained fixed, while he saw the earth itself slowly rotate beneath it. For a scaled-up performance, Arago gave him access to the vaulted Meridian Hall of the Observatory. Foucault invited journalists and scientists "to come and watch the rotation of the earth," using a pendulum eleven meters long. Word of this striking display reached President Bonaparte, who had a personal interest in scientific experimentation; he decided to lend his support.[25] Foucault reported, "At lightning speed the President's high influence flashed to the uppermost rungs of the administration"; and by mid-March the experiment had been moved to an even grander location, the Pantheon, with a wire sixty-seven meters long (see fig. 10.6).

An interesting aspect of this "experiment"—perhaps the most famous instance of public science of the nineteenth century—was that it didn't really prove much. By 1851, very few members of the public, let alone scientists, had any doubt that the earth rotated.[26] While the event did unleash a storm of mathematical proofs and explanations for Foucault's "effect," its success was due instead to its powerful effect on its audiences. Here, for the first time, was a full-bodied, immediate experience of a central article of scientific faith. The effect was heightened by its staging. The decision to locate it in the Pantheon was an overdetermined choice. The building is situated on one of the highest points in Paris, a stone's throw from the Observatory, the Sorbonne, and the Ecole Polytechnique. Its meaning has oscillated as well. Built as a chapel in the eighteenth century, it was transformed into a

FIG. 10.6. Foucault's pendulum, in *L'Illustration*, April 5, 1851. Courtesy of Rare Book and Manuscript Library, University of Pennsylvania.

national temple during the Revolution. Napoleon made it a cathedral; it was closed during the July Monarchy, but during the Revolution of 1848, some of the bloodiest fighting of the June Days took place on Rue Soufflot, just outside, and several revolutionaries occupied the building. During these standoffs, François Arago, as a member of the Executive Power Commission, had pleaded in person with the armed workers for a peaceful resolution; despite recognizing him as a long-standing ally, they rebuffed his overtures with the cry "But you have never been hungry, M. Arago. You have never known misery."[27]

Placing the demonstration in the Pantheon has also been seen as a belated rejoinder to the Catholic authorities who forced Galileo to deny the movement of the earth. It could also be seen as a continuation of Arago's republican popularization of science, reaching an audience of all classes: the drawing in *L'illustration* (fig. 10.6) showed workers' caps next to bourgeois top hats. Along these lines, the experiment was a state-sanctioned ritual to affirm, after the violence of the Revolution, the unity of the new republic under the power of reason, science, and technology.

For Bonaparte, a president criticized equally by monarchists and republicans, this scientific display in a temple of national glory was seized as a publicity coup of the highest order. Bonaparte had himself written, in *The Napoleonic Idea*, "The influence of a great human genius, similar to the influence of divinity, is a fluid that spreads like electricity, elating the imagination, making hearts palpitate, as it touches the soul before it persuades the mind."[28] Far from ruling out what Etienne Arago (François's playwright brother and the mayor of Paris after 1848) had warily called "imperial fetishism," Foucault's demonstration was a means of putting the quasi-electric fluid of "the Napoleonic idea" to work.[29] Foucault explained: "The plane of oscillation of the pendulum is not a material object. It does not belong to the support, or to the table, or to the circle. It belongs to space—to absolute space."[30] The fact that this immaterial and absolute space was made visible on such a grand scale only through the intervention of the president was widely reported. Louis-Napoleon Bonaparte had every interest in associating his name with Foucault's demonstration, since it showed the power of state-supported science and technology *to produce the absolute*—to harness and make experiential the order and grandeur of the cosmos.

Unlike Meyerbeer's spectacles, Foucault's pendulum experiment was not a symphony of blaring saxophones, pounding tympani, or haunting masses of sound. Nor did its audience greet it with thunderous applause. On the contrary, Foucault described the pendulum's effect as simultaneously pacific and overpowering: "The phenomenon develops calmly, it is fated, irresistible. . . . One feels, in seeing it born and growing, that it is not in the power of the experimenter either to hasten or hinder the manifestation. Any man placed in the presence of this fact remains for some thoughtful and silent moments, and generally he draws away, carrying with him a more pressing and more lively sentiment of our incessant mobility in space."[31] Whereas the Parisian "opera machine" had produced harmonies and cheers, Foucault's machine created a striking absence of sound: a resonant silence. It was a wordless poem justifying the works of the universe to man. Under the dome of the Pantheon, Foucault had crafted such perfectly controlled conditions that even he lacked the power "to hasten or hinder" its outcome. Human ingenuity had opened the door to the more-than-human sublime—greeted, appropriately, with silence.[32]

As if by fate, within nine months, after the coup d'état of 1851, President Bonaparte had himself declared Emperor Napoleon III. For Baudelaire, this act was above all a *coup de théâtre*, proving that "the first to arrive, by taking over the telegraph and national press, can govern a large nation."[33] Indeed, throughout the Second Empire, Napoleon III relied heavily on technologically astute theatrics to project and secure his authority. Sciences and technologies that had been deployed in the 1830s by Arago and other mechanical romantics to challenge the monarchy were deployed (and decried) in the Second Empire as tools of deception and domination. The demonstration of Foucault's pendulum can thus be read as a pivotal moment in this retrograde movement. On the upswing it displayed the power of reason over tradition and arbitrary rule, offering a challenge to the traditional alliance between religious superstition and monarchy. But then, when the lighting shifted and the backswing began, the demonstration could be read as an affirmation of the power of the absolute state and the necessary submission of subjects to a higher power.

In themselves, of course, machines neither liberate nor enslave. Yet their manifest powers can be yoked to many kinds of projects at different times and under different circumstances. Under France's succession of regimes in the first half of the nineteenth century, machines oscillated between two modes of action. Even while the injustices and brutalities of the new industrial order were being recognized by reformers and challenged by workers in strikes and uprisings, mechanical techniques of production, observation, transport, and communication were placed at the center of visions of liberation. These doubled machines were themselves the mechanical reflection of the workers, who claimed a greater stake in government on the basis of their labor but at the same time welcomed the promise of liberation from that labor. These paradoxical demands primed the explosion of utopian hopes in 1848, even as they eased the transition into the streamlined authoritarianism of the Second Empire.[34]

THE *APRÈS-COUP*

Thwarted by both republican and conservative factions, President Louis-Napoleon decided to make things easier on himself. On December 2, 1851, he occupied Paris with his own troops, locked up the opposition, shut down the National Assembly, and quickly seized control of the government, declaring himself emperor. According to one historical account of the subsequent events—Marx's *18th Brumaire*—the coup proved the futility and naïveté of the projects of reform and transformation in the 1830s and 1840s. It does not require a Baudelaire to point out that the Second Empire solidified the dominance of the bourgeoisie: its hypocritical morals, its sentimental taste, its commercial values. But this is not to suggest that the projects of the mechanical romantics came to a halt, nor that we should dismiss their previous efforts as benighted. Such projects continued, albeit with significant inflections, under the Second Empire. They included, not least, widespread advocacy of the notion that industrialization ushered in a new social order, requiring a new infrastructure; that engineers' models of networks, flow, and maximum rates of distribution of energy were essential to creating that order; and that some form of the Saint-Simonian vision of a mixture of state and private support for large-scale industry was crucial. Yet at the same time, romantic machines could become instruments of oppression and distraction. Most dramatically, Napoleon III's adviser Baron von Haussmann transformed Paris into a divided space of spectacle in which injustices and misery were largely hidden from view.[35]

Some of the actors we have met in early chapters were, like Balzac's donkey-zebra, able to change their stripes and find a place in the new taxonomical order. Others were not. Released from prison, Enfantin shaved his beard and continued to advocate on behalf of the development of the railroad; he became a successful adviser for plans of industrialization throughout the Mediterranean, especially Algeria. Already in the 1840s, Michel Chevalier had turned himself into a respected advocate of free trade and the development of France's manufactures; in the 1850s he became one of the emperor's chief advisers. Charles Dupin, with characteristic dexterity, was able to move from energetic supporter of the republic to distinguished defender of the empire. Arago died in 1852, blind, but with his dignity intact: Napoleon III exempted the great republican from having to swear an oath of loyalty to the emperor, and his funeral procession was a giant, citywide event in which republican and revolutionary values were commemorated and even revived.[36]

Most of the rest of those whom Michelet had in 1848 named for admission to the Academy of Moral Sciences wound up in prison or in exile, though not Michelet himself. Victor Hugo famously denounced *Napoleon le petit* from exile in Belgium and across the English Channel; his caustic comments about the emperor aptly sum up his reign. Eventually Hugo found himself on the Isle of Jersey, where his neighbor was the exiled Pierre Leroux. The two shared philosophical and poetic visions: Leroux expounded on his notion of the "circulus" as the best means of employing the human body's natural metabolism for the improvement of society, and Hugo was increasingly drawn to spiritualist circles while composing his titanic steam-fantasy *Les travailleurs de la mer*.

Auguste Comte did his best to win the ear of the emperor; after all, in his view, a dictatorship was the most efficient government for hastening the transition to the positive state. The version of positivism culled from Comte's *Course* by his one-time disciple the lexicographer Emile Littré became one of the leading philosophies of the Second Empire and the Third Republic, guiding the restructuring of the educational system and preparing for the admission of social science into the university. Karl Marx, having registered the increased prestige of experimental and exact sciences — and their claims to take over all domains of inquiry — shifted his attention from his romantic and philosophical musings on the "species being" and the "externalization of the senses" and took up the scientistic analysis that resulted in *Capital*. The conservative astronomer Leverrier, presented as a new Newton for predicting the location of Neptune (after following up on Arago's suggestions), replaced Arago at the Observatory, purging it of republicans and instituting a regime of oppressive discipline. At the same time, painting and literature increasingly assumed a new, restricted place in what Rancière calls the "distribution of the sensible": the arts were frequently understood either as mere entertainment or, in avant-garde expressions, as the inscription of subjectivity in which both form and content could be evacuated of their political or moral consequences.[37] Under the Second Empire, the state began to invest on a vast scale in technological research, especially for military purposes that would serve the ends of empire, conquest, and expansion — tendencies that have continued, with various adjustments, into the twentieth and twenty-first centuries.

The period opened up by the reactionary governments of the 1850s has been called "high modernity"; it has been defined by a polarization between subjects and objects, values and facts, arts and sciences. Those times are still said in important respects to be with us. And certainly we do not currently inhabit the unified worlds dreamed up by the mechanical romantics. Nevertheless, as this book has shown, the roots of many of high modernity's central traits — positivism, the global mission of industry and commerce, the notion of an isolated avant-garde, and the fascination for technologically produced phantasmagoric mass spectacles—can be traced back to projects designed to overcome these divides. And it may well be the case that the era of high modernity is now coming to a close.[38]

MODERNITIES RETRIEVED

In the early years of the twenty-first century, there are indications that the architecture of high modernity may be undergoing fundamental shifts, due both to its internal instabilities and to a pressing recognition of the impact of human activity on the earth. The central tenet of the mechanical romantics—that nature and human society are inextricably linked and that technology transforms us and our milieu, for good or for ill—is finding new echoes, as warnings about the dangers of the unchecked pursuit of economic growth and technological mastery grow more urgent.[39] Scholars and prophets from a range of perspectives argue that unless we see nature as part of the social fabric—and not as a remote wilderness opposed to humanity, a lost paradise, a hostile alien force, or a passive "standing reserve"—we may well destroy the very conditions of our existence.[40]

This book has shown that within romanticism—a cultural movement often seen as one of the earliest reactions against the prevalence of technology and instrumental reason, one that urged a return to a premodern nature—the advocacy and development of machines did not necessarily entail alienation, specialization, dehumanization, or disenchantment. To the contrary, technology, broadly understood, was seen as a means of overcoming divisions between people, metaphysical domains, and fields of knowledge. Properly administered, machines could empower individuals and groups as well as actualize virtual powers, both natural and supernatural. Romantic machines served as focal points, both material and symbolic, for utopian projects that made a place for technology and science. Science and technology were seen as the bearers of emotion, aesthetics, and individual needs, capable of flexing in response to multiple and constantly changing settings. Four themes advanced by the leftist technophile mystics of this period seem particularly resonant for the concerns of the present: the view of the human as a technological animal; the necessity of including human thought and action in the world picture; an emphasis on projects situated at the medium scale; and the idea of nature as modifiable, but only within limits.[41]

Mechanical romanticism provided the outlines of a vision of the human as a creature dependent on its ecological milieu, defined by its power to alter its surroundings by means of technology. The notion of "the technological animal" goes back at least to Benjamin Franklin's definition of "man as tool-user"; reclaimed by Marx, the notion continued through Saint-Simon's and Comte's reflections on the preeminent place of humanity in the animal series—because of our complexity and our capacity to alter our environment. In this view, the process of our species' evolution is externalized and socialized; we adapt as a collective by means of our tools. This strand of thought was later elaborated by Henri Bergson and Hannah Arendt in their conceptions of *homo faber* and in the works of Pierre Teilhard de Chardin, André Leroi-Gourhan, Gilbert Simondon, Marshall McLuhan, and later postcybernetic theorists of the human. By emphasizing technological codependence and emergence, and deemphasizing any permanent essence for either humans or machines, this way of thinking about science and technology makes room for the contradictions and demands of subjectivity, emotion, and a changing nature. As in Leroux's conception, humans become the organs of a wider process of growth, contributing to the earth's "organogenesis" and continuing the process of "naturing nature."

Further, much as Comte decentered the notion of "life" to include the organism's interactions with its milieu, many of the philosophies of that time privileged relations over substances or essences. Comte's positivism is one such epistemological de-centering, while Geoffroy's ecstatic materialism, and its reinterpretation by Leroux, provides an ontological version. A key component of philosophies based on relations is that the act of drawing entities together—including, notably, human acts of perceiving, ordering, and organizing—must be considered as part of the elements' realization. In other words, any attempt to depict the world, and especially to conceive of the cosmos as a whole (for instance, in creating cosmograms like those studied here) must include recognition of the human activity involved in shaping the world picture. As Comte argued in the *Subjective Synthesis*, *how* and *why* we know are indissociably part of *what* we know, and a valid picture of nature must include all three. This reflexive loop has been recognized as intrinsic to historical, cultural, and ecological thinking; it is inevitable in quantum physics; it has a role to play in neuroscience, evolutionary theory, and philosophy of science.[42]

Another contribution of the mechanical romantics worth retrieving for the present is the scale at which they imagined themselves to be operating: at the middle scale, between the local and the cosmic. In projects like Fourier's phalanstery, Leroux's commune, or Comte's intendancies, we see daily life organized at the level of a relatively self-sufficient, medium-sized locality. These schemes do not ignore global connections, nor do they reject larger-scale coordinated projects. Against the caricatures of intentional communities as either terrifyingly efficient or absurdly bedlamized, such experiments need not preclude either individualism or the embrace of technology and connections to the larger world.[43] The ideas advanced in the 1830s and 1840s regarding global associations of diversified, self-supporting, largely face-to-face communities run counter to the gargantuan scale of much of today's commerce, politics, and media. But at the same time, they refuse an antimodern retreat into fragmented isolationism. Such schemes of loose federalism may appear a plausible ideal for a world undergoing both accelerated connectivity and processes of devolution.

Finally, and perhaps most importantly, the central tenet of the mechanical romantics deserves to be restated. Despite the claims of those who defend the notion of a rigid natural and social order, the relentless innovations of technology and science have made it hard to deny that nature is both open to modification and inextricably bound to human acts of intellectual and physical reorganization. At the same time, nature's modifiability is not infinite. True reform requires an acknowledgment of limits, of possibilities that cannot be realized. One message to be derived from this historical episode, now more than 150 years past, is that "freedom" can only be imagined as a concomitant of the dependencies, responsibilities, and limits that join us to others and to the nonhuman world.

* * *

At the close of the letter that Charles Duveyrier, the "poet of God," sent to Lavocat, the editor of *Le Paris des cent et-un*, as an introduction to his ecstatic vision of the "New City of the Saint-Simonians," he wrote:

> I end this letter, which is already too long, by imploring you to use all of your influence with your readers to reanimate the virtues of courage and of hope that are so rare today—if only for a moment, the time it takes to read these pages. For, even if these pages are intelligible, they may yet appear as a dream, as a fantastic hallucination—especially if your elegant readership obstinately persists in that gullible disposition, in that FAITH which often is pushed to the point of SUPERSTITION, which consists in considering as an impossibility every great, every generous, every excellent thought that aims at the improvement of the condition of the people.[44]

The same might be said, more generally, of the superstition that relegates romanticism—that attempt to envision humans, in the range of their emotions, capacities, and failings, as part of a developing, infinitely resourceful, yet elusive nature—to the dustbin of the past. When romanticism is dismissed as a naïve and nostalgic flight from "reality," as narcissistic brooding, or as a precursor to totalitarianism, a vital tradition and a central source for imagining the future is lost. The diverse streams of romanticism that circulated in the early nineteenth century pushed readers to see themselves as part of the earth, and take inspiration from both external and internal nature; they prompted democratic action and a new vision of social equality. In the emblematic city of Paris, in the early stages of industrial modernity, the diverse strands of romanticism were compatible with an embrace—an ambivalent, cautious embrace—of science and technology. This linkage of romanticism and technology propelled new visions of social and natural harmony and suggested concrete solutions to metaphysical debates.

This is a book largely about predictions that never came true, prophecies that might seem even less likely now. Even so, a broader intention than nostalgia, lamentation, revisionism, or the collection of historical curiosities lies behind its recovery of themes—whether the enchantment of machines, the power of the social and religious imagination in the face of ideological stagnation, the technologizing of the arts, or the aestheticizing of the sciences—from the oddly pregnant situation of interrevolutionary Paris. It has sought to give a sympathetic hearing to thinkers and builders who saw artistic experiment, scientific research, introspection, technological innovation, and social justice as tendrils growing from a common stem, demanding to be grafted or soldered back together—as materials and methods out of which better worlds can be made. Their futures do not belong only to the past.

Acknowledgments

A friend of mine once explained the idiosyncrasies of another friend this way: "She's the product of extreme forces." The same can probably said of any book. Happily, one force at work on this book was the extreme generosity of many people. Simon Schaffer offered countless suggestions, revisionist readings, and a constant, sublime example. Bruno Latour's ideas opened paths followed here, and he has helped me through many thickets. George Stocking patiently set a confused undergrad on his contextualizing feet. Three friends read and commented on the entire manuscript: Cathy Gere, with insight and patience, held its hands as it took its first steps; Ken Alder made many precision modifications and has offered multifarious support over many years; Karen Russell asked the right questions and found the right words. In taming the bibliography and images, Alexander Jacobs's persistence and percipience made a colossal difference. The hard work and sharp eyes of Deanna Day, Crystal Biruk, Amy Paeth, and Will Kearney were invaluable in the final stages, as were comments of Bernadette Bensaude-Vincent on a late draft. Karen Darling and the University of Chicago Press, along with Lois Crum, Michael Koplow, Kailee Kremer, Abby Collier, Isaac Tobin, and Holly Knowles, guided me through the process of publication with grace.

The Department of History and Sociology of Science of the University of Pennsylvania provided supportive conditions. Each of my colleagues, past and current, has helped in many different ways, and I thank them all; Mark Adams, Nathan Ensmenger, Riki Kuklick, Rob Kohler, and Beth Linker were particularly generous with book advice. Sections dealing with music and aesthetics and the book's overall argument owe a great deal to conversations and collaborations with Emily Dolan, and to the participants of the seminar she and I taught on "Instruments of Music and Science." Thanks also to Warren Breckman, Martha Farah, Lynn Farrington, Pat Johnson, Sharrona Pearl, John Pollack, Wendy Steiner, and Peter Struck.

At the University of Chicago and the Franke Institute for the Humanities, James Chandler taught me Blake and saved my bacon on more than two occasions. Warm thanks also to the incomparable Laura Desmond and Rosa Desmond, Mai Vukich, Margot Browning, and Jessica Burstein, and to Raymond Fogelson, Jan Goldstein, John Kelly, Robert Richards, Jay Williams, and William Wimsatt. I am indebted to Northwestern University's Program for Science in Human Culture, where I had the good fortune to work with Francesca Bordogna, David Joravsky, Jessica Keating, Lyle Massey, Sarah Maza, Guy Ortolano, Shobita Parthasarathy, and Claudia Swan. Thanks also to the Society of Fellows in the Humanities of Columbia University and to my fellow fellows. The Max Planck Institute for the History of Science, Berlin, provided a fantastic place for research and conversation; many thanks to Lorraine Daston, Otto Sibum, and Fernando Vidal.

This book was decisively shaped by my time at the Department of History and Philosophy of Science at Cambridge University and at the Ecole Normale Supérieure, Paris. I wish to thank Pembroke College, Barbara Bodenhorn, John Forrester, Nick Hopwood, Marina Frasca-Spada, Peter Lipton, Jim Secord, and Mark Wormald. At the ENS, I owe particular thanks to Christian Baudelot and the Department of Social Sciences. Thanks also to the EHESS, the CSI at l'Ecole des Mines, the Centre Alexandre Koyré, and the CRHST at La Villette. I am grateful for the insights and support of Jean Bazin, Alban Bensa, Christine Blondel, Luc Boltanski, Eric Brian, Christophe Bonneuil, Pierre Bourdieu, Benoît de l'Estoile, Claude Imbert, Dominique Lestel, Deborah Levy-Bertherat, Max Marcuzzi, Dominique Pestre, Jacques Rancière, and Sophie Roux. The book reached its conclusion at the New York Public Library's Cullman Center for Scholars and Writers, under the care of the formidable Jean Strouse.

Over the years of this book's writing I've been helped by many other friends and colleagues: Aaron Davis, holy diver, lent his expertise in German translation and bibliographic survivalism; Keith Hart provided the perspective of cosmic entrepreneurship; Catherine Leblanc made the leap to France conceivable; Clare Carlisle helped put things in order; Annie Siddons and Peter Gates have been there through thick and thin; Anne Stevens offered sober commentary; and Tamara Barnett-Herrin and Jay Basu rule. For many species of kindness, my thanks goes to the following: Esther Allen, Nalini Anatharaman, Naomi Andrews, Karl Appuhn, Carol Armstrong, Babak Ashrafi, David Aubin, Hylda Berman, Josh Berson, Charlotte Bigg, Julien Bonhomme, Véronique Bontemps, Bob Brain, Sam Breen, Terrance Brown, Brujo de la Mancha, Beatrice Collier, Peter Collopy, Isabelle Combes, Rachel Cooper, Deb Coy, Annelle Curulla, David Charles, Debbie

Davis, James Delbourgo, Ulrike Decoene, Marie d'Origny, Susan Einbinder, Matthew Engelke, François Furstenberg, Rivka Galchen, Peter Galison, S. N. Goenka, Michael Golston, Matthieu Gounelle, David Graeber, Richard Gray, Hildegard Haberl, Jacob Hellman, Linda Henderson, Jessica Hines, James Hofmann, Myles Jackson, Andi Johnson, Minsoo Kang, Christopher Kelty, Eion Kenny, Françoise de Kermadec, the Kings of Madison (Julie, Justin, Tobias, and Ronan), Aden Kumler, Julia Kursell, Andrew Lakoff, Ginger Lightheart, Fabien Locher, Kathy Lubey, Lori Reese, Kevin Lambert, Hannah Landecker, Pearl Latteier, Nhu Le, Rebecca Lemov, Theresa Levitt, Dominique Linhardt, James Livingston, Deirdre Loughridge, Stéphane Madelrieux, Andreas Mayer, Richard McGuire, Matthieu Merygniac, Mara Mills, Kevin Moser, Cecile Nail, Fred Nocella, Jason Oakes, Claire O'Brien, Noara Omouri, Thomas Patteson, Annie Petit, Danton Voltaire Pereira de Souza, Sylvain Perdigon, Mary Pickering, Christelle Rabier, Joanna Radin, Lori Reese, Justine de Reyniès, Jessica Riskin, Agathe Robilliard, Henning Schmidgen, Dana Simmons, Dimitri Topitzes, Joerg Tuske, Jean-Christophe Valtat, Margarete Vöhringer, Adelheid Voskuhl, Jen Walshe, Simon Werrett, Lord Whimsy, Kristoffer Whitney, Norton Wise, Albena Yaneva, and Jason Zuzga. Thanks also to my mother, my sister, and my brother for everything, and for not asking too often when it would be done.

Some material in this book previously appeared in the following publications: "The Machine Awakens: The Science and Politics of the Fantastic Automaton," *French Historical Studies* 34 (2011): 88–123; "The Prophet and the Pendulum: Popular Science and Audiovisual Phantasmagoria around 1848," *Grey Room Quarterly* 43 (2011): 16–41; with Emily Dolan, "'A Sublime Invasion': Meyerbeer, Balzac, and the Paris Opera Machine," *Opera Quarterly*, March 29, 2011, doi: 10.1093/oq/kbr001; "La Technaesthétique: Répétition, habitude, et dispositif technique dans les arts romantiques," *Romantisme* 150 (2010): 63–73; "The Order of the Prophets: 'Series' in Early French Socialism and Social Science," *History of Science* 48 (2010): 315–42; "Even the Tools Will Be Free: Humboldt's Romantic Technologies," in *The Heavens on Earth: Observatories and Astronomy in Nineteenth Century Science and Culture*, ed. David Aubin, Charlotte Bigg, and Otto Sibum (Durham: Duke University Press, 2010), 253–85; "Electromagnetic Alchemy in Balzac's *Quest for the Absolute*," in *The Shape of Experiment*, ed. Henning Schmidgen and Julia Kursell (Berlin: Max-Planck preprint, 2007), 57–78; "The Daguerreotype's First Frame: François Arago's Moral Economy of Instruments," *Studies in History and Philosophy of Science* 38 (2007): 445–76.

With apologies to anyone I forgot, and with extreme gratitude to all.

NOTES

PREFACE

1. Balzac, *La recherche*, 717.
2. Claës's experiments involved the electric decomposition of various substances and attempts to produce artificially the conditions needed for plant growth. His chemistry was guided by concepts derived from the chemistry of Lavoisier, Davy, Berzelius, Dalton, and Stahl as well as the alchemy of Paracelsus and Agrippa, Pythagorean numerology, and a triadic unity of "matter, means, and result" akin to a number of schemes in *Naturphilosophie*. See Fargeaud, *Balzac*; Mertens, "Du côté d'un chimiste."
3. Balzac, *La recherche*, 719-20.
4. Ibid., 719.
5. Balzac, *Louis Lambert; Gambara*.
6. Fargeaud, *Balzac*, 98.
7. Balzac, *Le père Goriot*, 80.
8. Balzac, *La recherche*, 698, 736, 742, 771.
9. Ibid., 780.
10. Balzac, *La chef d'oeuvre inconnu*.
11. Balzac, *Les illusions perdues*; Bruce Tolley, "Balzac et les Saint-Simoniens," *Année Balzacienne* (1966): 49–66; Macherey, "Leroux."
12. Sand, *Autour*, 200.

CHAPTER ONE

1. On the persistence of romanticism, see Abrams, *Natural Supernaturalism*; Liu, *Local Transcendence*; Lacoue-Labarthe and Nancy, *Literary Absolute*; Chandler, *England in 1819*; Cavell, *In Quest of the Ordinary*; McGann, *Romantic Ideology*. On the romantic self, see Taylor, *Sources*; Siegel, *Idea of the Self*; on romantic views of nature, see, for example, Gusdorf, *Le savoir*. On romanticism and politics, see Berlin and Hardy, *Roots of Romanticism*; more sympathetically, Butler, *Romantics, Rebels, and Reactionaries*. Other texts about this period in France are cited below, but see in particular Pinkney, *Decisive Years in France*.

2. See also Collingwood, *Idea of Nature*.
3. On the pluralist heritage of romanticism—in contrast to readings that see it as a forerunner of totalitarianism, such as Isaiah Berlin (e.g., *Roots of Romanticism*)—see Taylor, "Importance of Herder."
4. On the first page of *The Romantic Conception of Life*, Robert Richards makes this opposition clear: "The Romantics attacked a significant bulwark of earlier thought, one that advanced mechanism as the engine of progress in science.... [They] replaced the concept of mechanism with that of the organic, elevating it to the chief principle for interpreting nature" (xvii). Richards goes on to show that nineteenth-century biology, and particularly that of Darwin, was deeply indebted to the romantic life sciences. The claim is deliberately provocative, as historians of biology have long held that Darwin's success was due to his avoidance of dubious concepts coming from theology, romanticism, or *Naturphilosophie*: in particular, to his refusal to explain adaptation with reference to a divine plan, teleology, or intentionality and in identifying the mechanism (i.e., natural selection) to explain the diversity of species (on Darwin as exemplary mechanistic thinker, see Barzun, *Darwin, Marx, Wagner*). Richards argues, instead, that Darwin's conception of nature was "organic and aesthetic" and points out that the term *machine* appears only once, in a vitalistic context, in *Origin of Species*. Richards, *Romantic Conception of Life*, 534. Richards suggests we recategorize Darwin as a romantic. Yet Darwin's affinity to romantic biology might equally be read as a challenge to the assumption of a categorical opposition between "romantic" and "mechanistic" views.
5. Wellek, "Romanticism Re-examined," 221. This was Wellek's response to A. O. Lovejoy, who helped invent the twentieth-century field of history of ideas with exemplary studies of the great chain of being and romanticism; Lovejoy was skeptical about the existence of a general essence of romanticism. See Lovejoy, "Meaning of Romanticism."
6. The argument that scientific progress can be measured by the spread of mechanistic explanations is a long-standing topos in histories of science and of "the mechanical philosophy" from the seventeenth century onward (see discussions in Hall, "Establishment"; Westfall, *Construction*; Butterfield, *Origins*; Dijksterhuis, *Mechanization*); this argument was challenged by Kant in relation to the science of living things in the *Critique of Judgment* and in France by Bichat and later Comte (see chapter 9 below). The related but distinct notion that scientific progress can be measured by the degree to which knowledge production is assisted by machines was laid out by Francis Bacon in the *Novum Organum* and *The New Atlantis*; some of its nineteenth-century instantiations are examined in Daston and Galison's *Objectivity*. Relationships between experimental machines and "the mechanical philosophy" have been explored in Shapin, Schaffer, and Hobbes, *Leviathan and the Air-Pump*; Dear, *Revolutionizing the Sciences*; Shapin, *Scientific Revolution*; Meli, *Thinking with Objects*.
7. See Mumford, *Technics and Civilization*; Mayr, *Authority, Liberty*; Adas,

Machines; Schaffer, "Enlightened Automata"; Riskin, "Defecating Duck"; Foucault, *Discipline*; Hughes, *Human-Built World*.

8. "Dissociation of sensibility" is Eliot's expression (*Points of View*, 71) referring to the inability of poets since Donne to join reason and feeling; see Snow, *Two Cultures*; for the science wars, see Gross and Levitt, *Higher Superstition*.

9. Such trends were satirized by an 1835 vaudeville play cowritten by Etienne Arago, brother of astronomer François Arago and collaborator with Honoré de Balzac: *Paris dans la comète*. Responding to the popular frenzy about the comet visible over Paris in 1834, the play claimed to be written entirely by steam. This period saw the formation of the technological infrastructure for mass entertainment. For related themes in the United States, see Marx, *The Machine in the Garden*; Nye, *American Technological Sublime*.

10. See Canguilhem, "Le vivant et son milieu," in *La connaissance de la vie*. According to Canguilhem, key points in the development of the concept "milieu" were articulated by Lamarck, Geoffroy Saint-Hilaire, Alexander von Humboldt, and Auguste Comte, all of whom are featured in the following chapters.

11. Oken, *Elements of Physiophilosophy*, 178–97; on Oken's influence in France, see Braunstein, "Comte, de la nature."

12. Such a view of the earth's development with the aid of humanity is discussed below, with regard to both Leroux and his follower Ange Guépin's writings and to Comte's concept of the earth as the Great Fetish. On Lamarck's view of nature's "production," see Jordanova, "Nature's Powers."

13. This is a point of connection with Daston and Galison's *Objectivity*, which traces changes undergone by the epistemic and ethical ideal of objectivity from roughly Linnaeus and the age of Goethe to the present and relates practices and technologies of observation and inscription to conceptions of the self. They show that objectivity has both a practical and an ethical dimension, regulating practices of knowledge as well as conceptions of the knowing subject. The period on which the present volume concentrates, 1815 to 1851, is a transitional no-man's-land between two phases in their account: first, the late-eighteenth-century ideal of "truth to nature," which guided the hand of natural historians as they sketched idealized images of specimens to capture the essence of a plant or animal, and second, the ideal of "mechanical objectivity," taking shape in the 1850s and 1860s, when scientists sought to avoid the corruption brought by individual bias, irregularity, or interpretation by using machines; the sensitive and intuitive observer of "truth to nature" gives way to the self-monitoring, repressively ascetic subject of "mechanical objectivity." Since this book's main topics are situated in the years between these two regimes, it is perhaps unsurprising that it presents epistemologies and conceptions of the self that combine aspects of both.

14. Important reference points in this tradition include Collingwood, *Idea of Nature*; Lovejoy, *Great Chain of Being*; Foucault, *Order of Things*; Daston and Park, *Wonders*. A striking commonality of these books is their discomfort in

attempting to characterize the period after 1800. The theme of nature as an organism subject to history is frequently presented as essential to the modern view of nature (see also Greene, *Death of Adam*); yet this generalization stands at odds with the portrayal of modernity in terms of a fixed, reductive, ultimately static "mechanical worldview" (what Heidegger called "enframing" in "Question"). The projects discussed below sought explicitly to bring these two perspectives into harmony, forging a unified view of the cosmos that embraced both the organic and the mechanical.

15. That this ethics was realized externally, and often with the aid of machines, represents an extension and transformation of Kant's view of human nature as both a free reason and a mechanically determined body—what Foucault called the "empirico-transcendental doublet" in *The Order of Things*.
16. See Tresch, "Cosmograms"; "Technological World-Pictures."
17. For a rousing panorama of the English case, see Richard Holmes, *Age of Wonder*. Perhaps because it is set approximately a generation earlier than the present book, Holmes's work rarely touches upon the questions of technology and politics that became so central to the French case.
18. See Johnson, *Birth of the Modern*.
19. "Notice communale."
20. See Jardin and Tudesq, *La France des notables*; Harvey, *Paris*.
21. Guillerme's *La naissance de l'industrie* details the social and environmental impact of new industry in Paris in the fifty years around 1800, a perspective that enriches the histories of workers' lives and modes of association found in Sewell, *Work and Revolution*; Scott, *Glassworkers of Carmaux*; Rancière, *Nights of Labor*. For a perspective attuned to both technical and social aspects of an earlier moment, see Hafter, *Women at Work in Preindustrial France*; and Hafter, "Cost of Inventiveness."
22. Michelet, *People*; see also Orr, *Headless History*.
23. Bezanson, "The Term Industrial Revolution."
24. See Spitzer, *Generation of 1820*. On the structures of patronage in French science from the Revolution to the July Monarchy, and Cuvier's skill in adapting to them, see Outram, *Georges Cuvier*.
25. See Goldstein, *Post-Revolutionary Self*, on "horizontal fragmentation," as the central problem for postrevolutionary theories of the self. See also Bénichou, *Le sacré de l'écrivain*; Rosanvallon, *Moment Guizot*. Bénichou and Rosanvallon argue that reestablishing a "spiritual power" to unite society was the definitive political issue of this period; where Bénichou focuses on the role attributed to the arts, Rosanvallon analyzes Guizot's notion (deriving in part from early exchanges with Auguste Comte) that social unity could best be assured by a meritocratic elite armed with more fluid instruments (including education and propaganda) to influence public opinion than a one-way state apparatus of command and obedience.
26. See discussions below of Saint-Simon, Comte, Leroux, and Pecquer; see also Charlton, *Secular Religions*; Berenson, *Populist Religion*. For a longer historical

perspective, see Noble, *Religion of Technology*; on the neglected importance of nineteenth-century "machinolatry," see Clark, "Should Benjamin?"
27. In particular, the French Enlightenment's dynamic materialism, vitalism, and emphasis on sensibility, as discussed in Riskin, *Science*; see also Reill, *Vitalizing Nature*. The present book aims to show that the emphasis on sensibility and sentiment that Riskin and Reill show as crucial to the Enlightenment (as a counterweight and occasional partner to mechanism) carried on into the age of industrialization; the result was mechanical romanticism.
28. See, however, Chandler and Gilmartin, *Romantic Metropolis*.
29. See Wise's analysis of the balance and related technologies in "Mediations."
30. See Kuhn, "Simultaneous Discovery"; Serres, "Turner Translates Carnot"; Rabinbach, *Human Motor*; Wise and Smith, "Work and Waste"; Vatin, *Le travail*.
31. This conception resonates with Henning Schmidgen's discussion of "nonmodern machines." Taking off from Deleuze and Guattari's discussions of "machinic assemblages" and Latour's notion of the "nonmodern," Schmidgen treats Donders's experimental arrangements as "time-based syntheses of diverse and distinct partial objects opening paths or passages from materiality to semiotics. In this perspective, physiological experiments are machines not simply because the historical focus is on instruments and technological systems; rather, they are machines because they combine technological components with parts of human and nonhuman organisms to form essentially precarious, but functioning, arrangements of flows and interruptions that are directed toward the production of semiotic events." Schmidgen, "Donders Machine," 214. Significantly, Donders's experimental setups share a genealogy with those of two of my protagonists, Ampère and Humboldt.
32. Engels, "Socialism: Utopian and Scientific"; "fantastic" in Marx and Engels, *Manifesto of the Communist Party*, 498–99; commodity fetishism in *Capital*, 319–29, all in Marx and Engels, *Marx-Engels Reader*.
33. Even William Pietz's remarkable three-part excavation of the term in "Problem of the Fetish" emphasizes the critical uses of the term and neglects its positive evaluations; he shows it primarily as a term of dismissal, not a feature of primitive life to be imitated, as it was by Comte.
34. Enfantin in Régnier, *Le livre nouveau*, 184–85; Gere, *Knossos*.
35. For Max Weber, "charismatic authority" was a mode of political legitimation directly opposed to traditional or bureaucratic authority, one that eventually subsided into the "routinization of charisma." Charismatic authority might flare up occasionally in modern societies, but Weber saw it harking back to earlier, prebureaucratic times—as a recurrence of modes of thought and experience that preceded the "disenchantment of the world." But despite the fact that machines are now often seen as impersonal, as linked to rationalization, and as instruments or even agents of disenchantment, machines were often perceived as vessels of charismatic recurrence. Weber's "disen-

chantment of the world" was derived from Schiller's *Aesthetic Education*, which spoke of the "de-divinization of the world"; see chapter 3 below. On the circulation of "charisma" in the late eighteenth century and its relation both to the gothic and to uncanny technologies, see Castle, *Female Thermometer*. Shapin, *Scientific Life*, downplays the tension between charismatic individualism and the bureaucratic imperatives of modern research.

36. On the mythic power of machines, see, for example, Marx, *Machine in the Garden*; Mumford, *Myth of the Machine*; Noble, *Religion of Technology*; Latour, *Aramis*. Useful histories of Western conceptions of the self, with particular emphasis on romanticism, include Taylor, *Sources*; Seigel, *Idea of the Self*; see also Schaffer, "Genius in Romantic Natural Philosophy," in Cunningham and Jardine, *Romanticism and the Sciences*; and discussions throughout Richards, *Romantic Conception of Life*, especially on Fichte. Helpful insights about the political consequences of the expressive and sensitive self of romanticism are found in Sennett, *Fall of Public Man*. On variable meanings of the techniques of introspection and self-knowledge, see Foucault, Gros, Ewald, and Fontana, *Hermeneutics of the Subject*.

37. Nevertheless, it is worth keeping in mind the words of Adorno on this point: "The cliché which claims that modern technology has fulfilled the fantasies of the fairy tales only ceases to be a cliché if one adds that the fulfillment of wishes rarely benefits those who make them. The right way of wishing is the most difficult art of all, and we are taught to unlearn it from childhood on.... Utopias have been realized only to disabuse human beings of any utopian desire and commit them all the more thoroughly to the status quo, to fate." Adorno, *Eingriffe*, translated in Hansen, "Benjamin and Cinema," 223.

38. Benjamin, "Work of Art," 147–48; "Motifs in Baudelaire."

39. Novalis, *Henry of Ofterdingen*, 24.

40. Benjamin, "Work of Art"; Hansen, "Benjamin and Cinema"; Buck-Morss, "Benjamin's Passagen-Werk."

41. Among works incorporating Benjamin's insights into more synthetic and linear overviews of nineteenth-century Paris, see Buck-Morss, *Dialectics of Seeing*; Harvey, *Paris*; Marrinan, *Romantic Paris*; Clark, *Image of the People*; Clark, *Absolute Bourgeois*; Ferguson, *Paris as Revolution*; Prendergast, *Paris and the Nineteenth Century*.

42. Gérard de Nerval, *Vers dorés* (1845), in *Oeuvres*, 1:739.

43. Peter Sloterdijk aligns Benjamin's *Arcades Project* with the notion of milieu, noting his discovery of the technologically aided transformation of external nature into an interior, domesticated space: "The 19th century citizen seeks to expand his living room into a cosmos and at the same time to impress the dogmatic form of a room on the universe." "Spheres Theory," 2.

44. This book thus sheds light on the political and technical background for the vitalist tradition recently evoked by Jane Bennett in *Vibrant Matter: A Political Ecology of Things*.

45. For example, see Chandler and Gilmartin, *Romantic Metropolis*; Ziolkowski, *German Romanticism*; Abrams, *Natural Supernaturalism*; Chandler, *Words-*

worth's *Second Nature*; Butler, *Romantics, Rebels, and Reactionaries*; Recht, *La lettre de Humboldt*; Levy-Bertherat, *L'artifice romantique*; Bénichou, *L'école*.

46. See especially Cunningham and Jardine, *Romanticism and the Sciences*; Canguilhem, "Machine et organisme" and "Aspects du vitalisme," in *La connaissance de la vie*; Lenoir, *Strategy of Life*, "Gottingen School," and "Eye as Mathematician"; Dettelbach, "Humboldtian Science" and "Face of Nature"; Galison, "Objectivity Is Romantic"; Wise, "Architectures for Steam"; Otis, *Networking*; Vatin, "Des polypes"; Limoges, "Milne-Edwards."

47. The rehabilitation of romanticism for the life sciences has been most notable in Richards's *Romantic Conception of Life* and its follow-up (if we grant Haeckel the status of "romantic"), Richards's *Tragic Sense of Life*. On the influence of romanticism on physics in Germany and England, see suggestions by Kuhn ("Simultaneous Discovery") and Rabinbach (with his notion of "transcendental materialism" in *Human Motor*); Wise, "Architectures for Steam"; the discussion of Grove in Morus, *When Physics Became King*; Winter, *Mesmerized*; Brain, Cohen, and Knudsen, *Ørsted*; Friedman and Nordmann, *Kantian Legacy* (see esp. Beiser, "Kant and *Naturphilosophie*"; and Friedman, "Electromagnetism"); along with contributions to the debate about *Naturphilosophie*'s influence by Stauffer, Gower, Williams, Caneva, Dettelbach, Strickland, and Friedman discussed below.

48. When scholars of romanticism do confront technology, it is often, still, as the object of deflationary irony. This is frequently the case in media studies, where analysis of the constitutive impact of technology on romantic expressions is often treated as an exposé of a scandalous secret—the unveiling of a sinister mechanism at the heart of the illusion of creativity, spontaneity, and inspiration. Kittler, *Discourse Networks*; Crary, *Techniques of the Observer*; Siegert, *Relays*. As notable exceptions, see Bowie, "Romantic Technology"; Fiorentini, *Observing Nature*; Brain, "Romantic Experiment."

49. See Ben-David and Freudenthal, *Scientific Growth*; and the landmark paper by Fox, "Rise and Fall of Laplacean Physics"; Herivel, "Aspects of French Theoretical Physics"; see historiographical discussion in Dörries, "Future of Science"; and skeptical notes in Belhoste, "Arago, les journalistes."

50. Dörries, "Future of Science."

51. See Bradley and Perrin, "Dupin's Visits."

52. Babbage, *Ninth Bridgewater Treatise*; see Schaffer, "Babbage's Intelligence"; Ashworth, "Calculating Eye"; Morus, *When Physics Became King*.

53. See Crosland and Smith, "Transmission of Physics"; Miller, "Revival of Physical Sciences."

54. See Desmond, *Politics of Evolution*; Lawrence, "Heaven and Earth"; Schweber, "Comte and the Nebular Hypothesis"; Wise and Smith, "Work and Waste"; Pancaldi, "Republic of Letters"; Wise, "Mediating Machines."

55. Hahn, *Pierre Simon Laplace*; Crosland, *Society of Arcueil*; Dhombres and Dhombres, *La naissance d'un pouvoir*; Fox and Weisz, *Organization of Science*; Belhoste, *La formation d'une technocratie*; Outram, *Georges Cuvier*.

56. Deleuze and Guattari's discussion of royal science versus nomadic science

(*Thousand Plateaus*) offers a stimulating alternative to the hopeless romanticism/soulless mechanism divide and helps illuminate the opposition that separated Laplaceans from what Caneva called "the etherians," the algebrists from Mongean engineers, and Cuvier from Geoffroy.

57. See Outram, *Georges Cuvier*; Corsi, *Age of Lamarck*; Foucault, *Order of Things*; Fox, "Rise and Fall of Laplacian Physics"; Hahn, *Pierre Simon Laplace*; Herivel, "Aspects of French Theoretical Physics"; Friedman, "Creation of a New Science"; Vermeren, *Victor Cousin*; Goldstein, *Post-Revolutionary Self*; however, see Goldstein's closing chapters on phrenology (including Comte's phrenological writings) as a marginal discourse of the self opposed to that of Cousin.

58. Crosland, "Popular Science"; Bensaude-Vincent, "Historical Perspective"; and other contributors to *Isis* of summer 2009 (Focus: Historicizing "Popular Science"); Belhoste, "Arago"; Levitt, *Shadow of Enlightenment*; Staum, "Physiognomy and Phrenology." Much more has been done on the period after 1850 and on English popular science: see, for example, the essays in Cantor and Shuttleworth, *Science Serialized*; Secord, *Victorian Sensation*; Winter, *Mesmerized*; Morus, *When Physics Became King*; Fyfe, *Science and Salvation*, to name just a few.

59. Oehler, *Le spleen contre l'oubli*; Clark, *Absolute Bourgeois*.

60. See Parent-Lardeur, *Paris au temps de Balzac*; Bourdieu, *Rules of Art*; Guillerme, *La naissance de l'industrie*; Moretti, *Atlas of the European Novel*.

61. Jardin and Tudesq, *Restoration and Reaction*; *La France des notables*; Price, *Revolution and Reaction*; Parent-Lardeur, *Paris au temps de Balzac*; Moretti, *Atlas of the European Novel*; Ferguson, *Paris as Revolution*; *Osiris* 18 (2003), a volume titled "Science in the City."

62. See Marx, *18th Brumaire of Louis Bonaparte*. One critical response is Price, *French Second Republic*. See also Sennett, *Fall of Public Man*, for discussion of Lamartine's role in the revolution.

63. A lively overview of such movements is found in Erik Davis, *TechGnosis*; on green architecture and urbanism, see, for instance, Wheeler and Bentley, *Sustainable Urban Development Reader*.

CHAPTER TWO

1. Dulong to Berzelius, October 2, 1820, in Caneva, "Ampère."
2. On Ampère and electrodynamics, see Blondel, *A.-M. Ampère*; Harman, *Energy, Force, and Matter*; Hesse, *Forces and Fields*; Hofmann, *André-Marie Ampère*; Whittaker, *History of the Theories*; Williams, "Ampère, André-Marie"; Darrigol, *Electrodynamics*.
3. Douglas, *Natural Symbols*; Caneva, "What Should We Do with the Monster?"
4. In a case study in the history of chemistry and electricity, Bachelard recounted how researchers had noted an unusual smell produced when oxygen is exposed to electric sparks. In 1839 Schonbein, a correspondent of François Arago, claimed to have identified the cause of this smell: ozone.

According to Bachelard, ozone soon suffered a "cosmic overvaluation": led by his adherence to a two-fluid theory of electricity, Schönbein suggested an analogous substance created by negative electricity, "antozone," which was then identified as an enabling cause for epidemics. Ozone and its phantom sister became central objects in a wide-scale though short-lived hygienic movement to map their appearance and absence as indicators of insalubrity, a latter-day version of Priestley's eudiometry (see Schaffer, "Measuring Virtue"). "In these conditions," Bachelard writes, "it would be a long and difficult task to bring into the laboratory this 'cosmic substance.'" Bachelard, *Le rationalisme appliqué*, 223. Ozone was an object entangled with too many disorganized phenomena and cultural expectations to reckon it according to its true proportions or ratio. In my presentation of Ampère's electromagnetism, however, I wish to suggest that these extrarational associations are crucial for understanding the intentions going into the scientific research of such objects.

5. Kuhn, "Mathematical Traditions"; Williams, "Kant, *Naturphilosophie*, and Scientific Method"; Caneva, "Ampère, the Etherians, and the Oersted Connexion"; Blondel, "Vision physique." In a more recent survey of the *Naturphilosophie*-physics connection (although one that does not deal explicitly with Ampère), Caneva distinguishes between the "unity of nature" pursued by *Naturphilosophie*—with its emphasis on dialectical processes and triads—and the unity brought to physics by ether theories. He rightly argues that there are different modes of "unification," and the idea of a "unity of forces" is anachronistic; further, ether physics remained hospitable to materialism and atomism. Caneva, "Physics and *Naturphilosophie*," 41. Nevertheless, the line between ether physics and *Naturphilosophie* was easily blurred in France in the 1820s and 1830s, as physicists including Ampère and Arago were frequently willing to engage with both *Naturphilosophie* and animal magnetism, which could be read in both materialist and spiritualist terms.

6. Fargeaud, *Balzac*, suggests that Ampère was a model for Claës, the obsessive hero of Balzac's *Quest for the Absolute*; in his discussion of ozone and antozone, Bachelard ironically alluded to this novel to signal Schönbein's irrationality.

7. Schelling, *Ideas*, 42. See Oersted, *Soul in Nature*.

8. See discussion and diagram in Richards, *Romantic Conception*, 132.

9. Schelling, *Ideas*, book 1, with chapters progressing from combustion to light, air, electricity, and magnetism.

10. See discussion in Friedman, "Kant—Naturphilosophie—Electromagnetism."

11. On *Naturphilosophie*, Schelling, and Oersted, see Brain, Cohen, and Knudsen, *Hans Christian Ørsted*; Stauffer, "Speculations"; Kuhn, "Simultaneous Discovery"; Caneva, "Ampère"; "Physics and *Naturphilosophie*"; Gower, "Speculation in Physics"; Levere, *Poetry Realized*.

12. The homogeneity of "Laplacean physics" and the starkness of the opposition between faithful Laplaceans and renegades like Arago, Fresnel, Ampère, and Fourier can appear somewhat overstated when we examine Laplace's

own scientific and political positions closely, such as his support of Fourier as perpetual secretary of the Academy; nevertheless, there was a concerted effort on the part of Arago, with considerable assistance from his allies, to create the image of a unified (and misguided) Laplacean program during the Restoration. The opposition between science in the Laplacean and Lagrangean mold versus that of the scientist-engineers allied with romantic philosophy corresponds remarkably well to the opposition between "royal science" and "nomadic science" in Deleuze and Guattari, *Thousand Plateaus*, 351–423.

13. See Crosland, *Society of Arcueil*; Fox, "Rise and Fall of Laplacian Physics"; Hahn, *Pierre Simon Laplace*.
14. Heilbron, "Some Connections"; Heilbron, *Electricity*.
15. Biot, *Précis*.
16. Fox, "Background to the Discovery."
17. Williams, "Faraday and Ampère."
18. See Buchwald, *Rise of the Wave Theory*; Levitt, *Shadow of Enlightenment*.
19. See Kuhn, "Mathematical Traditions," Frängsmyr, Heilbron, and Rider, *Quantifying Spirit*; and Frankel, "J. B. Biot," on the emergence of experimental physics.
20. See Fox, "Rise and Fall of Laplacian Physics"; Hofmann, *André-Marie Ampère*.
21. Fargeaud, *Balzac*, 179. Caneva, for example ("What Should We Do with the Monster?"), links Ampère's openness to electromagnetism to his belief in the "naturalness" of categories ("high-grid," in Mary Douglas's terms) and to his relative privileging of his own intuitions and codes of conduct over those of the group ("low-group"). In sum, Ampère was rigid and realist in terms of theory and nonconformist in terms of social groups; this outlook allowed him to believe in and make a place for an "undomesticated" or "raw" phenomenon like electromagnetism that was rejected outright by the dominant clique. Caneva's suggestion of a correlation between attitudes toward social and natural categories and social positioning is intriguing, although the explanation of Ampère's unusual embrace of electromagnetism in terms of his isolation and uniqueness is difficult to square with Caneva's subsequent identification of Ampère's "Oersted connection" and membership in the group of "etherians." Was the whole group nonconformist?
22. Sainte-Beuve and Antoine, "Ampère," in *Portraits littéraires*
23. Hofmann, *André-Marie Ampère*, 14–16. See Williams, "Ampère, André-Marie," on his family resemblance with leading *Naturphilosophen*.
24. Ampère's personal life was often of central interest in the nineteenth century. Henry James wrote on the sentimental lives of the father and the son and portrayed Isabelle Archer in *Portrait of a Lady* with "a volume of Ampère" on her lap. Others have given a greater share of attention to his religious struggles: Buche concentrated on his connections to *L'école mystique de Lyon*, placing him in community with that city's famous mesmerists of the eighteenth century, "the unknown philosopher"—the mystic Saint-

Martin—and especially the romantic social prophet Pierre-Simon Ballanche. The spiritual biography by Valson, *La vie et les travaux* (Valson also studied the Catholic mathematician Cauchy), presents Ampère (Père) as a case study in the proper relation between faith and reason, as an example of "the religious and divine sentiment in science." For Valson, Ampère's Christianity helped make him the blessed exception to the rule that scientific discoveries are usually made in early youth.

25. For Caneva, this contributed to his marginal status in Parisian scientific circles, freeing him to accept Oersted's unexpected claims.
26. Buche, *L'école mystique de Lyon*. His friends from Lyon believed "in the unity of the design of the world, in the presence of spirit in the least particle of matter, in the analogy of laws of man (the microcosm) and of the universe (the macrocosm). An immense desire for synthesis and for universal harmony directs their entire existence." Blondel and Descamps, "Avec Ampère," 24.
27. Viatte, *Sources occultes*.
28. See Segala, "Electricité animale"; Gauld, *History of Hypnotism*, 111–40, 163–78.
29. Robertson, in Pancaldi, *Volta*, 230–31.
30. Lamarck, *Histoire Naturelle*, 42–47. The political valence of debates over vitalism, materialism, and the mechanisms of life is mapped in Jacyna, "Medical Science and Moral Science"; Williams, *Physical and the Moral*; on contemporary British discussions, see Secord, *Victorian Sensation*.
31. Deleuze, *Histoire critique*, 81.
32. "Lettre à M. Ampère sur une classe particulière de mouvements musculaires," *Revue des Deux Mondes*, 2nd series (1833) : 258–66. See discussion in Bensaude-Vincent and Blondel, *Savants face à l'occulte*, 201.
33. Viatte, *Sources occultes*.
34. See Ampère to Bredin, August 18, 1811, 377; Ampère to Roux, late February 1806, 298, both in Ampère, *Correspondance*, vol. 1.
35. Appel, *Cuvier-Geoffroy Debate*; Arago, "Eloge d'Ampère"; Sainte-Beuve and Antoine, *Portraits littéraires*, 247; cf. Picavet, *Les idéologues*.
36. On Oersted's early visit, during which he met with Ampère, Arago, Clément, and Chevreul, as reported by the anti-Cuvierian anatomist Blainville (about whom, see chapters 8 and 9 in this book), see Caneva, "Ampère," 137nn72, 74; further notes are in Sainte-Beuve and Antoine, *Portraits littéraires*, 248.
37. Ampère, "Lettre de M. Ampère à M. le Comte Berthollet"; for discussion of Ampère's chemistry, see Williams, "Ampère."
38. Williams, "Kant, *Naturphilosophie*, and Scientific Method," 16–17.
39. Goethe, "Experiment as Mediator," 15; see Wetzels, "Art and Science."
40. Ampère, notes to lessons 25, 26, and 27, carton 16, chemise 261, Archive de l'Académie des Sciences, in Hoffman, "Ampère's Invention of Equilibrium Apparatus," 313.
41. Ampère, *Essai*, 1:ix.
42. Brown, "Electric Current." On Ampère's application of these concepts, see

Blondel, *Ampère et la création de l'électrodynamique*, 82–84. On his methods, see Blondel and Williams, "Ampère and the Programming of Research"; and Williams's reply in *Isis* 75, no. 3 (September 1984).

43. See Pickering, *Mangle of Practice*, with its notion of "a dance of agency."
44. Tricker, *Early Electrodynamics*, 145.
45. Gooding, *Experiment*; Steinle, "Experiments"; *Explorative Experimente*.
46. In teaching physics, this notion has been replaced by the more convenient "right hand rule": clutching a current-carrying wire with the right hand, with the outstretched thumb pointing in the direction of the current, the fingers curl in the direction in which the magnetic force circulates.
47. See Hofmann, "Ampère, Electrodynamics"; Williams, "Faraday and Ampère"; Steinle, "Experiments."
48. See Blondel, *Ampère et la création de l'électrodynamique*. Biot and Savart, his rivals, had nevertheless based their law of the attraction between a magnet and an electric current on empirical measures, in line with the program of exact physics Biot laid out in his textbook. Their work maintained the Laplacean assumption that electricity and magnetism were two different forces, although they were forced to recognize their interaction. Their law was later shown by Ampère's student (and Balzac's teacher) Félix Savary to be derivable from Ampère's more general law, as was Coulomb's electrostatic law. See ibid.
49. See description in Ampère, *Théorie*, 186–99.
50. On successive forms of this equation, linking the strength of the current in each wire and the distance between them, see Blondel, *Ampère et la creation*; Hofmann, *André-Marie Ampère*; or, briefly, Williams, "Ampère," 146.
51. Ampère, *Théorie*, 186.
52. See Maxwell's commentary on Ampère's research in *Treatise on Electricity and Magnetism*, 132–64. For Maxwell's defense of the necessity of an ether as medium of propagation, see 448–49.
53. Ampère, *Théorie*, 177. On 180 he makes a "hypothesis non fingo" argument à la Newton: "Whatever may be the physical cause to which one might like to relate the phenomena produced by this action, the formula that I have obtained will still remain the expression of the facts." See discussion in Blondel, "Ampère," 62.
54. Maxwell was quick to qualify his praise: "We can scarcely believe that Ampère really discovered the law of action by means of the experiments which he describes. We are led to suspect, what, indeed, he tells us himself, that he discovered the law by some process which he has not shewn us, and that when he had afterwards built up a perfect demonstration he removed all traces of the scaffolding by which he had raised it." Maxwell, *Treatise on Electricity and Magnetism*, 175–76.
55. Ampère, "Notes sur cet exposé," 213.
56. Ampère, "Idées."
57. Ampère to Roux-Bordier, February 21, 1821, in Ampère, *Correspondance*, 2:567.

58. Caneva, "Ampère," 128,
59. Blondel, "Vision physique," 126.
60. See Goldstein, *Console and Classify*.
61. See Dupotet, *Cours du magnétisme*; Georget, *Physiologie du système nerveux*; Rostan, "Magnétisme"; Deleuze and Rostan, *Instruction pratique*. Although there has been considerable recent interest in spiritualism, the occult, and theosophy from the 1860s to the early 1900s, the relations between magnetism and the exact sciences, in France at least, from 1784 (the first magnetism commission) to the middle of the 1800s, remain underexamined. See, however, Méheust, *Un voyant prodigieux*; Viatte, *Sources occultes*; and for England in this period, Winter, *Mesmerized*. For the second half of the century, see Bensaude-Vincent and Blondel, *Des savants*; Owen, *Place of Enchantment*; Noakes, "Spiritualism, Science, and the Supernatural."
62. See Méheust, *Somnambulisme*; Bertrand, *Du magnétisme animal*.
63. Cuvillers, *Archives du magnétisme animal*, 6:60, in Fargeaud, *Balzac*, 150.
64. Quoted in Bertrand, *Du magnétisme animal*, 518.
65. Ultimately, however, Bailly refused to recognize the existence of a magnetic fluid, citing a fear of exposing the Academy to "the ridicule which is attached to all of those who concern themselves with animal magnetism" and of being associated with "the jugglers" who have already taken advantage of the investigation. In Bertrand, *Du magnétisme animal*, 518.
66. Bertrand, *Le globe*, 2 (1825): 1000; see chapters 7 and 8 below.
67. Bertrand, *Du magnétisme animal*.
68. Ampère, *Essai*, 1:147.
69. Ampère, "Fragment sur l'origine de l'idée de causalité," in Barthelémy-Saint-Hilaire, *Philosophie*, 322.
70. Geoffroy Saint-Hilaire, *Notions synthétiques*, 92. See Appel, *Cuvier-Geoffroy Debate*. Similarly, J.-B. Dumas compared the process of "combustion" in animal respiration to that of a steam engine, while physiologist Milne-Edwards wrote of "the division of organic labor." See Simmons, "Waste Not, Want Not."
71. See Valson, *La vie et les travaux*, 341; Ampère, *Correspondance*, 3:636.
72. March 7, 1825: "M. Edwards lit une note sur les Contractions musculaires produites par le contact d'un corps solide avec les nerfs sans arc galvanique"; Magendie and Ampère were commissioners of the report. January 3, 1825: "M. Pelletan fils lit une note sur les phénomènes galvaniques qui accompagnent l'acupuncture. Cet écrit est renvoyé à MM. Duméril, Ampère et Magendie." *Procès-verbaux des séances de l'Académie des Sciences* (1825), 8: 170, 196.
73. Ampère, *Correspondance*, 2:616.
74. Dettelbach, "Humboldtian Science"; Cawood, "Magnetic Crusade."
75. Ampère, *Correspondance*, 2:567; on sunlight, see Sainte-Beuve and Antoine, *Portraits littéraires*, 565.
76. Ampère, *Correspondance*, 1:229; Valson, *La vie et les travaux*, 273.

77. Ampère in 1802 in Valson, *La vie et les travaux*, 167, 161.
78. Ampère in 1801, printed in Valson, *La vie et les travaux*, 373–93; reprinted in Blondel, *Ampère et la creation*, appendix.
79. Ballanche to Ampère, 1805, in Valson, *La vie et les travaux*, 210.
80. Journal de Bredin, December 8 and 15, 1805, in *Correspondance*, 1:294.
81. Ampère to Bredin, March 29, 1818, ibid., 1:243
82. In Valson, *La vie et les travaux*, 194–95. An *oraison* is a phrase or prayer constantly repeated by contemplatives.
83. Ibid., 400; Ampère, March 1, 1817, in *Correspondance*, 2:525.
84. Ampère, October 28, 1824, in *Correspondance*, 1:275.
85. See Goldstein, *Post-Revolutionary Self*.
86. An influential statement of this opposition was in Foucault's introduction to Canguilhem's *Normal and Pathological* (1991), which pointed to "the line that separates a philosophy of experience, of sense and of subject and a philosophy of knowledge, of rationality and of concept. On the one hand, one network is that of Sartre and Merleau-Ponty; and then another is that of Cavaillès, Bachelard and Canguilhem. In other words, we are dealing with two modalities according to which phenomenology was taken up in France" (8). See Janicaud, *Une généalogie*; Braunstein, "Bachelard, Foucault, Canguilhem."
87. Maine de Biran, *L'influence*, 113–78.
88. Ibid., 143.
89. Likewise, Jonathan Crary and Elizabeth Green Musselman argue that, in the domain of vision, perception in all its forms was often presented at this time as a form of hallucination. Crary, *Techniques of the Observer*; Musselman, *Nervous Conditions*. See chapter 5 below.
90. Maine de Biran, *L'influence*, 114, 63–64.
91. For Foucault, Destutt de Tracy was significant as a transitional figure between philosophy as the analysis of ideas (aligning them on the "table" of differences and identities) and the study of "man" as a producing, speaking, living (and dying) historical organism, thanks to Tracy's analysis of the physiological formation of ideas.
92. See Azouvi, *Maine de Biran*, especially chapter 5, "La science subjective," 207–83.
93. Maine de Biran, *Mémoire sur la décomposition*, 89. Elsewhere he wrote: "I found myself led to recognize (by a truly *explanatory hypothesis*) the existence of a *hyper-organic* force, a permanent substance which we call *soul* (*âme*) when it does not have the sentiment of itself in its acts, and *self* (*moi*) or individual person, when it has this sentiment, in the constant effort that it exerts while awake. The soul is to the organic center of movement [*motilité*] what the latter is to the nervous and muscular system. . . . In order for effort to arise, the soul must begin to act on the center with which it is united, and, just as there must be heterogeneity between the nervous and the muscular systems in order for movement to be *felt*, there must be heterogeneity of the two substances, movement and soul, in order for there to be a feeling of ef-

fort." Maine Biran to Ampère, October 21, 1805, in Ampère, *Correspondance*, 1:287–90, cited in Goldstein, *Post-Revolutionary Self*, 138. Note the essential heterogeneity of soul and body, as in Descartes, here analogized with the essential heterogeneity of nerves and muscles (needed for the perception of causality in *emesthèse*). On the hyperorganic, see Azouvi, *Maine de Biran*, 81–83.

94. Maine de Biran, *Correspondance philosophique*, 37–58, 122–61. Ampère crowed: "I have so much changed the ideas of Maine Biran on the subject of Kant that he told me this morning that Kant was the greatest metaphysician who has ever existed." Ampère to Roux, late February 1806, in Ampère, *Correspondance*, 1:298.

95. See Naville, *Maine de Biran*, 414. On his mysticism, see Huxley, *Themes and Variations*.

96. Later philosophers to recognize his influence include not only Bergson but Merleau-Ponty, whose phenomenology began from the experience of embodiment. See also Derrida, *On Touching*, 143–58.

97. Ampère in "Fragment sur l'origine de l'idée de la causalité," in Barthelémy-Saint-Hilaire, *Philosophie*, 322–23.

98. See article from *Le temps*, in Ampère, *Essai*, 1:lxv, with a fascinating discussion of Laplace and the perception of the words sung in opera, Necker cubes, and trompe l'oeil painting. Ibid., 1:lxvij.

99. Ampère to Bredin, October 10, 1805, in Ampère, *Correspondance*, 1:285.

100. Ampère to Maine de Biran, September 18, 1810, in Barthelémy-Saint-Hilaire, *Philosophie*, 353.

101. The idea of matter comes from these resistances; see Jean-Jacques Ampère in Barthelémy-Saint-Hilaire, *Philosophie*, 80.

102. Ampère, *Essai*, 1:lxiij; see Marcovich, "Théorie philosophique des rapports"; Merleau-Ponty, "*Essai sur la Philosophie des sciences d'Ampère.*"

103. Ampère to Roux, late February 1806, in Ampère, *Correspondance*, 1:298. See also Jean-Jacques Ampère in Barthelémy-Saint-Hilaire, *Philosophie*, 155.

104. Ampère, *Essai*.

105. Budding positivist E. Littré, in his introduction to the posthumous second volume of Ampère, *Essai*, 2:xcvj, notes, "A place will always be reserved [in future books on the sciences] for the name of M. Ampère and for his beautiful and simple law on electro-magnetism," implying that, on the contrary, the *Essai* is not likely to be reprinted. On natural versus artificial classifications, see Corsi, *Age of Lamarck*; and see Riskin, *Science*, on Guyton de Morveau.

106. Ampère, *Essai* (1843), 2:vii.

107. See discussion of D'Alembert's "mappe-monde" and Bacon's classifications in Robert Darnton, "Philosophers Trim the Tree of Knowledge."

108. Ampère, "Technologie," in *Essai*, 1:97; "Cybernétique," in *Essai*, 1:127. The fact that Ampère's concept of technology emerged within a taxonomy of human knowledge is one case of the intimate links in this period among natural history, activist epistemology, and machines. Another is a key text in the formation of the modern concept of "technology," the *Technonomie* of

Gérard-Joseph Christian, director of the Centre National des Arts et Métiers and a friend of Ampère's (see Alexander, *Mantra of Efficiency*; Guillerme and Sebestik, "Commencements"), which was a classification of production processes and their economic value.
109. Ampère, *Essai*, 2:80.
110. Ibid., 2:140. Norbert Wiener cited Ampère's "Cybernétique" in *Human Use of Human Beings*. While Ampère would have rejected twentieth-century cybernetics' ontological monism, nevertheless, its ambition—to create a new science to unite all other sciences, to guide technical interventions, and to reform the order of human society—was one Ampère shared. Clarke and Henderson, *From Energy to Information*.
111. See Frankel, "J. B. Biot."
112. See Morrell and Thackray, *Gentlemen of Science*; Cahan, *From Natural Philosophy*; Schaffer, "Scientific Discoveries"; Frankel, "J. B. Biot."
113. See Bénichou, "Le grand oeuvre de Ballanche"; McCalla, *Romantic Historiosophy*; Schwab, *Oriental Renaissance*; Busst, "Ballanche"; Sharp, "Metempsychosis"; Charlton, *Secular Religions*; Berenson, *Populist Religion*.
114. Bonnet is a key figure in Lovejoy's *Great Chain of Being*.
115. Ballanche, *Oeuvres complètes*, 29; McCalla, "Palingénésie Philosophique."
116. Ballanche, *Palingénésie sociale*, 208, in Ballanche, *Oeuvres complètes*.

CHAPTER THREE

1. See Botting, *Humboldt*; on French hostility, see Daumas, *Arago*; Dettelbach, "Romanticism and Resistance," 247–58.
2. Schelling, quoted in Humboldt, *Cosmos*, vol. 1, p.55.
3. Humboldt, *Cosmos* vol. 1, 23.
4. On the importance of the concept of "the aesthetic" for critical philosophy, see Eagleton, *Ideology of the Aesthetic*; Bürger, *Theory of the Avant-Garde*; Marcuse, *Aesthetic Dimension*; Rancière, *The Politics of Aesthetics*.
5. Eggli, *Schiller et le romantisme français*.
6. Humboldt, Heath, and Losonsky, *On Language*; the quote is from Heidegger, "The Way to Language," 405.
7. My conception of Humboldtian science is indebted to the Ph.D. dissertation and several articles of Michael Dettelbach, as well as his presentation of the second volume of the new edition of *Cosmos*; also, Cannon, *Science in Culture*; Botting, *Humboldt*; Bourguet, "La république des instruments," Kehlmann, *Measuring the World*. More recent works treating Humboldt in the USA—Sachs, *Humboldt Current*, and Walls, *Passage to Cosmos*—have enlarged the picture by presenting Humboldt as a founding figure of the environmental movement.
8. See Aubin, Bigg, and Sibum, *Heavens on Earth*.
9. Schlegel, *Kritische Fragmente*, quoted in Brain, "Romantic Experiment," 224.
10. On Schiller's literary and critical influence in France, see Eggli, *Schiller*.
11. Bourguet, "La république des instruments."

12. See Daston and Galison, *Objectivity*.
13. See Lenoir, "Gottingen School"; Rehbock, *Philosophical Naturalists*; Schlanger, *Les métaphores*; Richards, *Romantic Conception of Life*.
14. See Beiser, *Fate of Reason*.
15. More exactly, organisms had to be studied "as if" they were expressions of freedom; see Richards, *Romantic Conception*. On Kant's diverse impact on science, see L. P. Williams, "Kant, *Naturphilosophie*, and Scientific Method."
16. Kant and Ellington, *Grounding*, 53. See Foucault's discussion, in *Order of Things*, of Kant's presentation of man as the "empirico-transcendental doublet."
17. See Kant, *Critique of Pure Reason*, on the categories; Cassirer, *Kant's Life and Thought*, 170; Kant, *Critique of Judgment*, section 36; Deleuze, *Kant's Critical Philosophy*, 3–10.
18. "For we are constrained to think the pure will as something bound by law and hence 'objective,' but this objectivity belongs to a sphere totally distinct from that which is expressed in the spatiotemporal phenomenon. It is not a world of things we are assured of here, but one of free personalities; not a set of causally related objects, but a republic of self-sufficient subjects purposively united." Cassirer, *Kant's Life and Thought*, 247, 154.
19. See Clavier, *Kant*.
20. Kant, *Grounding*, 49. The notion of autonomy ("self-law") can be traced to Rousseau's general will ("obedience to the law one has prescribed for oneself is freedom," *Social Contract*, 78). It has roots in stoicism, Saint Paul (who speaks of the Gentiles as "a law unto themselves" (Romans, 2:14)), and Luther's "Freedom of a Christian Man," where an inner law of faith overrides laws of the state. See Schneewind, *Invention of Autonomy*.
21. It was this gap between mechanical causality and the view of nature as a purposive, organic whole that Schelling's *Naturphilosophie* sought to close: "If, finally, we gather up Nature into a single Whole, *mechanism*, that is, a regressive series of causes and effects, and *purposiveness*, that is, independence from mechanism, simultaneity of causes and effects, stand confronting each other. If we unite these two extremes, the idea arises in us of a purposiveness of the whole: Nature becomes a circle which returns into itself, a self-enclosed system." Schelling, *Ideas*, 40.
22. The discussion of the difference between a machine and an organism is in Kant, *Critique of Judgment* (Pluhar, trans, 1987), sections 64, 65.
23. Ibid., 253 (section 65).
24. Ibid., 316–17 (section 82): "The possibility of reconciling the two ways of presenting [how] nature is possible may very well lie in the supersensible principle of nature (nature outside as well as within us). For presentation in terms of final causes is only a subjective condition of the use of our reason, [which applies] when reason wants us to judge certain objects not merely as appearances but insists on referring these appearances themselves, along with their principles, to the supersensible substrate. Reason insists on

making that reference so that it can consider as possible that there can be certain laws unifying those appearances, laws that reason can conceive of only as arising from purposes." In all quotes from this work, italics and sections in brackets are in the translation.

25. Ibid., 324 (section 82).
26. Ibid., 319 (section 83).
27. Ibid., 319 (section 83).
28. Both quotes ibid., 320 (section 83).
29. Ibid., 321 (section 83). See Cassirer, *Kant's Life and Thought*, 333. See also Kant, "Perpetual Peace."
30. Kant, "What is Enlightenment," in Schmidt, *What is Enlightenment*, 63; see also Foucault, "What Is Enlightenment?" 32–50.
31. This impression was perpetuated by such posthumous accounts as De Quincey, "The Life of Immanuel Kant."
32. Schiller, *Aesthetic Education*, 3.
33. Willoughby and Wilkinson's translation of the letters systematically renders *Selbständigkeit* as "autonomy." Critical discussions of Schiller frequently assimilate this notion to Kant's use of the term autonomy, at the cost of some confusion. See Bürger, *Theory of the Avant-Garde*, 41–46, on Kant and Schiller; 6–14 on Adorno, Lukacs, and Marcuse. Bürger's argument rests upon tracing the theoretical source for the functional differentiation of art as a separate sphere of activity (the "autonomy" championed not only by "bourgeois ideology" but also by critical theorists such as Adorno) back to Schiller. Yet, as I will argue, Schiller's ideal of autonomy, perhaps paradoxically, involves explicit recognition of interdependence.
34. Schiller, *Aesthetic Education*, 34, 17.
35. Ibid., 55, 109, 189.
36. Ibid., 21, 213. On the influence of Kant's critic Reinhold on Schiller's formulation—at times quite varied—of autonomy, see Roehr, "Freedom and Autonomy in Schiller."
37. Schiller, *Aesthetic Education*, 215.
38. Ibid., 19, 21.
39. Ibid., 33.
40. We have here a version of the central paradox of Schiller's work. The aesthetic is presented as a means of leading us upward to the world of form, but at the same time it appears as a corrective to an excess of form; similarly, his view of truth as both "pure object" and as part of the world of sense may well appear contradictory. Lovejoy attributes these difficulties to Schiller's attempt to combine "the two Gods of Plato—the immutable and self-contained Perfection and the Creative Urge which makes for the unlimited realization in time of all the possible" into a single system: "Since they are essentially antithetic, in any actual juncture in experience one of them must in some degree be sacrificed to the other." Schiller's wish to have it both ways, for Lovejoy, resulted in incoherence, though his sense is that for Schiller and subsequent romantics, "plenitude has the last word": the principles

41. Schiller, *Aesthetic Education*, 217, 219.
42. Ibid., 21. Upon mention of the "political artist," we must note the influence of Schiller on Goebbels's *Michael*, although as I hope to have made clear, Schiller's utopia had nothing to do with fascism, and is aimed specifically against despotism. Under the Third Reich, Schiller's *Don Carlos* was banned because its performances became the occasion for protests against the Nazi regime. Cf. Eagleton, *Ideology of the Aesthetic*; Jones, "Schiller, Goebbels, and Paul de Man."
43. On Humboldt's aesthetics, see Finkelstein, "Conquerors"; and Dettelbach, introduction to *Cosmos*. Humboldt lets us see how the sciences participate in what Rancière calls "the distribution of the sensible," the political dimension of the aesthetic. Rancière, *Politics of Aesthetics*.
44. Humboldt, *Cosmos*, 1:36, 26, 25.
45. Ibid., 1:76, 77. On the visibility of scientific labor, see Cawood, "François Arago"; Blondel, "Electrical Instruments"; Holmes and Olesko, "Images of Precision"; see the longer history in Roberts, Schaffer, Dear, eds. *The Mindful Hand*.
46. Dettelbach, "Humboldtian Science"; and "Stimulations of Travel," esp. 53–55, which depicts Humboldt's understanding of the researcher as sensitive instrument, in harmony with the late Enlightenment's culture of sensibility and vitalism.
47. See Kant, *Critique of Pure Reason*, section titled "Transcendental Aesthetic" (153–92) on pure forms of intuition (space and time); "Transcendental Deduction" (219–66) on the categories, pure forms of understanding (quantity, quality, relation, modality).
48. See Daston and Galison, "Image of Objectivity"; *Objectivity*, chapter 3.
49. See Humboldt, *Expériences sur le galvanisme*; Dettelbach, "Stimulations of Travel," 55, "Face of Nature"; Schaffer, "Self-Evidence"; Richards, *Romantic Conception of Life*; on Ritter's self-experiments, see Strickland, "Ideology of Self-Knowledge."
50. The picture adorns the cover of Jardine and Cunningham, *Romanticism and the Sciences*.
51. See Humboldt to Pictet, June 22, 1798, p. 5; to Pictet, November 7, 1798, p. 7; to Forell, à Ortava (sur Ténériffe), June 24, 1799, pp. 23, 39, all in Humboldt, Delamétherie, and Hamy, *Lettres américaines*.
52. Daston and Galison, "Image of Objectivity," 83.
53. Daston, "Moral Economy of Science," 23.
54. Relevant discussions of the role of instruments in social and epistemological coordination include Sibum, "Reworking the Mechanical Value of Heat"; Latour, *Pandora's Hope*, esp. chapters 2, 3 and 6; Galison, *Einstein's Clocks*; see also contributions to *Culture Technique* 7 (1992) and *Osiris* 9 (1993), special issue titled "Instruments"; O'Connell, "Metrology."
55. Dettelbach, *Romanticism and Administration*, 133; Schaffer, "Astronomers Mark Time."

56. Humboldt, *Cosmos*, vol. 2, 355–56. This quote directly follows a passage insisting on the difference between the organs of the lower animals and the higher, suggesting that the artificial organs of humans are part of what define them as a species. See the insightful discussion of this passage in Kapp and Chamayou, *Principes*, 130; Kapp's work is discussed in an exceptionally suggestive footnote in "Machine et Organisme," in Canguilhem, *La connaissance de la vie*. Humboldt identifies organs and instruments throughout *Cosmos*: 108, 112, 179, 200, 332, and 353, on the telescope, "which exercised an influence similar to some great and sudden event."
57. See "Treatise on Nomadology," in Deleuze and Guattari, *Thousand Plateaus*, 387–467.
58. Jackson, *Harmonious Triads*, 45–74; Morell and Thackray, *Gentlemen of Science*, 509–17; Cannon, *Science in Culture*, 181–96.
59. Humboldt, *Des Lignes Isothermes*.
60. See Cawood, "Terrestrial Magnetism"; Pratt, *Imperial Eyes*, places Humboldt's natural history in the context of exploration and imperialism, as does Dettelbach, *Romanticism and Administration*. Unquestionably, the projects of surveying, mapping, and artistically evoking the Americas were enticements for colonial and imperial adventures, many of which had dire consequences for the inhabitants of these lands. It is possible, however, to read an image like the frontispiece of *Views of the Andes*, where a mythological figure representing Western classical culture raises a fallen Aztec god to his feet, not as paternalistic bad faith, but as a depiction of the late-Enlightenment ideal of freedom as reciprocity and interdependence. The success of Humboldt and those who followed him in instituting this ideal is another question.
61. Schiller, *An de Freude* (1785), translation adapted from Raptus Association for Music Appreciation website, http://raptusassociation.org/ode1785.html, accessed June 12, 2011.
62. Schiller to Körner, February 23, 1793, in Schiller, *Aesthetic Education*, 300 (italics mine).
63. Humboldt frequently plays a role in accounts of the institution of standards and new instrumental verification; in his correspondence he served as a human relay at the intersection of multiple networks of scientific practitioners. On methods discussed here, see Wise's transitional chapters in Wise, *Values of Precision*; Olesko, "Meaning of Precision"; Schaffer, "Astronomers Mark Time"; O'Connell, "Metrology"; Alder, *Measure of All Things*.
64. Humboldt rarely foregrounded the language of "objectivity" versus "subjectivity"; when he did so, it was not to disparage the latter in favor of the former, instead arguing that they are inevitably entwined: for instance, he wrote, "The objective world, conceived and reflected in us by thought, is subjected to the eternal and necessary conditions of our intellectual being." Humboldt, *Cosmos*, 1:76. Humboldt, like Schiller (and Hegel), saw the goal of science as an eventual fusion of the two terms. In the same paragraph Humboldt wrote, "Science only begins when the spirit takes possession of substance, when the attempt is made to subject the mass of experience to

rational knowledge; science is spirit turned towards nature. The external world exists for us only when we take it into ourselves and it forms itself into a view of nature." The existence of the "external" or "objective" world for us depended on humans "taking hold" of or "overpowering" (*bemächtigt*) external substance (*Stoffe*) in order to "form" it into a view of nature (*Naturanschauung gestaltet*), the same dynamic interaction between "Stoffe" and "Gestalt" that Schiller called the aesthetic state. Humboldt's conception of "objectivity" should thus be understood, perhaps paradoxically, as a mixture between "objectivity" and "subjectivity," a tension analogous to the paradox of autonomy in Schiller (see note 40 above), where the goal of the "moral state" was a mixture of itself "morality" (form) and its opposite, "sense" (substance). In both, crucially, the two contradictory terms were brought into harmony by concrete mediators: art and instruments.

65. Du Bois-Reymond in 1849, quoted in Finkelstein, "Conquerors," 179; on Humboldt as transitional figure to nineteenth-century laboratory science, see that essay, as well as his many cameos in Cahan, *Helmholtz*; Lenoir ("Eye as Mathematician") has argued that we see Helmholtz's findings as accounts of complex experimental arrangements of diverse interacting apparatuses, the kind of amalgamated system that Humboldt arranged on a global scale.

66. On Humboldt and the birth of environmentalism, see Sachs, *Humboldt Current*, Walls, *Passage to Cosmos*. Humboldt's deployment of instruments as an autonomous externalization of the categories is part of a more general movement of post-Kantian philosophy. While retaining the idea that knowledge is a function of the subject as much as of things, many displaced the categories from their seat in the transcendental ego: Schiller's view that the formal universals of art, politics, and science must emerge within collective material practices was one response to this crux; Schopenhauer relocated the constitutive ego in the will and physical drives, making representations of the world a function of physiology; Wilhelm von Humboldt located the structures of thought in culturally and historically variable, external, and shared languages. For a comparison of the energetic conceptions underwriting the work of the Humboldts, see Reill, "Science and the Construction"; on the stature of the Humboldts later in the century and their influence on Franz Boas's landmark concept of "culture," see Stocking, *Shaping of American Anthropology*; and Bunzl, "Franz Boas."

67. Humboldt in Botting, *Humboldt*, 268, 273.

CHAPTER FOUR

1. Laplace, *Mécanique céleste; Philosophical Essay*, 4; see Hacking, "Cracks in Nineteenth Century Determinism." On the nebular hypothesis, see chapter 9 below, as well as Merleau-Ponty, *Science de l'univers*. Bachelard sees a theological intent in Laplace's determinism, even as he claimed to Napoleon to have no need of the "hypothesis" of God: "the hypothesis of the mathematician who possesses a formula that would reunite the past and the

future of all movements is, in the very style of Laplace's writing, the 'God hypothesis.'" *L'activité rationaliste*, 212.

2. Porter, "Objectivity and Authority"; Weiss, *Making of Technological Man*; Shinn, *L'Ecole Polytechnique*.
3. Hugo, letter of January 1864, in Gosling, *Nadar*, 16. For good background on Arago, see Daumas, *Arago*; Audiganne, *Arago*; Sarda, *Les Arago*; Cawood, "Arago"; Grison, "François Arago"; and Levitt, *Shadow of Enlightenment*. On the symbolically charged intersection of the Observatory, see Aubin, "Fading Star." On links between balloons and engineering, see Gillispie, *Montgolfier brothers*; on Hugo's views of technology, see Charles, *La pensée technique*.
4. Hugo's frequent imagery of infinite space and multiple worlds found inspiration in his visits to the observatory, one of which is described in his *Le promontoire du songe*; see also Seghers, *Victor Hugo visionnaire*.
5. Daston and Galison, *Objectivity*; "The Image of Objectivity." It would also be possible to read the concerns of mechanical objectivity onto the obsession with precision to lower the tolerances for variation in machine parts and to reduce "play" in engineering, as discussed in Alder, *Engineering the Revolution*; and Otto Sibum, "Exploring the Margins of Precision"; and as shown in many essays in Wise's collection *Values of Precision*.
6. See Dhombres, "L'image 'scientiste,'" Shinn, *L'Ecole Polytechnique*, Weiss, *Making of Technological Man*. Pierre Bourdieu has presented the school as a key site for the formation and reproduction of the *habitus* of a technocratic elite in his *La noblesse d'état*. According to Isambert, *De la charbonnerie*, Etienne Arago, François's brother, held meetings of the Carbonari at the school.
7. Foucault, *Discipline*.
8. Carnot, "Discours," 2:100–101. For a discussion of Carnot's "art of war," see Carnot, *Révolution et mathématique*, 1:155.
9. Note the transitivity in republican discourse between soldier and citizen—Carnot declared, "Tout citoyen est soldat"—and between citizen and representative, since all are equally constitutive members of the *res publica*. See Pilbeam, *Republicanism in France*.
10. See chapter 9.
11. Gillispie, *Pierre-Simon Laplace*, 177.
12. When Laplace served as director in the 1820s, an increasing tendency toward "abstract mathematics" was noted by critics, including Comte. See also Belhoste, *Augustin-Louis Cauchy*. Arago's "Sur l'ancienne Ecole Polytechnique," in his *Oeuvres*, laments changes in the institution's character in the Restoration and affirms its practical and republican ideals.
13. See Kranakis, *Constructing a Bridge*; Graber, "Obvious Decisions"; Alder, *Engineering the Revolution*.
14. Grattan-Guinness, "Work for the Workers"; "'Ingénieur Savant'"; Daston, "Physicalist Tradition."
15. Bradley, "Facilities for Practical Instruction"; Hoskin, "Education"; Smeaton, "Early History."

16. See Grattan-Guinness, "Work for the Workers," 16.
17. See Grattan-Guinness, "'Ingénieur Savant.'"
18. Charles Dupin, *Développement de géométrie*, 72; quoted in Guillerme, "Network," 155.
19. Guillerme, "Network"; Mattelart, *La communication*.
20. Hachette, *Traité Elémentaire*.
21. See Hankins, *D'Alembert*.
22. Petit, " Sur l'emploi du principe." On Leibniz and *Forces Vives*, see Mary Terrall, *Man Who Flattened the Earth*; and Wise, "Mediating Machines."
23. Montgolfier, quoted in Vatin, *Le travail*, 63; Coulomb, *Théorie des machines simples*, 255, discussed in Vatin, *Le travail*, 36–56; on fatigue, see Rabinbach, *Human Motor*.
24. Navier, ed., *Architecture hydraulique*, 356; quoted in Grattan-Guinness, "Work for the Workers," 13; see discussion in M. N. Wise and Smith, "Work and Waste"; Brain, *Graphic Method*.
25. Coriolis, *Du calcul*, 91–95, 40–44, 33; quoted and discussed in Grattan-Guinness, "Work for the Workers," 16; Poncelet, *Introduction à la mécanique industrielle*, 95–96; Navier, ed., *Architecture Hydraulique*.
26. Poncelet, *Introduction à la mécanique*, 95–96; see also Poncelet, *Cours de Mécanique Industrielle*.
27. Guillerme and Sebestik, "Les commencements"; Alexander, *Mantra of Efficiency*, 36.
28. Goethe, in *Le globe*. September 6, 1828, Tome VI, 673.
29. On Athénée, see Staum, "Physiognomy and Phrenology."
30. Dupin's *Forces* was an application of Navier's idea of work as universal currency or exchange, which allowed the addition and comparison of a range of distinct reservoirs of productivity; a similar concept appears with the Saint-Simonian vision of a central bank coordinating production and consumption.
31. Fox, "Charles Dupin," 473.
32. Recall Smith's assertion in *The Wealth of Nations* that increasing the division of labor encourages innovation and its later contradiction, in which he argues that the repetition involved in much mechanical work makes any individual as stupid as it is possible to be.
33. Grattan-Guinness, "Work for the Workers."
34. Say, *Traité d'économie politique*.
35. Robert Fox uncovered the Clément-Desormes/Carnot connection, including Sadi's allusions to his lectures; see his commentary in Carnot, *The Motive Power of Heat*.
36. Cf. Rancière, *Nights of Labor*, on the breakdowns of communication between workers and those who claimed to speak for them.
37. Notably by Schaffer, "Babbage's Intelligence"; Wise and Smith, "Work and Waste"; Ashworth, "The Calculating Eye"; Musselman, *Nervous Conditions*; Erna Fiorentini, "Practices of Refined Observation"; Jackson, "Joseph von Fraunhofer."

38. See Schaffer, "Babbage's Intelligence"; *Cultural Babbage*.
39. See discussions in *Science Incarnate*, ed. Lawrence and Shapin, especially Schaffer, "Regeneration"; and Winter, "A Calculus of Suffering," on Ada Lovelace.
40. See Roberts, Schaffer, Dear, *The Mindful Hand*.
41. Smith, *The Body of the Artisan: Art and Experience in the Scientific Revolution*.
42. Gillispie, "Jacobin Philosophy of Science," in Clagget; Pinault, "Les mains de l'*Encyclopédie*." Other important points in this story include Erigena's radical elevation of labor and the mechanical arts; the gradual revaluation of the *vita activa* against the *vita contemplativa*; the argument from makers' knowledge, as found in Vico and Hobbes; and the parallels between Locke's theory of the origin of wealth in the mixture of labor with matter and his theories of the origins of ideas in experience, frequently cited as a precursor to Marx's theory of labor and the "species being." See Noble, *Religion of Technology*. On the specifically epistemological dimensions of the labor theory of knowledge, see Meli, *Thinking with Objects*, which brings to light the "minor" tradition of sixteenth- and seventeenth-century mechanics frequently neglected in histories written backward from Newton's universal laws. The industrial mechanics of the early nineteenth century retained much of the equipment of this tradition: pulleys, springs, and pendulums (Borda, Hachette), as well as its attention to hydrodynamics, friction, and elasticity. These are precursors to Gaston Bachelard's view that "labor" is fundamental to modern scientific knowledge, a view linked to his claims that science involves *phénoménotechniques*; it produces nonnatural "effects" by means of instruments that are "reified theorems." *L'activité rationaliste*, 84. In Canguilhem's suggestive (and biologically informed) reading of Bachelard, "Scientific proof is labor because it reorganizes the given, because it provokes effects without natural equivalents, because it constructs its organs." "L'histoire des sciences," 192.
43. See Carnot's theory expressed in his *Machines in General*; Poncelet's *Textbook*; Navier, *Mécanique industrielle*; and Hachette, *Traité élémentaire des machines*; Hachette taught the course on machines to all cadets at the Ecole Polytechnique until the restoration; the course was later taken over by Arago.
44. Dumas, *Traité de chimie*, vol. 1.
45. Arago, "Discourse on Electoral Reform"; see below. Similarly, see Weiss, *Making of Technological Man*, for an account of the Ecole Centrale's attempt to institutionalize a labor-based epistemology.
46. On the hierarchies workers themselves recognized — including the highly skilled and itinerant "sublimes" — see Rifkin and Thomas, *Voices of the People*, 104–11; and Harvey, *Paris*, 230–31.
47. Courses on political economy were later taught at the CNAM by Adolphe Blanqui and the Saint-Simonian Michel Chevalier.
48. Poncelet, *Cours de Mécanique Industrielle*, 95–96.
49. See Alder, "Tolerance."
50. In other words, French engineering, shaped by republican and revolutionary

traditions, sought to minimize class hierarchy in the 1820s and 1830s, yet the division between laborers and bourgeois sharply reemerged with the class-based struggles of the 1840s. See Sarah Maza, *The Myth of the Bourgeoisie*.

51. See Wise, "Mediating Machines"; Wise and Smith, "Work and Waste"; Wise, "Mediations," 197; *Values of Precision*, introduction.
52. Wise, "Mediations," 256. In *L'activité rationaliste* Bachelard described Laplace's science as essentially idealist (212); see also Arago's biographical notice on Laplace (*Oeuvres*, 3, 456–515) indicating disengagement from the work of experiment and observation.
53. See Wise and Smith, "Work and Waste"; Serres, "Turner Discovers Carnot"; Prigogine and Stengers, *Order out of Chaos*; Schaffer, "Nebular Hypothesis"; Hayles, *Chaos Bound*.
54. "Voulez-vous me confier la clef du sucre?" Arago, "Histoire de ma jeunesse," 58.
55. Arago, "Biographies des principaux astronomes," 485.
56. Arago in J. A. Barral, ed., *Oeuvres*, 3:33–44, quoted in Crosland, *Society of Arcueil*, 427.
57. Arago's prolixity compelled his English translators to warn: "The attentive reader, while pursuing with deep interest some of the more argumentative parts of the work, may, probably, sometimes be induced to think that his instructor has entered more into details than he need have done to establish the correctness of his position; but it will invariably be found that these excursive episodes terminate in some useful result." They attribute this verbosity to the demands of explaining science to an uninformed audience, yet such "excursive episodes" also occurred in Arago's reports to the Academy, suggesting that his conception of the proper style of science exceeded the standards of "utility" promoted by his British counterparts. Arago, *Popular Astronomy*, v.
58. See Arago, "Histoire de ma jeunesse."
59. Arago, "Eloge d'Ampère," II, 102.
60. Arago, "Discours de M. François Arago," xi–xvi.
61. Arago, "Rapport," 27; "Eloge de Condorcet," 119.
62. On the expressive self, see Taylor, *Sources*. One observer described Arago's performances in the *Edinburgh Review*, "The very moment he enters on his subject he concentrates on himself the eyes and the attention of all. He takes science as it were in his hands: he strips it of its asperities and its technical forms, and he renders it so clear, that the most ignorant are astonished, as they are charmed at the ease with which they understand its mysteries. There is something perfectly lucid in his demonstrations. His manner is so expressive that light seems to issue from his eyes, from his lips, from his very fingers" (October 1856, 314, quoted in Miller, *Discovering Water*, 124).
63. Dettelbach, "Humboldtian Science."
64. Reciprocally, see the abundant references to Arago in Humboldt's *Cosmos*.
65. Aubin, "The Fading Star"; on the republican daydreams of Arago's assistants, see Levitt, "'I Thought.'"

66. Arago, "De la scintillation," 36.
67. On the acrimonious rupture and subsequent polemics between Arago and Biot (the leading representative of the Laplace school), see the insightful and deeply researched work by Theresa Levitt, *Shadow of Enlightenment*.
68. Arago, in Daguerre et al., *Historique*, 43. Such parallels between the eye and optical apparatus have a long history. See Hankins and Silverman, *Instruments*, chapter 7, esp. p. 177.
69. See Grattan-Guinness, "Work for the Workers."
70. Levitt, "Biot's Paper." On reactions, especially by Guillermo Libri, see Fox, "Rise and Fall of Laplacian Physics"; Tobin, *Léon Foucault*.
71. The commitments of Arago's polity of science could be contrasted with those of Cuvier, against whom he supported Geoffroy. See Outram, *Georges Cuvier*.
72. See Condorcet, *The Nature and Purpose of Public Instruction*, in Condorcet, ed. Baker, *Condorcet*, 105–42.
73. Arago, *Eloge de Condorcet*.
74. Arago, "Carnot." Arago's analogy anticipates Durkheim and Mauss's isomorphism between taxonomy and social structure in *Primitive Classification*.
75. Arago, "James Watt," 438.
76. During the Restoration and the July Monarchy, claims of the importance of the Third Estate, inaugurated during the Revolution, were reclaimed by workers in support of their collective rights. See Sewell, *Work and Revolution*.
77. Arago, "James Watt," 431.
78. See Miller, *Discovering Water*, chapter 6, for a detailed discussion of the composition and reception of this text, including Arago's journey in Britain to gather sources, his difficult correspondence with Watt's son, and the course of his reception in England, where after his death his reputation was tarnished, according to Humboldt, in an "infamous manner" due to "party spirit"(125).
79. Arago, "James Watt," 491.
80. See ibid., 435, where he speaks of England's system of canals and railroads linking the entire nation for the rapid movement of goods and exclaims: "Voilà donc l'utopie des nouveaux économistes réalisée."
81. Arago, "*Réforme éléctorale*," 603, 614. See Heurtin, *L'espace public parlementaire*, on "le peuple toujours malheurese" as a recurrent topos in the rhetoric of the Chamber of Deputies of this period.
82. On the dialectic between individual and collective in French romanticism, see Sharp, "Metempsychosis"; and Breckman, "Politics."
83. Barger and White, *Daguerreotype*, 27; for comments on Arago's speech, see Benjamin, *Oeuvres II*, 17. Above all, see McCauley's insightful analysis of the politics of the speech in "François Arago."
84. Paper and digital reproductions of daguerreotypes thus fail to render their actual impact; they must be seen in person. A recent exposition at the Musée d'Orsay and the Metropolitan Museum of Art made it possible to come face to face with the daguerreotype's crystalline eeriness in an unprecedented collection of early plates. See Bajac, *Le daguerreotype français*.

85. See Pinson, *Speculating Daguerre*, on Daguerre's art and entrepreneurship.
86. Arago, "Rapport," 20.
87. Arago, "Rapport," 10–11.
88. Arago, "Rapport," 19. On Egyptomania, see Assmann, *Moses*, including discussion of Humboldt's frontispiece of *Isis*; on metals, mummies, and automata, see Nelson, *Secret Life*.
89. In his "Rapport," Arago anticipates using daguerreotypes as part of descriptive geometry, "to scale up to the exact dimensions of the highest and most inaccessible parts of edifices" (20).
90. Ibid.
91. See Bann, *Delaroche*. On the close ties between romantic aesthetics and early romanticism, see Nochlin, *Realism*; Galassi, *Before Photography*; Rosen and Zerner, *Romanticism and Realism*; Recht, *La lettre de Humboldt*; Armstrong, *Scenes*; McCauley, "Talbot's Rouen Window."
92. See chapter 5 on Bachelard's theory of "phénoménotechniques" as the production of effects, a notion that arose principally from his studies of immediately post-Laplacean electricity and chemistry; he argues that the chemistry of this time created an "artificial nature," a view that found support in Comte's positivism.
93. Levitt, in *Biot's Paper*, quotes very similar lines from Biot, who used exposures on paper to explore the action of "invisible radiation." But whether Arago is borrowing from his rival Biot or, instead, continuing Humboldtian field science's patient analysis and mapping of atmospheres (or milieus), the quote shows that Arago appreciated the daguerreotype in part as *an inscription of the invisible*, a dimension of reality otherwise inaccessible to humans. See Joel Snyder, "Visualization."
94. On the invisibility of scientific labor, see Shapin, "Invisible Technician"; on its visibility in the public representations of Arago, see Blondel, "Electrical Instruments."
95. Levitt, "Biot's Paper," 457; see also Levitt, *Shadow of Enlightenment*.
96. The variable meanings of photographic resemblance, including the extent to which it is assumed that the apparatus must functionally imitate human physiology, are beautifully discussed with regard to Brewster's stereoscope in Hankins and Silverman, *Instruments and the Imagination*, chapter 7.
97. Arago, "Rapport," 7, 17. Daguerre immediately had Arago's discourse printed up, along with instructions on the process, a testimonial by Gay Lussac, and his business address, as in Daguerre et al., *Historique*.
98. Arago, "Rapport," 16–17.
99. Arago, "Rapport," 27. As Levitt shows, in Foucault and Fizeau's experiments, the intensity of a given light source was established not by the brightness of the image produced, but by the time it took to develop to a standard degree, "to compare lights by their effects" (Arago, "Rapport," 24); see Levitt, "Biot's Paper," 470.
100. Schaffer, "Glass Works," 70, referencing Pinch, *Confronting Nature*, 212–14; on black boxes, see Latour, *Science in Action*.

101. Arago, "Rapport," 18–19.
102. Rothermel, "Images of the Sun"; Schaffer, "Where Experiments End"; Pang, "Stars."
103. See Bajac, "Le daguerrotype" [sic], 86, for an article by Alfred Donné of 1839: "Cette parole de M. Arago a dû rassurer bien du monde; mais il nous a néanmoins paru curieux de nous rendre compte par nous-mêmes de difficultés du Daguerrotype en nous mettant de suite à l'oeuvre et en suivant de point en point les renseignements si bien décrits par M. le secrétaire de l'Académie. . . . Les premières expériences ont donné un résultat à peu près nul." Donné was a physician and popular scientist writer who—along with the mathematician and book thief Gugliermo Libri—frequently antagonized Arago in the press; he was also one of Léon Foucault's early patrons, a fact that Arago did not hold against Foucault.
104. Arago, "Rapport," 22.
105. See, for example, Yeo, *Defining Science*; Gieryn, *Cultural Boundaries*. In "Moral Economy of Science," Daston suggests that such detachment is a general tendency of modern science. Scientific boundaries are permeable by social values, but we should expect primarily a "one-way" communication: "Although moral economies in science draw routinely and liberally upon the values and affects of ambient culture, the reworking that results usually becomes the peculiar property of science," 7.
106. Kohler, *Lords of the Fly*; Daston, "Moral Economy of Science"; Thompson, "Moral Economy."
107. Mauss, *Gift*. Studies of early modern science have made use of Maussian analyses of gift exchange, but few attempts have been made to address similar processes in the "rationalized" sciences, states, and public cultures of the nineteenth century. See Biagioli, *Galileo*; Smith, "Alchemy as a Language of Mediation."
108. Arago, "Rapport," 24.
109. Krauss, "Tracing Nadar"; see also Gunning, "Phantom Images."
110. Daguerre had earlier set scenery into motion with his Diorama (see chapter 5). Niépce also had a history of interest in technologies of motion and conversion: in 1811 Lazare Carnot reported to the Academy of Sciences on an engine that gathered its energy from the heat differential between itself and cold air; it was invented by none other than Nicéphore Niépce and his brother, as noted in Gillispie, *Lazare Carnot*.
111. See Eamon, "Technology as Magic."
112. Hugo, *Le promontoire du songe*.
113. "Bailly: Biography Read at the Public Sitting of the Academy of Sciences, 26 February 1844," in Arago, *Biographies*, 154. See Levitt, *Shadow of Enlightenment*, 106–16, on "Arago and the magnetizers," linking Arago's early interest and eventual disenchantment with occult powers (including Hugo's theory of "magnetic vision") to his quest for transparency.
114. See Buchwald, *Rise of the Wave Theory*; Cantor, *Theories of Ether*.
115. On Humboldt's tepid response to Laplace's nebular hypothesis, see

Merleau-Ponty, *La science de l'univers*, 183–209; Arago also wrote ambivalently, "Our illustrious compatriot never proposes himself anything vague, undecided ... One time, one single time, Laplace launched himself, like Kepler, like Descartes, like Leibniz, like Buffon, into the region of conjectures. His conception at this point was nothing less than a cosmogony" (Arago, "Laplace," 505).

CHAPTER FIVE

1. Bachelard, *Le matérialisme rationnel*; see also Williams, "Kant, *Naturphilosophie*, and Scientific Method."
2. Comte, *Discours sur l'ensemble*, 138.
3. Comte wrote that our scientific laws "represent the universal order as much as we need to know it"; "every phenomenon presupposes a spectator." *Système*, 4:175, 1:439. Also see Pickering, *Auguste Comte: An Intellectual Biography*, 175–181. For Bachelard on Comte, see *Etude sur l'évolution*.
4. Ampère in Castex, *Le conte fantastique*, 45.
5. Todorov defined the fantastic as the "hesitation experienced by a person who knows only the laws of nature, confronting an apparently supernatural event"; Freud analyzed Hoffmann's "The Sandman" as a key instance of *Das Unheimlich*, or "the uncanny," the eerie feeling produced when a new, alien object is recognized as familiar and invested with a primal charge. Freud, "Uncanny"; Todorov, *Fantastic*, 25; See also Castex, *Le conte fantastique*; Kessler, *Demons of the Night*; Siebers, *Romantic Fantastic*.
6. "Académie Royale de Musique. Première représentation. *La sylphide*, ballet-pantomime en deux actes," *Courrier des théâtres*, March 13, 1832, quoted in Meglin, 122.
7. Ampère, *Essai*, 2:75. Technaesthetics had the following subfields (following the four-part structure described in chapter 2): *terpnographie*: what is agreeable and the beauties that strike us at first glance; *terpnognosie*: the underlying ideas or sentiments, the aims of the author or creator; *comparative technaesthetics*: appreciation according to the laws of their composition, musical, visual, rhetoric; *philosophie des beaux-arts*: What is "beauty," and what is its origin? Is it arbitrary? Does it reside in the human heart, in eternal archetypes? What are the causes that make it arise at a certain time or to a certain people?
8. Ampère's friend Gérard-Joseph Christian at the Centre National des Art et Metiers had in 1819 drafted his *Plan de Technonomie*, which was "a science of industrial operations." Guillerme and Sebestik, *Les commencements*, 60. Interestingly, Ampère also linked his instrumental conception of the arts to a broader community with his reference to action "upon the will of others," just as Christian's "technonomie" considered not only specific machines but the means of producing them and their effects on social organization.
9. De Stael, *De l'Allemagne*, 463–64.
10. Cousin's version of the German idealism of Kant and Schelling strongly

influenced the predominant school of philosophical romanticism in the United States, the transcendentalists. See Joyaux, "Cousin and American Transcendentalism."

11. De Staël, *De l'Allemagne*; Cousin, *Du vrai, du beau*, 188–91; Jouffroy, *Cours d'esthétique*, 238.
12. Abrams, *Mirror and the Lamp*.
13. See Robertson, *Fantasmagoria*; also Pancaldi, *Volta*. Robertson's performances are detailed, as are the complex dialectics of illusion and demystification implied by "phantasmagoria," in Gunning, "Long and Short of It," 23–35.
14. On Charles as transitional figure from eighteenth century cabinets de physique and science as mass entertainment in the nineteenth century, see Hankins and Silverman, *Instruments and the Imagination*, 61–63. See also Oersted, *Luftskibet*.
15. See Pancaldi, *Volta*; Robertson's performances are detailed in Gunning, "Long and Short of It."
16. See Marina Warner's discussion of Robertson, as well as the relationship between inscription technologies, the imagination, and inquiries into the reality of spirits, in her *Phantasmagoria*.
17. See Goulet, "'Tomber dans le phénomène'" and *Optiques*.
18. Vigny, "La maison du berger," 28, in Grant, *French Poetry*.
19. Victor-Louis-Amédée Pommier, "Aurolatrie," in ibid., 30.
20. Hugo, *Les contemplations*, in ibid., 40.
21. Hugo, *Les contemplations*, in ibid., 26.
22. De Laprade, "Le nouvel âge," in ibid., 11.
23. Sand, *Les sept cordes de la lyre*, in ibid., 33.
24. "Car le mot, qu'on le sache, est un être vivant./La main du songeur vibre et tremble en l'écrivant . . . /Il sort d'une trompette, il tremble sur un mur,/Et Balthazar chancelle, et Jéricho s'écoule./Il s'incorpore au peuple, étant lui-même foule./Il est vie, esprit, germe, ouragan, vertu, feu;/Car le mot, c'est le Verbe, et le Verbe, c'est Dieu." Hugo, *Les contemplations*, 1:43; see discussion in Larthomas, "Théories linguistiques," 67–72.
25. This goes with a shift from language as primarily concerned with representation/imitation, or from the didactic capacities of language, to emphasis on its effect. Foucault hinted at this shift but focused on the historicity of language, which became prominent at this time, rather than its demiurgic aspect, and asserted an epistemic gap between literary and didactic/descriptive uses of language. See Foucault, *The Order of Things*, 286, 290–91, 382–84. Such a divide was far from apparent in the literary-flavored works of scientists and the technical/scientific disquisitions of poets and novelists of this time.
26. Schwab, *The Oriental Renaissance*.
27. See Bellanger et al., *Histoire générale*; de la Motte and Przyblyski, *Making the News*; Allen, *Popular French Romanticism*; Parent-Lardeur, *Paris au temps de Balzac*.
28. Méry and Nerval, *L'imager de Harlem*.

29. Michelet, *People*, 191, 11.
30. Legouvé, "De l'invention," 297–307.
31. Lamartine wrote: "Speech, through the procedure perfected by Gutenberg, will have become, through matter, just as immaterial as when it was merely thought; but this thought will have become universal as it bursts forth from the intelligence or the will of man." Lamartine, *Gutenberg*, 237.
32. Bénichou, *Les mages romantiques*.
33. Sainte-Beuve in Castex, *Le conte fantastique*, 53.
34. Hoffmann, *Golden Pot*, 102.
35. On puppets versus gods, see Kleist, "On the Marionette Theater."
36. See Baudelaire's "Litanies of Satan," in *Les fleurs du mal*, which concluded with a prayer: "Glory and praise to you, Satan, in the heights of Heaven, where you reigned, and in the depths of Hell, where you dream, defeated, in silence! Grant that one day, beneath the Tree of Science, my soul will rest at your side, at the moment when over your head, like a new temple, its boughs grow forth!" (*Oeuvres*, 1:125). For references to "la science" as implying primarily occult sciences in the works of Saint-Martin, Ballanche, and others, see Viatte, *Les sources occultes*. On the theological implications of artistic creation in English romanticism, see Cantor, *Creature and Creator*. On hallucination and somnambulism in French romanticism, see James, *Dream, Creativity, and Madness*.
37. Balzac, *Père Goriot*.
38. Griffiths, *Shivers down Your Spine*; Siegel, "Wagner and Hoffmann."
39. "Rapport" (1800), in Thompson, "Essai," 51; see Bigg, "Panorama."
40. Humboldt, *Cosmos*, 90–91.
41. Marrinan, *Romantic Paris*, 173–77.
42. Serres, "Turner Translates Carnot."
43. Crary, "Géricault, the Panorama." See Bigg, "Staging the Heavens"; Mannoni, *Great Art*; Oettermann, *Panorama*; Comment, *Painted Panorama*.
44. See Pinson, *Speculating Daguerre*.
45. Gernsheim and Gernsheim, *Daguerre*, 34.
46. Ibid., 31.
47. Ibid., 32.
48. Daguerre, *Historique*, 75.
49. Gernsheim and Gernsheim, *Daguerre*.
50. Ibid., 32.
51. Ibid., 35.
52. On Berlioz and fantastic culture, see Brittan, "Berlioz."
53. See Savart, *Mémoire des instruments à chordes*.
54. Berlioz, *Berlioz's Orchestration Treatise* (*Grand traité d'instrumentation*).
55. H. Berlioz, *The Art of Music*, 182. According to Jean-Michel Hasler, Berlioz was "above all one of the first to be seriously pre-occupied with a problem up until then considered as secondary by musicians: the *place* in which his music is interpreted"; in Wasselin and Serna, *Hector Berlioz*, 52. Also see Locke, *Music and the Saint-Simonians*.

56. Baroli, *Le train*, 98–99. Also see Holoman, *Berlioz*.
57. Winter, *Mesmerized*, 317.
58. Heinrich Heine, *Lutèce*, April 25, 1844, reprinted in Wasselin and Serna, eds., *Hector Berlioz*, 358.
59. Tobin, *Léon Foucault*; Daumas, *Arago*.
60. See Jackson, *Harmonious Triads*; Savart, *Mémoire des instruments*; "On the Acoustic Figures"; "Recherches."
61. Chevreul, *De la loi*, 172. See Vienot, "Michel-Eugène Chevreul," 4–14.
62. On Henry's influence, see Brain, "Representation on the Line."
63. Musselman, *Nervous Conditions*, 158. See especially chap. 5, which makes both the ubiquity of hallucinations and their submission to the scientist's will keys to the moral epistemology of early-nineteenth-century Britain.
64. Crary, *Techniques of the Observer*, 16.
65. Valtat, "La littérature hallucinée," 74–115.
66. Maine de Biran, *De l'influence de l'habitude*, 146.
67. See Boas, *French Philosophies*; Foucault, *Order of Things*, 240–43.
68. See Goldstein, *Console*: "Thus neither a monist nor a dualist inclination in philosophy, but rather a belief in reciprocal psychosomatic influence guided the therapies of early nineteenth-century French alienists," 267.
69. See Bertrand's lengthy analysis of the "unity of composition in organized beings" in his review of Geoffroy's "Cours de l'histoire naturelle des mammifères, professé au Jardin du Roi" appearing in the *Globe* January 17, 1829. Bertrand argues that "this doctrine was proposed by the author nearly at the beginning of his career," and though it has at last been acknowledged, the glory has gone to the Germans (37).
70. See discussion in Goblot, "Extase."
71. Bertrand, *Du magnétisme*, 309 and following. Literally, "extase" means a transport "out" of a "base" or "foundation"—in the case of mesmerism, a way of leaving the self while remaining within the body, and thus an ecstatic materialism. Littré cites Corneille for one of the early uses of the term: "Que dans ces transports extatiques/Où seul tu me feras la loi,/Tout hors de moi,/mais tout en toi,/Je te chante mille cantiques," in Corneille, *Oeuvres*, 3:612. On Bertrand's notion of *extase*, see Bertrand, *Du magnétisme*; Goldstein, *Hysteria*; Goblot, "Extase"; Leroux, "Alexandre Bertrand"; Hibberd, "'Dormez donc.'"
72. See Goldstein, *Hysteria*; Azouvi, *Maine de Biran*.
73. See de Tours, *Du hachich*; Gautier, "Le Club." Dumas's *Le Comte de Monte-Cristo* features a scene with hashish, and many of de Nerval's works, including *Aurélia* (1855), are structured as a series of visions whose strangeness suggests both pharmacology and madness.
74. De Tours, *Du hachich*, 25n1.
75. On Moreau de Tours as self-experimenter, see Karin Soldhju, *Selbstexperimente*.
76. This reading of Maine de Biran and his readers also suggests an alternative genealogy for the French philosophical tradition of "spiritualism."

Historians of philosophy have shown that the emphasis on embodiment in Merleau-Ponty's phenomenology owed a debt to Maine de Biran, as did Bergson's philosophy of life (via Félix Ravaisson), yet the reading provided here concerns another line of development, one that is as important as it is neglected: the view that human nature and its evolution are intimately bound up with tool use. Bergson's notion of *homo faber* (see Bergson, *L'évolution créatrice*; and Canguilhem, "Machine and Organism") was a major inflection point for this theme; it has subsequently been pursued, in various ways, by André Leroi-Gourhan, Teilhard de Chardin, Gilbert Simondon, Marshall McLuhan, Gilles Deleuze, Bruno Latour, and more recently in a phenomenological vein by Bernard Stiegler (*La technique et le temps*). The "physiospiritualism" of the students of Maine de Biran was a variation on the early nineteenth century's technological Lamarckism—the idea that humans remake themselves and their milieu by means of new, technological organs. On French spiritualism without this technological emphasis, see Janicaud, *Une généalogie*.

77. We might see this as another version of the post-Kantian "empirico-transcendental doublet" that Foucault described as foundational to the modern human sciences (*The Order of Things*, 318–22).
78. Opera provided audiences—aristocratic and bourgeois—with a "common emotional bond" and a "shared dream." Lacombe, "'Machine' and the State," 41.
79. See Cormac Newark, "Meyerbeer."
80. "Opéra," *Le figaro*, November 28, 1831 (Paris), in Wilberg, "*Mise en scène*," 299. See Coudroy, *La critique parisienne*. For a discussion of Meyerbeer and Balzac's physiospiritual machines, see Dolan and Tresch, "'Sublime Invasion.'"
81. Gunning, "The Cinema of Attractions," 58.
82. Meyerbeer's diaries from the time detail his fears that his musical and visual effects would be scooped by rivals; critics focused as much on new harmonies, sound effects, and staging as they did on plots and music. See Letellier, *Diaries of Meyerbeer*; Coudroy, *La critique parisienne*. Even today, this aspect of opera's appeal remains largely neglected by musicologists: visible, present, but often treated as inessential or even antithetical to "the work" as canonized by idealist music criticism. The same neglect of "merely sensuous" elements such as timbre, color, effect, or even orchestration, has marked criticism of orchestral works; Emily Dolan's forthcoming *Orchestral Revolution*, in contrast, demonstrates the essential role these "supplements" played in the reception of Haydn, Beethoven, and their contemporaries.
83. Balzac, "Avant-Propos," 9.
84. Balzac, "Gambara," 105.
85. Balzac, "Gambara," 125.
86. Such scenes of technological transubstantiation abounded in this period; not only the opera, but labor, tools, and a range of "romantic machines" were

seen as a means of enlivening matter with spirit. Further, Balzac's refigured, semi-secular Communion can be linked to contemporary reinvestments of fetishism (Comte's *Système*) and the cult of the printed word. See Bénichou, *Le sacré de l'écrivain*.

CHAPTER SIX

1. On French popular science of this time, see Bensaude-Vincent, "Un public pour la science"; "Sciences pour tous." See also Crosland, "Popular Science"; Belhoste, "Arago, les journalistes et l'Académie"; Staum, "Physiognomy and Phrenology"; Levitt, *Shadow of Enlightenment*; Parent-Lardeur, *Lire à Paris au temps de Balzac*.
2. Haines, "Athénée de Paris"; Staum, "Physiognomy and Phrenology."
3. On performances of popular science, with a fascinating emphasis on magic and illusionism, as well as a discussion of Robert-Houdin, see Sophie Lachapelle, "Science on Stage: Amusing Physics and Scientific Wonder at the Nineteenth-Century French Theatre," *History of Science* 47 (2009): 297–315. See also Morus, "Seeing and Believing Science"; Schaffer, "Natural Philosophy and Public Spectacle."
4. Spary, *Utopia's Garden*.
5. See Burkhardt, "La ménagerie," on the *ménagerie* from Lamarck to Frédéric Cuvier; see also Frédéric Cuvier, "Examen de quelques observations de M. Dugald-Stewart."
6. See Kaenel, "Le buffon de l'humanité: J. J. Grandville."
7. Lagueux, "Geoffroy's Giraffe"; Burkhardt, "Ethology."
8. Burkhardt, *Spirit of System*; Richards, *Darwin*; Jordanova, "Nature's Powers."
9. Elizabeth Williams, *Moral and Physical*; Chappey, *La société des observateurs de l'homme*.
10. See Jacyna's immensely valuable "Medical Science and Moral Science" and "Immanence or Transcendence." See also Temkin, *Double Face of Janus*, especially "Background to Magendie"; "Romanticism"; and "German and French"; Lesch, *Science and Medicine*.
11. Maret, *Essai sur le panthéisme*.
12. Williams, *Physical and the Moral*.
13. Damiron, *Essai sur l'histoire*. Cousin and Damiron claimed Maine de Biran as their precursor, although Bertrand denied their spiritualist interpretation. See Goblot, "Extase"; Vermeren, "Les têtes rondes du Globe."
14. Cousin, *Cours de l'histoire*, 18. Cousin has recently been shown as a key figure in the construction of nineteenth-century bourgeois subjectivity in Goldstein, *Post-Revolutionary Self*; and Vermeren, *Victor Cousin*; For a polemic against contemporary uses of the term "eclecticism," see Scott, "Against Eclecticism."
15. Jacyna, "Medical Science and Moral Science."
16. See Richards, *Romantic Conception of Life*; Coleman, *Cuvier*; Foucault, *Order of Things*.

17. See Lamarck, *Philosophie zoologique*; Jordanova, "Nature's Powers"; Canguilhem, "Le concept de milieu."
18. Geoffroy Saint-Hilaire, in Appel, *Cuvier-Geoffroy Debate*, 78–79.
19. Geoffroy Saint-Hilaire, in Le Guyader, *Geoffroy Saint-Hilaire*, 31.
20. Geoffroy Saint-Hilaire, *Preliminary Discourse* (1818), 31.
21. Geoffroy Saint-Hilaire, in Le Guyader, *Geoffroy Saint-Hilaire*, 118.
22. Appel, *Cuvier-Geoffroy Debate*; See also Bourdier, "Le prophète."
23. Geoffroy Saint-Hilaire, in Le Guyader, *Geoffroy Saint-Hilaire*, 34. See discussion of representative "types" in Daston and Galison, *Objectivity*.
24. Serres, *Recherches*, 85. ("Ritta-Christina" was at times written as "Rita-Cristina.") The term "organogénie" is borrowed from the Lorenz Oken and his *Elements of Physiophilosophy*, which was translated into French in 1821. Oken's text may also have encouraged Serres to ground Geofrroy's "unity of composition" in the animal series—a concept toward which, as Appel has pointed out, Geoffroy was originally hostile. On the relationship between unity of plan and the zoological series, see Le Guyader, "Le concept de plan d'organisation." Serres's interest in both Oken and the animal series was encouraged by Blainville, another major opponent of Cuvier. As we will see in chapter 9, Blainville's animal series underwrote Comte's views not only of biology but also of the hierarchy of the sciences. On Blainville and Oken, see Braunstein, "Comte, de la nature."
25. Serres, *Recherches* (1832), 10.
26. See Russell, *Form and Function*, 79–83. On Serres's medical research, see Williams, *Physical and Moral*, 233–41. On his ethnology and study of races, see Staum, *Labeling People*. On the wider influence of "organogenesis" on Sand, Michelet, and Leroux, see Bourdier, "Le prophète."
27. *Annales des sciences naturelles*, xii; *Recherches d'anatomie transcendante, sur les lois de l'organogénie* (1827), 85. Support for this notion was found in Oken; on the many forms similar concepts took and the controversies they spurred in German life science of this period, see Richards, *Romantic Conception*.
28. Geoffroy Saint-Hilaire, in Le Guyader, *Geoffroy Saint-Hilaire*, 5.
29. He also noted the environment's capacity to produce immediate changes in an organism in his discussion of the *teleosaurus*, a fossilized creature resembling modern crocodiles. Le Guyader, *Geoffroy Saint-Hilaire*, 94–95.
30. Rostand, "E. Géoffroy Saint-Hilaire et la tératogénèse"; Oppenheimer, "Some Historical Relationships." See Secord on contemporary electrical experiments to produce life, in Gooding et al, *The Uses of Experiment*.
31. Serres, in Le Guyader, *Geoffroy*, 239; Serres, "Eloge de Geoffroy Saint-Hilaire," in ibid. See Serres, "Organogénie," 24; Isidore Geoffroy Saint-Hilaire, "Teratologie," in *L'Encyclopédie nouvelle*, 8:24.
32. See Jordanova, "Nature's Powers."
33. See the history traced in Le Guyader, *Geoffroy Saint-Hilaire*; Appel, *Cuvier-Geoffroy Debate*.
34. Cuvier, "Nature," *Dictionnaire des sciences naturelles*.
35. This is Cuvier's official rejection of concepts akin to the *Bildungstrieb* of Blu-

menbach and Cuvier's teacher, Kielmeyer. See Richards, *Romantic Conception*, chapters 5 and 6.
36. Cuvier, "Nature," 263–64.
37. Ibid., 267.
38. Geoffroy, *Encyclopédie Moderne*, 28.
39. Ibid., 32–33.
40. Ibid., 31.
41. Ibid., 36.
42. Ibid., 59, 44. These views are expanded in his *Notions synthétiques*. On combustion and electrification, see esp. 92.
43. Geoffroy, *Encyclopédie Moderne*, 45.
44. See Geoffroy, "On the necessity for printed writings," a prospectus he distributed to members of the Academy April 5, 1830, reprinted in Le Guyader, 122–25.
45. See chapter 5; de la Motte and Przyblyski, *Making the News*; Avenal, *Histoire de la presse*.
46. Goethe, "Les naturalistes français"; see related texts reprinted in Tort, *La querelle des analogies*.
47. For a long time, Cuvier was given the victory in the debate; recent studies have shown the results to be more mixed, especially in light of appreciations of Geoffroy's influence in England and in Scotland—where his ideas reached Darwin during the latter's medical studies (see Desmond, *Politics of Evolution*). The notion of an "animal plan" was more recently rehabilitated by mathematician René Thom as a precursor to mathematical biology and part of a return to development-focused biology. After Thom, Deleuze and Guattari's *Thousand Plateaus* contained a lengthy discussion of Geoffroy, in which the "animal plan" was presented as an example of their central, if elusive, concepts of the "body without organs," the "abstract machine," and the "plane of composition," as well as the concept of "the virtual" that Deleuze developed in his studies of Bergson (Deleuze, *Difference and Repetition*; *Bergsonism*; see also DeLanda, *Intensive Science*, which interprets Deleuze in terms of the mathematics of complexity).
48. On the dialectic of wonder and habit, focused in part on the International Expositions, see Gunning, "Re-Newing Old Technologies"; Blondel, "Electrical Instruments"; Ory, *Les expositions*.
49. Dupin, *Tableau comparé*, 7. On the national expositions which preceded the international expositions, see Bouin and Chanut, *Histoire des Foires et des Expositions*; De Plinval de Guillebon, *Bibliographie Analytique des Expositions Industrielles*.
50. See Bezanson, "The Term Industrial Revolution."
51. Cited in Flachat, *L'industrie*, 17.
52. Neufchâteau, quoted in Chandler in his online series of essays on the Expositions, http://charon.sfsu.edu/publications/PARISEXPOSITIONS/1798 EXPO.html, consulted June 25, 2011; see also Chandler, "First Industrial Exposition"; *Catalogue détaillé*, 1798.

53. Chandler, "First Industrial Exposition"; see also Sewell, *Work and Revolution*.
54. Héricart-Ferrand de Thury and Migneron, *Rapport* (1828), 3.
55. Chandler, "First Industrial Exposition," 5.
56. Flachat, *L'Industrie: Exposition de 1834*; Dupin, *Rapport du Jury*.
57. *Rapport du jury central, 1839*.
58. *Rapport du jury central, 1839*, 1:lix.
59. Noted by Arthur Chandler, in his online series of essays on the Expositions, http://charon.sfsu.edu/publications/ParisExpositions/JulyMonarchyExpos .html, accessed June 25, 2011.
60. See Rubichon, *Du mécanisme*, chap. 4; Buret, *De la misère*.
61. Dupin, along with his frequent ally François Arago, exerted himself to show the mechanization of industry as beneficial in the long run for laborers. See Dupin's pamphlets collected at the Bibliothèque Nationale de France, including *The Effects of Popular Instruction in Reading, Writing and Mathematics, Geometry and Mechanics Applied to the Arts, on the Prosperity of France* (1826); *The Future of the Working Class* (1834); *The Influence of the Working Class on the Progress of Industry* (1835). On workers' Luddite reactions against mechanization, see Jarrige, "Le mauvais genre," and Bordeau, Jarrige, *Les Luddites*.
62. Cantorowicz, *The King's Two Bodies*.
63. Robert-Houdin, *Memoirs*, 151.
64. Ether (nitrous oxide or sulphuric ether) was introduced for surgery in the 1840s; its perception-altering qualities encouraged an identification with the protean ethers assumed by animal magnetism and post-Laplacean physics. Seldow, *Vie et secrets*; Metzner, *Crescendo of the Virtuoso*, esp. chapters 1 and 5 on Robert-Houdin and "the vogue of the automaton-builders." See also Lachapelle, "Magic."
65. Robert-Houdin, *Memoirs*, 8.
66. See "La cafetière" (1831), 53–63; "Omphale" (1834), 65–76, both in Gautier, *Contes et récits*. See also Siebers, *Romantic Fantastic*.
67. Erdan, *La France mistique* [sic]; Viatte, *Les sources occultes du romantisme*; Wilkinson, *Dream*.
68. One of Victor Frankenstein's sources, Henry Cornelius Agrippa, wrote: "The ancient priests made statues, and images, foretelling things to come, and infused into them the spirits of the stars. . . . They do always, and willingly abide in them, and speak, and do wonderful things by them." Agrippa, Freake, and Tyson, *Occult Philosophy*, 114. On Renaissance automata, see Nelson, *Secret Life*; Eamon, "Technology as Magic." See also Kang, *Sublime Dreams*.
69. "Au risque d'être athée/J'aime Pygmalion et j'aime Prométhée." *Ce siècle est en travail*, from *Libres paroles* (1847), in Grant, *French Poetry*, 47.
70. Esquiros, *Le magicien*, 55, 60; in "La pipe d'opium," a character named Esquiros proposes magnetic explanations for the narrator's marvelous experiences, in Gautier, *Contes et récits*. An analysis of Esquiros's career is provided by Andrews in *Socialism's Muse*, 139–48. After imprisonment for

writing an *Evangile du peuple* that deified the working class, he rose to prominence in the Second Republic and went into exile after 1851; unlike his wife and frequent collaborator before 1851, Adèle Esquiros, whom he abandoned and who died in poverty, Esquiros returned to prominence and prosperity in 1859. Andrews sees this trajectory as paralleling the relationship between early socialism and feminism over the mid-century: an early sense of shared destiny seemingly fulfilled in 1848, followed by rupture after 1851, with the feminist cause abandoned by socialism.

71. Nerval, *Les filles du feu*, 224–38, 247, 268, 249. Similarly, on the morality of creation in English romanticism, see Cantor, *Creature and Creator*.
72. See During, *Modern Enchantments*; Riskin, *Modern Magic*. Metzner, in *Crescendo of the Virtuoso*, also interprets Robert-Houdin's flirtations with the supernatural in this way.
73. One indication of the economic range of his audience was a poster (fig. 6.5) showing entry prices between 1 F 50 for the gallery and 4 F for *loges* plus half-price entries at certain hours. Although he moved to England after 1848, in the immediate aftermath of the revolution, Robert-Houdin handed out free tickets to the *Soirées* during the February uprising, which passed just outside his theater at the Palais Royal; it was a gesture of solidarity or, perhaps, self-defense.
74. See Méheust, "Enquète," 63–83.
75. On the dialectic in the eighteenth century between claims that automata simulate natural phenomena and that they show the impossibility of doing so, see Riskin, "Defecating Duck."
76. In contrast, see During, *Modern Enchantments*; Riskin, *Modern Magic*; Metzner, *Crescendo of the Virtuoso*; Metzner interprets Robert-Houdin along similar lines.
77. Robert-Houdin's stances were deliberately ambivalent. The poster reproduced in figure 6.5 emphasizes his status as *mécanicien* and *physicien*, while proclaiming the "seconde vue" of his son. Although his *Memoirs* revealed the trick behind his clairvoyance, nothing in the performance discouraged those inclined to see the act as genuine. And although he debunked many charlatans, in 1847 he testified to observing the medium Alexis Didier reading while blindfolded, without "any doubt as to Alexis' lucidity." See Méheust, "Enquète," 63–83.
78. Grandville, *Un autre monde*, 1, 10, 11.
79. See Schlanger, *Les métaphores*; Blanckaert, *Les politiques*.
80. Grandville, *Un autre monde*, 121.
81. Ibid., 91.
82. Ibid., 44.
83. Ibid., 47.
84. Ibid., 42
85. Ibid., 242.
86. See Rousseau, *Robert Macaire*.

87. Grandville, *Un autre monde*, 61.
88. Ibid., 91.
89. Ibid., 139, 141.
90. Ibid., 144, 149.
91. Fourier's writings have a similar quality of wavering between seriousness and a playful hysteria that verges on self-parody; for illuminating discussion, see Barthes, *Sade, Fourier, Loyola*.
92. Benjamin saw in Grandville's images a warning about the reification of human labor and the penetration of the commodity form into all aspects of life. See Benjamin, "Fourier or the Arcades."
93. Critical theory has recently directed attention to the religious themes in the social philosophy between Kant and Marx (in the United States, this has been most visible in Derrida, *Specters of Marx*). More interesting work along these lines has come from philosophers who were part of Louis Althusser's circle at the Ecole Normale Supérieure in the 1960s and whose reconsideration of the project of revolutionary political philosophy has led them to study authors whose thought, frequently of a theological bent, persisted, under erasure, in Marx. One of the key, if neglected, aspects of poststructuralism that followed the incomplete revolution of 1968 (and the decline of Althusser's influence) was a questioning of the supposed (Bachelardian) "epistemological break" represented by *Capital* and the accompanying claim to an objective social science that would be the province of intellectuals. Through a close study of the writings—and views on writing—of workers published in *La ruche populaire* and *L'atelier*, both of which were started by Saint-Simonians, Jacques Rancière has reconstructed the often ambivalent views workers held about the "emancipation" promised to them by bourgeois publicists and philosophers; he discusses Leroux in *On the Shores of Politics*. Pierre Macherey has explored works of the social philosophers (and philosophical novelists) from the decades before 1848 (including Leroux, Comte, the Saint-Simonians, and Sand, as well as Hegel and Spinoza); Etienne Balibar has reexamined earlier political theologies, including those of Spinoza and Locke; Alain Badiou has returned to the writings of Saint Paul to interrogate theological precursors to Marx's reimagining of the social body and the revolutionary event. To a large extent, their arguments concern the relation of politics, and especially radically egalitarian politics, to philosophy and religion; they also reveal the religious dimension of the industrial phase of the Spinoza-inspired "radical enlightenment" (Spinoza's eighteenth-century phase has been discussed by Margaret Jacobs and Jonathan Israel). One aim of part 3 of this book, therefore, is to extend these discussions by showing that the arrival of a revolutionary and reformist socialism before 1848 was not only an event within philosophy, politics, and religion, but one that both drew upon and shaped developments in technology and the sciences; a broadened conception of the political (and of the philosophical) must include not only aesthetic and religious considerations,

but a reckoning of the means (technical, institutional, conceptual) through which humans transform the world around them.

CHAPTER SEVEN

1. Carlyle, "Signs of the Times."
2. Shine, *Carlyle and the Saint-Simonians*.
3. Chevalier, *Politique industrielle: Système de la Méditerranée*, 36.
4. Quotes in this and the next five paragraphs are from Duveyrier, "La ville nouvelle," 252–74.
5. Romans 8:22, New American Standard Bible.
6. Saint-Simon, "Mémoire," 172. This seminal text was written as a sketch for a longer work; he sent it personally to various philosophers, physicians, and savants.
7. See Williams, *Moral and Physical*; Chappey, *La Société des Observateurs*; Staum, *Cabanis*.
8. Stéphane Schmitt argues that by making comparative anatomy the science that would unify physiology, medicine, and natural history, Vicq d'Azyr was a crucial precursor to both Cuvier and Geoffroy Saint-Hilaire; Schmitt, "From Physiology to Classification."
9. Pierre Musso aligns this view with what Canguilhem called "baroque physiology," an interest in the fluids of varying constitution and density that pass through the vessels and canals of varying width and length to produce the phenomena of life. Pierre Musso, *Télécommunications et philosophie des réseaux*; Taton, *Histoire générale*.
10. Saint-Simon, "Mémoire," 108, my italics.
11. Ibid., 177, my italics.
12. Although he makes references to the Scottish stadial historians, his sequence is closest to Condorcet's *Outline*. See Meek, *Social Science and the Ignoble Savage*, 172–78; Manuel, *New World*.
13. Saint-Simon, "Mémoire," 108, 50. Comte developed a similar theory of human evolution; see also Buchez, *Introduction*, in which scientific theories are seen as "new faculties acquired by the human species" and where Buchez states that the progress of civilization modifies individual psychology and physiological "organization" (151n). Pierre Musso connects Saint-Simon's emphasis on "tubes" and "canals" to his description of the Ecole Polytechnique as the "canal" for his ideas and the hydraulics studied by its cadets, including the "capacities" of various tubes.
14. Despite this advocacy of religious freedom, Benjamin Constant argued for the importance of religion—defined primarily as "sentiment"—as the basis of society; Helena Rosenblatt depicts his argument as a response to Saint-Simonianism, in Rosenblatt, "Re-evaluating."
15. Compare to the evolution of the concept of "Fraternité" in Sewell, *Work and Revolution*. See it also as an update on Bacon (whom Saint-Simon cites

frequently), where the development of science is a form of worship, and technological mastery—in agriculture and production of goods—is the means of charity.
16. See Manuel, *New World*; Szajkowski, "Jewish Saint-Simonians."
17. See Belhoste and Chatzis, "From Technical Corps"; Picon, "Générosité sociale." Charléty lists among the leading polytechnicians who were also Saint-Simonians Charles Lambert, engineer of mines; the Paulin brothers; Léon and Edmond Talabot, engineers; Lamé; Clapeyron; Capella; Bigot, captain of engineering; Lefrant; and Le Play, plus about one hundred polytechnicians with whom Enfantin was in correspondence. Charléty, *Essai*, 101.
18. *Doctrines*, 143; see Hacking, *Taming of Chance*, 111.
19. See Busst, "Ballanche," for an enlightening article.
20. See the *Encyclopédie* of Diderot and D'Alembert, where the entry on "Evolution" is exclusively military.
21. Charléty, *Essai*.
22. Abel Transon, *Le globe*, February 12, 1831, in Charléty, *Essai*, 113.
23. *Le globe*, June 2, September 8, 1831; Charléty, *Essai*, 127.
24. This three-part phrase was imprinted on the masthead of *Le globe* after the Saint-Simonians took it over. See *Le globe*, 1830–32.
25. Isambert, *De la charbonnerie*; Buchez's foundational *Science de l'histoire* is discussed in chapter 8 below; see also Tolley, "Balzac et les Saint-Simoniens."
26. This aspect of their social science was derived from the *Nouveau christianisme*, in which Saint-Simon compared the movement of history to a pump, rising and falling. Another source was Ballanche's view of the periodic and providential rise and decline of societies, and the discussion by Jouffroy (originally published in the *Globe*), "Comment les dogmes finissent."
27. Charléty, *Essai sur l'histoire du Saint-Simonianisme*, 111.
28. These divisions were borrowed from Bichat: see Manuel, "Equality to Organicism."
29. On "sensible empiricism," see Riskin, *Science*.
30. They situated Saint-Simon in a longer history of religions—a history informed, as Raymond Schwab and Philippe Régnier have shown, by the explosion of interest in the early nineteenth century in "oriental" languages, the translation of sacred literatures from around the world, and the nascent comparative study of diverse religious traditions. Leroux and Reynaud's *Encyclopédie Nouvelle* was at the center of this "oriental renaissance." See Régnier, "Le mythe oriental."
31. See Breckman, *The Left Hegelians*, on the Hegelian and Schellingian sources of this synthesis.
32. "Credo," in D'Allemagne, *Les Saint-Simoniens*, 105.
33. See Macherey, *Hegel ou Spinoza*.
34. D'Allemagne, *Les Saint-Simoniens*, "Tableau synoptique de la religion Saint-Simonienne," 95.

35. *Doctrine*, 343. Emphasis mine. See discussion of the Meckel-Serres Law, sometimes called the law of recapitulation, in chapter 8.
36. Carnot, "Sommaire du rapport fait par les directeurs de l'enseignement," in Enfantin and Saint-Simon, *Oeuvres*, 4:75.
37. Enfantin and Saint-Simon, *Oeuvres*, 4:81
38. Charléty, *Essai*, 114.
39. Saint-Simon. "Literary, Philosophical, and Industrial Opinions," in *Art in Theory, 1815–1900: An Anthology of Changing Ideas*, ed. Charles Harrison and Paul Wood (Oxford: Blackwell, 1998), 40.
40. Baudelaire in Kelly, ed. *Salon de 1848*, 105.
41. Locke, *Music*, 33
42. See Musso, *Télécommunications et réseaux*.
43. Lamé, *Vues Politiques*. See Bradley, "Franco-Russian Engineering Links"; on their optimization work, see Grattan-Guinness, *Convolutions*, 1204–19. See Saint-Simon, "Mémoire," with its language of fluids of variable fineness and meshworks of tubes and canals of varying dimensions. Pierre Musso has noted the resonance between this "baroque physiology" and polytechnicians' hydraulics-based analyses of flows, pressures, and the "capacities" of different vessels, all of which fed into the theory of routes.
44. Guillerme "Réseau."
45. Chevalier, *Politique industrielle*, 22 (my emphasis).
46. Ibid., 31.
47. Ibid., 36.
48. Ibid., 41, 56.
49. Ibid., 32, 56.
50. For instance, see the models in the CNAM collection: "Machine à vapeur type compound," Inventory no. 02566–0003; "Machine de Woolf avec détente dans deux cylindres successifs," Inventory no. 04061–0001; and the 14,350 kg. "Deux mouvements simultanés du tiroir et du piston des machines à vapeur de Woolf et de Watt," Inventory no. 03512–0000.
51. Carnot's efficiency equation was $T_1 - T_2/T_1$ = % efficiency. Thus for example making the heat source 100 and the cold source 99 would give 1% efficiency; increasing the difference between the two, making the hot source 100 and the cold source 1, would produce 99% efficiency.
52. Carnot's own cognizance of the implications of his theory has been debated. His abstract and yet simple way of treating the efficiency of engines, regardless of the material they used, was a major source for William Thomson's conception of an underlying energy that was conserved through conversions. At the same time, he recognized that the energy's *state* changed, leading to loss and dissipation—what Clausius called entropy. This point went beyond Carnot's *Réflexions*, which used Laplace's notion of heat as a distinct, imponderable fluid, "caloric." Yet unpublished notebooks show that Carnot doubted the existence of caloric. They also reveal that he had arrived at the concept of absolute zero, later foundational to the scale named after Thom-

son (Lord Kelvin). See Thomson, "Absolute Thermometric Scale"; Fox, ed., *Réflexions*; Redondi, *L'accueil des idées*. The energetic interpretation of the steam engine became the basis for Saint-Simonian Stéphane Flachat's rejection of the distinction in Babbage's *Economy of Machinery and Manufacturers* between machines that transmit force and those that produce it. Flachat argued instead that machines (including steam engines) never make force; they only transform it. Flachat and Bury, *Traité élémentaire*; Redondi, *L'accueil des idées*, 126–28.

53. Clapeyron, "Mémoire sur la puissance motrice de la chaleur." Clapeyron had already presented a report with the same title to the Academy of Sciences in 1832. The article was followed by one by Lamé concerned with another imponderable: "Lois de l'équilibre du fluide éthéré" (ibid., 191–288), which mathematically modeled the passage of light through successive layers of ether.
54. The first French description of Watt's indicator diagram was in 1828, in the Saint-Simonian journal *Le producteur*, in Redondi, *L'accueil des idées*, 104.
55. Clapeyron, "Mémoire."
56. Hankins and Silverman, *Instruments*; Chadarevian, "The Graphical Method"; Brain, "Representation on the Line."
57. Clot, *Emile Pereire*.
58. Chevalier and Enfantin in Régnier, *Le livre nouveau*, 184–85.
59. Enfantin, "Lettre du Père Enfantin," 70.
60. Enfantin in Carlisle, *Proffered Crown*, 116.
61. Moses, *French Feminism*; Andrews, *Socialism's Muse*. Andrews argues that the figure of the androgynous god which appears frequently in early socialism — not least among the Saint-Simonians — is an attempt to overcome the male-centered ideology of individualism; it is also a figure of difference within unity, or freedom in connection.
62. D'Allemagne, "Les Saint-Simoniens," 284.
63. Picon, "L'Utopie-spectacle d'Enfantin."
64. These séances were transcribed and are now stored at the Bibliothèque de l'Arsenal; Philippe Régnier has edited the transcriptions along with an insightful introduction, annotations, and biographical appendices, as *Le livre nouveau*. On Saint-Simonian science, see Picon, "La science saint-simonienne entre romantisme et technocratisme."
65. Régnier, *Le livre nouveau*, 79.
66. The passage continues with sacred mathematics: "GOD, whom the revolutionary mathematicians have vainly sought to chase out of their sanctuary and who always nevertheless remained there, uncovered or hidden, under the divine name of the INFINITE or under the deceptive veil of LIMITS, GOD will reappear in science, more splendid than ever, to animate all conceptions." Ibid., 75.
67. Ibid., 79, 70, 157.
68. Chevalier, *Politique industrielle*, 39; Régnier, *Le livre nouveau*, 174.
69. Régnier, *Le livre nouveau*, 156.

70. "Conversations avec le Père," in ibid., 174.
71. Saint-Simonian grammar also included consideration of the rhetorical modes of chivalric tales, sacred prophecy, the language of the lover and the military general, all of which were resources for the priest in inspiring and guiding an audience; this tool kit of persuasive speech reflected the philological obsessions of the "Oriental Renaissance" and the romantic emphasis on the world-changing "Verbe."
72. Régnier, *Le livre nouveau*, 164.
73. Ibid., 271.
74. Duveyrier in Régnier, *Les Saint-Simoniens en Egypte*, 38–40.
75. Duveyrier in Régnier, *Le livre nouveau*, 232.
76. "Les Colonies," in ibid., 251. Morsy, *Les Saint-Simoniens et l'Orient*; Jouve, *L'épopée des Saint-Simoniens*, Abi-Mershed, *Apostles of Modernity*; Lorcin, *Imperial Identities*, ch.5.
77. Rancière, *Nights of Labor*.

CHAPTER EIGHT

1. Gaubert, *Rénovation de l'imprimerie*.
2. Many variants on the composition machine were developed in this period, although Leroux laid unusual emphasis on the fact that his device would alleviate the boredom of printing by making it possible to read while setting type. Such devices were in many cases the target of hostility from male workers because they threatened to transform an occupation identified with adult males into one that could be undertaken by children and women; the feminine associations of the piano heightened this suggestion. See Jarrige, "Le mauvais genre," 209. On the mechanics and economics of printing in this time including composition machines, see Secord, *Victorian Sensation*, 116–23.
3. Leroux, "D'une nouvelle typographie," 267.
4. Ibid., 259.
5. Note the difference between this view of the benefits of mechanization via the organization of labor and collective ownership and Arago's argument (following J. B. Say) in favor of mechanization. Arago said the increase of production via mechanization would increase demand and thus eventually replace lost jobs, following a "natural law" of the market.
6. The best full-length study in English is Bakunin, *Pierre Leroux*. See Berenson, *Populist Religion*; Charlton, *Secular Religions*; Griffiths, *Jean Reynaud*, 280–85; in the latter work, see 279n8 on Arago's nomination of Reynaud to the Academy of Sciences and Reynaud's participation in a salon-cenacle along with Janin, Marrast, and the publisher Girardin.
7. Leroux, *D'une religion nationale*, 160.
8. Breckman, "Politics"; Macherey, *Comte*; Maret, *Essai sur le panthéisme*; Behrent, "Mystical Body of Society"; see Goldstein, *Post-Revolutionary Self*.

9. Vatin, "Des polypes"; Blanckaert, "La nature de la société."
10. Leroux, "Aux artistes," 138.
11. Leroux, "D'une nouvelle typographie," 274.
12. See chapter 6; Goblot, "Extase"; Trahard, *Le romantisme*; Goblot, *La jeune France*.
13. Breckman, "Politics."
14. See Leroux, "De la poésie de style"; "A propos du Werther"; "Aux artistes,"all in "Aux Philosophes."
15. See Evans, *Le socialisme romantique*, on Leroux's anticipation of Baudelaire and symbolist poetry.
16. Leroux, "Aux Artistes," 142.
17. See Bénichou, *Le temps des prophètes*.
18. Another of his targets was the eclectic philosophy of Victor Cousin, whose individualism was the Scylla to the Charybdis of Saint-Simonian "absolute socialism." See Goldstein's *Post-Revolutionary Self*.
19. Leroux, "De la doctrine," xvii.
20. Geoffroy quoted this passage of Leroux, as cited in Le Guyader, *Geoffroy Saint-Hilaire*, 282n14.
21. Leroux, "De la doctrine," lxviii. This view of the convergence of all the fields of science around the concept of progress at the level of fluids, molecules, organisms, species, the earth, and the stars parallels comparable developments in England, which have been given much greater attention by historians of science. Desmond's *Politics of Evolution* (esp. 41–56) considers the impact of Geoffroy's philosophical anatomy in Edinburgh and London; Secord's *Victorian Sensation* examines the changing conditions of print and reading that conditioned the diverse meanings given to the anonymously published *Vestiges of the Natural History of Creation*; Schaffer's "Nebular Hypothesis" examines the astronomical aspects of this movement. While exiled on the Isle of Jersey in the 1850s, Leroux familiarized himself with these developments; he undertook to have *Vestiges* translated into French, but after the publication of Darwin's *Origin of Species* in 1857, the project fell through. He wrote about *Vestiges* in 1858: "All the ideas that are getting noticed come from France. It's Lamarck, it's Geoffroy[;] these are a few of the authors who truly made this book, that is, who thought it." Leroux quoted in Bourdier, "Le prophète," 53.
22. See Schor, *George Sand*.
23. Leroux, "De la doctrine," xxxv.
24. See Breckman, "Politics"; Courtine's introduction to Leroux, *Discours de Schelling*.
25. Leroux, "De Dieu," 19.
26. See Robert Richards, *Romantic Conception of Life*; Gusdorf, *Science romantique*.
27. Leroux, "De Dieu," 20–21.
28. Desmond, *The Politics of Evolution*; Deleuze, *Sacher-Masoch*; Deleuze and Guattari, *A Thousand Plateaus*.

29. Le Guyader, *Geoffroy Saint-Hilaire*, 231; Viard, "Leroux et les romantiques."
30. Geoffroy's salon was host to the romantic elite; regulars included Alfred de Musset, the socialist journalist Alphonse Esquiros (author of *Le magicien*), and sculptor David d'Angers. François Arago's brother Etienne and Ampère's son Jean-Jacques also attended at times, along with Victor Hugo, Franz Liszt, and George Sand. On Geoffroy's appeals to medical students, see Le Guyader, *Geoffroy Saint-Hilaire*, 35, 62. A moving and helpfully contextualizing cinematic interpretation of Musset's work, and his life with Sand (*Les enfants du siècle*), was directed in 1999 by Diane Kurys; the costumes alone are worth the price of renting the movie.
31. *Scènes de la vie privée*, 269, 272.
32. Ibid., 273, 277, 286.
33. Balzac, "Avant-Propos," *La comédie humaine*, 8. Despite mentioning Cuvier's claim to have reconstructed a mammoth on the basis of a single bone, Balzac hews more closely to the notions of Geoffroy. See Somerset, "Naturalist in Balzac."
34. Somerset, "The Naturalist in Balzac."
35. Geoffroy Saint-Hilaire, *Notions synthétiques*, 103-4.
36. Buchez, *Introduction*. Buchez's fusion of the animal series, the embryonic series, geohistory, and human history earned him, unsurprisingly, a rebuke from Geoffroy's enemy Cuvier. See Isambert, *De la charbonnerie*, 32.
37. See Bourdier, "Le prophète"; Macherey, "Un roman panthéiste"; Schor, *George Sand*.
38. Sand, in unpublished version of *Lélia*, ed. Didier, 2: 232.
39. Geoffroy Saint-Hilaire, *Notions synthétiques*, 92.
40. See Truesdell, *Essays*; Vatin, *Economie politique*. Leibniz's notion of *vis viva*, transformed into "work," turned out to be a more profitable means of analyzing potential energy. On *virtualis* from Duns Scotus to Leibniz and D'Alembert, and the difference between *vis viva* and *virtual velocities*, see Hankins, *D'Alembert*.
41. Geoffroy to Sand, July 13, 1838, quoted in Bourdier, "Le Prophète," 62.
42. Gilles Deleuze treats the concept of the virtual in *Bergsonism* and in *Difference and Repetition*; in *A Thousand Plateaux*, he and Guattari explicate their homologous concept of the *plane of immanence* through an evocative reading of Geoffroy's "unity of plan" and "unity of composition," quoting liberally from Geoffroy's *Notions synthétiques*. Le Guyader commends this reading (and its precursor in René Thom's applications of catastrophe to animal development) in *Geoffroy Saint-Hilaire*.
43. The term references Geoffroy's Lamarckian roots, his skepticism toward traditional religion, and the works of Geoffroy and Leroux's close ally Alexandre Bertrand, who considered magnetism a state of "ecstasy" that accompanies a physiological modification.
44. See Daudin, *Cuvier et Lamarck*.
45. Foucault argued that Cuvier's conception of "life"—founded in a hierarchy of organic functions and the organism's adaptation to its conditions of

existence—was in fact more modern than that of a thinker like Lamarck, who, despite his transmutationism, still ranked life according to the animal series. Lovejoy's history of the idea of the great chain of being went only as far as Schiller and Schelling, seeing the concept's demise in the irrationalism of the romantics. Thus the combined—and erroneous—impression left by these two influential intellectual historians was that the chain of being was a thing of the past by the early nineteenth century. On its persistence, see Bynum, "Great Chain"; and "Seriality," a special issue of *History of Science*, 2010.

46. See Bénichou, *Les mages*; McCalla, *Romantic Historiosophy*; Schwab, *Oriental Renaissance*; Brusst, "Ballanche and the Saint-Simonians."
47. C. Bougle and Elie Halevy, eds., *Doctrine de Saint-Simon: Exposition* (Paris, 1924), 4n. On the impact of the Saint-Simonians on sociology, see Giddens, ed., *Durkheim on Politics*, 17.
48. Fourier, *Theory of Four Movements*, 292.
49. Ibid., 50.
50. Leroux, "Lettres sur le Fourierisme," 187.
51. Proudhon, *De la creation*, 1: 244.
52. Ibid., 1:300.
53. See Tresch, "The Order of the Prophets."
54. Appel, "Henri de Blainville."
55. Serres quoted Blainville's textbook of 1822 at length in his discussion of Ritta-Christina, approving of his definition of transcendental anatomy. This praise of Blainville—a leading French supporter of Oken and the major influence on Auguste Comte's view of biology—reveals the common cause between Blainville, a fierce Catholic traditionalist, and Serres, whose strong support of Geoffroy Saint-Hilaire and influence on Michelet and Leroux would tend to place him, at least in the 1830s, on the republican and socialist left; it suggests the existence of a coalition in the 1820s and early 1830s among Cuvier's enemies from multiple sides. Both Serres and Blainville relied on the notion of the animal series or scale of being, with support from Oken.
56. Serres approvingly quoted Isidore Geoffroy Saint-Hilaire's argument for "parallel series," which placed varieties of a single species at the same rank in the great chain instead of lining them up on the single scale: "Nature, since it repeats itself in the creation of the diverse parts of the same being, repeats itself again in the creation of diverse partial series out of which, in reality, the animal series is composed." From *Considérations sur les caractères employés en orinthologie* (1832), quoted in Serres, "Organogénie," 50n. Unity of plan also provided a framework for understanding abnormalities where more or less "formative force" (akin to Blumenbach's *Bildungstrieb*) in a given part led to deformities.
57. The Meckel-Serres law was central to Ernst Haeckel's interpretation of Darwin in the late nineteenth century. See Gould, *Ontogeny and Phylogeny*.
58. He also entertained doctrines of reincarnation and metempsychosis, as did Ballanche and Jean Reynaud, Leroux's former Saint-Simonian comrade and

the coeditor of *L'encyclopédie nouvelle*. See Sharp, "Metempsychosis"; Griffiths, *Jean Reynaud*.
59. Leroux, *De l'humanité* (1840), 203.
60. Leroux, *De l'humanité* (Paris, 1845), 205; Leroux is quoting himself from *L'égalité*, part 2, chap. 4.
61. It is tempting to find in the works of Leroux and his colleagues at the *Encyclopédie nouvelle* the roots of the modern discipline of comparative religion and to see it taking inspiration from the comparative anatomy being elaborated and debated at the Muséum d'Histoire Naturelle. Such an argument could certainly be made in the case of comparative literature: one of the earliest uses of this term, perhaps the first, was in a series of lectures given at the Collège de France by Jean-Jacques Ampère, son of Geoffroy's friend André-Marie Ampère, a central member of the *Encyclopédie nouvelle* network. The younger Ampère corresponded with two founders of German philology, Friedrich and August Schlegel. Edgar Quinet, a member of the *Encyclopédie nouvelle* circle and a close friend of Michelet, translated Herder's *Outlines of Philosophy of the History of Mankind* in 1827; Eugène Burnouf, one of the *Encyclopédie nouvelle*'s main contributors, translated many recently discovered Buddhist and Hindu texts. See Burnouf, "La science des religions"; Leroux, "Importance of Oriental Studies"; Schwab, *Oriental Renaissance*.
62. This slogan may be a response to Malthus, whom he quoted in 1846 thus: "A man who is born in a world already occupied, if the rich do not have need of his labor, has not the slightest right to claim any portion of food, and is truly an excess on the earth (*réellement de trop sur la terre*); at the great banquet of Nature, there is no place set for him." "L'humanité et le capital," 84. The importance of this rite for Leroux is underlined by his statement that he was not a "communist" but a "communionist." See Alexandrian, *Le socialisme romantique*.
63. Leroux, "L'humanité et le capital," 63.
64. Leroux, *D'une religion nationale*, 160.
65. See Maret, *Essai sur le panthéisme*. According to Maret, the gravest error made by all modern pantheists was to put mankind in the place of God. Maret was a follower of the former Saint-Simonian Buchez (see Brusst, "Ballanche and the Saint-Simonians"); the French renewal of earlier pantheist debates can thus be seen as arising from a squabble within the Saint-Simonian family. See Macherey, "Leroux"; and "Le Saint-Simonianisme et le panthéisme." As we saw in chapter 6, Cuvier denounced Geoffroy as a pantheist; Geoffroy publicly denied but privately affirmed his pantheistic sympathies, although, as suggested above, "ecstatic materialism" comes closer to his view. See Bourdier, "Le prophète." Jan Goldstein, in *Post-Revolutionary Self*, 248, discusses the opposition between Comte and Cousin; yet Leroux's hostility toward Cousin's slack eclecticism was even fiercer. He wrote a lengthy book titled *Refutation of Eclecticism*, which denounced the absence of synthesis and life in Cousin's philosophy. Cousin's student Jules Simon later wrote: "Whenever a philosopher declared that he was *not* a pantheist, — 'You lie,'

said [Leroux]; 'you are a pantheist, since you are a philosopher; and, besides, Cousin, whose slave you are, with your gown and your square cap, is undoubtedly a pantheist.'" Simon, *Victor Cousin*, 149.

66. Leroux's stance echoes that of the Saint-Simonian "credo" quoted in chapter 7. The crucial difference, however, was the Saint-Simonian emphasis on hierarchy, which meant submission to the will of the priest. Leroux's pantheism insisted on equality as the ground of order, placing on the side of the individual the weight that the Saint-Simonian's had placed on the side of the collective (and on the side of the elite they instituted as its enlightened directors).
67. Breckman, "Politics."
68. Leroux, *De l'humanité*, 203.
69. Leroux, "De Dieu," 76
70. Ibid., 76.
71. Ibid., 87.
72. Ibid., 89.
73. Geoffroy, "Preliminary Discourse" (1818), reprinted in Le Guyader, *Geoffroy Saint-Hilaire*, 30.
74. Leroux, *D'une religion nationale* (1846), 109–11, in Evans, *Le socialisme romantique*, 83.
75. Ibid.
76. Leroux, "D'une nouvelle typographie," 265, 275–78.
77. Leroux, "Science sociale," 52.
78. On the compagnonnage, see Sand, *Compagnon*, Sewell, *Work and Revolution*, Rancière, *Nights of Labor*.
79. Leroux, "De la Philosophie et du Christianisme," cited in Evans, *Le socialisme romantique*, 79.
80. Leroux, *D'une religion nationale*.
81. Leroux, *De l'humanité*, 129; discussed in Andrews, *Socialism's Muse*, 85.
82. On the circulus, see Laporte, *Histoire de la merde*, 296–302; Le Bras-Chopard, *L'égalité dans la différence*, 97–117; Lacassagne, "Victor Hugo"; Reid, *Paris Sewers*; Ceri, "Leroux and the circulus." Leroux's circulus has been vindicated by grassroots, do-it-yourself ecologists; see Jenkins, *Humanure Handbook*.
83. See the related discussion of Eugène Huzar's technological/ecological doomsday book of 1855, *La fin du monde par la science* (Paris: Dentu), in Fressoz, "Beck."
84. Leroux, "De la recherche," 84–85.
85. Ibid., 90.
86. Ibid., 82.
87. Ibid., 87–88.
88. Ibid., 89. On the *circulus*, see Reid, *Paris Sewers*; Simmons, *Waste Not, Want Not*; Griffiths, *Jean Reynaud*, 280. Today, the notion seems more than sane and a precursor to environmentalist "zero waste" ideals. On the "poop cycle," see Paul Glover's website at www.paulglover.org/.
89. Alexandrian, *Le socialisme romantique*, 88.

90. Without explanation, Leroux gives another article in the same series the title of Toussenel's *Les juifs, rois de l'époque* and goes on to trace the contemporary dominance of finance capital to Jews' being granted exclusive rights under Christian kings to lend money at interest. He makes a point of distinguishing the "Jewish spirit" from any Jewish individuals or the Jewish people and writes, as he does throughout his works, in praise of the world-historical significance of Judaism in the progress of humanity. It is not in the name of Christianity that he criticizes capitalism but in the name of a new, democratically based and modifiable religion, which would insist, according to an 1849 publication in *La revue sociale*, on "la liberté des sectes." Nevertheless, the denunciation of capitalism and "the Jewish spirit" in a single breath throws his entire project of a new religion of society into question and returns us to one of the paradoxes in the discourse of secularism: the trade-off between intolerance and social disunity, and the latter was often seen by earlier socialists as a greater danger than the former. Durkheim's advocacy for a secular religion in the Third Republic is another chapter in this story. For discussion of anti-Semitic themes in early socialism, see Bakunin, "National Socialists and Socialist Anti-Semites."
91. Quoted in Sewell, *Work and Revolution*, 274.
92. For discussion of related theories of technology and symbol use as externalized evolution in the works of Bergson (with his concepts of creative evolution and *Homo Faber*) and the archaeologist Leroi-Gourhan, see Schlanger, "'Suivre les gestes.'"
93. Guépin, "La terre et ses organes," 12.

CHAPTER NINE

1. For a comparable approach to the paper technologies of medicine and their function in structuring time, see Hess and Mendelsohn, "Case and Series." A detailed appreciation of the contents and absences of Comte's calendar is provided by one of the founders of modern history of science, George Sarton, in "Auguste Comte, Historian of Science"; his peroration begins, "Auguste Comte was a great man, one of the greatest of his time, even if he was crazy," 357.
2. See Heilbron, "Comte and Epistemology"; and *Rise of Social Theory*. In the *Avertissement* of the *Cours*, Comte compared his project to "what the English call Natural Philosophy." *Ecrits*, discussed in Braunstein, "Comte, de la nature." Simon Schaffer has traced the transition from natural philosophy into the modern disciplines, including the role played by charismatic founders of new fields, many of them in the period covered by this book, whose innovations established fields that could be pursued routinely by subsequent generations. See, for example, "Natural Philosophy" in *Ferment*; "Discoveries and the End of Natural Philosophy." A similar problematic is adopted in the title and introduction to the essay collection edited by Cahan, *From Natural Philosophy to the Sciences*. See Comte, *Catéchisme Positiviste*, 75.

3. On elements of this tradition, see Guchet, *Un humanisme technologique*; Schlanger, "Suivre les gestes."
4. A denunciation of the Saint-Simonians as proto-totalitarian is Iggers, *Cult of Authority*; Simondon likewise spoke of the Saint-Simonian vision of engineering as the imposition of human will on passive, malleable nature, in *Du mode d'existence*.
5. For background and analysis of Comte , see essays in Canguilhem, *Etudes*; Petit, *Auguste Comte*; Juliette Grange, *Auguste Comte*; Gouhier, *Jeunesse*; Krémer-Marietti, *Le positivisme*; Braunstein, "Comte, de la nature"; Karsenti, *Politiques de l'esprit*; Scharff, *Comte after Positivism*.
6. See Reisch, *How the Cold War Changed*; Galison, "Aufbau/Bauhaus"; Friedman, "History and Philosophy of Science"; Cat, Cartwright, and Chang, "Neurath"; Zammito, *Nice Derangement of Epistemes*. Scharff, in *Comte after Positivism*, examines Comte specifically in the light of logical positivism.
7. Many recent works on Comte do not assume a complete break between his "first career," in which his writings focused on scientific methods, and his "second career," which concentrated on religion and the emotions. The fundamental connection between these phases, I suggest, resides in Comte's views on technology and the milieu. I have been particularly helped in thinking through this continuity by Pickering's three-volume intellectual biography *Auguste Comte*, Petit's *Heurs et malheurs* and *Trajectoires positivistes*, Juliette Grange's ecological reading of Comte in *Auguste Comte*, the diverse articles on Comte (especially his popular writings) by Bernadette Bensaude-Vincent, and Michel Serres' occasional writings on Comte ("Auguste Comte auto-traduit"; "Paris 1800").
8. Comte's positivism was always propaganda. See Petit, "La diffusion"; Bensaude-Vincent, "L'astronomie populaire."
9. On Comte's careful attention to visual iconography, see Mary Pickering, "Comte et la culture visuelle," unpublished MS.
10. See Hadot and Davidson, *Philosophy as a Way of Life*; Foucault, *The Use of Pleasure*.
11. See Fedi, "Auguste Comte et la technique."
12. Comte, *Correspondence*, 1:6, "in the moment where all of our fellow citizens are hurrying after slavery and despotism."
13. See discussion of Musset in chapter 1.
14. Shinn, *L'Ecole Polytechnique*; Belhoste, *La formation d'une technocratie*.
15. Comte to Mill, July 22, 1842, in Pickering, *Auguste Comte*, 1:469.
16. Goffman, *Asylums*: A total institution is "a place of residence and work where a large number of like-situated individuals, cut off from the wider society for an appreciable period of time together, lead an enclosed, formally administered round of life."
17. Dhombres, "L'image 'Scientiste,'" 55. Blackboards were soon adopted at West Point and other engineering schools modeled on the Ecole Polytechnique.
18. Dhombres, "L'enseignement des mathématiques."
19. Comte, "Plan des travaux," 255.

20. See also his "Considérations sur le pouvoir spirituel" (1826), included in vol. 4 of *Système*.
21. Comte, "Plan des travaux," 255; see Serres, "Paris 1800."
22. Pickering, *Auguste Comte*, 1:368; *Système*, 1:2.
23. His madness and recovery are detailed in Pickering, *Auguste Comte*, 1:380–403; on pantheism, see 403.
24. Comte, *Cours*, (Leçon 1), 21. On the military art of the "coup d'oeil" and its relation to techniques of panoramic representation, see Bigg, "Panorama."
25. Comte, *Cours*, (Leçon 1), 21.
26. Comte, *Cours*, (Leçon 1), 21–22.
27. See discussions in Laudan, "Towards a Reassessment"; and Armstrong, *Scenes in a Library*. After reading Mill's *Logic*, Comte argued that the basis for induction was the assumption—inductively derived—of the uniformity of nature. The concept of "the great fetish" developed in the *Subjective Synthesis* was a recognition of this inevitable leap. Against the view that Comte was a naïve empiricist, scholars have recognized his view that phenomena are shaped by expectation and preexisting ideas (Michael Hawkins), by "fictitious" hypotheses (Laudan), and by the consideration of "unobservables" (Bensaude-Vincent, "Atomism").
28. Heilbron, "Comte and Epistemology." Comte's pluralist conception of the sciences is thus comparable to Ian Hacking's "Styles of Reasoning" (as Hacking acknowledges) and John Pickstone's "Ways of Knowing."
29. Comte used the term *social physics* until 1839, when he replaced it with *sociology*; the former term was used by Quetelet for his application of quantitative and statistical methods to the study of social phenomena, a move Comte rejected.
30. The primary meaning of the term *positive* as used by Comte is simply "observable"—although, as Bensaude-Vincent and Laudan have shown, Comte's sense of positive facts was not strictly limited to observables. In the *Système de politique positive*, vol. 1, published in 1848 as *Vue générale sur l'ensemble du positivisme*, he explained that "positive" refers to qualities of *reality, usefulness, certainty, precision*, and those that are *organic* (i.e., having "a social purpose") and *relative*, in the sense of "repudiating all absolute principles"; it is simply "the proper philosophical attitude, which is nothing, at bottom, but the generalization and systematization of good sense." *Système*, 1:57.
31. See Heilbron, "Comte and Epistemology"; Schaffer, "Natural Philosophy."
32. Comte's idea that the sciences appear as responses to astonishment and fear—and not as justifications of the ordinary course of events—was taken from Adam Smith's lectures on the history of astronomy. See discussion in Canguilhem, "Histoire des religions."
33. A helpful discussion of Comte's use of the "échelle" and the "série" in both biology and the classification of sciences is in Tort, "L'échelle encylopédique."
34. Comte, *Cours*, (Leçon 2), 54–55.
35. Comte, *Cours*, (Leçon 26), 425–26.

36. Comte, *Cours*, (Leçon 35), 569.
37. Ibid.
38. On the circularity of Comte's presentation, see Bensaude-Vincent, "La science populaire." On the methods of Comte's sociology, see Gane, *Auguste Comte*, e.g., esp. 71: "It is a scientific history, a rational and normal history, constructed through the elaboration of homogeneous sequences or series across the range of human activities." For a sympathetic view of Comte's understanding of sociology as analysis and intervention within public opinion, see Karsenti, *Politique de l'esprit*.
39. See Schweber, "Comte and Nebular Hypothesis," on the implications of astronomy for the life sciences.
40. Comte, *Course* (Leçon 40), 716.
41. Ibid. This point comes intriguingly close to arguments of Geoffroy Saint-Hilaire after his debate with Cuvier. Although it has been argued that the debate only indirectly concerned evolution, in fact Geoffroy's discussions of crocodile-like creatures whose fossils were found in Normandy were aimed precisely at arguing that new species have emerged gradually over time; these new species would have been produced by changes in their milieu, primarily through changes in temperature. See Serres, "Organogénie"; Le Guyader, *Geoffroy Saint-Hilaire*, 88–95.
42. See Schweber, "Comte and Nebular Hypothesis"; Schaffer, "The Science of Progress"; Secord, *Victorian Sensation*, on the implications of astronomy for the contemporary life sciences.
43. For a view of historical events as the intersection of multiple and distinct developmental temporalities, see Protevi, *Political Affect*.
44. Bensaude-Vincent, "Auguste Comte, la science populaire."
45. See Comte's 1819 review of an essay on "interior navigation" in England and France. *Ecrits de jeunesse*, 168.
46. See Hahn, *Anatomy of a Scientific Institution*; Crosland, *Society of Arcueil*; and analyses of the careers of savants in this period in Fox, Morell, Cardwell, *Patronage in 19th Century Science*; Outram, *Georges Cuvier*.
47. See Crosland, *Science under Control*; on the Academy's international status and importance in establishing priority, see Miller, *Discovering Water*.
48. Included in *Ecrits de jeunesse*. See in-depth discussion of the work and its reception in Schweber, "Auguste Comte and the Nebular Hypothesis."
49. Applying his formula for the diffusion of heat, Fourier argued that the heat gradient observed by miners—the deeper one goes into the earth, the hotter it gets—was consistent with the view of the earth as progressively cooling following its condensation out of hot nebulous matter. Elie de Beaumont extended this theory into geology, attributing the formation of strata and features including mountains and valleys to successive cracks in the earth's surface as it cooled. Comte also presented Fourier's theory of terrestrial heat with great enthusiasm in his section of the *Course* on thermology. See Fourier, *Théorie du chaleur terrestre*; and discussion in Merleau-Ponty, *La science*. For discussion see Lawrence, "Heaven and Earth."

50. Comte, "Cosmogonie positive," in *Ecrits de jeunesse*, 597.
51. Ibid.
52. Schweber (1980).
53. Pickering, *Auguste Comte*, 1:556.
54. See Babbage, *Reflections on the Decline*; note too how leading mathematical physicists looked to French models: Thompson's interest in Fourier and Regnault, Maxwell's interest in Ampère. See Wise, "Flow Analogy."
55. See Laplace, *Exposition*, 542 (book 5, chap. 6); Schaffer, "On Astronomical Drawing"; Merleau-Ponty, "Laplace as Cosmologist"; Arago, "Laplace"; Secord, *Victorian Sensation*.
56. For discussion see Schweber, "Comte and Nebular Hypothesis"; Schaffer, "Science of Progress"; Secord, *Victorian Sensation*.
57. Mill's *On Liberty* has been read as a rejoinder to what Mill saw as Comte's denial of the individual in favor of the collective.
58. Comte, *Ecrits de jeunesse*, 584.
59. Pickering, 1:547–60.
60. Comte, *Cours* (Leçon 40), 704. Comte accepted what has been termed the Meckel-Serres theory, in which the comparison of the phases of a single animal's development offers, "on a small scale, and, as it were, under one aspect, the whole series of the most marked organisms of the biological hierarchy." He said, "The primitive state of the highest organism must present the essential characters of the complete state of the lowest; and thus successively" (704–5). This relationship between the developmental series of the embryo and the elements of the animal series formed a clear analogy with Comte's view that each individual, and each science, must recapitulate the overall progress of humanity from the theological state to the positive state. See Canguilhem, *Comte et l'embryologie*; Clark, "Contributions of Meckel"; Lessertisseur and Jouffroy, "L'idée de série."
61. See Cuvier, *Le règne animal*; Daudin, *Cuvier et Lamarck*; Foucault, *Order of Things*.
62. Bichat, *Recherches physiologiques*. Comte also argues against seeing all of nature as alive: if that were so, the concept of "life" would have no meaning. Nevertheless, at several points in the *Course* he points out fundamental similarities between brute matter and organic matter, often in terms of shared obedience to more general laws: matter has and can acquire *habits*; the synchronizing of pendulum clocks on a shared pedestal appears as a form of imitation in the nonorganic world; the molecular activity of organic and inorganic substances differs largely by degree.
63. Comte, *Cours* (Leçon 40), 680. Martineau's extremely effective free translation and condensation of the *Cours* (*The Positive Philosophy of Auguste Comte*, of 1855) renders the crucial term *milieu* as "medium."
64. Comte, *Cours* (Leçon 40), 680.
65. Comte, *Cours* (Leçon 40), 682.
66. Comte, *Cours* (Leçon 40), 773–74.
67. Comte, *Cours* (Leçon 40), 678.

68. Comte, "Plan des travaux," 275.
69. Comte, *Cours* (Leçon 40), 681.
70. *Système*, 1:439. See Pickering, *Auguste Comte*, 3:175–81.
71. Canguilhem, *Normal and Pathological*: "Social organization is, above all, the invention of organs—organs to look for and receive information, organs to calculate and even make decisions.... As for the assimilation of social information by means of statistics being analogous to the assimilation of vital information by means of sense receptors, to our knowledge it is older. It was Gabriel Tarde, who, in 1890, in *Les lois de l'imitation*, was the first to attempt it" (253–54). Canguilhem makes a comparable point about technologies as the extension of organs in a genealogy going back to at least Descartes, in his essay "Machine et organisme," in *La connaissance de la vie*.
72. See Rouvre, *L'amoureuse histoire*; for details of Clotilde's literary ambitions and her relationship with Comte, including his tyrannical obsession with her and a horrendous deathbed scene, see Pickering, vol. 2.
73. Comte's religious shift took place more than a decade after the decline of the Saint-Simonian movement; during its rise he and his former associates frequently clashed; he was bashed in a long digression in the *Doctrine Saint-Simonienne*, 284–92, for failing to understand the importance of religion, art, and emotion in social organization; all critiques he subsequently addressed to himself.
74. Clotilde was one of three "guardian angels" of humanity, along with Comte's mother and his servant, "a remarkable woman belonging to the working class, who has deigned to devote herself to my material service without imagining that she was showing me an admirable type of moral perfection." The three represented the "three sympathetic virtues: attachment between equals, veneration for superiors, kindness to inferiors" (*Système* 1:12–13).
75. Comte, *Catechism*, 2. See Michel Chevalier, who uses the same etymology of "religion" to present the connections forged by the railroad as a religion, in *Politique industrielle*.
76. Comte, *Système*, 1:402. Despite his loathing of Victor Cousin, these modes restated Cousin's principles of the true, the beautiful, and the good.
77. Comte, *Système*, 1:36.
78. Comte, *Catéchisme Positiviste*, 36.
79. Comte, *Catéchisme Positiviste*, 38.
80. Comte, *Synthèse subjective*.
81. For discussion, see Pickering, vol. 3.
82. *Synthèse Subjective*, 22–26.
83. *Système*, 1:617. Comte's ecological themes are presented in Grange, *Auguste Comte*.
84. Comte, *Système*, 1:618.
85. My account of Comte's ecological views is indebted to Juliette Grange's *Auguste Comte*; for Comte and biopolitics, see Cohen, *A Body Worth Defending*.
86. This notion (and this chapter's subtitle) is taken from an undeveloped but suggestive aside in Mike Gane's *Auguste Comte*, 14; Laurent Fedi made it the

organizing principle of his study of Comte's astronomy in "Monde clos contre l'univers infini"; Bruno Latour's *Cogitamus* also returns to Koyré's phrase, arguing that the continuous and infinite universe announced by Descartes and Newton has now been replaced by a finite cosmos, or multiverse, which humans must compose—a proposition with strong continuities with the projects of the mechanical romantics.

87. Hamlet's worry "The time is out of joint" was a leitmotif in Derrida's probing of the messianic metaphysics underlyling Marx and Engels's *Communist Manifesto*; like Comte, Marx saw himself confronting a temporal dislocation, a case of the present's being somehow maladjusted to its own demands.

88. Frank Manuel suggested that most utopias of the eighteenth century were "stable and ahistorical, ideals out of time," while in the French Restoration they "became dynamic and bound to a long prior historical series. They should henceforth be called euchronias—good place becomes good time." "Toward a Psychological History," 104.

89. Schiller to Körner, February 23, 1793, in Schiller, *Aesthetic Education*, 300.

90. Andrews, "Utopian Androgyny," persuasively argues that the recurrent image among romantic socialists of beings that combined masculine and feminine traits was a challenge to the supremacy of a fundamentally male-centered individualism; see also Kofman, *Aberrations: Le devenir-femme d'Auguste Comte*.

91. Leroux, *D'une religion nationale*. See discussions in Evans, *Romantic Socialism*; Bakunin, *Leroux*; Bénichou, *Le Temps*; and Rancière, *Aux bords de la politique*.

92. See Sessions, *By Sword and Plow*. In *Lévy-Bruhl*, Frédéric Keck traces Lévy-Bruhl's conception of primitive thought as "mystical participation" back to Comte's writings on fetishism, and sets both in the context of French colonial expansion; yet despite Comte's emphasis on the biological foundations of society and endorsement of the civilizing mission of France, his view of a future world society defied the racial hierarchies of his age. Martin Staum, in *Labeling People*, shows the participation of several former Saint-Simonians, notably Comte's former pupil, Gustave D'Eichtahl, in new social science projects from ethnology to phrenology in which a correlation between racial differences and degrees of civilization was taken for granted. In "Universal Alliance," Naomi Andrews shows how for many projects of romantic socialism, the human universality and the ideal of "association" led to support of empire.

93. The question of freedom-in-connection—taken as more than simply a political question, but one touching upon questions of causality and agency more broadly—resonates with Francisco Varela's use of the concept of "autonomy" in the study of microorganisms, immune systems, consciousness, and "selves" of all kinds. Although a difference between an inside and an outside emerges through the formation of membranes, this distinction is necessarily fragile; it is both a real and conventional distinction. See Varela, *Principles of Biological Autonomy*; Varela, Thompson, and Rosch, *Embodied*

Mind; Thompson, *Mind in Life*; on Varela's scruples about extending his biological concepts to political formations, see Protevi, *Political Affect*, 43–45.

CHAPTER TEN

1. See the special 2007 issue of *Romantisme* titled "Les banquets"; Fortescue, *France and 1848*; Price, *French Second Republic*; Agulhon, *Les quarante-huitards*.
2. See the American and French Research on the Treasury of the French Language (ARTFL) project's 1848 online collection, containing more than one hundred pamphlets and periodicals from 1848 to 1851.
3. Baudelaire, *Mon coeur mis à nu*, in *Oeuvres*, 1.
4. See Petitier, *Michelet*.
5. Marx referred to Napoleon's coup d'état as tragedy repeated as farce in the first paragraph of *The 18th Brumaire of Louis Napoleon Bonaparte*.
6. On Foucault's science journalism, see Tobin, *Léon Foucault*, 80–94.
7. See Tobin, *Léon Foucault*, 117–32.
8. See Berlioz's invitation in Tobin, *Léon Foucault*, 130.
9. Gautier, in Coudroy, *La critique parisienne*, 151.
10. Although the performance was a huge success (ticket lines turned into riots, there were repeat performances throughout Europe, and for fifty years it was the most performed opera in the world), criticism was not all positive. Castil-Blaze: "Destined to overexcite the imagination, this equipment of mise en scène is an instrument of destruction, of ruin." For Gautier, *Le prophète* was a further indication of the "decadence of art": "We see our eyes overwhelmed, when instead the play and the music should leave the mind, the heart and the ear in a calm flatness." Both in Coudroy, *La critique parisienne*, 157–58.
11. Jan van Leyden was a follower of Thomas Müntzer, cited by Engels as a precursor of communism; see Negri and Fadini, "Materialism and Theology"; Hibberd, "*Le Prophète*," 153–79.
12. Meyerbeer's religious themes, and the readings he has supported, deserve much more than an endnote: from *Robert* to *The Huguenots* and *The Prophet*, he repeatedly staged religious conflict and the violent consequences of enthusiasm. The young Wagner greatly admired and imitated Meyerbeer, borrowing such techniques as the defining motif, risqué harmonies, and the mythical framework; he denied the influence in his notorious essay "On Jewishness in Music," which rejected Meyerbeer's emphasis on stage effects as unseemly and impure. Today, while musicology has confronted and denounced Wagner's anti-Semitism, it nevertheless continues Wagner's condemnation of Meyerbeer as unworthy of serious consideration, a perverse stance toward the acknowledged founder of one its main objects, the French grand opera. As late as the early twentieth century, composer Reynaldo Hahn could write that "people of my father's generation would rather have doubted the solar system than the supremacy of 'Le Prophète' over all other operas." Letellier, *Operas of Giacomo Meyerbeer*, 197.

13. Fulcher, "Meyerbeer and the Music of Society."
14. See pamphlet collection at the Bibliotheque Nationale and at the Centre National des Arts et Metiers on the Expositions.
15. See Chandler, "First Industrial Exposition."
16. *Rapport du jury central*, 66.
17. Ibid.
18. Archives Nationales, Dossier F 12 5005, Exposition des produits de l'industrie. 1849.
19. Dupin's speech underlines how the study of productive force, the emphasis on knowledge as the product of active humanity, and the recognition of the transformative and dynamic powers of the imponderables locked in chemical compounds—all testify to the new conception of the universe as a storehold of arguably limitless potential, which human technology has only begun to unleash and apply. Leroux's attack on Malthus clearly goes in this direction: instead of an assumption of inevitable scarcity, he argues that economics must begin with the idea of the limitless fecundity of nature. Chevalier—informed, like Leroux, by Saint-Simon's vision of the abundance of nature and a possible paradise on earth—thus introduces this new, endlessly receding horizon of economic growth into his political economic writings of the 1830s and 1840s. See Staum, "French Lecturers in Political Economy."
20. *Rapport du jury central*, 70.
21. *Le siecle* of July 6, 1849, quoted in *Arithmaurel inventé par MM. Maurel et Jayet*, 43.
22. Quoted in ibid., 29.
23. See Brewster, *Letters on Natural Magic*. For discussion of the shifting characteristics of the human in response to mechancial imitations, see Riskin, "Defecating Duck."
24. On Chladni's combined musical and physical research, see Jackson, *Harmonious Triads*.
25. Bonaparte had spent several years in prison for a failed coup in 1840, during which he read up on Saint-Simonianism and science and even conducted electrical experiments in his cell.
26. Laplace had written that a demonstration of this rotation by a more direct, "internal" means than by changes in the relation of the stars would be welcome. Aczel, *Pendulum*, 40; see Laplace, *Exposition*. See Eco, *Foucault's Pendulum*, whose climax takes place in the Muséum of the Conservatoire National des Arts et Métiers, where the pendulum was suspended amid other relics of nineteenth-century physics and engineering. Much like the fantastic authors of 1830s, though with heavy irony, Eco superimposed upon the wonders of modern instrumental rationality the quest for mystical truths found in the "perennial" tradition of the occult sciences.
27. Daumas, *Arago*, 271.
28. Louis-Napoléon Bonaparte, *Les idées Napoléoniennes*, 10; Bonaparte's concep-

tion of the "Napoleonic idea" participated directly in the thematics of vital, at times transcendent, fluids we have seen in fantastic literature, physics, animal magnetism, and the sciences of the living at this time—the quasi-concrete form of the "spiritual power" sought by many reformers. The quote begins, "The Napoléonic idea manifests itself in different ways in the various branches of human ingenuity: it will make agriculture flourish[;] it will invent new products[;] it will exchange new ideas with other countries." Thanks to John Purciarello for this source.

29. Mayor Etienne Arago told Bonaparte: "You will be forced to march toward the monarchy. What will push you forward is the ignorance of the masses and imperial fetishism!" (Quoted by another famous brother of François Arago, the blind traveler, critic, and historian Jacques Arago, in his *Histoire de Paris*, 2:372.)
30. Foucault, in *Journal des débats*, March 12, 1851, quoted in Aczel, *Pendulum*, 157.
31. Foucault, "Démonstration expérimentale," 520.
32. Pascal's famous words, in his *Pensées*, were written in response to the shattering of the musical spheres: "The silence of those infinite spaces terrifies me."
33. Baudelaire, *Oeuvres complètes*, 692.
34. See Rancière, *Nights of Labor*.
35. Clark, *Absolute Bourgeois*; Buck-Morss, *Dialectics of Seeing*.
36. See Hugo, *Napoleon le petit*.
37. See Rancière, *Le partage du sensible*; see also Clark, *Painting of Modern Life*; Barthes, *Writing Degree Zero*; Wilson, *Axel's Castle*; Huyssens, *Beyond the Great Divide*.
38. This discussion builds on analyses of the emergence of the modern disciplinary system, including those by Wolf Lepenies, *Between Literature and Science*; Peter Burger, *Theory of the Avant-Garde*; and Bruno Latour, *We Have Never Been Modern*. In relation to the argument in the latter, my claim has been that something like a "modern constitution" did indeed have its proponents (as sketched in the introduction by the hopeless/soulless divide) and that this constitution really began to take shape in modern terms after 1800, in the face of the combined changes brought by political and industrial revolutions (and Kant's noumenon/phenomenon division both notes and contributes to this solidification). Yet against this dawning consensus, this book has shown how the period of early industrialization, in the first half of the nineteenth century, was crowded with attempts to undo or redraw or redirect this constitution. Its production of "hybrids" of natures/cultures was not, however, done in secret or unknowingly (as claimed by *We Have Never Been Modern*): "romantic machines" were self-consciously celebrated for their ability to cross the gap between matter and mind, objects and subjects: they were recognized as hybrids, as instruments working against the looming division of science from humanities, objects from subjects. This "constitution," in my view, rightly characterizes much (but not all) of the intellectual activity and disciplinary framings of what we might call "high modernity"

(borrowing from James Scott's *Seeing like a State*), whose consolidation depended on the successful repression of the utopian uprisings of 1848.
39. Fressoz, "Beck."
40. See Cronon, "Trouble with Wilderness."
41. On the notion of alternative trajectories of modernization in postcolonial studies, see Timothy Mitchell, ed., *Questions of Modernity*; on alternative pasts, Renouvier, *Uchronie*; Carrère, *Le détroit de Behring*.
42. On placing consciousness back into the world-picture, see Varela, "A Science of Consciousness as if Experience Mattered"; Thompson, *Mind in Life*; Malabou, *What Should We Do with Our Brain?*
43. One critique of utopias, going as far back as Hawthorne's *Blithesdale Romance* (dealing with a Fourierist experiment), is that they fail to overcome individual differences and lack basic technical competence, as in post-Manson revisitations of 1960s and 1970s communes, including Boyle's *Drop City* and Moodyson's film *Together*; another critique is that planned communities are, on the contrary, all too successful in eradicating individuality, as in *We*, *1984*, and *Brave New World*. In contrast, see the interest in the small-scale but high-tech orientation of the *Whole Earth Catalog* (see Markoff, *What the Dormouse Said*; Turner, *From Counterculture to Cyberculture*); the scientifically calibrated "natural farming" techniques at Polyface Farm described in Pollan's *Omnivore's Dilemma*; and the Leroux-like solutions to industrial devolution advocated in McKnight and Block, *The Abundant Community*.
44. Duveyrier, "La ville nouvelle," 257.

Bibliography

Abi-Mershed, Osama. *Apostles of Modernity: Saint-Simonians and the Civilizing Mission in Algeria*. Palo Alto, CA: Stanford University Press, 2010.
Abrams, Meyer H. *The Mirror and the Lamp: Romantic Theory and the Critical Condition*. New York: Oxford University Press, 1971.
———. *Natural Supernaturalism: Tradition and Revolution in Romantic Literature*. New York: Norton, 1971.
Académie des Sciences. *Procès-verbaux des séances de l'Académie tenues depuis la fondation de l'Institut jusqu'au mois d'août 1835*. Publiés conformément à une décision de l'Académie par MM. les secrétaires perpétuels. Hendaye: Observatoire d'Abbadia, 1910–22.
Aczel, Amir D. *Pendulum: Léon Foucault and the Triumph of Science*. New York: Simon and Schuster, 2003.
Adas, Michael. *Machines as the Measure of Men: Science, Technology, and Ideologies of Western Dominance*. Ithaca, NY: Cornell University Press, 1990.
Adorno, Theodor W. *Eingriffe: Neun kritische Modelle*. Frankfurt am Main: Suhrkamp, 1963.
Agrippa, Henry Cornelius, James Freake, and Donald Tyson. *Three Books of Occult Philosophy*. Woodbury, MN: Llewellyn, 1992.
Agulhon, Maurice. *Le cercle dans la France bourgeoise, 1810–1848*. Paris: Armand Colin, 1977.
———. *Les quarante-huitards*. Paris: Gallimard, 1973.
Alder, Ken. *Engineering the Revolution: Arms and Enlightenment in France, 1763–1815*. Princeton, NJ: Princeton University Press, 1997.
———. "Making Things the Same: Representation, Tolerance, and the End of the Ancien Régime in France." *Social Studies of Science* 28, no. 4 (1998): 499–545.
———. *The Measure of All Things: The Seven-Year Odyssey and Hidden Error That Transformed the World*. New York: Free Press, 2002.
Alexander, Jennifer Karns. *The Mantra of Efficiency: From Waterwheel to Social Control*. Baltimore: Johns Hopkins University Press, 2008.

Alexandrian, Sarane. *Le socialisme romantique*. Paris: Seuil, 1979.

Allen, James Smith. *Popular French Romanticism: Authors, Readers, and Books in the 19th Century*. Syracuse, NY: Syracuse University Press, 1981.

Ampère, André-Marie. *Essai sur la philosophie des sciences, ou exposition analytique d'une classification naturelle de toutes les connaissances humaines*. Vols. 1–2. Paris, 1834, 1843.

———. "Idées de M. Ampère sur la chaleur et la lumière." *Bibliothèque universelle des sciences, belles lettres et arts* 45 (1832): 225–35.

———. "Lettre de M. Ampère à M. le comte Berthollet sur la détermination des proportions dans lesquelles les corps se combinent d'après le nombre et la disposition respective des molécules dont les parties intégrantes sont composes." *Annales de chimie* 90, no. 1 (1814): 43–86.

———. "Mémoire présenté à l'Académie royale des sciences, le 2 octobre 1820, où se trouve compris le résumé de ce qui avait été lu à la même Académie les 18 et 25 septembre 1820, sur les effets des courans électriques." *Annales de chimie et de physique*, 15 (1820): 59–74, 170–218.

———. "Notes sur cet exposé des nouvelles expériences relatives aux phénomènes produits par l'action electro-dynamique, faites depuis le mois de mars 1821." In *Recueil d'observations électrodynamiques, contenant divers mémoires, notices, extraits de lettres ou d'ouvrages périodiques sur les sciences, relatifs à l'action mutuelle de deux courants électriques, à celle qui existe entre un courant électrique et un aimant ou le globe terrestre, et à celle de deux aimants l'un sur l'autre*. Paris: Crochard, 1822.

———. *Théorie des phénomènes électro-dynamiques, uniquement déduite de l'expérience*. Paris: Méquignon-Marvis, 1826.

Ampère, André Marie, Louis Auguste Alphonse de Launay, and Société des Amis d'André-Marie Ampère. *Correspondance du grand Ampère*. Vols. 1–2. Paris: Librairie Gauthier-Villars, 1936–43.

Andrews, Naomi. *Socialism's Muse: Gender in the Intellectual Landscape of French Romantic Socialism*. Lexington Books, 2006.

———. "'The Universal Alliance of All Peoples': Romantic Socialists, the Human Family, and the Defense of Empire during the July Monarchy, 1830–1848." *French Historical Studies* 34, no. 3 (2011): 473–502.

———. "Utopian Androgyny: Romantic Socialists Confront Individualism in July Monarchy France." *French Historical Studies* 26, no. 3 (Summer 2003): 437–57.

Appel, Toby. *The Cuvier-Geoffroy Debate: French Biology in the Decades before Darwin*. Oxford: Oxford University Press, 1987.

———. "Henri de Blainville and the Animal Series: A Nineteenth Century Chain of Being." *Journal of the History of Biology* 13, no. 2 (1980): 291–319.

Arago, Étienne, Charles Dupeuty, and Michel-Nicolas Balisson de Rougemont. *Paris dans la comète*. Paris: Marchant, 1836.

Arago, François. "Biographies des principaux astronomes: Laplace." In Arago, *Oeuvres complètes*, vol. 3.

———. *Biographies of Distinguished Scientific Men*. Translated by William Henry Smyth, Baden Powell, and Robert Grant. Vol. 1. London: Ticknor and Fields, 1859.

———. "Carnot, biographie lue en séance publique de l'Académie des Sciences." In Arago, *Oeuvres complètes*, vol. 1.

———. "De la scintillation." In Arago, *Oeuvres complètes*, vol. 7.

———. "Discours de M. François Arago." Preface to E. Salverte, *Des sciences occultes, ou essai sur la magie, les prodiges et les miracles*. Paris: Ballière, 1843.

———. "Discours sur la réforme éléctorale." In Arago, *Oeuvres complètes*, vol. 12.

———. "Eloge d'Ampère." In Arago, *Oeuvres complètes*, vol. 2.

———. "Eloge de Condorcet." In Arago, *Oeuvres complètes*, vol. 2.

———. "Histoire de ma jeunesse." In Arago, *Oeuvres complètes*, vol. 1.

———. "James Watt, biographie lue en séance publique de l'Académie des Sciences." In Arago, *Oeuvres complètes*, vol. 1.

———. *Oeuvres complètes de François Arago*, edited by J. A. Barral. 17 vols. Paris: Gide et J. Baudry, 1854–62.

———. *Popular Astronomy*. Translated by R. Grant and W. H. Smyth. London: Longman, Brown, Green, and Longmans, 1855.

———. "Rapport fait au nom de la commission chargée de l'examen du projet de loi tendant à accorder . . . pour la cession faite par eux du procédé servant à fixer les images de la chambre obscure." In Louis-Jacques-Mandé Daguerre, *Historique et description des procédés du daguerréotype et du diorama*. 1839. Reprint, Paris: Giroux, 2003.

Arago, Jacques. *Histoire de Paris, ses révolutions, ses gouvernements et ses événements de 1841 à 1852, comprenant les sept dernières années du règne de Louis-Philippe et les quatre premières de la République*. Paris: Dion-Lambert, 1853.

Armstrong, Carol. *Scenes in a Library: Reading the Photograph in the Book, 1843–1875*. Cambridge, MA: MIT Press, 1998.

Ashworth, William J. "The Calculating Eye: Bailly, Herschel, Babbage, and the Business of Astronomy." *British Journal for the History of Science* 27, no. 4 (December 1, 1994): 409–41.

Assmann, Jan. *Moses the Egyptian: The Memory of Egypt in Western Monotheism*. Cambridge, MA: Harvard University Press, 1997.

Aubin, David. "The Fading Star of the Paris Observatory in the Nineteenth Century: Astronomers' Urban Culture of Circulation and Observation." *Osiris* 18 (2003): 79–100.

———. "Orchestrating Observatory, Laboratory, and Field: Jules Janssen, the Spectroscope, and Travel." *Nuncius* 17 (2003): 143–62.

Aubin, David, Charlotte Bigg, and Heinz Otto Sibum, eds. *The Heavens on Earth: Observatories and Astronomy in Nineteenth-Century Science and Culture*. Durham, NC: Duke University Press, 2010.

Audiganne, Armand. *François Arago: Son génie et son influence*. Paris: Garnier, 1857.

Avenal, Henri. *Histoire de la presse française depuis 1789 jusqu'à nos jours*. Paris: E. Flammarion, 1900.

Azouvi, François. *Maine de Biran: La science de l'homme*. Paris: Vrin, 1995.

Babbage, Charles. *The Ninth Bridgewater Treatise*. London: J. Murray, 1837.

———. *On the Economy of Machinery and Manufactures*. London: Charles Knight, 1832.

———. *Reflections on the Decline of Science in England and on Some of Its Causes*. London: Fellowes, 1830.

Bachelard, Gaston. *L'activité rationaliste de la physique contemporaine*. Paris: Presses Universitaires de France, 1951.

———. *Etude sur l'évolution d'un problème de physique: La propagation thermique dans les solides*. Paris: Vrin, 1927.

———. *Le matérialisme rationnel*. Paris: Presses Universitaires de France, 1953.

———. *Le nouvel esprit scientifique*. Presses Universitaires de France, 1983.

———. *Le rationalisme appliqué*. Paris: Presses Universitaires de France, 1949.

Bajac, Quentin. *Le daguerréotype français: Un objet photographique: Paris, Musée d'Orsay, 13 Mai–17 Août 2003: New York, the Metropolitan Museum of Art, 22 Septembre 2003–4 Janvier 2004*. Paris: Musée d'Orsay, 2003.

Bakunin, Jack. "National Socialists and Socialist Antisemites." *Patterns of Prejudice* (1977): 29–33.

———. *Pierre Leroux and the Birth of Democratic Socialism, 1797–1848*. New York: Revisionist Press, 1976.

Ballanche, Pierre Simon. *Oeuvres complètes*. Geneva: Slatkine Reprints, 1967.

Balzac, Honoré de. *La chef d'oeuvre inconnu*. In *La comédie humaine*, vol. 10. Paris: Bibliothèque de la Pléiade, 1979.

———. *Gambara*. In *La comédie humaine*, vol. 10. Paris: Bibliothèque de la Pléiade, 1979.

———. *Les illusions perdues*. In *La comédie humaine*, vol. 1. Paris: Bibliothèque de la Pléiade, 1976.

———. *Louis Lambert*. In *La comédie humaine*, vol. 11. Paris: Bibliothèque de la Pléiade, 1980.

———. *Le père Goriot*. In *La comédie humaine*, vol. 1. Paris: Bibliothèque de la Pléiade, 1976.

———. *La recherche de l'Absolu*. In *La comédie humaine*, vol. 10. Paris: Gallimard, Bibliothèque de la Pléiade, 1979.

Bann, Stephen. *Paul Delaroche: History Painted*. Princeton, NJ: Princeton University Press, 1997.

Barger, Susan, and William White. *The Daguerreotype: Nineteenth Century Technology and Modern Science*. Washington, DC: Smithsonian Institution Press, 1991.

Baroli, Marc. *Le train dans la littérature française*. Paris: Ecole technique d'imprimerie Notre Famille, 1963.

Barthelémy-Saint-Hilaire, Jules. *La philosophie des deux Ampères*. 2nd ed. Paris: Didier, 1870. First ed. published in 1866.

Barthes, Roland. *Sade, Fourier, Loyola*. Baltimore: Johns Hopkins University Press, 1997.

———. *Writing Degree Zero*. New York: Hill and Wang, 1968.

Barzun, Jacques. *Darwin, Marx, Wagner: Critique of a Heritage*. Garden City, NY: Doubleday, 1958.

Baudelaire, Charles. *Oeuvres complètes*. Vol. 1. Paris: La Pléiade, 1990.

Behrent, Michael C. "The Mystical Body of Society: Religion and Association in Nineteenth-Century French Political Thought." *Journal of the History of Ideas* 69, no. 2 (2008): 219–43.

Beiser, Frederick C. *The Fate of Reason: German Philosophy from Kant to Fichte*. Cambridge, Mass.: Harvard University Press, 1987.

———. "Kant and *Naturphilosophie*." In *The Kantian Legacy in Nineteenth-Century Science*, edited by M. Friedman and A. Nordmann. Cambridge, MA: MIT Press, 2006.

Belhoste, Bruno. "Arago, les journalistes et l'Académie des sciences dans les années 1830." In *La France des années 1830 et l'esprit de réforme*, edited by Patrick Harismendy. Rennes: Presses Universitaires de Rennes, 2006.

———. *Augustin-Louis Cauchy: A Biography*. New York: Springer-Verlag, 1991.

———. *La formation d'une technocratie: L'école Polytechnique et ses élèves de la Révolution au Second Empire*. Paris: Belin, 2003.

Belhoste, Bruno, and Konstantinos Chatzis. "From Technical Corps to Technocratic Power: French State Engineers and Their Professional and

Cultural Universe in the First Half of the 19th Century." *History and Technology* 23, no. 3 (2007): 209–25.

Belhoste, Bruno, Amy Dalmedico, and Antoine Picon. *La formation polytechnicienne, 1794–1994*. Paris: Dunod, 1994.

Bélidor, Bernard Forest de. *Architecture hydraulique*. Paris, 1819.

Bellanger, Claude, Jacques Godechot, Pierre Guiral, and Fernand Terrou, eds. *Histoire générale de la presse française*. Vol. 2, *De 1815–1871*. Paris: Presses Universitaires de France, 1969.

Ben-David, Joseph, and Gad Freudenthal. *Scientific Growth: Essays on the Social Organization and Ethos of Science*. Berkeley: University of California Press, 1991.

Bénichou, Paul. *L'école du désenchantement*. Paris: Gallimard, 1992.

———. "Le grand oeuvre de Ballanche." *Revue d'histoire littéraire de la France* 75 (September–October 1975): 736–48.

———. *Les Mages Romantiques*. Paris: Gallimard, 1988.

———. *Le sacré de l'écrivain, 1750–1830: Essai sur l'avènement d'un pouvoir spirituel laïque dans la France moderne*. Paris: J. Corti, 1973.

———. *Le temps des prophètes: Doctrines de l'âge romantique*. Paris: Gallimard, 1977.

Benjamin, Walter. *The Arcades Project*. Belknap Press of Harvard University, 2002.

———. "Fourier or the Arcades." In Benjamin, *The Arcades Project*.

———. *Oeuvres II: Poésie et révolution*. Paris: Editions Denoël, 1955.

———. "On Some Motifs in Baudelaire." In *Illuminations: Essays and Reflections*, edited by Hannah Arendt. New York: Schocken, 1968.

———. "The Work of Art in the Age of Mechanical Reproduction." In *Illuminations: Essays and Reflections*, edited by Walter Benjamin, Hannah Arendt, and H. Zohn. New York: Schocken, 1935.

Bensaude-Vincent, Bernadette. "L'astronomie populaire: priorité philosophique et projet politique." *Revue de synthèse* 4, no. 1 (1991): 49–60.

———. "Atomism and Positivism: A Legend about French Chemistry." *Annals of Science* 56 (1999): 81–94.

———. "A Historical Perspective on Science and Its 'Others.'" *Isis* 100, no. 2 (June 1, 2009): 359–68.

———. "Présentation." *Romantisme* 19, no. 65 (1989): 3–5.

———. "Un public pour la science: L'essor de la vulgarisation au XIXe siécle." *Réseaux* 58 (1993): 47–66.

———. "La science populaire d'un philosophe: Auguste Comte." *Corpus: Revue de Philosophie* 4 (1987): 143–67.

———, ed. "Sciences pour tous." *Romantisme* 65, special issue (1989).

Bensaude-Vincent, Bernadette, and Bruno Bernardi, eds. "Rousseau et la chimie." *Corpus: Revue de Philosophie* 36, special issue (1999).

Bensaude-Vincent, Bernadette, and Christine Blondel. *Des savants face à l'occulte: 1870–1940*. Paris: La Découverte, 2002.

Berenson, Edward. *Populist Religion and Left-Wing Politics in France*. Princeton, NJ: Princeton University Press, 1984.

Bergson, Henri. *L'évolution créatrice*. Paris: Félix Alcan, 1908.

Berlin, Isaiah, and Henry Hardy. *The Roots of Romanticism*. Princeton, NJ: Princeton University Press, 1999.

Berlioz, Hector. *The Art of Music and Other Essays*. Translated by E. Csicsery-Rónay. Bloomington: Indiana University Press, 1998.

———. *Berlioz's Orchestration Treatise: A Translation and Commentary*. Translated and edited by Hugh Macdonald. Cambridge: Cambridge University Press, 2002.

Bertrand, Alexandre. *Du magnétisme animal en france et des jugements qu'en ont porté les sociétés savantes*. Paris: J. B. Baillière, 1826.

Bezanson, Anna. "The Early Use of the Term Industrial Revolution." *Quarterly Journal of Economics* 36, no. 2 (1922): 343–49.

Biagioli, Mario. *Galileo, Courtier: The Practice of Science in the Culture of Absolutism*. Chicago: University of Chicago Press, 1993.

Bichat, Xavier. *Recherches physiologiques sur la vie et la mort*. Paris: Delahays, 1806.

Bigg, Charlotte. "The Panorama, or La nature à coup d'œil." In *Observing Nature—Representing Experience: The Osmotic Dynamics of Romanticism, 1800–1850*, edited by Erna Fiorentini. Berlin: Reimer Verlag, 2007.

———. "Staging the Heavens: Popular Observatory Science in the Nineteenth Century." In *The Heavens on Earth: Observatories and Astronomy in Nineteenth-Century Science and Culture*, edited by David Aubin and Charlotte Bigg. Durham, NC: Duke University Press, 2010.

Biot, Jean-Baptiste. *Précis élémentaire de physique expérimentale*. Paris: Déterville, 1824.

Blanckaert, Claude. "La nature de la société: Organicisme et sciences sociales au XIXe siècle." In *Une histoire des sciences humaines*, edited by C. Blanckaert and L. Mucchielli. Paris: L'Harmattan, 2004.

———. *Les politiques de l'anthropologie: Discours et pratiques en France (1860–1940)*. Paris: L'Harmattan, 2001.

Blondel, Christine. "Ampère, le Newton de l'électricité." *SABIX* 34 (2004): 57–64.

———. *Ampère et la création de l'électrodynamique*. Paris: Bibliotheque Nationale, 1982.

———. *André-Marie Ampère*. Paris: Ecole des hautes études en sciences sociales, 1978.

———. "Electrical Instruments in Nineteenth Century France, between Makers and Users." *History and Technology* 13 (1997): 157–82.

———. "Vision physique "éthérienne," mathématisation "laplacienne": L'electrodynamique d'Ampère." *Revue d'histoire des sciences* 42, nos. 1–2 (1989), 123–37.

Blondel, Christine, and Philippe Descamps. "Avec Ampère, le courant passe...." *Cahiers de sciences et vie* 67 (2002): 20–27.

Blondel, Christine, and L. Pearce Williams. "Ampère and the Programming of Research." *Isis* 76, no. 284 (1985): 559–61.

Boas, Franz. *The Shaping of American Anthropology, 1883–1911*. Edited by George W. Stocking. New York: Basic Books, 1974.

Boas, George. *French Philosophies of the Romantic Period*. Baltimore: Johns Hopkins University Press, 1925.

———. "Maine de Biran." *Philosophical Review* 34, no. 5 (1925): 477–90.

Bonaparte, Louis-Napoléon. *Les idées Napoléoniennes*. Paris: Amyot, 1860.

Bonnet, Charles. *La Palingénésie philosophique ou Idées sur l'état passé et sur l'état futur des êtres vivants, ouvrage destiné à servir de supplément aux derniers écrits de l'auteur et qui contient principalement le précis de ses recherches sur le christianisme*. Amsterdam: M.-M. Rey, 1769.

Botting, Douglas. *Humboldt and the Cosmos*. New York: Harper and Row, 1973.

Bouin, Philippe, and Christian-Philippe Chanut. *Histoire Française des Foires et des Expositions Universelles*. Paris: Baudouin, 1980.

Bourdeau, Vincent, François Jarrige, and Julien Vincent. *Les luddites: Bris de machines, économie politique et histoire*. Maisons-Alfort: Editions è®e, 2006.

Bourdier, Franck. "Le prophète Geoffroy Saint-Hilaire, George Sand, et les Saint-Simoniens." *Histoire et nature* 1 (1973): 47–66.

Bourdieu, Pierre. *La noblesse d'état: Grandes écoles et esprit de corps*. Paris: Les Editions de Minuit, 1989.

———. *The Rules of Art: Genesis and Structure of the Literary Field*. Translated by Susan Emanuel. Stanford, CA: Stanford University Press, 1995.

Bourguet, Marie-Noëlle. *Déchiffrer la France: La statistique départementale à l'époque napoléonienne*. Paris: Editions des archives contemporaines, 1989.

———. "La république des instruments: Voyage, mesure et science de la nature chez Alexandre de Humboldt." In *Marianne Germania: Deutsch-Französischer Kulturtransfer im europäischen Kontext 1789–1914*, edited by Etienne François, Marie-Claire Hoock-Demarle, Reinhart Meyer-Kalkus, and Michael Werner. Leipzig: Leipziger Univ.-Verl., 1998.

Bowie, Andrew. "Romantic Technology." *Radical Philosophy* 72 (July–August 1995), www.radicalphilosophy.com/default.asp?channel_id=2188&editorial_id=10774.

Bown, Nicola, Carolyn Burdett, and Pamela Thurschwell, eds. *The Victorian Supernatural*. Cambridge: Cambridge University Press, 2004.

Bradley, Margaret. "The Facilities for Practical Instruction in Science during the Early Years of the Ecole Polytechnique." *Annals of Science* 33 (1976): 425–46.

———. "Franco-Russian Engineering Links: The Careers of Lamé and Clapeyron, 1820–1830." *Annals of Science* 38, no. 3 (1981): 291–312.

———. "Scientific Education versus Military Training: The Influence of Napoleon Bonaparte on the 'Ecole Polytechnique.'" *Annals of Science* 32 (1975): 415–49.

Bradley, Margaret, and Fernand Perrin. "Charles Dupin's Study Visits to the British Isles, 1816–1824." *Technology and Culture* 32, no. 1 (1991): 47–68.

Brain, Robert Michael. "The Graphic Method: Inscription, Visualization, and Measurement in Nineteenth-Century Science and Culture." Ph.D. thesis, UCLA, 1996.

———. "Representation on the Line: Graphic Recording Instruments and Scientific Modernism." In *From Energy to Information: Representation in Art, Science, and Literature*, edited by B. Clarke and L. Henderson. Stanford, CA: Stanford University Press, 2002.

———. "The Romantic Experiment as Fragment." In *Hans Christian Ørsted and the Romantic Legacy in Science: Ideas, Disciplines, Practices*, edited by Robert M. Brain, Robert S. Cohen, and Ole Knudsen. Dordrecht, Netherlands: Springer, 2007.

Brain, Robert, Robert S. Cohen, and Ole Knudsen. *Hans Christian Ørsted and the Romantic Legacy in Science: Ideas, Disciplines, Practices*. Dordrecht, Netherlands: Springer, 2007.

Braunstein, Jean-François. "Bachelard, Canguilhem, Foucault: Le style français en épistémologie." In *Les philosophes et la science*, edited by P. Wagner. Paris: Gallimard, 2002.

———. "Comte, de la nature à l'humanité." In *Philosophies de la nature*, edited by O. Bloch. Paris: Publications de la Sorbonne, 2000.

Breckman, Warren. *The Left Hegelians: Marx, the Young Hegelians, and the Origins of Radical Social Theory: Dethroning the Self*. Cambridge: Cambridge University Press, 2001.

———. "Politics in a Symbolic Key: Pierre Leroux, Romantic Socialism, and the Schelling Affair." *Modern Intellectual History* 2 (2005): 61–86.

Brewster, David. *Letters on Natural Magic, Addressed to Sir Walter Scott.* 7th ed. 1832. Reprint, London: William Tegg, 1856.

Brittan, Francesca. "Berlioz and the Pathological Fantastic: Melancholy, Monomania, and Romantic Autobiography." *19th-Century Music* 29, no. 3 (2006): 211–39.

Brown, Theodore. "The Electric Current in Early Nineteenth-Century French Physics." *Historical Studies in the Physical Sciences* 1 (1969): 61–103.

Brush, Stephen G. *The Kind of Motion We Call Heat: A History of the Kinetic Theory of Gases in the Nineteenth Century.* Amsterdam: North-Holland, 1986.

Buche, Joseph, and Edouard Herriot. *L'école mystique de Lyon, 1776–1847: Le grand Ampère, Ballanche, Cl.-Julien Bredin, Victor de Laprade, Blanc Saint-Bonnet, Paul Chenavard.* Lyon: Rey, 1935.

Buchez, Philippe Joseph. *Introduction à la philosophie de l'histoire.* Paris, 1842.

Buchwald, Jed Z. *The Rise of the Wave Theory of Light: Optical Theory and Experiment in the Early Nineteenth Century.* Chicago: University of Chicago Press, 1989.

———. *Scientific Practice: Theories and Stories of Doing Physics.* Chicago: University of Chicago Press, 1995.

Buchwald, Jed Z., and Sungook Hong. "Physics." In *From Natural Philosophy to the Sciences: Writing the History of Nineteenth-Century Science*, edited by D. Cahan. Chicago: University of Chicago Press, 2003.

Buck-Morss, Susan. "Benjamin's Passagen-Werk: Redeeming Mass Culture for the Revolution." *New German Critique* 29 (Spring–Summer 1983): 211–40.

———. *The Dialectics of Seeing: Walter Benjamin and the Arcades Project.* Cambridge, MA: MIT Press, 1989.

Bunzl, Matti. "Franz Boas and the Humboldtian Tradition: From Volksgeist and Nationalcharakter to an Anthropological Concept of Culture." In *Volksgeist as Method and Ethic: Essays on Boasian Ethnography and the German Anthropological Tradition*, edited by George W. Stocking. Madison: University of Wisconsin Press, 1996.

Buret, Eugène. *De la misère des classes laborieuses en Angleterre et en France.* Paris, 1849.

Bürger, Peter. *Theory of the Avant-Garde.* Translated by Michael Shaw. Reprint, Minneapolis: University of Minnesota Press, 1984.

Burkhardt, Richard W. "Ethology, Natural History, the Life Sciences, and the Problem of Place." *Journal of the History of Biology* 32, no. 3 (1999): 489–508.

———. "La Ménagerie et la vie du Muséum." In *Le Muséum au premier siécle de son histoire*, edited by C. Blanckaert. Paris: Muséum d'Histoire Naturelle, 1997.

———. *The Spirit of System: Lamarck and Evolutionary Biology*. Cambridge, MA: Harvard University Press, 1995.

Burnouf, Emile. "La science des religions, sa méthode et ses limites." *Revue des deux mondes*, December 15, 1864.

Busst, A. J. L. "Ballanche and Saint-Simonism." *Australian Journal of French Studies* 9 (1972): 291–92.

Butler, Marilyn. *Romantics, Rebels, and Reactionaries: English Literature and Its Background, 1760–1830*. New York: Oxford University Press, 1982.

Butterfield, Herbert. *The Origins of Modern Science*. New York: Simon and Schuster, 1997.

Bynum, William F. "The Great Chain of Being after Forty Years: An Appraisal." *History of Science* 13 (1975): 1–28.

Cahan, David, ed. *From Natural Philosophy to the Sciences: Writing the History of Nineteenth Century Science*. Chicago: University of Chicago Press, 2003.

———. *Hermann von Helmholtz and the Foundations of Nineteenth-Century Science*. Berkeley: University of California Press, 1994.

Caneva, Kenneth. "Ampère, the Etherians, and the Oersted Connexion." *British Journal for the History of Science* 14, pt. 2, no. 44 (1980): 121–38.

———. "Physics and *Naturphilosophie*: A Reconnaissance." *History of Science* 35 (1997): 35–107.

———. "What Should We Do with the Monster? Electromagnetism and the Psychosociology of Knowledge." In *Sciences and Cultures: Anthropological and Historical Studies of the Sciences*, edited by Everett Mendelsohn and Yehuda Elkana. Dordrecht: D. Reidel, 1981.

Canguilhem, Georges. *La connaissance de la vie, deuxième édition, revue et augmentée*. Paris: Vrin, 1965.

———. *Etudes d'histoire et de philosophie des sciences*. Paris: Vrin, 1983.

———. "Histoire des religions chez Auguste Comte." In Canguilhem, *Etudes*.

———. "L'histoire des sciences dans l'oeuvre épistémologique de Gaston Bachelard." In Canguilhem, *Etudes*.

———. "Machine and Organism." In *Zone 6: Incorporations*, translated by Mark Cohen and Randall Cherry, edited by Jonathan Crary and Sanford Kwinter. Cambridge, MA: Zone Books, 1992.

———. *The Normal and the Pathological*. Cambridge, MA: Zone Books, 1991.

———. "Le vivant et son milieu." In *La connaissance de la vie*. Paris: Vrin, 1992.

Cannon, Susan Faye. *Science in Culture: The Early Victorian Period*. New York: Science History, 1978.

Cantor, G. N., and M. J. S. Hodge, eds. *Conceptions of Ether: Studies in the History of Ether Theories, 1740–1900*. Cambridge: Cambridge University Press, 1981.

Cantor, Geoffrey, and Sally Shuttleworth, eds. *Science Serialized: Representations of the Sciences in Nineteenth-Century Periodicals*. Cambridge, MA: MIT Press, 2004.

Cantor, Paul A. *Creature and Creator: Myth-Making and English Romanticism*. Cambridge: Cambridge University Press, 1984.

Carlisle, Robert B. *The Proffered Crown: Saint-Simonianism and the Doctrine of Hope*. Baltimore: Johns Hopkins University Press, 1987.

Carlyle, Thomas. "Signs of the Times." In *A Carlyle Reader: Selections from the Writings of Thomas Carlyle*, edited by G. B. Tennyson. Cambridge: Cambridge University Press, 1986.

Carnot, Lazare. "Discours contre l'obéissance passive." In Carnot, *Révolution et mathématique*.

———. *Essai sur les machines en général*. Paris, 1786.

———. *Mémoire sur la théorie des machines*. Paris, 1780.

———. *Révolution et mathématique*. Edited by J. P. Charnay. 2 vols. Paris: L'Herne, 1984–1985.

Carnot, Sadi. *Réflexions sur la puissance motrice du feu*. Edited by Robert Fox. Paris: Vrin, 1978.

Carrère, Emmanuel. *Le détroit de Behring*. Paris: P.O.L., 1986.

Carroy, Jacqueline. "Observation, expérimentation et clinique de soi: Haschich, folie, rêve et hystérie au XIXe siècle." In *L'envers de la raison: Alentour de Canguilhem*, edited by Pierre F. Daled. Vrin: Paris, 2008.

Cassirer, Ernst. *Kant's Life and Thought*. New Haven, CT: Yale University Press, 1981.

Castex, Pierre-Georges. *Le conte fantastique en France de Nodier à Maupassant*. 1951. Reprint, Paris: José Corti, 1987.

Castle, Terry. *The Female Thermometer: Eighteenth-Century Culture and the Invention of the Uncanny*. New York: Oxford University Press, 1995.

Cat, Jordi, Nancy Cartwright, and Hasok Chang. "Otto Neurath: Politics and the Unity of Science." In *The Disunity of Science: Boundaries, Contexts, and Power*, edited by Peter Galison and David J. Stump. Stanford, CA: Stanford University Press, 1996.

Cavell, Stanley. *In Quest of the Ordinary: Lines of Skepticism and Romanticism.* Chicago: University of Chicago Press, 1988.

Cawood, John. "François Arago, homme de science et homme politique."*La recherche* 16, no. 172 (1985): 1464–71.

———. "The Magnetic Crusade: Science and Politics in Early Victorian Britain." *Isis* 70, no. 4 (December 1979): 492–518.

———. "Terrestrial Magnetism and the Development of International Collaboration in the Early Nineteenth Century." *Annals of Science* 34, no. 6 (1977): 551–88.

Chadarevian, Soraya de. "Graphical Method and Discipline: Self-Recording Instruments in Nineteenth-Century Physiology." *Studies in the History and Philosophy of Science* 24, no. 2 (1993): 267–91.

Chandler, Arthur. "Expositions of the July Monarchy" Paris, 1834, 1839, 1844.

———. "The First Industrial Exposition: L'exposition publique des produits de l'industrie française, Paris, 1798." *World's Fair* 10, no. 1 (2000).

Chandler, James. *England in 1819: The Politics of Literary Culture and the Case of Romantic Historicism*. Chicago: University of Chicago Press, 1998.

———. *Wordsworth's Second Nature: A Study of the Poetry and Politics*. Chicago: University of Chicago Press, 1984.

Chandler, James, and Kevin Gilmartin, eds. *Romantic Metropolis: The Urban Scene of British Culture, 1780–1840*. Cambridge: Cambridge University Press, 2005.

Chappey, Jean-Luc. *La société des observateurs de l'homme (1799–1804): Des anthropologues au temps de Bonaparte*. Paris: Société des Etudes Robespierristes, 2002.

Charles, David. *La pensée technique dans l'oeuvre de Victor Hugo: Le bricolage de l'infini*. Paris: Presses Universitaires de France, 1997.

Charléty, Sébastien. *Essai sur l'histoire du Saint-Simonisme*. Paris: Hachette: 1896.

Charlton, David G. *The French Romantics*. 2 vols. Oxford: Cambridge University Press, 1984.

———. *Secular Religions in France, 1815–1870*. London: Oxford University Press, 1963.

Chevalier, Michel. *Politique industrielle: Système de la Méditerranée (Articles extraits du Globe)*. Paris, 1832.

Chevreul, Michel Eugène. *De la loi du contraste simultané*. Paris: Pitois-Levrault, 1839.

———. *Lettre à M. Ampère sur une classe particulière de mouvements musculaires.* Paris: P. Renouard, 1833.

Christian, Gérard-Joseph. *Vues sur le système générale des opérations industrielles, ou Plan de technonomie.* Paris: Huzard, 1819.

Clapeyron, Emile. "Mémoire sur la puissance motrice de la chaleur." *Journal de l'Ecole Polytechnique* 14, no. 23 (1834): 153–90.

Clark, Owen E. "The Contributions of J. F. Meckel, the Younger, to the Science of Teratology." *Journal of the History of Medicine and Allied Sciences* 24, no. 3 (1969): 310–22.

Clark, T. J. *The Absolute Bourgeois: Artists and Politics in France, 1848–1851.* Berkeley: University of California Press, 1999.

———. *Image of the People: Gustave Courbet and the 1848 Revolution.* London: Thames and Hudson, 1973.

———. *The Painting of Modern Life: Paris in the Art of Manet and His Followers.* New York: Knopf, 1985.

———. "Should Benjamin Have Read Marx?" *Boundary 2* 30, no. 1 (2003): 31–49.

Clarke, Bruce, and Linda Dalrymple Henderson. *From Energy to Information: Representation in Science and Technology, Art, and Literature.* Stanford, Calif.: Stanford University Press, 2002.

Clavier, Paul. *Kant: Les idées cosmologiques.* Paris: Presses Universitaires de France, 1997.

Clot, Louis. *Emile Pereire.* Paris, 1856.

Cohen, Ed. *A Body Worth Defending: Immunity, Biopolitics, and the Apotheosis of the Modern Body.* Durham, NC: Duke University Press, 2009.

Coleman, William. *Georges Cuvier, Zoologist: A Study in the History of Evolution Theory.* Cambridge, MA: Harvard University Press, 1964.

Collingwood, R. G. *The Idea of Nature.* Oxford: Clarendon Press, 1945.

Comment, Bernard. *The Painted Panorama.* New York: Henry N. Abrams, 2000.

Comte, Auguste. *Catéchisme positiviste ou sommaire exposition de la religion universelle, en onze entretiens systématiques entre une femme et un prêtre de l'humanité.* Paris: Thunot, 1852.

———. *Correspondance générale et confessions.* Paris: J. Vrin, 1973.

———. *Cours de philosophie positive.* Vols. 1 and 2. Edited by Michel Serres, François Dagognet, and M. A. Sinaceur. Paris: Hermann, 1998.

———. *Discours sur l'ensemble du positivisme.* Edited by Annie Petit. 1848. Reprint, Paris: Flammarion, 1998.

———. *Ecrits de jeunesse (1816–1828), suivis du mémoire sur la cosmogonie de Laplace, 1835.* Paris: Mouton, 1970.

———. *La philosophie des sciences*. Edited by Juliette Grange. Paris: Gallimard, 1996.

———. "Plan des travaux scientifiques nécessaires pour réorganizer la société." In *Auguste Comte: Philosophie des sciences*, edited by Juliette Grange. Paris: Gallimard, 1996.

———. *The positive philosophy of Auguste Comte*. Translated by Harriet Martineau. New York: Blanchard, 1855.

———. *Synthèse subjective, ou système universelle des conceptions propres à l'état normal de l'humanité*. Paris: Dalmont, 1856.

———. *Système de politique positive ou Traité de sociologie, instituant la religion de l'humanité*. 4 vols. Osnabrück: O. Zeller, 1967.

Condorcet, J. A. N. *Condorcet: Selected Writings*. Edited by K. M. Baker. Indianapolis: Bobbs-Merrill, 1976.

Constant, Benjamin. *De la religion, considérée dans sa source, ses formes et ses développements*. Vol. 2. Paris: A. Leroux et C. Chantpie, 1823.

Coriolis, G. *Du calcul de l'effet des machines*. Paris: Carilian-Goeury, 1829.

Corneille, Pierre. *Oeuvres complètes de P. Corneille*. Vol. 3. Paris: Lefèvre, 1838.

Corsi, Pietro. *The Age of Lamarck: Evolutionary Theories in France, 1790–1830*. Berkeley: University of California Press, 1988.

Coudroy, Marie-Hélène. *La critique parisienne des "grands opéras" de Meyerbeer*. Saarbrücken: Lucie Galland, 1988.

Coulomb, C. "Mémoire sur la force des hommes." In Coulomb, *Théorie des machines simples*.

———. *Théorie des machines simples, en ayant égard au frottement de leurs parties et a la roideur des cordages*. Paris: Bachelier, 1821.

Cousin, Victor. *Cours de l'histoire de la philosophie*. Paris: Didier Libraire-éditeur, 1841.

———. *Du vrai, du beau et du bien*. Paris: Didier, 1854.

Crary, Jonathan. *Techniques of the Observer: On Vision and Modernity in the Nineteenth Century*. Cambridge, MA: MIT Press, 1990.

Cronon, William. "The Trouble with Wilderness; or, Getting Back to the Wrong Nature." *Environmental History* 1 (1996): 7–28.

Crosland, Maurice Pierre. "Popular Science and the Arts: Challenges to Cultural Authority in France under the Second Empire." *British Journal for the History of Science* 34 (2001): 301–22.

———. *The Society of Arcueil: A View of French Science at the Time of Napoleon I*. Cambridge MA: Harvard University Press, 1967.

Crosland, Maurice, and Crosbie Smith, "The Transmission of Physics from France to Britain, 1800–1840." *Historical Studies in the Physical Sciences* 9 (1978): 1–61.

Crossley, Ceri. "Pierre Leroux and the *Circulus*: Soil, Socialism, and Salvation in Nineteenth-Century France." In *Histoires de la Terre: Earth Sciences and French Culture, 1740–1940*, edited by Louise Lyle and David McCallam. Amsterdam: Rodopi, 2008.

Cunningham, Andrew, and Nicholas Jardine, eds. *Romanticism and the Sciences*. Cambridge: Cambridge University Press, 1990.

Cuvier, Frédéric. "Examen de quelques observations de M. Dugald-Stewart, qui tendent a détruire l'analogie des phénomènes de l'Instinct avec ceux de l'habitude." *Memoires du Muséum d'histoire naturelle* 10 (1823): 241–60.

Cuvier, Georges. "Nature." *Dictionnaire des sciences naturelles*, xxxiv. Paris: Levrault, 1825.

———. *Le règne animal, distribué d'après son organisation, pour servir de base à l'histoire naturelle des animaux et d'introduction à l'anatomie comparée*. 5 vols. Paris: Déterville, 1829–30.

Cuvillers, Baron Henin de. *Archives du magnétisme animal*. 8 vols. Limoges, 1868.

Daguerre, Louis-Jacques-Mandé, François Arago, Joseph Louis Gay-Lussac, and Nicéphore Niépce. *Historique et description des procédés du Daguerréotype et du Diorama*. Paris: A. Giroux, 1839.

D'Allemagne, Henri. "Les Saint-Simoniens, 1827–1837." In M. Wallon, "Les Saint-Simoniens et les chemins de fer." Ph.D. diss., University of Paris, 1908.

Damiron, Philibert. *Essai sur l'histoire de la philosophie en France, au XVII siècle*. Vol. 2. 1846. Paris: Nabu, 2010.

Darnton, Robert. "Philosophers Trim the Tree of Knowledge: The Epistemological Strategy of the *Encyclopédie*." In *The Great Cat Massacre: And Other Episodes in French Cultural History*. New York: Basic Books, 2009.

Darrigol, Olivier. *Electrodynamics from Ampère to Einstein*. New York: Oxford University Press, 2000.

Darwin, Charles. *The Origin of Species*. Edited by Philip Appelman. New York: Norton, 2002.

Daston, Lorraine. "The Moral Economy of Science." *Osiris* 10 (1995): 3–24.

———. "Objectivity and the Escape from Perspective." *Social Studies of Science* 22, no. 4 (November 1992): 597–618.

———. "The Physicalist Tradition in Early Nineteenth Century French Geometry." *Studies in History and Philosophy of Science, Part A* 17, no. 3 (1986): 269–95.

Daston, Lorraine, and Peter Galison. "The Image of Objectivity." *Representations* 40 (1992): 81–128.

———. *Objectivity*. Cambridge: Zone Books, 2007.
Daston, Lorraine, and Katharine Park. *Wonders and the Order of Nature, 1150–1750*. New York: Zone Books; Cambridge, MA: MIT Press, 1998.
Daudin, Henri. *Cuvier et Lamarck: Les classes zoologiques et l'idée de série animale*. Paris: Editions des Archives Contemporaines, 1983.
Daumas, Maurice. *Arago: La jeunesse de la science*. 1943. Reprint, Paris: Belin, 1987.
Davis, Erik. *TechGnosis: Myth, Magic, and Mysticism in the Age of Information*. New York: Harmony Books, 1998.
Dear, Peter. *Revolutionizing the Sciences: European Knowledge and Its Ambitions, 1500–1700*. Princeton, NJ: Princeton University Press, 2001.
De la Motte, Dean, and Jeannene Przyblyski, eds. *Making the News: Modernity and the Mass Press in Nineteenth-Century France*. Amherst: University of Massachusetts Press, 1999.
DeLanda, Manuel. *Intensive Science and Virtual Philosophy*. New York: Continuum Books, 2002, 2004.
Deleuze, Gilles. *Bergsonism*. Cambridge, MA: Zone Books, 1988.
———. *Difference and Repetition*. 1968. Reprint, New York: Continuum Books, 2004.
———. *Kant's Critical Philosophy: The Doctrine of the Faculties*. Minneapolis: University of Minnesota Press, 1984.
Deleuze, Gilles, and Félix Guattari. *A Thousand Plateaus: Capitalism and Schizophrenia*. Minneapolis: University of Minnesota Press, 1987.
Deleuze, Gilles, and Leopold von Sacher-Masoch. *Masochism: Coldness and Cruelty and Venus in Furs*. Translated by Jean McNeil. Cambridge, MA: MIT Press, 1991.
Deleuze, Joseph-Philippe-François. *Histoire critique du magnétisme animal*. Paris: Mame, 1813.
———. *Instruction pratique sur le magnétisme animal*. Paris: J.-G. Dentu, 1825.
De Musset, Alfred. *La confession d'en enfant du siècle*. Paris: Bibliothèque Charpentier, 1985.
De Plinval de Guillebon, Régine. *Bibliographie analytique des expositions industrielles et commerciales en France depuis l'origine jusqu'à 1867*. Dijon: L'Echelle de David, 2006.
De Quincey, Thomas. *Last Days of Immanuel Kant and Other Writings*. Edinburgh: A. and C. Black, 1863.
Derrida, Jacques. *On Touching: Jean-Luc Nancy*. Stanford, CA: Stanford University Press, 2005.
———. *Specters of Marx: The State of the Debt, the Work of Mourning, and the New International*. London: Routledge, 1994.

Desmond, Adrian. *The Politics of Evolution: Morphology, Medicine, and Reform in Radical London*. Chicago: University of Chicago Press, 1992.

De Staël, Germaine. *De l'Allemagne*. Paris: Garnier-Flammarion, 1967. (Paris: H. Nicolle, 1910. Reprint, Paris: Dido, 1845.)

De Tours, Moreau. *Du hachisch et de l'aliénation mentale: Etudes psychologiques*. Paris: Fortin, Masson, 1845.

Dettelbach, Michael. "The Face of Nature: Precise Measurement, Mapping, and Sensibility in the Work of Alexander Von Humboldt." *Studies in History and Philosophy of Science Part C: Studies in History and Philosophy of Biological and Biomedical Sciences* 30, no. 4 (1999): 473–504.

———. "Humboldtian Science." In *Cultures of Natural History*, edited by N. Jardine, J. A. Secord, and E. Spary. Cambridge: Cambridge University Press, 1996.

———. "Humboldt's Instruments." In *"The Passage to Cosmos*: A Symposium," edited by Susan Castillo, Michael Dettelbach, Jorge Cañizares-Esguerra, and Laura Dassow Walls. *Studies in Travel Writing* 15, no. 1 (2011): 61–75.

———. Introduction to *Cosmos: A Sketch of a Physical Description of the Universe*, vol. 2, by Alexander von Humboldt. Baltimore: Johns Hopkins University Press, 1997.

———. "Romanticism and Administration: Mining, Galvanism, and Oversight in Alexander von Humboldt's Global Physics." Ph.D. diss., Cambridge University, 1992.

———. "Romanticism and Resistance: Humboldt and "German" Natural Philosophy in Natural Philosophy in Napoleonic France." In *Hans Christian Ørsted and the Romantic Legacy in Science: Ideas, Disciplines, Practices*, edited by Robert M. Brain, Robert S. Cohen and Ole Knudsen. Dordrecht, Netherlands: Springer, 2007.

———. "The Stimulations of "Travel: Humboldt's Physiological Construction of the Tropics." In *Tropical Visions in an Age of Empire*, edited by Felix Driver and Luciana Martins. Chicago: University of Chicago Press, 2005.

Dhombres, Jean. "L'enseignement des mathématiques par la méthode révolutionnaire: Les leçons de Laplace à l'Ecole Normale de l'an III." *Revue d'histoire des sciences et de leurs applications* 33 (1980): 315–48.

———. "L'image 'scientiste' de l'Ecole Polytechnique." In *La formation polytechnicienne (1794–1994)*, edited by B. Belhoste, A. Dahan, and A. Picon. Paris: Dunod, 1996.

Dhombres, Jean, and Nicole Dhombres. *La naissance d'un nouveau pouvoir: Sciences et savants en France, 1793–1824*. Paris: Payot, 1989.

Diderot, Denis, and Jean le Rond d'Alembert, eds. *Encyclopédie, ou diction-*

naire raisonné des sciences, des arts et des métiers. Lausanne: Chez la Société typographique, 1778–81.

Dijksterhuis, E. J. *The Mechanization of the World Picture.* 1961. Reprint, Oxford: Clarendon Press, 1989.

Doctrine Saint-Simonienne: Exposition. Paris: Librairie Nouvelle, 1854.

Dolan, Emily. *The Orchestral Revolution: Haydn and the Technologies of Timbre, 1750–1810.* Forthcoming.

Dolan, Emily, and John Tresch. "'A Sublime Invasion': Meyerbeer, Balzac, and the Paris Opera Machine." *Opera Quarterly*, March 29, 2011, doi: 10.1093/oq/kbr001.

Dörries, Matthias. "Visions of the Future of Science in Nineteenth-Century France: 1830–1871." Ph.D. diss., Freie Universität, Berlin, 1997.

Douglas, Mary. *Natural Symbols: Explorations in Cosmology.* London: Routledge, 1996.

Dumas, Alexandre. *Le Comte de Monte-Cristo.* London: Chapman and Hall, 1844–46.

Dumas, Jean-Baptiste. *Traité de chimie appliquée aux arts.* 3 vols. Paris: Bechet Jeune, 1828–46.

Dupin, Charles. *Développement de géométrie, avec les applications à la stabilité des vaisseaux, aux déblais et remblais, aux défilements, à l'optique pour faire suite à la géométrie déscriptive et à la géométrie analytique de G. Monge.* Paris, 1813.

———. *Forces productives et commerciales de la France.* Paris: Bachelier, 1827.

———. *Rapport du jury central sur les produits de l'industrie française exposés en 1834.* 3 vols. Paris: Imp. Royale, 1836.

———. *Tableau comparé de l'instruction populaire avec l'industrie des départemens, d'après l'exposition de 1827, présenté dans la seconde séance du cours de géométrie et de mécanique appliquées aux arts, professé pour les ouvriers, par le Bon Charles Dupin, le 23 décembre 1827.* Paris: Impr. de J. Tastu, 1828.

Dupotet, Jules. *Cours de magnétisme animal.* Paris, 1834.

During, Simon. *Modern Enchantments: The Cultural Power of Secular Magic.* Cambridge, MA: Harvard University Press, 2002.

Durkheim, Emile, and Marcel Mauss. *Primitive Classification.* Chicago: University of Chicago Press, 1963.

Duveyrier, Charles. "'La ville nouvelle, ou le Paris des Saint-Simoniens.'" In *Paris, ou le livre des cent-et-un*, edited by C. Ladvocat, vol. 9. Stuttgart: Ladvocat, 1832.

Eagleton, Terry. *The Ideology of the Aesthetic.* Oxford: Blackwell, 1990.

Eamon, William. "Technology as Magic in the Late Middle Ages and the Renaissance." *Janus* 70 (1983): 171–212.

Eco, Umberto. *Foucault's Pendulum*. Translated by William Weaver. New York: Houghton Mifflin Harcourt, 1989.

Eggli, Edmond. *Schiller et le romantisme français*. Paris: Gamber, 1927.

Eliot, Thomas Stearns. "Metaphysical Lyrics and Poems of the Seventeenth Century: Donne to Butler." *Times Literary Supplement*, October 20, 1921.

———. *Points of View*. London: Faber and Faber, 1941.

Enfantin, Barthélémy-Prosper. "Lettre du Père Enfantin à Charles Duveyrier, sur la vie éternelle." In *Lettres sur la vie éternelle*, edited by Antoine Picon. Paris: Le Corridor bleu, 2004.

Erdan, Alexandre. *La France mistique [sic]: Tableau des excentricités [sic] religieuses de ce temps*. Amsterdam, 1858.

Esquiros, Henri-François-Alphonse. *L'évangile du people*. Paris: Le Gallois, 1840.

———. *Le magicien*. Paris: L. Dessesart, 1838.

———. *Paris ou les sciences, les institutions, au XIXème siècle*. Paris, 1847.

Evans, David Owen. *Le socialisme romantique: Pierre Leroux et ses contemporains*. Paris: Librairie Marcel Rivire, 1948.

Exposition des produits de l'industrie. Dossier F 12 5005a, 1849 (dossier 1), Archives Nationales de France.

Fargeaud, Madeleine. *Balzac et 'La recherche de l'Absolu.'* Paris: Hachette, 1968.

Fedi, Laurent. "Auguste Comte et la technique." *Revue d'Histoire des Sciences* 53, no. 2 (2000): 265–93.

———. *Comte*. Paris: Les Belles Lettres, 2000.

———. "Le monde clos contre l'univers infini: Auguste Comte et les enjeux humains de l'astronomie." *La mazarine*, no. 13, June 2000, 12–15.

Ferguson, Priscilla P. *Paris as Revolution: Writing the Nineteenth-Century City*. Berkeley: University of California Press, 1994.

Finkelstein, Gabriel. "'Conquerors of the Künlün'? The Schlagintweit Mission to High Asia, 1854–57." *History of Science* 38 (2000): 179–218.

Fiorentini, Erna. *Observing Nature—Representing Experience: The Osmotic Dynamics of Romanticism, 1800–1850*. Berlin: Reimer, 2007.

Flachat, Stéphane. *L'industrie: Exposition de 1834*. Paris: L. Tenré, 1834.

Flachat, Stéphane, and Jean-Baptiste Marie Bury. *Traité élémentaire de mécanique industrielle: Résumé des traités de Christian, Poncelet, d'Aubuisson, Coriolis, Hachette, Lanz et Bétancourt, Ch. Dupin, Borgnis, Guenyveau, Leblanc, etc., etc*. Paris: L. Tenré et H. Dupuy, 1835.

Fortescue, William. *France and 1848: The End of Monarchy*. Routledge, 2005.

Foucault, Léon. "Démonstration expérimentale du mouvement de rotation de la terre." In *Recueil de travaux scientifiques*, vol. 1. Paris: Gauthier-Villars, 1878.

Foucault, Michel. *Discipline and Punish: The Birth of the Prison*. New York: Random House, 1975.

——. Introduction to *Le normal et le pathologique*, edited by G. Canguilhem. Paris: Presses Universitaires de France, 2005.

——. *The Order of Things: An Archaeology of the Human Sciences*. New York: Vintage, 1994.

——. *The Use of Pleasure*. Vol. 2 of *The History of Sexuality*. New York: Vintage, 1990.

——. "What Is Enlightenment?" In *The Foucault Reader*, edited by Paul Rabinow. New York: Pantheon Books, 1984.

Foucault, Michel, Frédéric Gros, François Ewald, and Alessandro Fontana. *The Hermeneutics of the Subject: Lectures at the Collège de France, 1981–1982*. New York: Picador, 2005.

Fourier, Charles. *The Theory of the Four Movements*. Edited by G. Jones and I. Patterson. Cambridge: Cambridge University Press, 1996.

Fourier, Jean Baptiste Joseph. *Théorie analytique de la chaleur*. Paris, 1822.

Fox, Robert. "The Background to the Discovery of Dulong and Petit's Law." *British Journal for the History of Science* 4, no. 1 (1968): 1–22.

——. "Charles Dupin." In *Les professeurs du CNAM*. Paris: CNAM, 1999.

——. *The Culture of Science in France, 1700–1900*. Aldershot, UK: Variorum, 1992.

——. "The Rise and Fall of Laplacian Physics." *Historical Studies in the Physical Sciences* 4 (1976): 89–136.

Fox, Robert, et al. *The Patronage of Science in the Nineteenth Century*. Leyden: Noordhoff International Publishing, 1976.

Fox, Robert, and George Weisz, eds. *The Organization of Science and Technology in France, 1808–1914*. Cambridge: Cambridge University Press, 1980.

Frängsmyr, Töre, John L. Heilbron, and Robin E. Rider, eds. *The Quantifying Spirit in the Eighteenth Century*. Berkeley: University of California Press, 1990.

Frankel, Eugene. "J. B. Biot and the Mathematization of Experimental Physics in Napoleonic France." *Historical Studies in the Physical Sciences* 8 (January 1, 1977): 33–72.

Fressoz, Jean-Baptiste. "Beck Back in the 19th Century: Towards a Genealogy of Risk Society." *History and Technology* 23, no. 4 (2007): 333–50.

Freud, Sigmund. "The Uncanny." In *The Uncanny*, by Freud, translated by David McLintock. New York: Penguin, 2003.

Friedman, Michael. "History and Philosophy of Science in a New Key." *Isis* 99, no. 1 (March 2008): 125–34.

———. "Kant—Naturphilosophie—Electromagnetism." In *Hans Christian Ørsted and the Romantic Legacy in Science*, edited by Robert Michael Brain, Robert S. Cohen, and Ole Knudsen. Dordrecht: Springer Netherlands, 2007.

———. "Remarks on the History of Science and the History of Philosophy." In *World Changes: Thomas Kuhn and the Nature of Science*, edited by Paul Horwich. Cambridge, MA: MIT Press, 1993.

Friedman, Michael, and Alfred Nordmann, eds. *The Kantian Legacy in Nineteenth-Century Science*. Cambridge, MA: MIT Press, 2006.

Friedman, Robert Marc. "The Creation of a New Science: Joseph Fourier's Analytical Theory of Heat." *Historical Studies in the Physical Sciences* 8 (1977): 73–99.

Fulcher, Jane. "Meyerbeer and the Music of Society." *Musical Quarterly* 67, no. 2 (1981): 213–29.

Fyfe, Aileen. *Science and Salvation: Evangelical Popular Science Publishing in Victorian Britain*. Chicago: University of Chicago Press, 2004.

Galassi, Peter. *Before Photography: Painting and the Invention of Photography*. New York: Museum of Modern Art, 1981.

Galison, Peter. "Aufbau/Bauhaus." *Critical Inquiry* 16, no. 4 (1990): 709–52.

———. *Einstein's Clocks and Poincaré's Maps: Empires of Time*. New York: Norton, 2003.

———. "Objectivity Is Romantic." *ACLS Occasional Paper* 47 (2000): 15–43.

Gane, Mike. *Auguste Comte*. Routledge, 2006.

Garber, Elizabeth. *The Language of Physics: The Calculus and the Development of Theoretical Physics in Europe, 1750–1914*. Boston: Birkhäuser, 1999.

Gaubert, Etienne Robert. *Rénovation de l'imprimerie. Notice sur le Gérotype, ou machine à distribuer et à composer en Typographie, Rapport à l'Académie des Sciences, le 5 décembre 1842*. Chez l'Inventeur: Paris, 1843.

Gauld, Alan. *A History of Hypnotism*. Cambridge: Cambridge University Press, 1992.

Gautier, Théophile. "Le Club des Hachichins." In *Romans, contes et nouvelles*, vol. 1. Paris: Gallimard, 2002.

———. *Contes et récits fantastiques*. Paris: LGF, 1991.

Geoffroy Saint-Hilaire, Etienne. "Discours Preliminaire." In *Philosophie anatomique*. Paris: Méquignon-Marvis, 1818.

———"Nature." In *Encyclopédie moderne; ou, Dictionnaire abregée des lettres et des arts*. Paris: Mongé Ainé, 1829.

———. *Notions synthétiques, historiques et physiologiques de philosophie naturelle.* Paris: Démain, 1838.

———. "Rapport sur l'ouvrage de M. Buchez, entitulé 'Philosophie de l'histoire.'" *Revue encyclopédique* 59 (1833): 210–21.

Geoffroy Saint-Hilaire, Isidore. *Essais de zoologie générale; ou, Memoires et notices sur la zoologie générale, l'anthropologie, et l'histoire de la science.* Paris: Librairie encyclopédique Roret, 1841.

Georget, Etienne-Jean. *De la physiologie du système nerveux, et spécialement du cerveau.* Paris, 1821.

Gere, Cathy. *Knossos and the Prophets of Modernism.* Chicago: University of Chicago Press, 2009.

Gernsheim, Helmut, and Alison Gernsheim. *L. J. M. Daguerre: The History of the Diorama and the Daguerreotype.* London: Dover, 1968.

Giddens, Anthony, ed. *Durkheim on Politics and the State.* Palo Alto, CA: Stanford University Press, 1986.

Gieryn, Thomas. *Cultural Boundaries of Science: Credibility on the Line.* Chicago: University of Chicago Press, 1999.

Gillispie, Charles C. "The *Encyclopédie* and the Jacobin Philosophy of Science: A Study in Ideas and Consequences." In *Critical Problems in History of Science*, edited by Marshall Clagget. Madison: University of Wisconsin Press, 1959.

———. *Lazare Carnot Savant: A Monograph Treating Carnot's Scientific Work, with Facsimile Reproduction of His Unpublished Writings on Mechanics and the Calculus, and an Essay concerning the Latter by A. P. Youschkevitch.* Princeton, NJ: Princeton University Press, 1971.

———. *The Montgolfier Brothers and the Invention of Aviation, 1783–1784.* Princeton, NJ: Princeton University Press, 1983.

———. *Pierre-Simon Laplace, 1749–1827: A Life in Exact Science.* Princeton, NJ: Princeton University Press, 1997.

Goblot, Jean-Jacques. *Aux origines du socialisme français: Pierre Leroux et ses premiers écrits.* Lyon: Presses Universitaires de Lyon, 1977.

———. "Extase, hystérie, possession: Les théories d'Alexandre Bertrand." *Romantisme* 9, no. 24 (1979): 53–59.

———. *La jeune France libérale: Le Globe et son groupe libérale, 1824–1830.* Paris: Plon, 1995.

Goethe, Johann Wolfgang von. "The Experiment as Mediator between Object and Subject." In *Scientific Studies*, edited and translated by Douglas Miller. New York: Suhrkamp, 1988.

———. "Les naturalistes français, ou Méditations de Goethe sur la marche et le caractère philosophique des sciences naturelles à Paris." In *Paris; ou, Le livre des cent-et-un*, edited by C. Ladvocat, vol. 2. Paris: Ladvocat, 1832.

———. *Scientific Studies*. Translated by D. Miller. New York: Suhrkamp, 1988.

Goffman, Erving. *Asylums: Essays on the Social Situation of Mental Patients and Other Inmates*. New York: Doubleday, 1961.

Goldberg Moses, Claire. *French Feminism in the 19th Century*. Albany: State University of New York Press, 1984.

Goldstein, Jan. *Console and Classify: The French Psychiatric Profession in the Nineteenth Century*. Cambridge: Cambridge University Press, 1987.

———. *Hysteria Complicated by Ecstasy: The Case of Nanette Leroux*. Princeton, NJ: Princeton University Press, 2009.

———. *The Post-Revolutionary Self: Politics and Psyche in France, 1750–1850*. Cambridge, MA: Harvard University Press, 2005.

Golinski, Jan. *Science as Public Culture: Chemistry and Enlightenment in Britain, 1760–1820*. Cambridge: Cambridge University Press, 1992.

Gooding, David. *Experiment and the Making of Meaning: Human Agency in Scientific Observation and Experiment*. Dordrecht: Kluwer, 1990.

Gooding, David, Trevor Pinch, and Simon Schaffer. *The Uses of Experiment: Studies in the Natural Sciences*. Cambridge: Cambridge University Press, 1993.

Gosling, Nigel. *Nadar*. London: Secker and Warburg, 1976.

Gouhier, Henri G. *La jeunesse d'Auguste Comte et la formation de positivisme*. 3 vols. Paris: Vrin, 1933.

Gould, Stephen Jay. *Ontogeny and Phylogeny*. Cambridge, MA: Belknap Press of Harvard University Press, 1967.

Goulet, Andrea. *Optiques: The Science of the Eye and the Birth of Modern French Fiction*. Philadelphia: University of Pennsylvania Press, 2006.

———. "'Tomber dans le phénomène': Balzac's Optics of Narration." *French Forum* 26, no. 3 (2001): 43–70.

Gower, Barry. *Scientific Method: An Historical and Philosophical Introduction*. London: Routledge, 1996.

———. "Speculation in Physics: The History and Practice of *Naturphilosophie*." *Studies in the History and Philosophy of Science* 3 (1973): 301–56.

Graber, Frédéric. "Obvious Decisions: Decision Making among French Ponts-et-Chaussées Engineers around 1800." *Social Studies of Science* 37, no. 6 (2007): 935–60.

Grandville, J. J. *Un autre monde*. Paris: H. Fournier, 1844.

———. *Scènes de la vie privée et publique des animaux*. Paris: Hetzel, 1840.

Grange, Juliette. *Auguste Comte: La politique et la science.* Paris: Presses Universitaires de France, 1996.

———. *Saint-Simon (1760–1825).* Paris, Ellipses, 2005.

Grant, Elliot Mansfield. *French Poetry and Modern Industry, 1830–1870.* Cambridge, MA: Harvard University Press, 1927.

Grattan-Guinness, Ivor. *Convolutions in French Mathematics 1800–1840.* Vol. 2. Basel: Birkhäuser, 1990.

———. "The 'Ingénieur Savant,' 1800–1830: A Neglected Figure in the History of French Mathematics and Science." *Science in Context* 6 (1993): 405–33.

———. "Work for the Workers: Advances in Engineering Mechanics and Instruction in France, 1800–1830." *Annals of Science* 41, no. 1 (1984): 1–33.

Greene, John C. *The Death of Adam: Evolution and Its Impact on Western Thought.* Ames: Iowa State University Press, 1959.

Griffiths, Alison. *Shivers down Your Spine.* New York: Columbia University Press, 2008.

Griffiths, David Albert. *Jean Reynaud, encyclopédiste de l'époque romantique, d'après sa correspondance inédite.* Paris: Rivière, 1965.

Grison, Emmanuel. "François Arago et l'Ecole Polytechnique." *SABIX* 4 (May 1989): 1–28.

Gross, Paul R., and N. Levitt. *Higher Superstition: The Academic Left and Its Quarrels with Science.* Baltimore: Johns Hopkins University Press, 1998.

Guchet, Xavier. *Pour un humanisme technologique: Culture, technique et société dans la philosophie de Gilbert Simondon.* Paris: Presses Universitaires de France, 2010.

Guépin, Ange. "La terre et ses organes." *La revue sociale* 8 (1850): 10–12.

Guillerme, André. *La naissance de l'industrie à Paris: Entre sueurs et vapeurs, 1780–1930.* Seyssel: Champ Vallon, 2007.

———. "Network: Birth of a Category in Engineering Thought during the French Restoration." *History and Technology* 8 (1992): 151–66.

Guillerme, Jacques, and Jan Sebestik. "Les commencements de la technologie." *Thalès* 12 (1966): 1–110.

Gunning, Tom. "The Cinema of Attractions: Early Film, Its Spectator, and the Avant-Garde." In *Early Film*, edited by Thomas Elsaesser and Adam Barker. London: BFI, 1989.

———. "The Long and Short of It: Centuries of Projecting Shadows from Natural Magic to the Avant-Garde." In *The Art of Projection*, edited by S. Douglas and C. Eamon. Ostfildern: Hatje Cantz, 2009.

———. "Phantom Images and Modern Manifestations: Spirit Photography, Magic Theater, Trick Films, and Photography's Uncanny." In *Fugitive Im-

ages: From Photography to Video, edited by P. Petro. Bloomington: Indiana University Press, 1995.

———. "Re-Newing Old Technologies: Astonishment, Second Nature, and the Uncanny in Technology from the Previous Turn-of-the-Century." In *Rethinking Media Change*, edited by D. Thorburn and H. Jenkins. Cambridge, MA: MIT Press, 2003.

Gusdorf, Georges. *Fondements du savoir romantique*. Vol. 9 of *Les sciences humaines et la pensée occidentale*. Paris: Payot, 1982.

———. *Le savoir romantique de la nature*. Paris: Payot, 1985.

Hachette, Jean-Nicolas-Pierre. *Traité élémentaire des machines*. Paris, 1811.

Hacking, Ian. "Nineteenth Century Cracks in the Concept of Determinism." *Journal of the History of Ideas* 44, no. 3 (July 1, 1983): 455–75.

———. *The Taming of Chance*. Cambridge: Cambridge University Press, 1990.

Hadot, Pierre, and Arnold Ira Davidson. *Philosophy as a Way of Life: Spiritual Exercises from Socrates to Foucault*. Oxford: Blackwell, 1995.

Hafter, Daryl M. "The Cost of Inventiveness: Labor's Struggle with Management's Machine." *Technology and Culture* 44, no. 1 (2003): 102–13.

———. *Women at Work in Preindustrial France*. University Park: Pennsylvania State University, 2007.

Hahn, Roger. *The Anatomy of a Scientific Institution: The Paris Academy of Sciences, 1666–1803*. Berkeley: University of California Press, 1971.

———. *Pierre Simon Laplace, 1749–1827: A Determined Scientist*. Cambridge, MA: Harvard University Press, 2005.

Haines, Barbara. "The Athénée de Paris and the Bourbon Restoration." *History and Technology* 5, no. 2 (1988): 249–71.

———. "The Inter-Relations between Social, Biological, and Medical Thought, 1750–1850: Saint-Simon and Comte." *British Journal for the History of Science* 11, no. 1 (1978): 19–35.

Hall, Marie Boas. "The Establishment of the Mechanical Philosophy." *Osiris* 10 (1952): 412–541.

Hankins, Thomas. *Jean D'Alembert: Science and the Enlightenment*. Oxford: Oxford University Press, 1970.

Hankins, Thomas L., and Robert J. Silverman. *Instruments and the Imagination*. Princeton, NJ: Princeton University Press, 1999.

Hansen, Miriam. "Benjamin, Cinema, and Experience: 'The Blue Flower in the Land of Technology.'" *New German Critique* no. 40 (1987): 179–224.

Harari, Josué. "The Pleasures of Science and the Pains of Philosophy: Balzac's Quest for the Absolute." *Yale French Studies* 67 (1984): 135–63.

Harman, P. M. *Energy, Force, and Matter: The Conceptual Development of Nineteenth-Century Physics*. Cambridge: Cambridge University Press, 1982.

Harvey, David. *Paris, Capital of Modernity*. New York: Routledge, 2003.

Hawkins, M. J. "Reason and Sense Perception in Comte's Theory of Mind." *History of European Ideas* 5, no. 2 (1984): 149–63.

Hayles, Katherine N. *Chaos Bound: Orderly Disorder in Contemporary Literature and Science*. Ithaca, NY: Cornell University Press, 1990.

Heidegger, Martin. *Basic Writings: From Being and Time (1927) to the Task of Thinking (1964)*. New York: Harper and Row, 1977.

———. *The Question concerning Technology*. New York: Harper Perennial, 1982.

———. "The Way to Language." In *Basic Writings*, translated and edited by David Krell. New York: HarperCollins, 1993.

Heilbron, Johan. "Auguste Comte and Modern Epistemology." *Sociological Theory* 8, no. 2 (1990): 153–62.

———. *The Rise of Social Theory*. Translated by Sheila Gogol. Minneapolis: University of Minnesota Press, 1995.

Heilbron, John L. *Electricity in the 17th and 18th Centuries: A Study of Early Modern Physics*. Berkeley: University of California Press, 1979.

———. "Some Connections among the Heroes." *Revue d'histoire des sciences* 54, no. 1 (2001): 11–28.

Héricart-Ferrand de Thury, Louis, and Pierre-Henri Migneron. *Rapport sur les produits de l'industrie française*. Paris: Imprimerie Royale, 1823.

Herivel, J. W. "Aspects of French Theoretical Physics in the Nineteenth Century." *British Journal for the History of Science* 3, no. 2 (December 1, 1966): 109–32.

Hess, Volker, and J. Andrew Mendelsohn. "Case and Series: Medical Knowledge and Paper Technology, 1600–1900." *History of Science* 48, no. 161 (September–December 2010): 287–314.

Hesse, Mary B. *Forces and Fields: The Concept of Action at a Distance in the History of Physics*. Westport, CT: Greenwood, 1970.

Heurtin, Jean-Philippe. *L'espace public parlementaire: Essai sur les raisons du législateur*. Paris: Presses Universitaires de France, 1999.

Hibberd, Sarah. "'Dormez donc, mes chers amours': Hérold's La Somnambule (1827) and Dream Phenomena on the Parisian Lyric Stage." *Cambridge Opera Journal* 16, no. 2 (2004): 107–32.

———. "*Le Prophète*: The End of History?" In *French Grand Opera and the Historical Imagination*. Cambridge: Cambridge University Press, 2009.

Hoffmann, E. T. A. *The Golden Pot and Other Tales*. Oxford: Oxford University Press, 2000.

Hofmann, James R. "Ampère, Electrodynamics, and Experimental Evidence." *Osiris* 3 (1987): 45–76.

———. "Ampère's Invention of Equilibrium Apparatus: A Response to Experimental Anomaly." *British Journal for the History of Science* 20, no. 3 (1987): 309–41.

———. *André-Marie Ampère*. Cambridge: Cambridge University Press, 1996.

Holmes, Frederic L., and Kathryn M. Olesko. "The Images of Precision: Helmholtz and the Graphical Method in Physiology." In *The Values of Precision*, edited by M. Norton Wise. Princeton, NJ: Princeton University Press, 1997.

Holmes, Richard. *The Age of Wonder*. London: Harper Press, 2008.

Holoman, D. Kern. *Berlioz*. Cambridge, MA: Harvard University Press, 1989.

Hoskin, Keith. "Education and the Genesis of Disciplinarity: The Unexpected Reversal." In *Knowledges: Historical and Critical Studies in Disciplinarity*, edited by E. Messer-Davidow, D. Shumway, and D. Sylvan. Charlottesville: University Press of Virginia, 1993.

Hughes, Thomas Parke. *Human-Built World: How to Think about Technology and Culture*. Chicago: University of Chicago Press, 2004.

Hugo, Victor. "Ce qu'on entend sur la montagne." In *Les feuilles d'automne*. Paris: Hauman, 1832.

———. *Les contemplations*. Paris: Michel Lévy, 1856.

———. *Napoléon le petit*. London: Jeffs, 1862.

———. *Le promontoire du songe*. Paris: Belles Lettres, 1993.

———. *Victor Hugo, visionnaire: Illustrations et poèmes*. Compiled by Pierre Seghers. Paris: R. Laffont, 1983.

Humboldt, Alexander von. *Cosmos: A Sketch of the Physical Description of the Universe*. Translated by E. C. Otté. 2 vols. Baltimore: Johns Hopkins University Press, 1997.

———. *Des lignes isothermes et la distribution de la chaleur sur le globe*. Paris: V. H. Perronneau, 1817.

———. *Experiences sur le galvanisme, et en général sur l'irritation des fibres musculaires et nerveuses*. Paris: Didot jeune, 1799.

———. *Kosmos: Entwurf einer physischen Weltbeschreibung*. Stuttgart: Cotta, 1845.

Humboldt, Alexander von, Jean-Claude Delamétherie, and Ernest Théodore Hamy. *Lettres américaines d'Alexandre de Humboldt*. Paris: E. Guilmoto, 1904.

Humboldt, Wilhelm von, Peter Heath, and Michael Losonsky. *On Language:*

On the Diversity of Human Language Construction and Its Influence on the Mental Development of the Human Species. Cambridge: Cambridge Univ. Press, 1999.

Huxley, Aldous. *Themes and Variations*. London: Chatto and Windus, 1950.

Huyssen, Andreas. *After the Great Divide: Modernism, Mass Culture, Postmodernism*. Bloomington: Indiana University Press, 1986.

Iggers, George G. *The Cult of Authority: The Political Philosophy of the Saint-Simonians*. New York: Springer, 1970.

Isambert, François André. *De la charbonnerie au Saint-Simonisme: Etude sur la jeunesse de Buchez*. Paris: Les Editions de Minuit, 1966.

Jackson, Myles. *Harmonious Triads: Physicists, Musicians, and Instrument Makers in Nineteenth-Century Germany*. Cambridge, MA: MIT Press, 2006.

———. *Spectrum of Belief: Joseph von Fraunhofer and the Craft of Precision Optics*. Cambridge, MA: MIT Press, 2000.

Jacyna, L. S. "Immanence or Transcendence: Theories of Life and Organization in Britain, 1790–1835." *Isis* 74 (1983): 311.

———. "Medical Science and Moral Science: The Cultural Relations of Physiology in Restoration France." *History of Science* 25 (1987): 111–46.

James, Tony. *Dream, Creativity, and Madness in Nineteenth-Century France*. Cambridge: Oxford University Press, 1995.

Janicaud, Dominique. *Une généalogie du spiritualisme français: Aux sources du bergsonisme, Ravaisson et la métaphysique*. La Haye: Nijhoff, 1969.

Jardin, André, and André Jean Tudesq. *La France des notables*. Paris: Editions du Seuil: 1973.

———. *Restoration and Reaction, 1815–1848*. Cambridge: Cambridge University Press; Paris: Editions de la Maison des Sciences de L'Homme, 1983.

Jarrige, François. "Le mauvais genre de la machine." *Revue d'histoire moderne et contemporaine* 54 (2007): 193–221.

———. "Les ouvriers parisiens et la question des machines au début de la Monarchie de Juillet." In *La France des années 1830 et l'esprit de réforme*, edited by P. Harismendy. Rennes: PUR, 2006.

Jenkins, Joseph. *The Humanure Handbook: A Guide to Composting Human Manure*. 2nd ed. Jenkins, 1999.

Johnson, Paul. *The Birth of the Modern: World Society, 1815–1830*. New York: HarperCollins, 1991.

Jones, Michael T. "Schiller, Goebbels, and Paul de Man: The Dangers of Comparative Studies." *Mosaic* 32 (1999): 53–72.

Jordanova, Ludmilla. "Nature's Powers: A Reading of Lamarck's Distinction between Creation and Production." In *History, Humanity, and Evolution*, edited by J. R. Moore. Cambridge: Cambridge University Press, 1989.

Jouffroy, Théodore. "Comment les dogmes finissent." In *Mélanges philosophiques*. Geneva: Slatkine, 1979.

———. "Cours d'esthétique." In Jouffroy, *Mélanges philosophiques*. Geneva: Slatkine, 1979.

Jouve, Bernard. *L'épopée saint-simonienne: Saint-Simon, Enfantin et leur disciple Alexis Petit de Suez au pays de George Sand*. Paris: Guénégaud, 2001.

Joyaux, Georges J. "Victor Cousin and American Transcendentalism." *French Review* 29, no. 2 (December 1, 1955): 117–30.

Kaenel, Philippe. "Le buffon de l'humanité: La zoologie politique de J.-J. Grandville (1803–1847)." *Revue de l'art* 74, no. 1 (1986): 21–28.

Kang, Minsoo. *Sublime Dreams of Living Machines: The Automaton in the European Imagination*. Cambridge, MA: Harvard University Press, 2011.

Kant, Immanuel. *Critique of Judgment*. Translated by Werner S. Pluhar. Indianapolis: Hackett, 1987.

———. *To Perpetual Peace: A Philosophical Sketch*. Translated by Ted Humphrey. Indianapolis: Hackett, 2003.

Kant, Immanuel, and James W. Ellington. *Grounding for the Metaphysics of Morals; with On a Supposed Right to Lie Because of Philanthropic Concerns*. Indianapolis: Hackett, 1993.

Kant, Immanuel, Paul Guyer, and Allen W. Wood. *Critique of Pure Reason*. Cambridge: Cambridge University Press, 1998.

Kantorowicz, Ernst. *The King's Two Bodies: A Study in Mediaeval Political Theology*. Princeton, NJ: Princeton University Press, 1957.

Kapp, Ernst, and Grégoire Chamayou. *Principes d'une philosophie de la technique*. Paris: Vrin, 2007.

Karsenti, Bruno. *Politique de l'esprit: Auguste Comte et la naissance de la science sociale*. Paris: Hermann, 2006.

Keck, Frédéric. *Lucien Lévy-Bruhl: entre philosophie et anthropologie, contradiction et participation*. Paris: Editions CNRS, 2008.

Kehlmann, Daniel. *Measuring the World*. New York: Vintage Books, 2007.

Kelly, David, ed. *Salon de 1848 de Charles Baudelaire*. Oxford: Clarendon Press, 1975.

Kerr, Greg. "'Nous avons enlacé le globe de nos réseaux . . .' Spatial Structure in Saint-Simonian Poetics." In *Histoires de la Terre: Earth Sciences and French Culture 1740–1940*, edited by Louise Lyle and David McCallam. Amsterdam: Rodopi, 2008.

Kessler, Joan C. *Demons of the Night: Tales of the Fantastic, Madness, and the Supernatural from Nineteenth-Century France*. Chicago: University of Chicago Press, 1995.

Kittler, Friedrich A. *Discourse Networks, 1800/1900*. Stanford, CA: Stanford University Press, 1990.

Kleist, Heinrich von. "On the Marionette Theater." In *German Romantic Criticism*. Translated by Christian-Albrecht Gollub, edited by A. L. Willson. New York: Continuum, 1982.

Kofman, Sarah. *Aberrations: Le devenir-femme d'Auguste Comte*. Paris: Aubier, 1978.

Kohler, Robert. *Lords of the Fly: Drosophila Genetics and the Experimental Life*. Chicago: University of Chicago Press, 1994.

Koyré, Alexandre. *From the Closed World to the Infinite Universe*. Baltimore: Johns Hopkins University Press, 1968.

Kranakis, Eda. *Constructing a Bridge: An Exploration of Engineering Culture, Design, and Research in Nineteenth-Century France and America*. Cambridge, MA: MIT Press, 1996.

Krauss, Rosalind. "Tracing Nadar." *October* 5 (1978): 29–47.

Krémer-Marietti, Angele. *Le positivisme*. Paris: Presses Universitaires de France, 1982.

Kuhn, Thomas S. "Mathematical versus Experimental Traditions in the Development of Physical Science." In *The Essential Tension: Selected Studies in Scientific Tradition and Change*. Chicago: University of Chicago Press, 1977.

———. "Simultaneous Discovery." In *Critical Problems in History of Science*, edited by Marshall Clagget. Madison: University of Wisconsin Press, 1959.

———. *The Structure of Scientific Revolutions*. Chicago: University of Chicago Press, 1962.

Kurys, Diane. *Les enfants du siècle*. Studio Canal, 1999. Feature Film.

Lacassagne, Jean-Pierre. "Victor Hugo, Pierre Leroux et le Circulus." *Bulletin de la Faculté des Lettres de Strasbourg* 48 (1970): 389–400.

Lachapelle, Sofie. "From the Stage to the Laboratory: Magicians, Psychologists, and the Science of Illusion." *Journal of the History of the Behavioral Sciences* 44, no. 4 (Fall 2008): 319–34.

Lacombe, Hervé. "The 'Machine' and the State." In *The Cambridge Companion to Grand Opera*, edited by D. Charlton. Cambridge: Cambridge University Press, 2003.

Lacoue-Labarthe, Philippe, and Jean-Luc Nancy. *The Literary Absolute: The Theory of Literature in German Romanticism*. Albany: State University of New York Press, 1988.

———. *Retreating the Political*. London: Routledge, 1997.

Lagueux, Olivier. "Geoffroy's Giraffe: The Hagiography of a Charismatic Mammal." *Journal of the History of Biology* 36 (2003): 225–47.

Lamarck, Jean Baptiste Pierre Antoine de Monet de. *Histoire naturelle des animaux sans Vertèbres*. Paris: J. B. Baillière, 1835.

———. *Philosophie zoologique*. Paris: Flammarion, 1997.

Lamartine, Alphonse. *Gutenberg, inventeur de l'imprimerie*. Paris: Hachette, 1853.

Lamé, Gabriel. *Vues politiques et pratiques sur les travaux publics de France*. Paris: Imprimerie d'Evérat, 1832.

Laplace, Pierre-Simon. *Exposition du système du monde*. 2 vols. Paris: Bachelier, 1836.

———. *Mécanique Céleste*. 5 vols. Paris: J. B. M Duprat, 1799–1825.

———. *A Philosophical Essay on Probabilities*. Translated by Frederick Wilson Truscott and Frederick Lincoln Emory. New York: Dover, 1951.

Laporte, Dominique. *Histoire de la merde*. Paris: Christian Bourgois, 1978.

Larthomas, Pierre. "Théories linguistiques de l'école romantique: Le cas de Victor Hugo." *Romantisme* 24, no. 86 (1994): 67–72.

Latour, Bruno. *Aramis; or, The Love of Technology*. Cambridge, MA: Harvard University Press, 1996.

———. *Cogitamus: Six letters sur le humanisme scientifique*. Paris: La Découverte, 2010.

———. *Politics of Nature: How to Bring the Sciences into Democracy*. Translated by Catherine Porter. Cambridge, MA: Harvard University Press, 2004.

———. *We Have Never Been Modern*. Translated by Catherine Porter. Cambridge, MA: Harvard University Press, 1993.

Laudan, Larry. "Towards a Reassessment of Comte's 'Méthode Positive.'" *Philosophy of Science* 38, no. 1 (1971): 35–53.

Lawrence, Christopher, and Steven Shapin, eds. *Science Incarnate: Historical Embodiments of Natural Knowledge*. Chicago: University of Chicago Press, 1998.

Lawrence, Philip. "Heaven and Earth: The Relation of the Nebular Hypothesis to Geology." In *Cosmology, History, and Theology*, edited by Wolfgang Yourgrau and Allen D. Breck. New York: Plenum Press, 1977.

Le Bras-Chopard, Armelle. *De l'égalité dans la différence: Le socialisme de Pierre Leroux*. Paris: Presses de Sciences Po, 1986.

Le Globe 1824–1832. Geneva: Slatkine Reprints, 1974–78.

Legouvé, Ernest. "De l'invention de l'imprimerie." *Revue de Paris* 5 (1829): 297–307.

Le Guayader, Hervé. "Le concept de plan d'organisation: Quelques aspects de son histoire." *Revue d'histoire des sciences* 53 no. 3–4 (2008): 339–80.

———. *Geoffroy Saint-Hilaire: A Visionary Naturalist*. Chicago: University of Chicago Press, 2004.

Lenoir, Timothy. "The Eye as Mathematician." In *Hermann Von Helmholtz and the Foundations of Nineteenth-Century Science*, edited by David Cahan. Berkeley: University of California Press, 1993.

———. "The Göttingen School and the Development of Transcendental Naturphilosophie in the Romantic Era." *Studies in History of Biology* 5 (1981): 111–205.

———. *The Strategy of Life: Teleology and Mechanics in Nineteenth-Century German Biology*. Dordrecht, Holland: D. Reidel, 1982.

Lepenies, Wolf. *Between Literature and Science: The Rise of Sociology*. Cambridge: Cambridge University Press, 1992.

Leroux, Pierre. "Alexandre Bertrand." In Leroux and Reynaud, *Encyclopédie nouvelle*, vol. 2.

———. "Aphorismes." In *Doctrine de l'humanité*, edited by P. Leroux. Paris: Boussac, 1848.

———. *Aux philosophes, aux artistes, aux politiques: Trois discours et autres texts*. Paris: Payot 1994.

———. "De Dieu ou de la vie considérée dans les êtres particuliers et dans l'être universel." In *La revue indépendante*, edited by P. Leroux, G. Sand, and L. Viardot. Paris, 1842.

———. "De la doctrine du progrès continu." *Revue encyclopédique* 1 (1833): i–lxxi.

———. "De la poésie de style." In *Oeuvres de Pierre Leroux (1825–1850)*, vol. 1. Paris, 1850.

———. "De la recherche des biens matériels (quatrième article). De l'humanité et le capital." *Revue sociale, ou Solution Pacifique du Problème du Prolétariat* 1 (March 1846): 81–90.

———. *De l'humanité, de son principe, et de son avenir, où se trouve exposée la vraie définition de la religion et où l'on explique le sens, la suite, et l'enchaînement du mosiasme et du christianisme*. 2 vols. Paris: Perrotin, 1845.

———. "De l'influence philosophique des etudes orientales." *Revue encyclopédique* (April 1832): 73–82.

———. *Discours de Schelling a Berlin: Du cours de philosophie de Schelling: Du Christianisme*. Paris: J. Vrin, 1982.

———. "D'une nouvelle typographie." *Revue indépendante* 25 (January 1843): 262–91.

———. *D'une religion nationale, ou du culte*. New ed. Paris: Boussac, 1848.

———. "Lettres sur le Fourierisme, IVème Lettre: Le plagiat de Fourier." *Revue sociale* 12 (1846): 187.

———. "Science sociale: Analyse des fonctions, pour servir à l'intelligence du principe d'organisation appelé Triade." *Revue sociale* 3, nos. 2–4 (1847): 48–55.

Leroux, Pierre, and J. Reynaud, eds. *Encyclopédie nouvelle; ou, Dictionnaire philosophique, scientifique, littéraire et industriel, offrant le tableau des connaissances humaines au dix-neuvième siècle*. Paris: Gosselin, 1838–41. Reprint, Geneva: Slatkine Reprints, 1991.

Leroux, Pierre, and George Sand. *Histoire d'une amitié (d'après une correspondance inédite 1836–1866): Pierre Leroux et George Sand*. Edited by J-P Lacassagne. Paris: Klincksieck, 1973.

Lesch, John E. *Science and Medicine in France: The Emergence of Experimental Physiology, 1790–1855*. Cambridge, MA: Harvard University Press, 1984.

Lessertisseur, J., and F. K. Jouffroy. "L'idée de série chez Blainville." *Revue d'histoire des sciences* 32, no. 1 (1979): 25–42.

Letellier, Robert, ed. and trans. *The Diaries of Giacomo Meyerbeer*. 4 vols. Madison, NJ: Fairleigh Dickinson University Press, 1999–2004.

———. *The Operas of Giacomo Meyerbeer*. Madison, NJ: Fairleigh Dickinson University Press, 2006.

Levere, Trevor Harvey. *Poetry Realized in Nature: Samuel Taylor Coleridge and Early Nineteenth-Century Science*. Cambridge: Cambridge University Press, 1981.

Levitt, Theresa. "Biot's Paper and Arago's Plates: Photographic Practice and the Transparency of Representation." *Isis* 94 (2003): 456–76.

———. "'I Thought This Might Be of Interest . . .': The Observatory as Public Enterprise." In *The Heavens on Earth: Nineteenth Century Observatory Sciences*, edited by D. Aubin, C. Bigg, and O. Sibum. Durham, NC: Duke University Press, 2010.

———. *The Shadow of Enlightenment: Optical and Political Transparency in France, 1789–1848*. Oxford: Oxford University Press, 2009.

Lévy-Bertherat, Ann-Déborah. *L'artifice romantique: De Byron à Baudelaire*. Paris: Klincksieck, 1994.

Limoges, Camille. "Milne-Edwards, Darwin, Durkheim and the Division of Labor: A Case Study in Reciprocal Conceptual Exchanges between the Social and Natural Sciences." In *The Relations between the Natural Sciences and the Social Sciences*, edited by I. B. Cohen. Princeton, NJ: Princeton University Press, 1994.

Littré, Emile. *Dictionnaire de la langue française*. Paris: L. Hachette, 1863.

Liu, Alan. *Local Transcendence: Essays on Postmodern Historicism and the Database*. Chicago: University of Chicago Press, 2008.

Locke, Ralph. *Music, Musicians, and Saint-Simonians*. Chicago: University of Chicago Press, 1986.
Lorcin, Patricia M. E. *Imperial Identities: Stereotyping, Prejudice and Race in Colonial Algeria*. London: I. B. Tauris, 1999.
Lovejoy, Arthur. *The Great Chain of Being: A Study of the History of an Idea*. Cambridge, MA: Harvard University Press, 1976.
———. "The Meaning of Romanticism for the Historian of Ideas." *Journal of the History of Ideas* 2, no. 3 (1941): 257–78.
Lyle, Louise, and David McCallam, eds. *Histoires de la Terre: Earth Sciences and French Culture, 1740–1940*. Amsterdam: Rodopi, 2008.
Macherey, Pierre. "Un chapitre de l'histoire du panthéisme: La religion Saint-Simonienne et la réhabilitation de la matière." In *Philosophie de la nature*, edited by Olivier Bloch. Paris: Sorbonne, 2000.
———. *Comte: La philosophie et les sciences*. Paris: Presses Universitaires de France, 1989.
———. *Hegel ou Spinoza*. Paris: F. Maspero, 1979.
———. "Leroux dans la querelle du panthéisme." *Les Cahiers de Fontenay* 36–38 (2003): 215–22.
———. "Un roman panthéiste: Spiridion." In *A quoi pense la littérature?* Paris: Presses Universitaires de France, 1990.
Maine de Biran, Pierre. *Correspondance philosophique avec Ampère*. Vol. 13–1 of *Oeuvres*. Edited by André Robinet. Paris: Vrin, 1993.
———. *L'influence de l'habitude sur la faculté de juger*. Paris: Henrichs, 1803.
———. *Mémoire sur la décomposition de la pensée*. Vol. 3 of *Oeuvres*. Edited by François Azouvi. Paris: J. Vrin, 1988.
Malabou, Catherine. *What Should We Do with Our Brain?* Translated by Sebastian Rand. New York: Fordham University Press, 2008.
Mannoni, Laurent. *Great Art of Light and Shadow: Archaeology of the Cinema*. Translated by Richard Crangle. Chicago: University of Chicago Press, 2000.
Manuel, Frank. "From Equality to Organicism." *Journal of the History of Ideas* 17, no. 1 (1956): 54–69.
———. *The New World of Henri Saint-Simon*. Cambridge, MA: Harvard University Press, 1956.
———. *The Prophets of Paris*. Cambridge, MA: Harvard University Press, 1962.
———. "Toward a Psychological History of Utopia." In *Utopias and Utopian Thought*. Boston: Beacon Press, 1967.
Marcovich, André. "La théorie philosophique des rapports d'André-Marie Ampère." *Revue d'histoire des sciences* 30, no. 2 (1977): 119–23.

Marcuse, Herbert. *The Aesthetic Dimension: Toward a Critique of Marxist Aesthetics*. Boston: Beacon Press, 1978.
Maret, Henri. *Essai sur le panthéisme dans les sociétés modernes*. Paris, 1845.
Markoff, John. *What the Dormouse Said: How the 60s Counterculture Shaped the Personal Computer*. New York: Viking Penguin, 2005.
Marrinan, Michael. *Romantic Paris: Histories of a Cultural Landscape, 1800–1850*. Stanford, CA: Stanford University Press, 2009.
Marx, Karl, and Friedrich Engels. *The Marx and Engels Reader*, edited by R. Tucker. New York: Norton, 1978.
Marx, Leo. *The Machine in the Garden: Technology and the Pastoral Ideal in America*. New York: Oxford University Press, 1964.
———. "Technology: The Emergence of a Hazardous Concept." *Technology and Culture* 51, no. 3 (2010): 561–77.
Mattelart, Armand. *L'invention de la communication*. Paris: Editions la Découverte, 1994.
Maurel and Jayet. *Arithmaurel, inventé par MM. Maurel et Jayet: Rapport à l'Académie et opinion des journaux sur l'arithmaurel*. Lille: Lefort, 1849. Bibliothèque Nationale de France, code FRBNF30911155.
Mauss, Marcel. *The Gift: The Form and Reason for Exchange in Archaic Societies*. New York: Norton, 1990.
Maxwell, James Clerk. *A Treatise on Electricity and Magnetism*. Vol. 2. New York: Dover, 1979.
Mayr, Otto. *Authority, Liberty, and Automatic Machinery in Early Modern Europe*. Johns Hopkins Studies in the History of Technology, vol. 8. Baltimore: Johns Hopkins University Press, 1986.
Maza, Sarah C. *The Myth of the French Bourgeoisie: An Essay on the Social Imaginary, 1750–1850*. Cambridge, MA: Harvard University Press, 2003.
McCalla, Arthur. "Palingénésie Philosophique to Palingénésie Sociale: From a Scientific Ideology to a Historical Ideology." *Journal of the History of Ideas* 55, no. 3 (1994): 421–39.
———. *A Romantic Historiosophy: The Philosophy of History of Pierre-Simon Ballanche*. Leiden, Netherlands: Brill, 1998.
McCauley, Anne. "François Arago and the Politics of the French Invention of Photography." In *Multiple Views: Logan Grant Essays on Photography, 1983–89*, edited by D. P. Younger. Albuquerque: University of New Mexico Press, 1991.
———. *Industrial Madness: Commercial Photography in Paris, 1848–1871*. New Haven, CT: Yale University Press, 1994.
———. "Talbot's Rouen Window: Romanticism, Naturphilosophie, and

the Invention of Photography." *History of Photography* 26, no. 2 (2002): 124–31.

McGann, Jerome J. *The Romantic Ideology: A Critical Investigation*. Chicago: University of Chicago Press, 1983.

McKnight, John, and Peter Block. *The Abundant Community: Awakening the Power of Families and Neighborhoods*. San Francisco: Berrett-Koehler, 2010.

Meek, Ronald L. *Social Science and the Ignoble Savage*. Cambridge: Cambridge University Press, 1976.

Meglin, Joellen A. "Behind the Veil of Translucence: An Intertextual Reading of the Ballet Fantastique in France, 1831–1841. Part One. Ancestors of the Sylphide in the Conte Fantastique." *Dance Chronicle* 27, no. 1 (2004): 67–129.

Méheust, Bernard. "Enquête sur la recontre entre l'illusioniste Robert-Houdin et la somnambule Alexis Didier: Y a-t-il des faits attestès?" In *Des savants face à l'occulte: 1870–1940, science et société*, edited by B. Bensaude Vincent and C. Blondel. Paris: Dècouverte, 2002.

———. *Somnambulisme et médiumnité: 1784–1930*. Le Plessis-Robinson: Institut Synthélabo, 1998.

———. *Un voyant prodigieux: Alexis Didier (1826–1886)*. Paris: Les Empêcheurs de penser en rond, 2003.

Meli, Domenico Bertoloni. *Thinking with Objects: The Transformation of Mechanics in the Seventeenth Century*. Baltimore: Johns Hopkins University Press, 2006.

Merleau-Ponty, Jacques. "L'essai sur la philosophie des sciences d'Ampère." *Revue d'histoire des sciences* 30, no. 2 (1977): 113–18.

———. "Laplace as Cosmologist." In *Cosmology, History, and Theology*, edited by W. Yourgrau and A. Breck. New York: Plenu, 1977.

———. *La science de l'univers a l'âge du postivisme: Etude sur les origines de la cosmologie contemporaine*. Paris: Vrin, 1983.

Mertens, Joost. "Du côté d'un chimiste nommé Thilorier." *L'année balzacienne* 1, no. 4 (2003): 251–63.

Méry, Joseph, and Gérard de Nerval. *L'imager de Harlem; ou, La découverte de l'imprimerie*. Paris: La Libraire Théatrale, 1851.

Metzner, Paul. *Crescendo of the Virtuoso: Spectacle, Skill, and Self-Promotion in Paris during the Age of Revolution*. Berkeley: University of California Press, 1998.

Michelet, Jules. *The People*. Urbana: University of Illinois Press, 1973.

Miller, David Philip. *Discovering Water: James Watt, Henry Cavendish, and the Nineteenth Century "Water Controversy."* Burlington, VT: Ashgate, 2004.

———. "The Revival of the Physical Sciences in Britain, 1815–1840." *Osiris* 2 (January 1, 1986): 107–34.

Mitchell, Timothy. *Questions of Modernity.* Vol. 11 of *Contradictions in Modernity.* Minneapolis: University of Minnesota Press, 2000.

Montgolfier, Joseph. "Mémoire sur la possibilité de substituer le belier hydraulique à l'ancienne machine de Marly." *Journal de l'Ecole Polytechnique* 7 (1808): 289–318.

Moreau, Jacques-Joseph. *Du hachisch et de l'aliénation mentale: Etudes psychologiques.* Paris: Fortin, Masson, 1845.

Morrell, Jack, and Arnold Thackray. *Gentlemen of Science: Early Years of the British Association for the Advancement of Science.* Oxford: Clarendon Press, 1981.

Moretti, Franco. *Atlas of the European Novel, 1800–1900.* London: Verso, 1998.

Morsy, Magali, ed. *Les Saint-Simoniens et l'Orient: Vers la modernité.* Aix-en-Provence: Edisud, 1990.

Morus, Iwan Rhys. "Seeing and Believing Science." *Isis* 97 (2006): 101–10.

———. *When Physics Became King.* Chicago: University of Chicago Press, 2005.

Mumford, Lewis. *The Myth of the Machine.* New York: Harcourt, Brace and World, 1967.

———. *Technics and Civilization.* New York: Harcourt, Brace and World, 1963.

Musselman, Elizabeth Green. *Nervous Conditions: Science and the Body Politic in Early Industrial Britain.* Albany: State University of New York Press, 2006.

Musso, Pierre. *La religion du monde industriel, analyse de la pensée de Saint-Simon.* La Tour D'Aigues: Editions de l'Aube, 2006.

———. *Saint-Simon et le Saint-Simonisme.* Series "Que sais-je?" Paris: Presses Universitaires de France, 1992.

———. *Télécommunications et philosophie des réseaux, la postérité paradoxale de Saint-Simon.* Paris: Presses Universitaires de France, 1997.

Navier, Claude-Louis-Marie-Henri. *Résumé des leçons de mécanique données á l'Ecole polytechnique.* Paris: Carilian-Goeury et Vor Dalmont, 1841.

Naville, Ernest de. *Maine de Biran: Sa vie et ses Pensées.* Paris: Didier, 1877.

Negri, Antonio, and Gabriele Fadini. "Materialism and Theology: A Conversation." *Rethinking Marxism* 20, no. 4 (2008): 665–72.

Nelson, Victoria. *The Secret Life of Puppets*. Cambridge, MA: Harvard University Press, 2001.
Nerval, Gérard de. *Les filles du feu suivi de Aurélia*. Paris, Livre de Poche: 1968.
——. *Oeuvres Complètes*. Edited by Jean Guillaume and Claude Pichois. Vol. 1. Paris: Gallimard, 1989.
Newark, Cormac. "Metaphors for Meyerbeer." *Journal of the Royal Musical Association* 127, no. 1 (2002): 23–43.
Noakes, Richard. "Spiritualism, Science, and the Supernatural in Mid-Victorian Britain." In *The Victorian Supernatural*, edited by Nicola Bown, Carolyn Burdett, and Pamela Thurschwell. Cambridge: Cambridge University Press, 2004.
Noble, David. *The Religion of Technology: The Divinity of Man and the Spirit of Invention*. New York: Knopf, 1997.
Nochlin, Linda. *Realism*. New York: Penguin, 1971.
"Notice Communale." http://cassini.ehess.fr/cassini/fr/html/fiche.php?select_resultat=26207, accessed September 1, 2010.
Novalis [Friedrich von Hartenberg]. *Henry of Ofterdingen*. Cambridge, MA: John Owen, 1842.
Nye, David E. *American Technological Sublime*. Cambridge, MA: MIT Press, 1994.
O'Connell, J. "Metrology: The Creation of Universality by the Circulation of Particulars." *Social Studies of Science* 23, no. 1 (1993): 129.
Oehler, Dolf,. *Le spleen contre l'oubli, juin 1848: Baudelaire, Flaubert, Heine, Herzen*. Paris: Payot, 1996.
Oersted, Hans-Christian. *Luftskibet*. Copenhagen, 1836.
Oettermann, Stephan. *The Panorama: History of a Mass Medium*. Cambridge: Zone Books, 1997.
Oken, L. *Elements of Physiophilosophy*. Translated by A. Turk. London: Ray Society, 1848.
Olesko, Kathryn M. "The Meaning of Precision: The Exact Sensibility in Early Nineteenth-Century Germany." In *The Values of Precision*, edited by M. Norton Wise. Princeton, NJ: Princeton University Press, 1997.
Oppenheimer, Jane M. "Some Historical Relationships between Teratology and Experimental Embryology." *Bulletin of the History of Medicine* 42 (1968): 124–28.
Orr, Linda. *Headless History: Nineteenth-Century French Historiography of the Revolution*. Ithaca, NY: Cornell University Press, 1990.
Ørsted. *See* Oersted.

———. *The Soul in Nature: With Supplementary Contributions*. Translated by Leonora Horner and Joanna B. Horner. London: H. G. Bohn, 1852.

Ory, Pascal. *Les expositions universelles de Paris: Panorama raisonnée, avec des aperçus nouveaux et des illustrations par les meilleurs auteurs*. Paris: Editions Ramsay, 1982.

Otis, Laura. *Networking: Communicating with Bodies and Machines in the Nineteenth Century*. Ann Arbor: University of Michigan Press, 2002.

Outram, Dorinda. *Georges Cuvier: Vocation, Science, and Authority in Post-Revolutionary France*. Manchester: Manchester University Press, 1984.

———. "Uncertain Legislator: Georges Cuvier's Laws of Nature in Their Intellectual Context." *Journal of the History of Biology* 19, no. 3 (October 1, 1986): 323–68.

Owen, Alex. *The Place of Enchantment: British Occultism and the Culture of the Modern*. Chicago: University of Chicago Press, 2004.

Pancaldi, Giuliano. "The Republic of Letters in Transition: William Thomson and Natural Philosophy ca. 1850." In *The Global and the Local: The History of Science and the Cultural Integration of Europe*, edited by Michal Kokowski. Krakow: Polish Academy of Arts and Sciences, 2007.

———. *Volta: Science and Culture in the Age of Enlightenment*. Princeton, NJ: Princeton University Press, 2003.

Pang, Alexander. "'Stars Should Henceforth Register Themselves': The Rhetoric and Reality of Early Astrophotography." *British Journal for the History of Science* 31 (1997): 177–201.

Parent-Lardeur, Françoise. *Lire à Paris au temps de Balzac: Les cabinets de lecture à Paris, 1815–1830*. Paris: Editions de l'Ecole des hautes études en sciences sociales, 1981.

Paris; or, The Book of the Hundred-and-One. 3 vols. London: Whittaker, Treacher, 1833.

Petit, Alexis. "Sur l'emploi du principe des forces vives dans le calcul de l'effet des machines." *Annales de chimie et physique* 2, no. 8 (1818): 287–305.

Petit, Annie, ed. *Auguste Comte, trajectoires positivistes, 1798–1998*. Paris: L'Harmattan, 2003.

———. "La diffusion des savoirs comme devoir positiviste." *Romantisme* 19, no. 65 (1989): 7–26.

———. "Heurs et malheurs du positivisme." Ph.D. diss., University of Paris I, 1995.

Petitier, Paule, ed. *Michelet et "la question sociale."* Littérature et Nation. Tours: L'Université François-Rabelais, 1997.

Picavet, François. *Les idéologues: Essai sur l'histoire des idées et des théories sci-*

entifiques, philosophiques, religieuses, etc. en France depuis 1789. New York: B. Franklin, 1971.

Pickering, Andrew. *The Mangle of Practice: Time, Agency, and Science*. Chicago: University of Chicago Press, 1995.

Pickering, Mary. *Auguste Comte: An Intellectual Biography*. Cambridge: Cambridge University Press, 1993.

Picon, Antoine. "Générosité sociale et aspirations technocratiques: Les polytechniciens Saint-Simoniens." *"Pour mémoire": Revue du Comité d'Histoire* 2 (April 2007): 106–14.

———. "La science saint-simonienne entre romantisme et technocratisme." In *L'utopie en questions*, edited by Michele Riot-Sarcey. Saint-Denis: Presses Universitaires de Vincennes, 2001.

———. "L'utopie-spectacle d'Enfantin: De la Retraite de Ménilmontant au procès et à l'Année de la Mère.'" In *Le siècle des Saint-Simoniens, du Nouveau Christianisme au Canal de Suez*, edited by N. Coilly and P. Régnier. Paris: Bibliothèque Nationale de France, 2006.

Pietz, William. "The Problem of the Fetish. I." *RES: Anthropology and Aesthetics* 9 (April 1, 1985): 5–17.

———. "The Problem of the Fetish. II: The Origin of the Fetish." *RES: Anthropology and Aesthetics* 13 (April 1, 1987): 23–45.

———. "The Problem of the Fetish. IIIa: Bosman's Guinea and the Enlightenment Theory of Fetishism." *RES: Anthropology and Aesthetics* 16 (October 1, 1988): 105–24.

Pilbeam, Pamela. *French Socialists before Marx: Workers, Women, and the Social Question in France*. Montreal: McGill-Queen's Press, 2000.

———. *Republicanism in Nineteenth Century France, 1814–1871*. Basingstoke: Palgrave, 1995.

Pinault, Madeleine. "Les mains de l'*Encyclopédie*." In "L'instrument." *Corps écrit* 35 (September 1990): 27–32.

Pinch, Trevor. *Confronting Nature: The Sociology of Solar Neutrino Detection*. Dordrecht: D. Reidel, 1986.

Pinkney, David H. *Decisive Years in France, 1840–1847*. Princeton, NJ: Princeton University Press, 1986.

———. *The French Revolution of 1830*. Princeton, NJ: Princeton University Press, 1972.

Pinson, Stephen. *Speculating Daguerre: Art and Enterprise in the Work of L. J. M. Daguerre*. Chicago: University of Chicago Press, 2011.

Pollan, Michael. *The Omnivore's Dilemma: A Natural History of Four Meals*. New York: Penguin, 2005.

Poncelet, Jean Victor. *Cours de mécanique industrielle*. Paris, 1829.

———. *Introduction à la mécanique appliqué aux machines*. Metz, France: Gauthier-Villars, 1829.

Porter, Theodore. "Objectivity and Authority: How French Engineers Reduced Public Utility to Numbers." *Poetics Today* 12 (1991): 245–65.

———. *Trust in Numbers: The Pursuit of Objectivity in Science and Public Life*. Princeton, NJ: Princeton University Press, 1996.

Pratt, Mary Louise. *Imperial Eyes: Travel Writing and Transculturation*. London: Routledge, 1992.

Prendergast, Christopher. *Paris and the Nineteenth Century*. Oxford: Blackwell, 1992.

Price, Roger. *The French Second Republic: A Social History*. Ithaca, NY: Cornell University Press, 1972.

———. *Revolution and Reaction: 1848 and the Second French Republic*. London: C. Helm; New York: Barnes and Noble Books, 1975.

Prigogine, Ilya, and Isabelle Stengers. *Order out of Chaos: Man's New Dialogue with Nature*. London: Flamingo, 1984.

Protevi, John. *Political Affect*. Minneapolis: University of Minnesota Press, 2009.

Proudhon, Pierre. *De la creation de l'ordre dans l'humanité*. 2 vols. 2nd ed. Paris: Garnier Freres, 1849.

Purrington, Robert D. *Physics in the Nineteenth Century*. New Brunswick, NJ: Rutgers University Press, 1997.

Rabinbach, Anson. *The Human Motor: Energy, Fatigue, and the Origins of Modernity*. Berkeley: University of California Press, 1992.

Rancière, Jacques. "The Community of Equals." In *On the Shores of Politics*, edited by J. Rancière. London: Verso, 1995.

———. *The Nights of Labor: The Workers' Dream in Nineteenth Century France*. Philadelphia: Temple University Press, 1989.

———. *The Politics of Aesthetics: The Distribution of the Sensible*. Translated by Gabriel Rockhill. London: Continuum, 2004.

Rapport du jury central en 1839: Exposition des produits de l'industrie française. 3 vols. Paris: Bouchard-Huzard, 1839.

Rapport du jury central sur les produits de l'agriculture et de l'industrie exposés en 1849. 3 vols. Paris: Imprimerie nationale, 1850.

Raspail, F. V. *Nouveaux coups de fouet scientifiques*. Paris: Meilhac, 1831.

Recht, Roland. *La lettre de Humboldt: Du jardin paysager au daguerréotype*. Paris: Christian Bourgois, 1989.

Redondi, Pietro. *L'accueil des idées de Sadi Carnot et la technologie française de 1820 à 1860: De la légende à l'histoire*. Paris: Vrin, 1980.

Régnier, Philippe, ed. *Le livre nouveau des Saint-Simoniens: Manuscrits d'Emile*

Barrault, Michel Chevalier, Charles Duveyrier, Prosper Enfantin, Charles Lambert, Léon Simon et Thomas-Ismayl Urbain (1832–1833). Tusson, Charente, France: Du Lérot, 1991.

———. "Le mythe oriental des Saint-Simoniens." In *Les Saint-Simonians et l'Orient*, edited by M. Morsy. Aix en Provence: Edisud, 1989.

———. *Les Saint-Simoniens en Egypte, 1833–1851*. Cairo: Banque de l'Union Européenne, 1989.

Rehbock, Philip F. *The Philosophical Naturalists: Themes in Early 19th-Century British Biology*. Madison: University of Wisconsin Press, 1983.

Reid, Donald. *Paris Sewers and Sewermen: Realities and Representations*. Cambridge, MA: Harvard University Press, 1991.

Reill, Peter Hans. "Science and the Construction of the Cultural Sciences in Late Enlightenment Germany: The Case of Wilhelm Von Humboldt." *History and Theory* 33, no. 3 (1994): 345–66.

———. "Vitalizing Nature and Naturalizing the Humanities in the Late Eighteenth Century." *Studies in Eighteenth-Century Culture* 28 (1999): 361–81.

———. *Vitalizing Nature in the Enlightenment*. Berkeley: University of California Press, 2005.

Reisch, George. *How the Cold War Changed Philosophy of Science: To the Icy Slopes of Logic*. Cambridge: Cambridge University Press, 2005.

Renouvier, Charles. Uchronie. In *L'Utopie dans l'histoire*, edited by Emmanuel Carrère. Paris: P.O.L., 1986.

Richards, Robert J. *Darwin and the Emergence of Evolutionary Theories of Mind and Behavior*. Chicago: University of Chicago Press, 1989.

———. *The Romantic Conception of Life: Science and Philosophy in the Age of Goethe*. Chicago: University of Chicago Press, 2002.

———. *The Tragic Sense of Life: Ernst Haeckel and the Struggle over Evolutionary Thought*. Chicago: University of Chicago Press, 2008.

Rifkin, Adrian, and Roger Thomas. *Voices of the People: The Social Life of "La Sociale" at the End of the Second Empire*. London: Routledge, 1988.

Riskin, Jessica. "The Defecating Duck; or, The Ambiguous Origins of Artificial Life." *Critical Inquiry* 29, no. 4 (2003): 599–633.

———. *Genesis Redux: Essays in the History and Philosophy of Artificial Life*. Chicago: University of Chicago Press, 2007.

———. *Modern Magic*. New York: Basic Books, in press.

———. *Science in the Age of Sensibility: The Sentimental Empiricists of the French Enlightenment*. Chicago: University of Chicago Press, 2002.

Robert-Houdin, Jean.-Eugène. *Memoirs of Robert-Houdin, Ambassador, Author, and Conjurer*. Vol. 1. London: Chapman and Hall, 1859.

Roberts, Lissa, Simon Schaffer and Peter Dear, eds. *The Mindful Hand: Inquiry and Invention from the Late Renaissance to Early Industrialisation.* Amsterdam: Koninklijke Nederlandse Akademie van Wetenschappen, 2007.

Robertson, Etienne-Gaspard. *Mémoires récréatifs, scientifiques et anecdotiques d'un physicien-aéronaute.* Vol. 1, *La Fantasmagorie.* Langres: Clima, 1985.

Roehr, Sabine. "Freedom and Autonomy in Schiller." *Journal of the History of Ideas* 64, no. 1 (2003): 119–34.

Rosanvallon, Pierre. *Le moment Guizot.* Paris: Gallimard, 1985.

Rosen, Charles, and Henri Zerner. *Romanticism and Realism: The Mythology of Nineteenth-Century Art.* New York: Viking Press, 1984.

Rosenblatt, Helena. "Re-evaluating Benjamin Constant's Liberalism: Industrialism, Saint-Simonianism, and the Restoration Years." *History of European Ideas* 30, no. 1 (March 2004): 23–37.

Rostan, Léon. "Magnétisme." *Dictionnaire de médecine et de chirurgie pratique.* Vol. 13. Paris, 1825.

Rostand, Jean. "E. Geoffroy Saint-Hilaire et la tératogénèse expérimentale." *Revue d'histoire des sciences* 17 (1964): 41–50.

Rothermel, Holly. "Images of the Sun: Warren De La Rue, George Biddell Airy, and Celestial Photography." *British Journal for the History of Science* 26 (1993): 137–69.

Rousseau, James. *Physiologie du Robert Macaire.* Paris: Laisné, 1842.

Rousseau, Jean-Jacques. *On the Social Contract with Geneva Manuscript and Political Economy.* Edited by R. D. Masters. Translated by J. R. Masters. New York: St. Martin's Press, 1978.

Rouvre, Charles de. *L'amoureuse histoire d'Auguste Comte et de Clotilde de Vaux.* Paris: Calmann Lévy, 1820.

Rubichon, Maurice. *Du mécanisme de la société en France et en Angleterre.* Paris, 1833.

Russell, E. S. *Form and Function: A Contribution to the History of Animal Morphology.* Chicago: University of Chicago Press, 1982.

Sachs, Aaron. *The Humboldt Current: A European Explorer and His American Disciples.* Oxford: Oxford University Press, 2007.

Sainte-Beuve, Charles-Augustin, and Gérald Antoine. *Portraits littéraires.* Paris: Laffont, 1993.

Saint-Simon, Claude-Henri. "Mémoire sur la science de l'homme (1813)." In Saint Simon, *Oeuvres*, vol. 6.

——. *Oeuvres de Claude-Henri de Saint Simon.* Paris: Editions Anthropos, 1966.

Sand, George. *Autour de la table.* Paris: Michel Levy, 1876.

———. *Le compagnon du tour de France*. Brussels: Compagnie de Editeurs Réunis, 1840.

———. *Lélia*. Edited by Béatrice Didier. Meylan: Editions de l'Aurore, 1987.

Sarda, François. *Les Arago: François et les autres*. Paris: Tallandier, 2002.

Sarton, George. "Auguste Comte, Historian of Science: With a Short Digression on Clotilde de Vaux and Harriet Taylor." *Osiris* 10 (1952): 328–57.

Savart, Félix. *Mémoire des instruments à chordes et à archet*. Paris: Librairie encyclopédique de Roret, 1819.

———. "On the Acoustic Figures Produced by the Vibrations Communicated through the Air to Elastic Membranes." *Edinburgh Journal of Science* 2 (1825): 296–301.

———. "Recherches sur les usages de la membran du tympan et de l'oreille externe." *Journal de physiologie expérimentale* 4 (1824): 183–219.

Say, Jean-Baptiste. *Traité d'économie politique; ou, Simple exposition de la manière dont se forment, se distribuent, et se composent les richesses*. Paris: Guillaumin, 1841.

Schaffer, Simon. "Astronomers Mark Time: Discipline and the Personal Equation." *Science in Context* 2, no. 1 (1988): 115–45.

———. "Babbage's Intelligence: Calculating Engines and the Factory System." *Critical Inquiry* 21, no. 1 (1994): 203–27.

———. "Enlightened Automata." In *The Sciences in Enlightened Europe*, edited by William Clark, Jan Golinski, and Simon Schaffer. Chicago: University of Chicago Press, 1999.

———. "Glass Works: Newton's Prisms and the Use of Experiment." In *The Uses of Experiment: Studies in the Natural Sciences*, edited by David Gooding, Trevor Pinch and Simon Schaffer. Cambridge: Cambridge University Press, 1993.

———. "Measuring Virtue: Eudiometry, Enlightenment, and Pneumatic Medicine." In *The Medical Enlightenment of the Eighteenth Century*, edited by Andrew Cunningham and Roger French. Cambridge: Cambridge University Press, 1990.

———. "Natural Philosophy." In *The Ferment of Knowledge*, edited by George Sebastian Rousseau and Roy Porter. Cambridge: Cambridge University Press, 2008.

———. "Natural Philosophy and Public Spectacle in the Eighteenth Century." *History of Science* 21 (1983): 1–43.

———. "The Nebular Hypothesis and the Science of Progress." In *History, Humanity, and Evolution: Essays for John C. Greene*, edited by James Moore. Cambridge: Cambridge University Press, 1989.

———. "On Astronomical Drawing." In *Picturing Science, Producing Art*, edited by Peter Galison and Caroline Jones. New York: Routledge, 1998.

———. "Scientific Discoveries and the End of Natural Philosophy." *Social Studies of Science* 16, no. 3 (1986): 387–420.

———. "Self Evidence." *Critical Inquiry* 18, no. 2 (1992): 327–62.

———. "Where Experiments End: Tabletop Trials in Victorian Astronomy." In *Scientific Practice: Theories and Stories of Doing Physics*, edited by J. Z. Buchwald. Chicago: University of Chicago, 1995.

Scharff, Robert. *Comte after Positivism*. Cambridge: Cambridge University Press, 2002.

Schatzberg, Eric. "*Technik* Comes to America: Changing Meanings of Technology before 1930." *Technology and Culture* 47, no. 3 (2006): 486–512.

Schelling, Friedrich W. J. *Ideas for a Philosophy of Nature*. Cambridge: Cambridge University Press, 1988.

Schiller, Friedrich. *On the Aesthetic Education of Man, in a Series of Letters*. New Haven, CT: Yale University Press, 1954.

Schivelbusch, Wolfgang. *The Railway Journey: The Industrialization and Perception of Time and Space*. Berkeley: University of California Press, 1987.

———. *Tastes of Paradise: A Social History of Spices, Stimulants, and Intoxicants*. New York: Vintage, 1993.

Schlanger, Judith E. *Les métaphores de l'organisme*. Paris: L'Harmattan, 1971.

Schlanger, Nathan. "'Suivre les gestes, éclat par éclat'—la chaîne opératoire d'André Leroi-Gourhan." In *Autour de l'homme: Contexte et actualité d'André Leroi-Gourhan*, edited by Françoise Audouze and Nathan Schlanger. Antibes: APDCA, 2004.

Schmidgen, Henning. "The Donders Machine: Matter, Signs, and Time in a Physiological Experiment, ca. 1865." *Configurations* 13, no. 2 (2007): 211–56.

Schmidt, James, ed. *What Is Enlightenment? Eighteenth-Century Answers and Twentieth-Century Questions*. Berkeley: University of California Press, 1996.

Schmitt, Stéphane. "From Physiology to Classification: Comparative Anatomy and Vicq d'Azyr's Plan of Reform for Life Sciences and Medicine (1774–1794)." *Science in Context* 22, no. 2 (2009): 145–93.

Schneewind, J. B. *The Invention of Autonomy: A History of Modern Moral Philosophy*. Cambridge: Cambridge University Press, 1998.

Schor, Naomi. *George Sand and Idealism*. New York: Columbia University Press, 1983.

Schwab, Raymond. *The Oriental Renaissance: Europe's Rediscovery of India and the East, 1680–1880*. Translated by Gene Patterson-Black and Victor Reinking. New York: Columbia University Press, 1987.

Schwartz, Vanessa. *Spectacular Realities: Early Mass Culture in Fin-de-Siècle Paris*. Berkeley: University of California Press, 1998.

Schweber, Silvan S. "Auguste Comte and the Nebular Hypothesis." In *Presence of the Past: Essays in Honor of Frank Manuel*, edited by R. T. Bienvenu and M. Feingold. Dordrecht: Springer, 1990.

———. "Darwin and the Political Economists: Divergence of Character." *Journal of the History of Biology* 13, no. 2 (October 1, 1980): 195–289.

"The Science of Religion." *Anthropological Review* 9 (1865): 89–120.

Scott, Joan W. "Against Eclecticism." *Differences* 16 (2005): 114–37.

———. *The Glassworkers of Carmaux: French Craftsmen and Political Action in a Nineteenth-Century City*. Cambridge, MA: Harvard University Press, 1974.

Secord, James. "Extraordinary Experiment: Electricity and the Creation of Life in Victorian England." In *The Uses of Experiment: Studies in the Natural Science*, edited by David Gooding, Trevor Pinch, and Simon Schaffer. Cambridge: Cambridge University Press, 1989.

———. *Victorian Sensation: The Extraordinary Publication, Reception, and Secret Authorship of "The Vestiges of the Natural History of Creation."* Chicago: University of Chicago Press, 2000.

Segala, Marco. "Electricité animale, magnétisme animal, galvanisme universel: A la recherche de l'identité entre l'homme et la nature." *Revue d'histoire des sciences* 54, no. 1 (2001): 71–84.

Seigel, Jerrold E. *The Idea of the Self: Thought and Experience in Western Europe since the Seventeenth Century*. New York: Cambridge University Press, 2005.

Seldow, Michel. *Vie et secrets de Robert-Houdin*. Paris: Fayard, 1971.

Sennett, Richard. *The Fall of Public Man*. New York: Knopf, 1975.

Serres, Etienne. "Organogénie." In Leroux and Reynaud, *Encyclopédie nouvelle*, vol. 8.

———. *Recherches d'anatomie transcendante sur les lois de l'organogénie appliquée à l'anatomie pathologique*. Paris : Thuau, 1827.

Serres, Michel. "Auguste Comte auto-traduit dans l'Encyclopédie." In *La Traduction: Hermes III*. Paris: Editions de Minuit, 1974.

———. "Paris 1800." In *A History of Scientific Thought: Elements of a History of Science*, edited by M. Serres. Cambridge, MA: Blackwell, 1995.

———. "Turner Translates Carnot." In *Hermes: Literature, Science, Philosophy*. Translated by Josué Harari. Baltimore: Johns Hopkins University Press, 1982.

Sessions, Jennifer E. *By Sword and Plow: France and the Conquest of Algeria*. Ithaca, NY: Cornell University Press, 2011.

Sewell, William. "La confraternité des prolétaires: Conscience de classe sous la Monarchie de juillet." *Annales* 36 (1981): 650–71.

———. "Visions of Labor." In *Work in France: Representations, Meaning, Organization, and Practice*, edited by S. Kaplan and C. Koepp. Ithaca, NY: Cornell University Press, 1986.

———. *Work and Revolution in France: The Language of Labor from the Old Regime to 1848*. Cambridge: Cambridge University Press, 1980.

Shapin, Steven. "The Invisible Technician." *American Scientist* 77 (1989): 554–63.

———. *The Scientific Life: A Moral History of a Late Modern Vocation*. Chicago: University of Chicago Press, 2008.

———. *The Scientific Revolution*. Chicago: University of Chicago Press, 1996.

Shapin, Steven, Simon Schaffer, and Thomas Hobbes. *Leviathan and the Air-Pump: Hobbes, Boyle, and the Experimental Life, Including a Translation of Thomas Hobbes, "Dialogus Physicus De Natura Aeris," by Simon Schaffer*. Princeton, NJ: Princeton University Press, 1985.

Sharp, Lynne. "Metempsychosis and Social Reform: The Individual and the Collective in Romantic Socialism." *French Historical Studies* 27, no. 2 (2004): 349–79.

Shine, Hill. *Carlyle and the Saint-Simonians: The Concept of Historical Periodicity*. New York: Octagon Books, 1971.

Shinn, Terry. *L'Ecole Polytechnique: 1794–1914*. Paris: Presses de la Fondation nationale des sciences politiques, 1980.

Sibum, H. Otto. "Exploring the Margins of Precision." In *Instruments, Travel, and Science: Itineraries of Precision from the Seventeenth to the Twentieth Century*, edited by Marie-Noëlle Bourguet, Christian Licoppe, and H. Otto Sibum. London: Routledge, 2002.

———. "Reworking the Mechanical Value of Heat: Instruments of Precision and Gestures of Accuracy in Early Victorian England." *Studies in History and Philosophy of Science* 26, no. 1 (1995): 73–106.

Siebers, Tobin. *The Romantic Fantastic*. Ithaca, NY: Cornell University Press, 1984.

Siegal, L. "Wagner and the Romanticism of E. T. A. Hoffmann." *Musical Quarterly* 51, no. 4 (1965): 597–613.

Siegel, Jerrold. *The Idea of the Self: Thought and Experience in Western Europe since the Seventeenth Century*. Cambridge: Cambridge University Press, 2005.

Siegert, Bernhard. *Relays: Literature as an Epoch of the Postal System*. Stanford, CA: Stanford University Press, 1999.

Simmons, Dana. "Waste Not, Want Not: Excrement and Economy in Nineteenth-Century France." *Representations* 96 (2006): 73–98.

Simon, Jules. "La philosophie de Pierre Leroux: Les articles du *Globe* et la réfutation de l'éclectisme." *Revue des deux mondes*, April 15, 1899.

———. *Victor Cousin*. Translated by Melville B. Playfair and Edward Playfair. Chicago: McClurg, 1888.

Simondon, Gilbert. *Du mode d'existence des objets techniques*. 1958. Reprint, Paris: Aubier, 1989.

Sloterdijk, Peter. *Sphères*. Paris: Pauvert, 2002.

———. "Spheres Theory: Talking to Myself about the Poetics of Space." *Harvard Design Magazine*, Spring–Summer 2009, 1–8.

Smeaton, William. A. "The Early History of Laboratory Instruction in Chemistry at the Ecole Polytechnique and Elsewhere." *Annals of Science* 10, no. 3 (1954): 224–33.

Smith, Pamela H. "Alchemy as a Language of Mediation at the Habsburg Court." *Isis* 85, no. 1 (March 1, 1994): 1–25.

———. *The Body of the Artisan: Art and Experience in the Scientific Revolution*. Chicago: University of Chicago Press, 2004.

Snelders, H. A. M. "Romanticism and Naturphilosophie and the Inorganic Natural Sciences, 1797–1840: An Introductory Survey." *Studies in Romanticism* 9, no. 3 (1970): 193–215.

Snow, C. P. *The Two Cultures and the Scientific Revolution*. New York: Cambridge University Press, 1959.

Snyder, Joel. "Visualization and Visibility." In *Picturing Science, Producing Art*, edited by P. Galison and C. Jones. New York: Routledge, 1998.

Soldhju, Katrin. *Selbstexperimente: Die Suche nach der Innenperspektive und ihre epistemologischen Folgen*. Munich: Fink, 2011.

Somerset, Richard. "The Naturalist in Balzac: The Relative Influence of Cuvier and Geoffroy Saint-Hilaire." *French Forum* 27, no. 1 (2002): 81–111.

Spary, Emma. *Utopia's Garden: French Natural History from Old Regime to Revolution*. Chicago: University of Chicago Press, 2000.

Spitzer, Alan. *The French Generation of 1820*. Princeton, NJ: Princeton University Press, 1987.

Spufford, Francis, and Jenny Uglow, eds. *Cultural Babbage*. London: Faber and Faber, 1997.

Stauffer, R. C. "Speculations and Experiment in the Background of Oersted's Discovery of Electromagnetism." *Isis* 48 (1957): 33–50.

Staum, Martin. "Cabanis and the Science of Man." *Journal of the Behavioral Sciences* 10, no. 1 (1974): 135–37.

——. "French Lecturers in Political Economy." *History of Political Economy* 30, no. 1 (1998): 95–120.

——. *Labeling People: French Scholars on Society, Race, and Empire, 1815–1848*. Montreal: McGill Queen's University Press, 2003.

——. "Physiognomy and Phrenology at the Paris Athénée." *Journal of the History of Ideas* 56, no. 3 (1995): 443–62.

Steinle, Friedrich. "Experiments in History and Philosophy of Science." *Perspectives on Science* 10, no. 4 (2002): 408–32.

——. *Explorative Experimente: Ampère, Faraday und die Ursprünge der Elektrodynamik*. Stuttgart: Steiner, 2005.

Strickland, Stuart. "The Ideology of Self-Knowledge and the Practice of Self-Experimentation." *Eighteenth-Century Studies* 31, no. 4 (1998): 453–71.

Szajkowski, Zosa. "The Jewish Saint-Simonians and Socialist Antisemites in France." *Jewish Social Studies* 9, no. 1 (1947): 33–60.

Taton, Réné. *Histoire générale des sciences*. Paris: Presses Universitaires de France, 1957.

Taylor, Charles. "The Importance of Herder." *Philosophical Arguments* (1995): 79–99.

——. *Sources of the Self: The Making of the Modern Identity*. Cambridge: Cambridge University Press, 1989.

Temkin, Owsei. *The Double Face of Janus and Other Essays in the History of Medicine*. Baltimore: Johns Hopkins University Press, 2006.

Terrall, Mary. *The Man Who Flattened the Earth: Maupertuis and the Sciences in the Enlightenment*. Chicago: University of Chicago Press, 2002.

Thompson, E. P. "The Moral Economy of the English Crowd in the Eighteenth Century." *Past and Present* 50 (1971): 76–136.

Thompson, Evan. *Mind in Life: Biology, Phenomenology, and the Sciences of Mind*. Cambridge, MA: Belknap Press, 2007.

Thompson, Patrice. "Essai d'analyse des conditions du spectacle dans le Panorama et le Diorama." *Romantisme* 12, no. 38 (1982): 47–64.

Thomson, William. "On an Absolute Thermometric Scale Founded on Carnot's Theory of the Motive Power of Heat, and Calculated from the Results of Regnault's Experiments on the Pressure and Latent Heat of Steam." *Philosophical Magazine* 33 (1848): 313–17.

Tobin, William. *The Life and Science of Léon Foucault: The Man Who Proved the Earth Rotates*. Cambridge: Cambridge University Press, 2003.

Todorov, Tzvetan. *The Fantastic: A Structural Approach to a Literary Genre*. Ithaca, NY: Cornell University Press, 1975.

Tolley, Bruce. "The 'Cénacle' of Balzac's Illusions Perdues." *French Studies* 15, no. 4 (1961): 324–37.
Tort, Patrick. "L'échelle encylopédique: Auguste Comte et la classification des sciences." In Patrick Tort, *La pensée hiérarchique et l'évolution*. Paris: Aubier, 1983.
———. *La querelle des analogies*. Paris: Editions d'Aujourd'hui, 1983.
Trahard, Pierre. *Le romantisme défini par "Le globe."* Paris: Les Presses françaises, 1925.
Tresch, John. "Cosmogram." In *Cosmograms*, edited by Melik Ohanian and Jean-Christophe Royoux. New York: Lukas and Sternberg, 2005.
———. "Technological World-Pictures: Cosmic Things and Cosmograms." *Isis* 98, no. 1 (March 2007): 84–99.
Tricker, R. A. R. "Ampère as a Contemporary Physicist." *Contemporary Physics* 3, no. 6 (1962): 453–69.
———. *Early Electrodynamics: The First Law of Circulation*. New York: Pergamon Press, 1965.
Truesdell, Clifford. *Essays in the History of Mechanics*. New York: Springer-Verlag, 1968.
Valson, Claude-Alphonse. *La vie et les travaux d'André-Marie Ampère*. Lyon: E. Vitte, 1897.
Valtat, Jean-Christophe. *La littérature halluciné: Entre pathologie et technologie (1800–1900)*. Habilitation, Université Clermont-Ferrand II, 2009.
Varela, Francisco. "A Science of Consciousness as If Experience Mattered." In *Towards the Science of Consciousness: The Second Tucson Discussions and Debates*, edited by Stuart R. Hameroff, Alfred W. Kaszniak, and Alwyn C. Scott. Cambridge, MA: MIT Press, 1997.
Vatin, François. "Des polypes." In *Trois essais sur les origines de la pensée sociologique*. Paris: La Découverte–MAUSS, 2005.
———. *Economie politique et économie naturelle chez Antoine-Augustin Cournot*. Paris: Presses Universitaires de France, 1998.
———. *Le travail: Economie et physique, 1780–1830*. Paris: Presses Universitaires de France, 1993.
Verin, Hélène. *La gloire des ingénieurs: L'intelligence technique du XVIe au XVIIIe siècle*. Paris: Albin Michel, 1993.
Vermeren, Patrice. "Les têtes rondes du *Globe* et la nouvelle philosophie de Paris." *Romantisme* 88 (1995): 23–34.
———. *Victor Cousin: Le jeu de la philosophie et de l'état*. Paris: L'Harmattan, 1995.
Viard, Jacques. "George Sand et Michelet, disciples de Pierre Leroux." *Revue d'histoire littéraire de la France* 75 (1975): 749–73.

———. "Leroux 'ouvrier typographe,' carbonaro et fondateur du *Globe*." *Romantisme* 28 (1980): 239–54.

———. "Pierre Leroux et les romantiques." *Romantisme* 12, no. 36 (1982): 27–50.

Viatte, Auguste. *Les sources occultes du romantisme*. 2 vols. Paris: H. Champion, 1928.

Vienot, Françoise. "Michel-Eugène Chevreul: From Laws and Principles to the Production of Colour Plates." *Color Research and Application* 27, no. 1 (2002): 4–14.

Voskuhl, Adelheid. "Motions and Passions: Music-Playing Women Automata and the Culture of Affect in Late Eighteenth-Century Germany." In *Genesis Redux: Essays in the History and Philosophy of Artificial Life*, edited by Jessica Riskin. Chicago: University of Chicago Press, 2007.

Warner, Marina. *Phantasmagoria: Spirit Visions, Metaphors, and Media into the Twenty-First Century*. Oxford: Oxford University Press, 2006.

Wasselin, Christian, and Pierre-René Serna, eds. *Hector Berlioz*. Paris: Editions de l'Herne, 2003.

Weber, Max. "Science as Vocation." In *The Vocation Lectures*. Indianapolis: Hackett, 2004.

Weiss, John. *The Making of Technological Man: The Social Origins of French Engineering Education*. Cambridge, MA: MIT Press, 1982.

Wellek, René. "Romanticism Re-examined." In *Concepts of Criticism*. New Haven, CT: Yale University Press, 1963.

Westfall, Richard. *The Construction of Modern Science: Mechanisms and Mechanics*. Cambridge: Cambridge University Press, 1977.

Wetzels, Walter D. "Art and Science: Organicism and Goethe's Classical Aesthetics." In *Approaches to Organic Form: Permutations in Science and Culture*, edited by Frederick Burwick. Dordrecht: D. Reidel, 1987.

Wheeler, Stephen, and Timothy Beatley, eds. *The Sustainable Urban Development Reader*. London: Routledge, 2008.

Whitehead, Alfred N. *Science and the Modern World*. New York: Free Press, 1967.

Whittaker, E. T. *A History of the Theories of Aether and Electricity*. New York: Harper, 1960.

Wiener, Norbert. *The Human Use of Human Beings: Cybernetics and Society*. Garden City, NY: Doubleday, 1954.

Wilberg, Rebecca. "The *mise en scène* at the Paris Opéra—Salle Le Peletier (1821–1873) and the Staging of the First French Grand Opéra: Meyerbeer's Robert Le Diable." Ph.D. diss. Brigham Young University, 1990.

Wilkinson, Lynn R. *The Dream of an Absolute Language: Emanuel Swedenborg and French Literary Culture*. Albany: State University of New York Press, 1996.

Williams, Elizabeth A. *The Physical and the Moral: Anthropology, Physiology, and Philosophical Medicine in France, 1750–1850*. Cambridge: Cambridge University Press, 1994.

Williams, L. Pearce. "Ampère, André-Marie." In *Dictionary of Scientific Biography*, edited by Charles Coulston Gillispie and American Council of Learned Societies. New York: Scribner, 1970.

———. "Faraday and Ampère: A Critical Dialogue." In *Faraday Rediscovered: Essays on the Life and Work of Michael Faraday, 1791–1867*, edited by David Gooding and Frank A. J. L. James. Melville, NY: American Institute of Physics, 1989.

———. "Kant, *Naturphilosophie*, and Scientific Method." In *Foundations of Scientific Method in the Nineteenth Century*, edited by Ronald N. Giere and Richard S. Westfall. Bloomington: Indiana University Press, 1973.

———. *Michael Faraday*. New York: Basic Books, 1966.

———. "What Were Ampère's Earliest Discoveries in Electrodynamics?" *Isis* 74 (1983): 492–508.

Williams, Rosalind H. *Notes on the Underground: An Essay on Technology, Society, and the Imagination*. Cambridge, MA: MIT Press, 1990.

———. "Opening the Big Box." *Technology and Culture* 48, no. 1 (2007): 104–16.

Wilson, Edmund. *Axel's Castle: A Study in the Imaginative Literature of 1879–1930*. New York: Scribner's, 1931.

Winter, Alison. *Mesmerized: Powers of Mind in Victorian Britain*. University of Chicago Press, 2000.

Wise, M. Norton. "Architectures for Steam." In *The Architecture of Science*, edited by Peter Galison and Emily Thompson. Cambridge, MA: MIT Press, 1999.

———. "The Flow Analogy to Electricity and Magnetism, Part I: William Thomson's Reformulation of Action at a Distance." *Archive for History of Exact Sciences* 25, no. 1 (1977): 19–70.

———. "Mediating Machines." In *World Changes: Thomas Kuhn and the Nature of Science*, edited by P. Horwich. Cambridge, MA: MIT Press, 2008.

———. "Mediations: Enlightenment Balancing Acts; or, The Technology of Rationalism." In *World Changes: Thomas Kuhn and the Nature of Science*, edited by P. Horwich. Cambridge, MA: MIT Press, 1993.

———, ed. *The Values of Precision*. Princeton, NJ: Princeton University Press, 1997.

Wise, M. Norton, and C. Smith. "Work and Waste: Political Economy and Natural Philosophy in Nineteenth Century Britain." *History of Science* 27 (1989): pt. 1, 263–301; pt. 2, 391–449; pt. 3, 28; (1990): 221–61.

Wise, M. Norton, and Elaine M. Wise. "Staging an Empire." In *Things that Talk*, edited by Lorraine Daston. Cambridge, MA: Zone Books, 2003.

Wittman, Richard. "Space, Networks, and the Saint-Simonians." *Grey Room* 40 (2010): 24–49.

Yeo, R. *Defining Science: William Whewell, Natural Knowledge, and Public Debate in Early Victorian Britain*. Cambridge: Cambridge University Press, 1993.

Zammito, John. *A Nice Derangement of Epistemes: Post-Positivism in the Study of Science from Quine to Latour*. Chicago: University of Chicago Press, 2004.

Ziolkowski, Theodore. *German Romanticism and Its Institutions*. Princeton, NJ: Princeton University Press, 1992.

Index

Page numbers in italics represent figures.

Absolute: Balzac's *The Quest for the Absolute*, xii–xv, 244, 317 n 2; Schelling and, 30–31
Academy of Political and Moral Sciences, 291, 306
Academy of Sciences (Académie des Sciences), 29, 36; Arago's position as permanent secretary, 34, 106, 109, 267; Arago's speeches to, 106–7, 341 n 62; Comte's presentation of his "Cosmogonie positive," 266–69; and the Cuvier-Geoffroy debate, 167
Adorno, Theodor, 294, 322 n 38
aesthetic, Schiller's conception of, 63, 65, 70, 71–76, 84, 335 n 40; art and the aesthetic state, 72–73; and freedom, 73–75, 79–80, 285; Humboldt's response to, 76–77, 79–80, 83–84; and ideal of autonomy (*Selbständigkeit* ["self-standingness"]), 71, 75, 334 n 33; and Kantian objectivity, 74–75; and the natural sciences, 75; and Nazi regime, 335 n 42; refashioning of Kant's philosophy, 63, 70, 71–75, 337 n 66, 337 n 68; and the Roman bust *Juno Ludovisi*, 72, 77; and the state, 73–75; and the third drive (the "play drive"), 63, 71
Agrippa, Heinrich Cornelius, 353 n 68
Aladdin's Magic Lamp (opera), 137
alchemy. *See* Renaissance alchemy
Alice in Wonderland (Carroll), 176
Althusser, Louis, 355 n 93
Ampère, André-Marie, 29–59; and animal magnetism, 36–37, 46–48; and Balzac's hero Claës, xiii, 30, 35, 325 n 6; Catholicism/Christian faith, 49–51, 327 n 24; "Classification of the Sciences," 39–40, 47, 54–57, 55, 56, 58–59, 331 n 108; electromagnetism experiments, 40–45, 43, 328 nn 53–54; *Essay on the Philosophy of Sciences*, 39, 54, 127–28, 331 n 105; and ether physics, 30, 44–45; exploratory experimentation, 38–40; and Geoffroy Saint-Hilaire, 47–48, 121–22; and Humboldt, 41, 48, 62; and Kantian philosophy, 38, 52, 53, 58, 331 n 94; and Laplacean physics, 30, 32–35, 44–45, 328 n 48; and Lyon circle, 36, 62, 326–27 n 24, 327 n 26; and Maine de Biran, 22, 31–32, 36, 51–53, 58, 129, 145–46; and Oersted's discovery of electromagnetism, 22, 29, 30–31, 33, 35, 38–39, 45; outsider status and nonconformism, 35, 326 n 21; and Paris social and intellectual milieu, 36, 61–62; personal life, 35–36, 326 n 24; and post-Kantian German *Naturphilosophie*, 22, 30–32, 35, 38–40, 45, 58, 325 n 5; and post-Laplacean physics, 20, 30, 31, 33–35, 44–51, 58, 120–22, 325–26 n 12; refusal to link electricity with the mind/thought, 31–32, 48–51; and romantic conceptions of the power of language, 133; scientific monomania and breadth of scientific interests, 35; and "la technesthétique," 57, 127–28, 146, 332 n 109, 345 n 7, 345 n 8; *Théorie mathématique des phénomènes électromagnetiques*, 42; theory of knowledge, 39–40, 53–57; "*troponomique*," 39–40, 56, 56. *See also* electromagnetism
Ampère, Jean-Jacques, 36, 126, 362 n 30, 364 n 61
Andersonian Institute (Glasgow), 98
Andrews, Naomi, 213, 285, 353–54 n 70, 359 n 61, 372 n 90, 372 n 92
animal magnetism (mesmerism), xiii–xiv; and Ampère's philosophy of electromagnetism, 31, 36–37, 46–48; Arago's interest in, 119–20; and Berlioz's symphonies, 143; Bertrand's theories, 46–47, 146; and Enfantin (Saint-Simonian), 212; and

animal magnetism (mesmerism) (*continued*)
    the imponderable fluids, xiii–xiv, 36–37;
    and Robert-Houdin's *Soirées fantastiques*,
    171–73, *173*, 175
animal series, 236–39; and Comte, 237,
    256–57, 261, 270–73, 370 n 60; and
    Cuvier's division of animal kingdom,
    160–61, 237, 362–63 n 45; foundational
    notion of the progressive series/serial
    organization, 237–39; and Geoffroy
    Saint-Hilaire, 162–63, 226, 236–39,
    242–43, 351 n 24; and Lamarck, 362–63
    n 45; and Leroux's virtual "humanity,"
    226, 239, 242–43; and Saint-Simon's physi-
    ological tableau, 195–97, 201, 205–6; and
    teratology experiments (embryological),
    163–64, 238–39; and unity of type/unity
    of composition, 162–63, 238–39, 351 n 24;
    and Vicq d'Azyr's comparative anatomy,
    195–96, 236–37. *See also* Cuvier-Geoffroy
    debate
Annales de chimie et de physique, 40
Annales des sciences naturelles, 38
Appel, Toby, 238
Arago, Etienne: and Balzac, xiii, 319 n 9;
    on Foucault's pendulum and imperial
    fetishism, 304, 375 n 29; and Geoffroy
    Saint-Hilaire's salon, 362 n 30
Arago, François, 89–122; and the Academy
    of Sciences, 34, 106, 109, 223–24, 267,
    269; advocacy on behalf of workers and
    inventors, 90, 109, 110–11, 360 n 5; and
    Ampère's electromagnetism experi-
    ments, 29, 41; and the "Arithmaurel"
    invention, 300; astronomical and meteo-
    rological instruments and experiments,
    107–9, *108*; and Balzac, xiii; in Cham-
    ber of Deputies, 90, 109, 111–12; and
    CNAM, 92, 95, 97–99, 101; and Comte,
    259, 266–67, 269; and the daguerreotype,
    90–92, 108, 112–20, 295, 343 n 89, 343
    n 93; Dantan's statue of, *119*; death, 306;
    and the Ecole Polytechnique, 33, 89, 93,
    95–97; and the "engineer-scientists,"
    92, 95–105; and Foucault's arc lamp,
    295; and Foucault's pendulum, 302; and
    Geoffroy Saint-Hilaire, 106, 121–22, 232;
    Histoire de ma jeunesse, 106, 341 n 57; and
    Humboldt, 64, 86, 92, 106–7; influence
    and achievements, 90, 269; interest in
    properties of light and sound, 143–44;
    and the labor theory of knowledge,
    102–5; and Laplace's nebular hypothesis,
    121, 266–67, 344–45 n 115; lectures on
    "Popular Astronomy," 106, 109, 154,
    269; and Leroux's pianotype invention,
    223–24; memorial obituaries, 109–10;
    and *Naturphilosophie*, 92; "On Electoral
    Reform," 111, 117; and Paris Observa-
    tory, 80, 106, 107, 109, 154, 269; poetic
    sensibility, 106–7; and post-Laplacean
    physics, 33–34, 89–92, 95, 105–9, 120–22,
    267, 325–26 n 12; and representation of
    science in the press, 109; republican
    politics and ideals, 90, 109–12; science
    and magic, 119–20; and science in the
    Second Empire, 17–18, 306; and the
    Second Republic, 90, 291, 292; and social
    impact of steam technology, 110–11; and
    the workers' revolution (1848), 90, 290,
    303–4
Archives du magnétisme animal, 46
arc lamp invention (self-regulating), 295–98,
    296
Arendt, Hannah, 309
"Arithmaurel" (calculating machine),
    300–301, *301*
artiste, L', 140
association: Leroux and, 225; the Saint-
    Simonians and universal association, 192,
    204, 205–8, 208, 225. *See also* freedom-in-
    connection
Association Polytechnique, 99, 103, 109
astronomy: Arago lectures on, 106, 109, 154,
    269; Arago's instruments and experi-
    ments, 107–9, *108*; and Comte's hierarchy
    of the sciences, 261–63, 265. *See also*
    "nebular hypothesis"
atheism, 158–59
Athénée, 98, 154, 203–4, 265–66
auras and auratic experience (Benjamin),
    14–16
automata: and genre of fantastic literature,
    134–35, 173; Hoffmann's "Automata,"
    134–35; and Robert-Houdin's *Soirées
    fantastiques*, 154, 171–76, *172*, *173*
autre monde, Un (Grandville), 27, *123*, 154, 155,
    176–85, *180*, *182*, *183*, *189*
Azouvi, François, 52

Babbage, Charles, 18, 81, 100–101, 171, 300;
    Economy of Machines and Manufactures,
    100
Bachelard, Gaston: on electricity and cosmic
    substances, 30, 324–25 n 4; and the labor
    theory of knowledge, 340 n 42; and

INDEX   433

Laplace's determinism, 337 n 1; theory of *phénoménotechniques*, 125, 343 n 92
Bacon, Francis, 101, 318 n 6
Badiou, Alain, 355 n 93
Bailly, Jean Sylvain, 46, 329 n 65
balances, 103
Balibar, Etienne, 355 n 93
Ballanche, Pierre-Simon, 36, 59, 232, 237; and Ampère, 36, 49, 59, 326–27 n 24; and palingenesis, 59, 198, 237; and the Saint-Simonians, 203; vision of history, 59
Balzac, Honoré de, xii–xvii; and the Cuvier-Geoffroy debate, 232–34, 362 n 33; and dialectic between realist and fantastic techniques, 131; and the Grandville-illustrated *Scenes from the Private and Public Life of Animals*, 155, 156, 232–34; *The Human Comedy* (and "Avant Propos"), xii, xiv, 233; *Lost Illusions* (*Les illusions perdues*), xv–xvi, 134; *Louis Lambert*, 234; and Meyerbeer's *Robert le diable*, 151–52, 294; the milieu (social and physical), xiv–xvi, 4; and Moreau de Tours's experiments in simulated madness (Club des Hachichins), 147; and new technologies for visual and auditory illusions, 131, 151–52, 349–50 n 86; Parisian novels, xii–xvii; *Le père Goriot*, xiv, 135–36; and the publishing industry/printing technology, xv–xvi, 133–34; *The Quest for the Absolute* (and the character Claës), xii–xiii, xv, 30, 35, 134, 244, 317 n 2, 325 n 6; story "Gambara," xiv–xv, 151–52, 349–50 n 86
"baroque physiology" (Georges Canguilhem), 356 n 9, 358 n 43
Barrault, Jean-Louis, 204, 205
Barrot, Odilon, 290–91
Baudelaire, Charles, 141, 147, 291; and Bonaparte's 1851 coup, 305; "Correspondances," 15; "Litanies of Satan" in *Les fleurs du mal*, 347 n 36; *Les paradis artificiels*, 147; on photography, 114; *Salon of 1846*, 205
Bazard, Claire, 204, 213
Bazard, Saint-Armand, 200, 213
Ben-David, Joseph, 17
Bénichou, Paul, *Le Sacré de l'écrivain*, 10–11, 320 n 26
Benjamin, Walter: *Arcades Project*, 15, 322 n 44; and Grandville's images, 15, 16, 355 n 92; and notion of the milieu, 322 n 44; and Paris intellectual scene, 7; on "the aura" of a work of art, 14–16; "The Work of Art in the Age of Mechanical Reproduction," 14–15
Bérard, Frédéric, 159
Bergson, Henri: conception of *homo faber*, 309, 348–49 n 76; and Maine de Biran, 331 n 96, 348–49 n 76
Berlioz, Hector, 141–43; *Chant des chemins de fer*, 205; and Foucault's inventions for stagecraft and lighting, 296; innovative instruments, 141; new genre of "program music," 141; and the optical illusions of the diorama/panorama, 143; *Requiem*, 142–43; and the Saint-Simonians' industrial religion, 143, 205; "Song of the Railroad," 143; *Symphonie fantastique*, 141; *Te Deum* staged at the 1855 International Exposition, 143; *Treatise of Instrumentation*, 141; view of the orchestral space, 141–43, 347 n 55
Berthollet, Claude-Louis, 33, 64
Bertrand, Alexandre: and animal magnetism, 46–47, 146, 228, 362 n 43; and the "feuilleton scientifique," 109; and Maine de Biran's physiospiritualism, 129, 146; science writings for the *Globe*, 46, 146, 153, 228; as supporter of Geoffroy Saint-Hilaire, 146, 228, 348 n 69; on trance states and possessions (in mesmerism), 146, 348 n 71
Bichat, Xavier: and Comte's view of the animal series, 270; definition of life, 270; and narrative of scientific progress, 318 n 6; medical advances of, 9; and Saint-Simon's physiological thought, 195; vitalism, 158, 159, 195
"biocracy," 246–47, 281, 282–84
biology, discipline of, 260, 261, 263, 269–70
Biot, Jean-Baptiste: and Ampère's electromagnetism experiments, 44, 328 n 48; battles with Arago, 115, 343 n 93; and Laplacean physics, 32, 33, 44, 328 n 48; *Précis élémentaire de physique expérimentale*, 33
Blainville, Henri de, 232; and the animal series, 237, 238, 261, 270, 351 n 24, 363 n 55; and Comte, 259, 270–71; definition of life, 270–71
Blanc, Louis: *History of Ten Years, 1830–1840*, 290; "The Organization of Labor," 290; and the Second Republic, 291–92, *292*; and the workers' revolution (1848), 289, 290–91
Blondel, Christine, 45, 327 n 26
Boehme, Jacob, 49, 145

Bonaparte, Louis-Napoleon (Napoleon III): conception of "the Napoleonic idea," 304, 374–75 n 28, 375 n 29; coup d'état of 1851 (launching the Second Empire), 8–9, 25, 305–6; and 1849 Exposition Nationale, 299–301; and Foucault's pendulum, 302–5; scientific interests, 302, 374 n 25; the Second Republic and romantic machines, 293–94, 299–301, 302–5, 374–75 n 28

Bonnet, Charles, 59, 160–61, 198, 236

Bonpland, Aimé, 64, 64

Bourdieu, Pierre, 338 n 6

Bourguet, Marie-Noëlle, 65

Breckman, Warren, 242

Bredin, Claude-Julien, 36, 38

Brewster, David, 100, 268, 300

British Association for the Advancement of Science, 81, 100

British science: and convergence of fields of science around concept of progress, 361 n 21; and the "decline thesis," 18; imitations of the Expositions Nationales, 170, 298; and industrial epistemology, 100–101; influence on Paris intellectual scene, 11; and the "nebular hypothesis," 266, 268

Broussais, François-Joseph-Victor, 9, 158, 259

Brown, Thomas, 145

Buche, Joseph, 326 n 24

Buchez, Philippe-Joseph: and the cenacle in Balzac's *Lost Illusions*, xvi; and Geoffroy Saint-Hilaire, 234, 362 n 36; on new scientific theories and progress of civilization, 356 n 13; *Philosophy of History*, 250; and the Saint-Simonians, 200, 213, 237

Buffon, Comte de (Georges-Louis Leclerc), 35

Bulletin des élèves de l'Ecole Polytechnique, 210

Bürger, Peter, 334 n 33

Burnouf, Eugène, 133, 364 n 61

Cabanis, Jean-Pierre, 36, 158, 195

Cabet, Etienne, 185, 289, 292

Caneva, Kenneth, 30, 45, 325 n 5, 326 n 21

Canguilhem, Georges, 4, 319 n 10, 330 n 86, 340 n 42, 356 n 9, 371 n 71

Carbonari, 9, 226, 245

Carlyle, Thomas, 7, 191–92, 204

Carnot, Hippolyte: and Comte, 259; as editor of *Revue encyclopédique*, 227; and Geoffroy Saint-Hilaire, 232; and the Saint-Simonians, 199–200, 203, 204, 208, 213, 229

Carnot, Lazare, 344 n 110; Arago's memorial obituary, 109, 110; and the Ecole Polytechnique, 93–95, 96, 101, 200; *Essay on Machines in General*, 96

Carnot, Sadi, 99, 137; and Clement-Desormes, 38, 99; engine and four-stage cycle of conversion, 208–12, 211, 358 n 51–52; *Reflexions on the Motive Power of Heat*, 99, 208–12

Cassirer, Ernst, 333 n 18

Castil-Blaze (François-Henri-Joseph Blaze), 373 n 10

Catholic Church: Ampère's faith, 49–51, 327 n 24; and the animal series, 237; and debates about fixity or mutability of species, 158–59; the mass and Berlioz's *Symphonie Fantastique*, 141; and mechanical romantics' interest in new spiritual power, 10–11, 197–98; and the Saint-Simonians, 196, 197–98, 214. *See also* Ballanche, Pierre-Simon; Buchez, Philippe-Joseph; Chateaubriand, François René de; de Bonald, Joseph

Cavaignac, Louis-Eugène, 292–93

Cawood, John, 82

Chamber of Deputies, 90, 109, 111–12

Chambers, Robert, 268

Chaptal, Jean-Antoine, 168–69

"charismatic authority" (Weber), 14, 321 n 36

Charles, Jacques-Aléxandre-César, 130

Charles X, reign of, 9

Chateaubriand, François René de, 125–26

Chevalier, Michel: and the *Globe*, 203; Mediterranean vision, 206–8, 217, 375 n 20; and the Saint-Simonians, 206–8, 212, 217; and the Saint-Simonian temple, 218–19; during the Second Empire, 306; *Système de la Méditerranée*, 206–8

Chevreul, Eugène, 37, 144

Chladni, Ernst, 144, 302

Chopin, Frédéric, 141, 234

Christian, Gérard-Joseph, 97–98, 331 n 108, 345 n 8

Christian Society of Lyon, 49

Clapeyron, Emile: applications of "optimization theory" to routes and engines, 206; article on Sadi Carnot's four-stage cycle of conversion, 210–12, 359 n 53; as Saint-Simonian, 206

classical mechanism: the "clockwork universe," 5, 11–12; Laplacean physics and the "standard view," 32–35; and "mechanical objectivity," 92, 319 n 13, 338 n 5; modern opposition between romanticism

and, 1–3, 318 n 4; narrative of scientific progress, 2, 318 n 6; and photography/daguerreotypes, 92; romantic poets on, 131–32; and shift of central cosmological model to the steam engine, 103–4
classification of species. *See* animal series; Cuvier-Geoffroy debate
Clément-Desormes, Nicolas, 38, 99
Club des Hachichins, 147, 174
CNAM. *See* Conservatoire National des Arts et Métiers (CNAM)
Coleridge, Samuel Taylor, 7, 38, 129–30
Collège de France, 50, 54, 266
color experiments/color theory, 144–45
Commission of the Académie des Sciences (1784), 37
Commission on Animal Magnetism (Academy of Medicine), 46, 120
comparative religion, 240, 364 n 61
Comte, Auguste, 125, 154, 185–86, 251, 253–86; and the animal series (animal hierarchy), 237, 256–57, 261, 270–73, 370 n 60; and Arago, 259, 266–67, 269; concept of life, 270–71, 273, 309, 370 n 62; *Course of Positive Philosophy*, 253–54, 255, 259–66, 268, 269, 273, 274, 275, 281, 307; *The Discourse on the Positive Spirit*, 273; and the Ecole Polytechnique, 19, 93, 257–59, 265–69, 274, 283; exclusion from established science, 266, 269; "first career" (focusing on philosophy of the sciences), 273–74, 367 n 7; "Fundamental Opuscule" of 1822 (on the "Plan of Intellectual Works Necessary for the Re-Organization of Society"), 258–59; and Maine de Biran's philosophy, 51, 330 n 86; and mechanical reductionism, 255, 261; mental breakdowns, 259, 273–74; and the nebular hypothesis, 121, 266–69; *Plan for Scientific Works*, 272; *Positive Catechism*, 274; religious shift and "second career," 256, 273–74, 367 n 7, 371 n 73; and the Saint-Simonians, 198, 203, 258–59, 274, 278, 371 n 73; and the social organism's interaction with its milieu, 255, 270–71, 272–73, 309; sociology (social physics), 260, 262, 263–64, 265, 368 n 29, 369 n 38; *The Subjective Synthesis*, 281, 309; and technaesthetics, 125, 295; theory of the human as technological animal, 255, 257, 273. *See also* Comte's positivism; Comte's Religion of Humanity

Comte's positivism, 51, 125, 368 n 30; and astronomy, 261–63, 265; and biology, 260, 261, 263, 269–70; and concept of "biocracy" and great chain of being, 246–47, 281, 282–84; definition of positivism, 368 n 30; foundational "laws" and intellectual organization of the sciences, 254–55, 259–65, 368 n 27; the hierarchy of the sciences, 260–65, *264*; influence and legacy of, 255–56, 283; introductory lectures and audiences, 259; knowledge/reason and the positive age, 260; the law of three states, 259–65, 275–76, 368 n 32; the Positivist Calendar, 253–54, *254*, 277, *278*; presentation of "Cosmogonie positive" to Academy of Sciences, 266–69; and the Religion of Humanity, 256, 273–84; as response to natural philosophy, 254, 261, 366 n 2; during the Second Empire and Third Republic, 307; slogan of "Order and Progress," 265, 274–75; and sociology (social physics), 260, 262, 263–64, 265, 368 n 29, 369 n 38; *Système de politique positive*, 255, 273, 274–75, 281, 368 n 30

Comte's Religion of Humanity, 256, 273–84; "altruism" supplanting "egotism," 255, 276–77; and Clotilde de Vaux, 256, 273–74, 277, *279*, 371 n 74; concept of "biocracy" and great chain of being, 246–47, 281, 282–84; and the dead (dependence of the living on the dead), 277; doctrine, 275–76, 281; and "fetishism," 13, 275–76, 281; "intendancies" and division of spiritual and temporal powers, 278–81, 285–86; and the Positivist Calendar, 253–54, *254*, 277, *278*; the Positivist Temple, 277–78, 279, *279*, *280*; practices and spiritual exercises, 256, 276–77; *psychotechnics* (prayers and mental training), 276; and racial hierarchies, 372 n 92; redefinition of religion as "linkage," 274; regime (sociocracy), 278–81; "subjective method," 275, 281; *The Subjective Synthesis*, 281, 309; and women/feminism, 274; worship (culte), 276–78, 281
Condillac, Étienne Bonnet de, 58
Condorcet, Marquis de, 109–10
Conservatoire National des Arts et Métiers (CNAM), 21, 97–99, 208; and Christian's "technonomie," 97–98, 331 n 108, 345 n 8; Dupin's lectures, 98–99, 103; Dupin's reforms, 98; and the "engineer-scientists," 92, 95, 97–99, 102, 169–70; and the Expositions Nationales, 168, 169–70; professorial chairs, 99
Considerant, Victor, 237, 288–89, 291, *292*

Constant, Abbé, 173
Constant, Benjamin, 356 n 14; *Histoire des religions*, 228
Coriolis, Gaspard-Gustave, 97
cosmograms, 6, 309; and Ampère's "Classification of the Sciences," 58–59; Grandville's *Un autre monde*, 177; Humboldt's works, 86; and Leroux's *Encyclopédie Nouvelle*, 227, 239; and the Saint-Simonian temple, 218–19
Cosmos (Humboldt), 62–63, 76–77, 80–81, 86, 122, 136
Coulomb, Charles Augustin de, 33, 44, 96
Cousin, Victor, 11, 20, 36, 63, 159, 202; eclecticism, 128, 159–60, 364 n 65; and Leroux, 20, 231, 361 n 18, 364 n 65; and Maine de Biran's physiospiritualism, 51, 129, 145; "spiritualism," 51, 128, 129, 145, 345–46 n 10
Crary, Jonathan, 78, 137, 330 n 89
Cuvier, Frédéric, 51, 155–56
Cuvier, Georges, 19–20, 51, 160–67, 237; and Ampère, 38; classification of animals based on functional systems, 160–61, 237; conception of life, 237, 362–63 n 45; dictionary entry on "Nature," 164–65; view of nature as fixed, stable order, 166. *See also* Cuvier-Geoffroy debate
Cuvier-Geoffroy debate, 154, 157, 160–67, 232, 352 n 47; and the Academy of Sciences, 167; Balzac's dramatization in Grandville's *Scenes from the Private and Public Life of Animals*, 232–34, 362 n 33; Cuvier's classifications based on functional systems, 160–61, 237; Cuvier's dictionary entry on "Nature," 164–65; form vs. function, 161–62; Geoffroy and the transformation of organisms, 162–66, 351 n 29, 369 n 41; Geoffroy's appeals to the press (publicity), 167; Geoffroy's concept of the "unity of composition," 162–63, 226, 234, 238–39, 351 n 24; Geoffroy's dictionary entry on "Nature," 165; Geoffroy's revival of Lamarckian transmutationist theories, 156–58, 161, 162; Geoffroy's supporters, 38, 106, 167, 228; Geoffroy's teratology experiments (embryological), *163*, 163–64; and opposing views of role of humans and technology in nature, 166
cybernetics, 300, 332 n 110; and Ampère's "La Cybernétique," 57, 332 n 110; Watt indicator diagram as cybernetic feedback machine, 210–11

Daguerre, Louis-Jacques-Mandé: background and work, 137; daguerreotype invention, 112, 117; dioramas, 112, 137–40, *138*, *139*, 344 n 110
daguerreotypes, 90–92, *91*, 112–20, *113*; aesthetic effects, 114; as an "artificial eye," 108; announcement of the invention, 90–92, 112–15; Arago and, 90–92, 108, 112–20, 295, 343 n 89, 343 n 93; association with classical machines/mechanism, 92; Foucault and Fizeau's first solar daguerreotype, 295–96; inscription of invisible phenomena unfolding over time, 115, 343 n 93; and the labor theory of knowledge, 115, 118; and the "moral economy of science," 117–18; opacity/mystery involved in, 119–20; practical uses in surveying/descriptive geometry, 113–14, 343 n 89; putative transparency, 115–17, 343 n 96; and question of the "natural" in fantastic mechanistic spectacles, 140–41; scientific uses and applications, 114–15; as technologies of time and transformation, 118–20
D'Alembert, Jean le Rond, 35, 96, 236
Damiron, Philibert, 146, 159
Dantan, Jean-Pierre (Dantan Jeune), 118, *119*
Darwin, Charles: and Comte's cosmogony, 268; and the romantic life sciences, 318 n 4; and transcendental anatomy of Geoffroy Saint-Hilaire, 20, 268, 352 n 47
Darwin, Erasmus, 112
Daston, Lorraine, 79, 85, 92, 117, 318 n 6, 319 n 13, 344 n 105
David, Félicien, 205, 215, *216*
David, Jean-Louis, 136, 137
David d'Angers, Pierre-Jean, 362 n 30
Davy, Humphry, 7, 38, 113
de Berry, Duc (Charles Ferdinand d'Artois), 197
de Bonald, Joseph, 10, 158, 195
Debord, Guy, 294
"decline thesis" and French science, 17–21
Degerando, Joseph-Marie, 36
D'Eichthal, Adolphe, 259
D'Eichtahl, Gustave, 191, 204, 258, 372 n 92
Delacroix, Eugène, 136–37, 147; *Barque of Dante*, 137; *The Death of Sardanapalus*, 137
Delambre, Jean Baptiste Joseph, 34
Delaroche, Paul, 114
Deleuze, Gilles, 81, 232, 236, 323 n 57, 348–49 n 76, 352 n 47, 362 n 42
Deleuze, J. P. F., 37

Della Porta, Giovanni Battista, 113, 118, 120
Démar, Claire, 289
de Musset, Alfred, 10, 234, 362 n 30
De Quincey, Thomas, 141
Deroin, Jeanne, 289
Derrida, Jacques, 372 n 87
Descamps, Philippe, 327 n 26
Descartes, René, 49, 54
Desmond, Adrian, 361 n 21
de Staël, Germaine, 11, 63, 128
Destutt de Tracy, Antoine Louis Claude, 36, 52, 158, 330 n 91
de Vaux, Clotilde, 256, 273–74, 277, 279, 371 n 74
Diderot, Denis, 31, 35, 101
Didier, Alexis, 354 n 77
dioramas, Daguerre's, 112, 137–40, *138*, *139*, 344 n 110; and Berlioz's fantastic symphonies, 143; *Midnight Mass*, 138, *139*; and question of the "natural" in mechanistic spectacles, 140–41; *View of Mont Blanc Taken from the Valley of Chamonix*, 138–40
Dolan, Emily, 349 n 82
Donders, Franciscus, 321 n 32
Donné, Alfred, 153–54, 295, 344 n 103
Dörries, Matthias, 18
Douglas, Mary, 29
drugs, hallucinogenic/psychotropic, 141, 147
Dubois, Paul, 226
Du Bois-Reymond, Emil, 85
Dulong, Pierre Louis, 34, 110, 120
Dumas, Alexandre, 126, 147
Dumas, Jean-Baptiste, 101, 120, 232
Dupin, Charles, 98; and the Expositions Nationales, 168, 169, 171, 298–300, 353 n 61; *Forces productives de la France*, 98, 339 n 30; and French statistics, 98; and Geoffroy, 232; and the labor theory of knowledge, 103; lectures and reforms at CNAM, 98–99, 103; during the Second Empire, 306; and study of work at the Ecole Polytechnique, 95
Dupotet, Jules, 46
Duveyrier, Charles, 192–94, 204, 218, 310
dynamometers, 97, 118

eclecticism, Cousin's, 128, 159–60, 364 n 65
Eco, Umberto, *Foucault's Pendulum*, 375 n 27
Ecole de Génie at Metz, 95, 97
Ecole de Médecine, 195
Ecole des Mines, 95
Ecole Normale Supérieure, 97, 355 n 93
Ecole Polytechnique, 18, 21, 93–97, *94*; and Ampère, 36; and Arago, 33, 89, 95–97; chemistry laboratory, 94–95, *96*; and Comte, 19, 93, 257–59, 265–69, 274, 283; the "engineer-scientists," 95–97, 198, 199–200, 204, 205–8; founding and history, 93–95, 338 n 6; and Laplace, 93, 258, 338 n 12; Monge's courses in descriptive geometry, 34, 94; pedagogical instruction and schedules, 257–58; practical knowledge ethos, 93–95; and republican agitation, 93, 102; and the Saint-Simonians, 198, 199–200, 204, 205–8, 220, 258, 356 n 13, 357 n 17; sciences of routes and work, 95–97, 206–8; and the series (serial organization), 237
ecstatic materialism, 236, 362 n 43
Edwards, Humphrey, 208
Eglise du Midi, 204
Egypt: and the Saint-Simonians, 196, 220; survey expedition under Napoleon, 107, 113, 156, 161
electromagnetism: Ampère's experiments, 40–45, *43*, 328 nn 53–54; Ampère's findings about attraction/repulsion of electric currents, 41–42, 328 n 48; Ampère's philosophy of, 45–51, 58; and animal magnetism (mesmerism), 31, 36–37, 46–48; the "*bonhomme d'Ampère*," 41, 328 n 46; cosmological associations, 29–30, 324–25 n 4; equilibrium experiments, 42–44, *43*; "etherian" theories, 30, 31, 44–45; Geoffroy Saint-Hilaire's theories connecting electricity and thought, 47–48; and Laplacean physics, 30, 32–35, 44–45, 328 n 48; and medical uses of electricity, 47–48; Oersted's 1820 discovery, 22, 29, 30–31, 33, 35, 38–39, 45; the term "circuit," 41; and terrestrial magnetism (magnetic crusade), 23, 41, 48, 336 n 60
Elie de Beaumont, Jean-Baptiste, 19, 232, 369 n 49
embryology: and the animal series, 238–39, 370 n 60; organogenesis and Geoffroy Saint-Hilaire's teratology experiments, *163*, 163–64, 238–39; and the Saint-Simonians' pantheistic religion, 203; theory of recapitulation (the Meckel-Serres Law), 238–39, 370 n 60. *See also* Serres, Etienne
Emerson, Ralph Waldo, 7
Encyclopédie nouvelle (Leroux and Reynaud), 227, 239, 357 n 30, 364 n 61
Enfantin, Barthélémy-Prosper, 13, *199*, 212–16; biography, 200–202; the New Bible (*Le livre nouveau*) and systematization of Saint-Simonian doctrine, 212,

Enfantin, Barthélémy-Prosper (*continued*) 215–17; proclamations of free love, 213, 214, 229; reflections on female oppression, 213; reflections on the metaphysics of sympathy, 212–13; and the Saint-Simonians' religion, 200–201, 212–16, 285; and Saint-Simonians' retreat to Ménilmontant, 214–15; during the Second Empire, 306. *See also* Saint-Simonians (religion of)

Engels, Friedrich, 12–13

"engineer-scientists": and Arago, 92, 95–105; and British industrial epistemology, 100–101; at CNAM, 92, 95, 97–99, 101, 169; and Comte, 267; at the Ecole Polytechnique, 95–97, 198, 199–200, 204, 205–8; and the Expositions Nationales, 154, 169–70; and the labor theory of knowledge, 100–105; and the Saint-Simonians, 198, 199–200, 204, 205–8, 220, 258, 356 n 13, 357 n 17; and sciences of routes and work (including "optimization"), 95–97, 206–8

England. *See* British science

Esquirol, Jean-Etienne, 147, 259

Esquiros, Adèle, 353–54 n 70

Esquiros, Alphonse, 353–54 n 70; and Geoffroy Saint-Hilaire's salon, 362 n 30; *Le magicien*, 174, 353–54 n 70

ether, 353 n 64; electromagnetism and the "etherian" theories of the anti-Laplaceans, 30, 31, 44–45; Robert-Houdin's *Soirées fantastiques*, 171–73, *173*, 175

"evolution," 198–99

Expositions Nationales des Produits de l'Industrie Française, 154, 167–71, *170*, 298–301; Berlioz's *Te Deum* and the 1855 Festival of Industry, 143; Bonaparte and the 1849 Exposition, 299–301; Champs Elysées site, 170; display of the "Arithmaurel," 300–301, *301*; displays of daguerreotypes, luxury goods, machines, 170, 171; displays of industrial machinery (steam engines), 169–70; and Dupin, 168, 169, 171, 298–300, 353 n 61; and the engineer-scientists, 154, 169–70; goal of linking animal power and industry, 299; goal of linking industrial transformation and social revolution, 168, 171, 353 n 61; imitators, 170, 298; as legacy of the 1789 French Revolution, 168–69; and Louis-Philippe, 170–71; Louvre courtyard site, 169; and Napoleon, 168–69; Place de la Concorde sites, 170, *170*; and the Saint-Simonians, 169–70; during the Second Republic, 298–301; and visions of unlimited growth, 299–301, 374 n 19; and the workers' revolution (1848), 299

Fabre d'Olivet, Antoine, 49, 173

Fantasmagoria (Robertson), 130–31

fantastic literature, genre of, 126, 134–35, 173, 345 n 5

fantastic spectacles (popular displays of animals and machines), 153–87; and debates over the formation of species, 157–60; deliberate ambivalence/ambiguities, 174–75, 354 n 77; the Expositions Nationales, 154, 167–71, *170*, 298–301; Grandville's *Scenes from the Private and Public Life of Animals*, 155, *156*; Grandville's *Un autre monde*, 154, 155, 176–85, *180*, *182*, *183*; the Muséum d'Histoire Naturelle and Ménagerie at the Jardin des Plantes, 154, 155–60; reflections on animals as models for human society, 155; Robert-Houdin's *Soirées fantastiques*, 154, 171–76, *172*, *173*, 354 n 73, 354 n 77; and science as mass entertainment, 153–55; and suggestions of divine or diabolical mastery over creation, 172

fantastic spectacles (visual and auditory illusions), 125–52, 171–76; as allegories of cosmic order, 125–27, 130–31, 153; altered states of consciousness, 146–47; automata, 134–35, 154, 171–76, *172*, *173*; and Bachelard's *phénoménotechniques*, 125, 343 n 92; and Balzac, 131, 151–52; Berlioz's fantastic symphonies, 141–43, *142*; color experiments/color theory, 144–45; Daguerre's dioramas, 112, 137–40, *138*, *139*, 143, 344 n 110; dialectic of doubt and certainty, 126–27; Egyptian occult science and the Tarot, 173; Foucault's arc lamp invention, 295–98, *296*; Foucault's pendulum, 302–5, *303*; and genre of fantastic literature, 126, 134–35, 173, 345 n 5; grand operas, 148–52, 349 n 82, 349–50 n 86; hallucinogenic/psychotropic drug use, 141, 147; and the imagination, 130–31; and mechanics of human perception and vision, 144–45; Meyerbeer's opera *Le prophète*, 296–98, *297*, 373 n 10; Meyerbeer's opera *Robert le diable*, 149–52, *150*, 294; and the "natural," 140–41; panorama craze, 135–37, 143; and physiospiritualism of Maine de Biran, 52–53, 129, 143–48, 348–49 n 76; and the positive sciences, 126–27; power of language and

INDEX 439

the printed word, 132–33, 134–35, 346 n 25; and Renaissance-era alchemy and hermeticism, 173–74; Robert-Houdin's *Soirées fantastiques*, 154, 171–76, *172*, *173*; Robertson's *Fantasmagoria*, 130–31; Satanic and Mephistophelean images, 135, 141, 347 n 36; stagecraft, 141–43, 148–52, 171–76, 294–98, 302, 354 n 77; studies of light and sound, 143–44; technaesthetics, 125–31, 295, 345 n 7

Faraday, Michael, 7, 38–39, 41

feminism: and Comte's Religion of Humanity, 274; and the Saint-Simonians, 213, 359 n 61; and socialism, 213, 353–54 n 70, 372 n 80

"fetishism," 13, 275–76, 281, 321 n 34

"feuilleton scientifique," 109, 153–54, 295

Fichte, Johann Gottlieb, 202

Fizeau, Hippolyte, 116, 144, 295–96, 343 n 99

Flachat, Eugène, 206

Flachat, Stéphane, 169, 206, 358–59 n 52

Flaubert, Gustave, 291

Foucault, Léon: and Arago, 109, 116, 144, 295; arc lamp invention, 295–98, *296*; and the "feuilleton scientifique," 109, 153–54, 295; first solar daguerreotype, 295–96; and Meyerbeer's opera *Le prophète*, 296–98; pendulum, 302–5, *303*; photometric observations, 116, 144, 295, 343 n 99

Foucault, Michel: and Cuvier's conception of life, 237, 362–63 n 45; the "empirico-transcendental doublet," 270, 285, 320 n 15, 349 n 77; on historicity of language, 346 n 25; and philosophies of Comte and Maine de Biran, 330 n 86; and philosophy of Destutt de Tracy, 330 n 91

Fourier, Charles, 259; cosmological predictions, 15; and Grandville's *Un autre monde*, 177, 181; ideal community (the Phalanstery), 237; and nebular hypothesis, 266; and serial organization, 237–38

Fourier, Joseph, 107, 161, 259; anti-Laplacean physics, 34, 325–26 n 12; heat studies, 34, 44, 266, 369 n 49

Fox, Robert, 208

Frankfurt school, 294

Franklin, Benjamin, 37, 101–2, 130, 308–9

freedom-in-connection, 372 n 93; Chevalier's vision of industry, 206–7; as goal of pre-1848 social thought, 284–85, 286; Humboldt's conception of objectivity and freedom/autonomy, 65, 79–80, 85, 336 n 64; Kantian reason and ethical freedom, 66–67, 69–70, 333 n 18, 333 n 24; and Leroux's project for organizing the social milieu, 225–26, 241–48, 285; Saint-Simonians and universal association, 192, 204, 205–8, 208, 225; Schiller's aesthetic and, 73–75, 79–80, 285; Schiller's "Ode to Joy" ("Ode to Freedom"), 73–74, 83

French Enlightenment, 11, 321 n 28

French Revolution. *See* Revolution of 1789

Fresnel, Augustin, 29, 34, 48, 120, 325–26 n 12

Freud, Sigmund, 345 n 5

Galison, Peter, 79, 92, 318 n 6, 319 n 13

Gall, Franz Joseph, 154, 158

Galvani, Luigi, 37

Gane, Mike, 369 n 38

Garnier-Pages, Louis Antoine, 292

Gaubert, Etienne Robert, 223

Gauss, Carl Friedrich, 85

Gautier, Théophile, 126, 128, 147, 173, 297, 373 n 10

Gay-Lussac, Joseph Louis, 64

Geoffroy Saint-Hilaire, Etienne, 19–20, 161–67, 232; and Ampère, 47–48; and Balzac, xiv, 232–34, 362 n 33; concept of the "unity of composition" (unity of type), 162–63, 178–79, 226, 234, 238–39; dictionary entry on "Nature," 165; ecstatic materialism, 236, 362 n 43; electricity theories, 47–48; "Geoffroy's giraffe," 156–57, 169, 232; influence on Darwin, 20, 268, 352 n 47; and Lamarckian transmutationism, 156–58, 161, 162; and Leibnizian language of "virtuality," 162–63, 236, 239, 242–43; and Leroux, xvi, 226, 231–32, 235–39, 242–43; materialist monism, 235–36; and Napoleon's Egypt expeditions/geographical surveys, 107, 156, 161; and *Naturphilosophie*, 121–22, 164, 165–66, 231–32; *Notions synthétiques*, 234; and organogenesis, 4–5, 163, 167, 239, 244; and pantheism, 364 n 65; and post-Laplacean physicists, 106, 121–22, 232; and publicity, 167; salon, 232, 362 n 30; and Sand, 232, 234–35, 239, 362 n 30; and Serres's animal series, 162–63, 226, 236–39, 242–43, 351 n 24; supporters within the scientific community, 38, 167, 228, 232; teratology experiments, *163*, 163–64, 238–39; and transformation of organisms through technological modification, 162–66, 351 n 29, 369 n 41. *See also* Cuvier-Geoffroy debate

Geoffroy Saint-Hilaire, Isidore, 156, 163, 232, 363 n 56

geometry, descriptive: and daguerreotypes, 113–14, 343 n 89; and Monge, 34, 94, 215–16; and the Saint-Simonians, 215–16, 359 n 66
geophysics, Humboldt's: and Arago, 64, 107; global expeditions and collection of data, 63–64, *64*, 79; global maps of isothermic lines, 81–82; and instruments, 63, 65–66, 76–85, 115, 121; mapping projects, 63, 81–84, 336 n 60; terrestrial magnetism (magnetic crusade), 23, 41, 48, 336 n 60
Georget, R. J., 46
Gere, Cathy, 13
Géricault, Théodore, *Raft of the Medusa*, 137
German philosophy: Leroux and the reception in France, 228, 231; and mechanical romanticism in Paris, 7, 11; "spiritualist aesthetics," 128–29. *See also* Kant, Immanuel/Kantian philosophy; *Naturphilosophie* (German natural philosophy); Schelling, Friedrich; Schiller, Friedrich
Globe, 36, 126; articles on religion, 228; Bertrand's science writings, 46, 146, 153, 228; as channel for German philosophy entering France, 228; and Leroux, 203, 221, 225, 226–28; after the Revolution of 1830, 203; and the Saint-Simonians, 203, 204
Goebbels, Joseph, *Michael*, 335 n 42
Goethe, Johann Wolfgang von, 38, 39, 98, 159, 228; and the Cuvier-Geoffroy debate, 167; *Faust*, 134; *Sorrows of Young Werther*, 228
Goldberg Moses, Clair, 213
Goldstein, Jan, 364 n 65
Gooding, David, 41
Grandville, J. J.: *Un autre monde (Another World)*, 154, 155, 176–85, *180*, *182*, *183*; depictions of new industrial technology, 181–82, *183*, 184; "The Metamorphoses of Sleep," 179, *180*; neo-Gods and retellings of myths, 182–84; *Scenes from the Private and Public Life of Animals*, 155, *156*, 232–33; "The Steam Concert," 182, *183*, 184; transpositions of the human and animal worlds, 155, *156*, 178, 232–33
Grattan-Guinness, Ivor, 95
Guattari, Félix, 81, 323 n 57, 352 n 47, 362 n 42
Guépin, Ange, 250, 265
Guizot, François, 289, 290–91
Gunning, Tom, 150
Gutenberg, Johannes, 133–34, 347 n 31

Hachette, Jean-Nicolas-Pierre, 96, 266
Hahn, Reynaldo, 373 n 12

Hasler, Jean-Michel, 347 n 55
Haussmann, Baron von, 8, 220, 306
Hegel, G. W. F., 71, 74, 241
Heilbron, John, 33
Heine, Heinrich, 143
Helmholtz, Hermann von, 85–86
Henry, Charles, 144
Henry, Joseph, 7
Hermes Trismegistus, 145
Herschel, John, 100, 112, 268–69
Hetzel, Pierre-Jules (pseud. P. J. Stahl), 155, 177, 184, 233
hierarchy of animals. *See* animal series
hierarchy of the sciences, Comte's, 260–65, 264
"high modernity" (and romantic machines), 307–8, 375 n 38
Hoëne-Wronski, Count, xii
Hoffmann, E. T. A., 126, 134–35; "Automata," 134–35
Hofmann, James, 39
homo faber, 309, 348–49 n 76
Horkheimer, Max, 294
Hugo, Victor, 61, 126, 232; and Arago, 89–90, 119; and Geoffroy Saint-Hilaire's salon, 232, 362 n 30; *Notre Dame de Paris*, 133; poem cycle *Les voix intérieures*, 131–32; poetry of machines, 131–32; poetry on power of the word/language, 132; during the Second Empire, 306–7; *Les travailleurs de la mer*, 307
Humboldt, Alexander von, 61–87, 259; and Ampère, 41, 48, 62; and Arago, 64, 86, 106–7; conception of objectivity and freedom, 65, 79–80, 85, 336 n 64; conception of scientific knowledge, 79–81; concept of the aesthetic, 62–63, 76–77; correspondence, 78–79, 81; *Cosmos*, 62–63, 76–77, 80–81, 86, 122, 136; electricity and terrestrial magnetism, 48; *Essai sur la géographie des plantes*, 82; galvanism and animal magnetism, 37, 78; geophysical instruments, 63, *64*, 65–66, 76–85, 115, 121; and German *Naturphilosophie*, 62, 63–65, 82; global science and geophysical mapping projects, 4, 63–64, *64*, 79, 81–84, *83*, 107, 336 n 60; influence and role in international scientific community, 63, 81–82, 85–87; and institution of scientific standards and instrumental verification, 336 n 63; and the "nebular hypothesis," 266–67; and panorama craze, 136; and Parisian social and intellectual scene, 62; and post-Kantian epistemology, 77, 78,

79, 337 n 66; and post-Kantian philosophy, 65, 66–70, 77, 78, 79, 337 n 66; and post-Laplacean physics, 120–22; response to Schiller's aesthetic, 76–77, 79–80, 83–84; *Tableau of Equatorial Plants*, 86
Humboldt, Wilhelm von, 63, 132–33, 337 n 66

imagination, romantic conception of, 130–31
imponderable fluids, xiii–xiv, 4; and animal magnetism (mesmerism), xiii–xiv, 36–37; Coulomb's model, 33; and Lamarck, xiv, 37; Laplacean physics and the standard view, 32–35
industrial mechanics, 96–97
Industrial Revolution, 9, 168
Ingres, Jean-Auguste-Dominique, 137
Institut de Statistiques, 98

Jacobins, 35, 49
Jacquard loom, 169, 171, 300
James, Henry, *Portrait of a Lady*, 326 n 24
Jardin des Plantes, 155–60, *156*, *157*, 178
Jouffroy, Théodore, 63, 128, 146, 159, 226, 228
Journal des débats, 116, 153, 295
July Monarchy, 9, 169
Jussieu, Bernard de, 38, 54

Kant, Immanuel/Kantian philosophy, 65, 66–70, 318 n 6; and Ampère, 38, 52, 53, 58, 331 n 94; categorical imperative, 67; and civil society, 69–70; *Critique of Judgment*, 65, 67–69, 70, 128; *Critique of Practical Reason*, 66, 69–70, 74; *Critique of Pure Reason*, 38, 66, 78; faculty of judgment, 67–68; and humankind's two natures, 66–70, 320 n 15; and Humboldt, 65, 66–70, 77, 78, 79, 337 n 66; influence of, 66; and Maine de Biran, 38, 52, 331 n 94; *Metaphysics of Nature*, 31; objectivity, 66–70, 74–75, 333 n 18; reason and ethical freedom, 66–67, 69–70, 333 n 18, 333 n 24; and Schelling, 31, 231; and Schiller, 63, 70, 71–75, 337 n 66; "What is Enlightenment?," 70; the will and notion of autonomy ("self-law"), 67, 333 n 20
Kardec, Allan, 36
Keck, Frédéric, 372 n 92
Kielmeyer, Carl Friedrich, 160
Kittler, Friedrich, 323 n 48
Kohler, Robert, 117
Koreff, Ferdinand, 36
Koyré, Alexandre, 283

Kuhn, Thomas, 256

labor theory of knowledge, 100–105, 121, 340 n 42; and aesthetics, 128, 130; Clapeyron's reworking of Carnot cycle, 211; and the daguerreotype/photography, 115, 118; and the "engineer-scientists," 100–105; and Saint-Simonians, 103, 211
labor theory of value, 102–3
Lagrange, Joseph-Louis: and balances, 103; concept of "virtual velocities," 96, 104, 236; and Ecole Polytechnique, 93; *Mécanique analytique*, 93
Lamarck, Jean-Baptiste/Lamarckianism: and animal magnetism, 36–37; and the animal series, 362–63 n 45; and concept of the milieu, 4–5; Deism, 236; Geoffroy Saint-Hilaire's revival of transmutationist theories, 156–58, 161, 162; and physio-spiritualism of Maine de Biran, 348–49 n 76; transmutationism, 156–58, 161–62, 362–63 n 45; *Zoological Philosophy* and the giraffe, 156–57
Lamartine, Alphonse, 126, 134, 291, 292, 347 n 31
Lambert, Charles, 215
Lamé, Gabriel, 206, 210, 258, 266
Lamennais, Hugues-Félicité Robert de, 49
La Mettrie, Julien Offray de, 31
Langlois, M., 138–39
language: discipline of philology, 133; and the Oriental Renaissance, 133, 357 n 30, 360 n 71; romantic conceptions of, 132–33, 134–35, 346 n 25, 360 n 71. *See also* printing technology and literary production
Laplace, Pierre-Simon/Laplacean physics, 20, 89–92, 325–26 n 12, 375 n 27; and Ampère's philosophy of electromagnetism, 45–51, 58; classical mechanism and determinism, 32–35, 89, 337 n 1; and the Ecole Polytechnique, 93, 258, 338 n 12; and electromagnetism research, 30, 32–35, 44–45, 328 n 48; *Mécanique céleste*, 32–33, 178; and mesmerism, 37; and Napoleon, 32–33, 89; and the nebular hypothesis, 121, 266–67, 344–45 n 115; and Newtonian physics, 32–35, 89; and the "second scientific revolution," 30; and Society of Arcueil, 33, 62, 64, 105. *See also* post-Laplacean physics
Laprade, Victor de, *The New Age*, 132
Latour, Bruno, 348–49 n 76, 372 n 86, 375 n 38
Laurent-Pichat, Léon, "Ce siècle est en travail," 174

442 INDEX

Lavoisier, Antoine, 94, 103
Ledru-Rollin, Alexandre Auguste, 290–91, 292
Legouvé, Ernest, 134
Leibniz, Gottfried, 162–63, 215, 236, 242, 362 n 40
Leroi-Gourhan, André, 309, 348–49 n 76
Leroux, Pierre, 185, 223–51, *227*, *249*, 284; advocacy on behalf of workers, 249–50; and the animal series, 226, 236–39, 242–43; and the cenacle in Balzac's *Lost Illusions*, xvi; and character in Grandville's *Un autre monde*, 177; *circulus* invention (for recycling human waste), 246–47, 248, 307; and Cousin, 20, 231, 361 n 18, 364 n 65; and democratic socialism, 225–26, 285; on division of labor and progressive role of machinery, 245–46; doctrine of progress and convergence of fields of science, 229–30; *Encyclopédie nouvelle*, 227, 239, 357 n 30, 364 n 61; and Geoffroy Saint-Hilaire's transcendental anatomy, xvi, 226, 231–32, 235–39, 242–43; and the *Globe*, 203, 221, 225, 226–28; *L'humanité*, 239–41; "L'humanité et le capital," 247; and the incessant creation of the universe, 230, 250; and Maine de Biran's physiospiritualism, 129, 146; and the milieu, 225–26, 241–48; and pantheism, 159, 186, 229, 241, 364 n 65–66; pianotype invention, 223–25, *224*, 245, 360 n 2; on property, capital, and wealth, 246, 247, 366 n 90; and the reception of German philosophy in France, 228, 231; refutation of Malthus's law of nature, 247–48, 375 n 20; relations between humans and nature, 246–48; and *Revue encyclopédique*, 227, 229–30; as romantic printer and encyclopedist, 226–32; and the Saint-Simonians, 203, 204, 213, 221, 225, 228, 240–41, 246–47, 285, 365 n 66; and Sand, 224, 227, 234, 245, *292*; and Schelling's philosophy, 231; on scientific instruments and experimental devices, 243–44; during the Second Empire, 307; and the Second Republic, 291, *292*; theory of the symbol, 228–29, 242, 245; and virtual humanity, 226, 239–41, 242–43
Leverrier, Urbain Jean Joseph, 18, 307
Levitt, Theresa, 115, 343 n 99
Liebig, Justus, 94, 248
life, conceptions of: Bichat's, 270; Blainville's, 270–71; Comte's, 270–71, 273, 309, 370 n 62; Cuvier's, 237, 362–63 n 45

Linnaeus, Carl, 38
Liszt, Franz, 137, 141, 232, 234, 362 n 30
literary production. *See* printing technology and literary production
Littré, Emile, 307, 331 n 105
Louis XVI, execution of, 8
Louis XVIII, Restoration of, 8–9, 10
Louis-Philippe: and Daguerre's diorama *Mont Blanc*, 140; and Humboldt, 62; and industrial displays of the Expositions Nationales, 170–71; Revolution of 1830 and "July Monarchy," 9; and Robert-Houdin's writing automaton, 171; and the workers' revolution (1848), 290–91
Lovejoy, A. O., 335 n 40, 362–63 n 45; *Great Chain of Being*, 1, 335 n 40
Luxembourg Commission, 291
Lyell, Charles, 19

Macaire, Robert, 181, *292*
Machereau, Philippe Joseph, 215
Macherey, Pierre, 241, 355 n 93
magnetism. *See* animal magnetism (mesmerism); electromagnetism
Maine de Biran, Pierre: and Ampère, 22, 31–32, 36, 51–53, 58, 129, 145–46; and Comte's positivist philosophy, 51; conversation circle of, 51, 362 n 30; and electromagnetism, 22, 31–32, 36; and Kantian philosophy, 38, 52, 331 n 94; lasting influence, 52–53, 331 n 96; *On the Influence of Habit*, 51–52; physiospiritualism and methodology of introspection, 51–53, 129, 145–48, 348–49 n 76; and the willful self as noncorporeal starting point for experience (the "hyperorganic"), 52, 330–31 n 93. *See also* physiospiritualism
Malthus, Thomas, 11, 247–48, 364 n 62, 375 n 20
Manifesto of the Communist Party, The (Marx and Engels), 13, 201, 372 n 87
Manuel, Frank, 356 n 12, 357 n 16, 357 n 28, 372 n 88
Maret, Henri, 241, 364 n 65
Marie, Pierre, 292
Marriman, Michael, 137
Martin, John, 136, 143
Marx, Karl, 283, 355 n 93, 372 n 87; and Bonaparte's 1851 coup, 306; *Capital*, 13, 307, 355 n 93; *Manifesto of the Communist Party*, 13, 201, 372 n 87; and mechanical romanticism, 12–13, 289, 307; and Paris intellectual scene, 7; during the Second Empire, 307

materialism: ecstatic, 236, 362 n 43; of Geoffroy Saint-Hilaire, 235–36, 362 n 43; monism, 158–59, 235–36
Mauss, Marcel, 117, 344 n 107
Maxwell, James Clerk, 18–19, 44, 328 n 54
McLuhan, Marshall, 309, 348–49 n 76
"mechanical objectivity," 92, 319 n 13, 338 n 5
mechanical romanticism, xi–xii, 1–26; concept of the "milieu," 4–5, 319 n 10, 322 n 44; contributions worth retrieving today, 25–26, 308–11, 375 n 38; fantastic spectacles, 125–52, 153–87, 294–95, 302; and France's history of social and political turmoil, 8–9; and "high modernity," 307–8, 375 n 38; the impact of romantic philosophy on science/epistemology, 17–21, 323 n 49; and new concept of nature, 5, 319 n 14; and new theory of knowledge, 5–6; and nineteenth-century Paris, xi–xii, 3–11; and the physical sciences, 22–23, 29–59, 61–87, 89–122, 287; and post-Laplacean science, 120–22; and Second Empire, 306–7; and Second Republic, 9, 24, 291–305; and social reformers, 23–24, 119–221, 185–87, 223–51, 253–86, 288, 355 n 93; view of the human as technological animal, 255, 308–9. *See also* romantic machines
mechanism. *See* classical mechanism
Meckel-Serres Law, 238–39, 370 n 60
Medical Faculty of Paris, 160
Mehmet Ali, viceroy of Egypt, 170
Mendelssohn-Bartholdy, Felix, 81
Merleau-Ponty, Maurice, 330 n 86, 331 n 96, 348–49 n 76
Merton, Robert, 18
Mesmer, Franz, 37
mesmerism. *See* animal magnetism (mesmerism)
Meyerbeer, Giacomo, 349 n 82, 373 n 12; critics, 298, 373 n 12; opera *Le prophète*, 296–98, *297*, 373 n 10; opera *Robert le diable*, 149–52, *150*, 294, 349 n 82
Michelet, Jules, 239, 240, 291, 298; *Le peuple (The People)*, 8, 134
milieu, concept of, 4–5, 319 n 10; Arago and, 107–9; Balzac and, xiv–xvi, 4; Benjamin and, 322 n 44; Comte and, 255, 270–71, 272–73, 309; Lamarck and, 4–5; Leroux and, 225–26, 241–48, 285; and Newton's ether, 4; and organogenesis, 4–5; origins of, 4–5, 319 n 10
Mill, J. S., 204, 257, 268
Milne-Edwards, Henri, 17

Monge, Gaspard: descriptive geometry, 34, 94, 215–16; and the Ecole Polytechnique, 34, 93–95, 101; and the 1806 Exposition Nationale, 169
monism, materialist, 158–59, 235–36
Montgolfier, Joseph, 96
moral economies and science, 117, 344 n 105, 344 n 107
Moreau de Tours, Joseph, 129, 146–47
Morse, Samuel, 7, 41
Musée Ampère, 40
Muséum d'Histoire Naturelle, 21, 154, 155–60; and Balzac's "Natural History of Society," 155; exotic species and comparative anatomy studies, 156–57, *157*; and Grandville's *Scenes from the Private and Public Life of Animals*, 155, *156*; and Grandville's *Un autre monde*, 178; and the Ménagerie of the Jardin des Plantes, 155–60, *156*, *157*, 178; and metaphysical debates about fixity or mutability of species, 157–60
Musselman, Elizabeth Green, 330 n 89
Musso, Pierre, 356 n 9, 356 n 13, 358 n 43
mystical religiosity of Lyon, 11, 36

Napoleon III. *See* Bonaparte, Louis-Napoleon (Napoleon III)
Napoleon Bonaparte (Napoleon I), 8, *94*; Egypt survey expedition, 107, 113, 156, 161; and French science, 6, 17–18, 32–33, 89; and industrial fairs following the Revolution, 168–69; and Laplace, 32–33, 89
"Napoleonic idea," 304, 374–75 n 28, 375 n 29
national, Le, 133, 290
National Institute of Sciences and Arts, 6, 32, 89
Naturphilosophie (German natural philosophy): and Ampère's electromagnetism experiments, 22, 30–32, 35, 38–40, 45, 58, 325 n 5; and Geoffroy Saint-Hilaire, 121–22, 164, 165–66, 231–32; and Humboldt, 62, 63–65, 82; and Kantian philosophy, 38, 52, 53, 58
Navier, Claude-Louis, 97, 259, 266, 339 n 30
Nazi regime and Schiller's utopia, 335 n 42
"nebular hypothesis," 5, 121, 266–69, 344–45 n 115, 369 n 49
neopositivism, 256
Nerval, Gérard de, 118, 126, 133, 147; *Aurélia*, 174, 348 n 73; *Vers dorée*, 15
network theory, 95–96
Neufchâteau, François de, 168

Newtonian physics, 4; and Ampère's electromagnetism experiments, 44–45, 328 n 53; Saint-Simon's goal of unifying knowledge under principle of gravitation, 195
Niépce, Nicéphore, 112, 117, 344 n 110
Nodier, Charles, 126
Novalis, 7, 14–15, 16

objectivity: and Comte's *Course of Positive Philosophy*, 275; De Quincey's use of the term, 141; Humboldt's conception of freedom/autonomy and, 65, 79–80, 85, 336 n 64; Kantian, 66–70, 74–75, 333 n 18; mechanical, 92, 319 n 13, 338 n 5
Oersted, Hans Christian, 130, 144; discovery of electromagnetism, 22, 29, 30–31, 33, 35, 38–39, 45
Oken, Lorenz, 7, 81; and the animal series, 238, 363 n 55; notion of polarity, 165; and organogenesis, 4–5, 163, 351 n 24
opera, grand, 148–52, 349 n 82; appeal of, 150–51, 349 n 82; and Balzac's "Gambara," 151–52, 349–50 n 86; Meyerbeer's *Le prophète*, 296–98, *297*, 373 n 10; Meyerbeer's religious themes, 373 n 12; Meyerbeer's *Robert le diable*, 149–52, *150*, 294; scenes of technological transubstantiation, 152, 349–50 n 86; stage lighting and Foucault's electric lamp, 296–98
"optimization theory," 206
organicism, 2, 11–12
organogenesis, 4–5, 163, *163*, 239, 244, 351 n 24; and Geoffroy Saint-Hilaire, 4–5, 163, 167, 239, 244, 351 n 24; Leroux's doctrine of progress and the incessant creation of the universe, 230, 250; and teratology, *163*, 163–64, 238–39. See also Oken, Lorenz
Oriental Renaissance: and ancient ritual languages, 133, 357 n 30, 360 n 71; and Leroux and Reynaud's *Encyclopédie Nouvelle*, 357 n 30; and the Saint-Simonians, 202, 357 n 30, 360 n 71. See also comparative religion; Schlegel, August; Schlegel, Friedrich
ozone and electricity's cosmological associations, 324–25 n 4

Paganini, Nicolo, 141
palingenesis, 59, 198, 237. See also Ballanche, Pierre-Simon; Bonnet, Charles
panoramas, 135–37, 143; and Balzac's *Père Goriot*, 135–36; and Berlioz's symphonies, 143; and Daguerre's dioramas, 137–40;

David's, 136, 137; Delacroix's *Barque of Dante*, 137; Delacroix's *The Death of Sardanapalus*, 137; Humboldt on pedagogical utility of, 136; and the mechanics of natural perception, 136
pantheism: and Cousin's eclecticism, 364 n 65; and Geoffroy Saint-Hilaire, 364 n 65; Leroux's participation in the controversy over, 159, 186, 229, 241, 364 nn 65–66; perspectives on organisms displayed at the Muséum d'Histoire Naturelle, 159; and romantic symbolism, 229; Saint-Simonian religion, 159, 194, 202–3, 213, 220, 364 nn 65–66; and Spinoza, 159, 202–3, 241
Pantheon: and Foucault's pendulum, 302–5; and the workers' revolution (1848), 293, *293*, 303–4
Paracelsus, 101
Paris, nineteenth-century: classic sites of French science, 21; the *France des notables*, 21; and history of social and political turmoil, 8–10; romanticism and mechanism, xi–xii, 3–11; and the Saint-Simonians' "New City," 192–94, 218; size/population, 8; social and intellectual scene, 7–8, 21, 36, 61–62; social geography/social ecology, 7–8, 21
Paris Observatory, xiii, *108*; and Arago, 80, 106, 107, 109, 154, 269; Arago's lectures on popular astronomy, 106, 109, 154, 269; and Humboldt's geophysical maps, 82; Leverrier's leadership during Second Empire, 18, 307; rooftop meteorological and atmospheric measures, 107, *108*; and studies of terrestrial magnetism (magnetic crusade), 48
pendulum, Foucault's, 302–5, *303*; and Bonaparte, 302–5; Pantheon location, 302–5; as public science (stagecraft and audiences), 302
Perdonnet, Auguste, 206
Pereire, Emile, 211–12
Petit, Alexis Thérèse, 34, 110, 120
phénoménotechniques, 125, 343 n 92. See also technaesthetics
philology, 133. See also Oriental Renaissance
photography: Arago's emphasis on mystery involved in, 118–20; association with classical machines/mechanism, 92; Baudelaire on, 114; and labor theory of knowledge, 115, 118; as process dependent on humans, 118; putative transparency, 115–17, 343 n 96. See also daguerreotypes

INDEX   445

phrenology, 154, 158
physiologies, literary genre of, 15, 155
physiospiritualism, 51–53, 129, 143–48, 348–49 n 76; and altered states of consciousness, 146–47; and Ampère, 51–53, 129, 146; and Bertrand, 129, 146; and complex relationship between mind and body, 145; and Cousin's "spiritualism," 51, 129, 145; and Leroux, 129, 146; Maine de Biran's theory of perception, 51–53, 129, 145–48, 348–49 n 76; and Moreau de Tours, 129, 147; the nature of perception and defects of human vision, 145; and new field of psychiatry, 147; and Ravaisson, 129, 146, 147; reflections on technology, 148
pianotype, 223–25, 224, 245, 360 n 2
Pickering, Andrew, and mangle of practice, 328 n 43
Picon, Antoine, 215
Pietz, William, 321 n 34
Poe, Edgar Allan, 7
Poisson, Siméon-Denis, 32, 44, 266, 269
Poncelet, Jean-Victor, 97, 102, 232
Ponts et Chaussées, school of, 95
positivism: Comte's definition, 368 n 30; Comte's positivist gospel, 254–55, 259–65, 368 n 27; Comte's Religion of Humanity, 256, 273–84; the fantastic arts and the positive sciences, 125–27; neopositivism, 256; and spiritualism, 51; twentieth-century, 255–56; Vienna Circle, 255–56. *See also* Comte's positivism
post-Laplacean physics, 20, 33–35, 44–45, 89–92, 120–22, 325–26 n 12; Ampère, 20, 30, 31, 33–35, 44–51, 58, 120–22, 325–26 n 12; Arago, 33–34, 89–92, 95, 105–9, 120–22, 267, 325–26 n 12; and electromagnetism, 30, 31, 45–51, 58; ether physics, 30, 31, 44–45; Geoffroy Saint-Hilaire, 106, 121–22, 232; Humboldt, 120–22; and the nebular hypothesis, 266–67. *See also* Dulong, Pierre Louis; Fourier, Joseph; Fresnel, Augustin; Petit, Alexis Thérèse
poststructuralism, 355 n 93
presse, La, 133
Prevost, Pierre, 137
Prigogine, Ilya, 103
printing technology and literary production, 131–35, 180–81; Balzac's *Lost Illusions* and the ambivalent power of the press, xv–xvi, 133–34; daily and weekly journals, 133; the *feuilleton* (serial novel), 133; glorification of the press and Gutenberg, 133–34,

347 n 31; and Grandville's *Un autre monde*, 180–81; labor-saving devices, 224–25, 245, 360 n 2, 360 n 5; Leroux's pianotype invention, 223–25, 224, 245, 360 n 2; and publicity, 167, 180–81; and "puffery," 181; romantic conceptions of the power of language and the printed word, 132–33, 134–35, 346 n 25; typography and notion of series, 238
producteur, Le (journal of the Saint-Simonians), 203, 258
Prony brake, 97
prophète, Le (Meyerbeer opera), 296–98, 297, 373 n 10
Proudhon, Pierre-Joseph: and character in Grandville's *Un autre monde*, 177; and notion of progressive series, 238; *On the Creation of Order in Humanity*, 238; and the Second Republic, 291, 292; and the workers' revolution (1848), 289
psychiatry, 146–47
publicity (*publicité*): and Cuvier-Geoffroy debate, 167; Grandville's *Un autre monde* and critique of, 180–81
"puffery," 177, 181
Puységur, Marquis de (Armand-Marie-Jacques de Chastenet), 37, 46, 146

Quest for the Absolute, The (Balzac), xii–xiii, xv, 30, 35, 134, 244, 317 n 2, 325 n 6
Quinet, Edgar, 133, 232, 291, 364 n 61

Rabinach, Anson, 103
Radcliffe, Anne, 126
railroads, 206–8, 211, 220
Rancière, Jacques, 221, 307, 355 n 93
Ravaisson, Félix, 129, 147
Récamier, Julie, 21, 36
réforme, La, 290
Regnault, Victor, 210
Régnier, Philippe, 357 n 30
Renaissance alchemy: and imagery of fantastic spectacles, 173–74, 353 n 68; and the labor theory of knowledge, 101; and the photographic process, 119–20
republican politics and ideals, 9; and Arago, 90, 109–12; and labor theory of knowledge, 102; and the Saint-Simonians, 197, 200, 203; and social impact of steam technology, 110–11
Restoration of Louis XVIII, 8–9, 10; and the Expositions Nationales, 167–71; goals of demarcating between "positive science" and superstition/imagination, 125–26

Revolution of 1789, 8; Arago and tradition connecting the sciences to, 109–10; and the "children of the century," 9–10; festivities, industrial fairs, and expositions following, 168–69; and Leroux's doctrine of progress, 229–30

Revolution of 1830, 8–9, 203

Revolution of 1848 (workers' revolution), 9, 24, 103, 288–91; the economic downturn, 289–90; key agitators (mechanical romantics), 289, 290–91; and the labor theory of knowledge, 103; and the Pantheon, 293, *293*, 303–4; and "Social Workshops," 290. *See also* Second Republic

Revue des deux mondes, 48

Revue du progrès, 290

Revue encyclopédique, 227, 229–30

Révue indépendante, 224, 227

Reynaud, Jean: and Geoffroy Saint-Hilaire, 232; and Leroux, 227; and the Saint-Simonians, 204, 213, 229; and the Second Republic, 291

Ricardo, David, 11

Richards, Robert, *The Romantic Conception of Life*, 318 n 4

Ritter, Johann, 7, 39

Robert-Houdin, Eugène: audiences, 354 n 73; automaton on display at the 1844 Exposition Nationale, 171; *Memoirs*, 172, 354 n 77; *Soirées fantastiques*, 154, 171–76, *172*, *173*, 354 n 73, 354 n 77; stagecraft and deliberate ambivalence/ambiguities, 174–75, 354 n 77; theater near the Palais Royal, 171, 354 n 73

Robert le diable (Meyerbeer opera), 149–52, *150*, 294, 349 n 82

Robertson, Etienne-Gaspard, 36–37, 130–31

Robespierre, Maximilien, 168

Rodrigues, Olinde, 200, 214, 258

romanticism: conception of the imagination, 130–31; Darwin and the romantic life sciences, 318 n 4; and the fantastic/fantastic arts, 125–27, 130–31; German, 7, 11, 128–29; opposition with mechanism, 1–3, 318 n 4; and organicism, 2, 11–12. *See also* mechanical romanticism

romantic machines, xi, 11–16; and active participation of the observer, 12; and "charismatic authority" of objects, 14, 321 n 36; as extensions of human senses, 5, 80–81, 118, 121; and "fetishism," 13, 321 n 34; and the routinization of enchantment/enchantment of routine, 14; and shifts in key arguments of critical theory, 12–16; as technologies of time and transformation, 118–20. *See also* mechanical romanticism

Rosanvallon, Pierre, 320 n 26

Rostan, Léon, 46

Rousseau, Jean-Jacques, 67, 74, 333 n 20

routes, science of, 95–97, 206–8

Royer-Collard, Pierre-Paul, 36, 51

Russian Institute of Roads, 206

Sainte-Beuve, Charles-Augustin, 11, 35, 134

Saint-Martin (mystic of Lyon), 36, 145, 326–27 n 24

Saint-Simon, Count (Henri de Rouvroy), 185, 194–99, 205; *Catéchisme des industriels*, 197; and character in Grandville's *Un autre monde*, 177; *Du système industriel*, 197; goal of unifying knowledge under the principle of Newtonian gravitation, 195; "Letters from a Citizen of Geneva," 195; and liberal politics, 197; "Mémoire sur la science de l'homme," 195–97, 202, 205–6; and New Christianity (*Le nouveau christianisme*), 197–98, 200, 202, 356–57 n 15, 357 n 26; "Parable," 197; *Le savant, l'industriel, l'artiste*, 205; physiological tableau of human society, 195–97, 201, 205–6; and work of the anatomist Vicq d'Azyr, 195–96

Saint-Simonians, 185, 186, 191–221, 284; army of industrial salvation, 198, 199–205; *avant-garde* of artists, 205; Berlioz and, 143, 205; Carlyle's assessment, 191–92; and the cenacle in Balzac's *Lost Illusions*, xvi; central bank plan, 201–2, 339 n 30; Chevalier's Mediterranean vision, 206–8, 217; the collected *Doctrine de Saint-Simon*, 201–3, 237; Comte and, 198, 203, 258–59, 274, 278, 371 n 73; descriptive geometry and sacred mathematics, 215–16, 359 n 66; Duveyrier and "The New City of the Saint-Simonians," 192–94, 218–20, 310; engineering science, 216–17; and engineer-scientists, 198, 199–200, 204, 205–8, 220, 258, 356 n 13, 357 n 17; and "evolution," 198–99; and the Expositions Nationales, 169–70; industrial projects, 206–8, 211, 220; journal *Le producteur*, 203, 258; and the labor theory of knowledge, 103, 211; law of history's progression, 201; Leroux and, 203, 204, 213, 221, 225, 228, 240–41, 246–47, 285, 365 n 66;

machines as mediators between spirit and world, 192, 208–12; music, 205, 215, *216*; new scientific principles, 215–17; optimization of technical systems, 206–8; and the Oriental Renaissance, 202, 357 n 30, 360 n 71; and railroads, 206–8, 211, 220; and republican ideals, 197, 200, 203; Sadi Carnot's engine and four-stage cycle of conversion, 208–12, 358 nn 51–52; Saint-Simon's industrial vision, 194–99, 205; Saint-Simon's "Mémoire sur la science de l'homme," 195–97, 202, 205–6; science of routes and transportation of goods, 206–8; and series/serial organization, 237, 238; social reform projects, 200–202, 204, 220–21; symbols, 191–92; and universal association, 192, 204, 205–8, 208, 225. *See also* Saint-Simonians (religion of)

Saint-Simonians (religion of), 200–205, 212–20, 360 n 71; Enfantin and, 200–201, 212–16, 285; Enfantin's free love proclamations and challenges to marriage, 213, 214, 229; evangelism through direct contact and preaching, 203–5; and external threats to the movement, 214; Félicien David's musical contributions, 205, 215, *216*; and feminine liberation, 213; "houses of association," 204; Ménilmontant retreat, 214–15; missions to various cities, 204; the New Bible (*Le livre nouveau*), 212, 215–17, 360 n 71; New Christianity (*Le nouveau christianisme*), 197–98, 200, 202, 357 n 26; new "Genesis" (retold story of creation), 217–18; new scientific principles, 215–17; pantheism, 159, 194, 202–3, 213, 220, 364 nn 65–66; priests and priestesses, 201–2, 213–14, 278, 285; public lectures and sermons, 203–5; temple-building preparations, 215, 221; the Temple-Woman, 193, 218–20, *219*; uniforms, rituals, and parades, 214–15

Salverte, Eusèbe, 106

Sand, George, xvii, 137; *Le compagnon du tour de France*, 234, *235*; and Geoffroy Saint-Hilaire, 232, 234–35, 239, 362 n 30; and the Grandville-illustrated *Scenes from the Private and Public Life of Animals*, 155, *156*; *Indiana*, 234; *Lélia*, 234–35; and Leroux, 224, 227, 234, 245, *292*; *Les sept cordes de la lyre*, 132, *292*

Satan figures and Mephistophelean images, 135; Baudelaire's "Litanies of Satan" in *Les fleurs du mal*, 347 n 36; Berlioz's *Symphonie Fantastique*, 141; Robert-Houdin's *Soirées fantastiques*, 172

Savart, Félix, 44, 144, 328 n 48

Savary, Félix, xiii, 328 n 48

Sax, Adolphe, 141

Say, J. B., 99, 110–11

"Say's Law," 99

Scenes from the Private and Public Life of Animals (Grandville), 155, *156*, 232–33

Schaffer, Simon, 318–19 nn 6–7, 322 n 36, 323 n 52, 325 n 4, 332 n 112, 335 n 45, 335 n 49, 335 n 55, 336 n 63, 339 n 37, 340 nn 38–39, 341 n 53, 343 n 100, 344 n 102, 350 n 3, 361 n 51, 366 n 2, 368 n 31, 369 n 42, 370 nn 55–56

Schelling, Friedrich, 7; closing gap between mechanical causality and nature as organic whole, 30, 62, 147, 231, 333 n 21; and Geoffroy Saint-Hilaire's philosophy, 165; and German *Naturphilosophie*, 30–31, 62, 165, 231, 333 n 21; and Humboldt, 62; *Ideas for a Philosophy of Nature*, 30–31; and Kant, 31, 231; Leroux and philosophy of, 231; and notion of the absolute identity between mind and matter, 30, 147, 231; and Oersted's discovery of electromagnetism, 30–31; and Ravaisson's work, 147; and the Saint-Simonians' pantheism, 159, 202–3

Schiller, Friedrich: and the aesthetic, 63, 65, 70, 71–76, 84, 335 n 40; conception of freedom, 73–75, 79–80, 285; Humboldt's response to aesthetic of, 76–77, 79–80, 83–84; ideal of autonomy (*Selbständigkeit* ["self-standingness"]), 71, 75, 334 n 33; and Kantian objectivity, 74–75; and Nazi regime, 335 n 42; "Ode to Joy"/"Ode to Freedom," 73–74, 83; *On the Aesthetic Education of Man*, 63, 65, 71–76, 129, 321–22 n 26; refashioning of Kant's philosophy, 63, 70, 71–75, 337 n 66, 337 n 68

Schlegel, August, 133, 364 n 61

Schlegel, Friedrich, 64–65, 132–33, 364 n 61

Schmidgen, Henning, 321 n 31

Schmitt, Stéphane, 356 n 8

Schoechler, Victor, 292

Schonbein, Christian Friedrich, 324–25 n 4

Schopenhauer, Arthur, 337 n 66

Schwab, Raymond, 133, 357 n 30

Schweber, Silvan, 268

Scott, Walter, 126

Second Empire: and the arts, 307; and Bonaparte's 1851 coup, 8–9, 25, 305–6; Haussmann's planning, 8, 220, 306; mechanical romantics and romantic machines, 306–7; as start of "high modernity," 307–8; politicized science and the "decline thesis," 18

Second Republic, 9, 24, 291–305; Bonaparte and French science, 293–94, 299–301, 302–5, 374–75 n 28; the change of regime and provisional government, 9, 90, 291–93; early goals of suffrage and relief for unemployed and workers, 291–93; the 1849 Exposition, 298–301; Executive Power Commission, 291–92; National Workshops, 291–92, 299; paradoxical presentations of machinery and labor, 9, 24, 294–305; political clubs, newspapers, and intellectuals, 291; stagecraft and spectacle, 294–95, 302; utopian projects and proposals, 291, *292*; visions of unlimited growth, 299, 374 n 19

Secord, James, 361 n 21

Séguier, Pierre-Armand, 90

Selbständigkeit ("self-standingness"), 71, 75, 334 n 33

sensory perception and processing, 51–53, 136, 330 n 89; fantastic spectacles and illusions, 141, 144–45, 147; and hallucinogenic/psychotropic drug use, 141, 147; and Kant's distinction between aspects of nature, 66; Maine de Biran's physiospiritualism, 51–53, 129, 145–48; and panoramas, 136; romantic machines as extensions of human senses, 5, 80–81, 118, 121; visual epistemology/human vision, 145

serial organization, 237–39. *See also* animal series

Serres, Etienne, 9, 363 nn 55–56; and the animal series, 363 nn 55–56; embryology and theory of recapitulation (the Meckel-Serres Law), 238–39, 370 n 60; and Geoffroy Saint-Hilaire, 228, 232, 238–39, 363 nn 55–56; and organogenesis, 163, 239, 244, 351 n 24; *Researches on Transcendental Anatomy*, 163; and teratology, 163–64, 238–39

Serres, Michel, 103, 137

Sewell, William, 221

Shelley, Mary, 7

siècle, Le, 133

Siegert, Bernhard, 323 n 48

Simon, Jules, 364 n 65

Simondon, Gilbert, 309, 348–49 n 76

Sloterdijk, Peter, 322 n 44

Smith, Adam, 11, 339 n 32, 368 n 32

Smith, Pamela, 101

social reform projects and philosophies of utopian thinkers and romantic socialists, 20, 23–24, 185–87, 284–86, 288, 355 n 93. *See also* Buchez, Philippe-Joseph; Comte, Auguste; Fourier, Charles; Leroux, Pierre; Marx, Karl; Proudhon, Pierre-Joseph; Saint-Simonians

Société Chrétienne de Lyon, 36

Society of Arcueil, 33, 62, 64, 81–82, 105

Society of Naturalists and Natural Philosophers (Humboldt), 18, 81

sociology: and Comte's hierarchy of the sciences, 260, 262, 263–64, 265, 368 n 29, 369 n 38; and Comte's ranking of humans in the animal series, 272–73; and Comte's Religion of Humanity, 275–76, 281; and future progress of the human organism, 281; and the historical method, 272–73; temporal complexity of, 263–64

Soirées fantastiques (Robert-Houdin), 154, 171–76, *172*, *173*, 354 n 73, 354 n 77

species, classification and variability of, 157–60. *See also* animal series; Cuvier-Geoffroy debate; transmutationism

Spinoza, Baruch: and pantheism, 159, 202–3, 241; and "radical enlightenment," 355 n 93

spiritualism: Cousin's, 51, 128, 129, 145, 345–46 n 10; de Bonald's traditionalist spiritualism, 158; German "spiritualist aesthetics," 128–29; and physiospiritualism, 51, 129, 145, 348–49 n 76; and positivism, 51. *See also* physiospiritualism

Spurzheim, Johann, 154

Stahl, P. J. (Pierre-Jules Hetzel), 155, 177

statistics, French, 98

Staum, Martin, 372 n 92

steam technology and steam engines: Arago and safety, social impact of, 110–11; Chevalier's Mediterranean vision, 217; displays at the Expositions Nationales, 169–70; Grandville's *Un autre monde* ("The Steam Concert" image), 181–82, *183*, 184; and the Industrial Revolution, 9; and the labor theory of knowledge, 104; shift of central cosmological model from the classical machine, 103–4; Woolf compound steam engine, *209*

Steinle, Friedrich, 39

Stengers, Isabelle, 103

Stiegler, Bernard, 348–49 n 76

Sturm, Jacques Charles François, 269
Swedenborg, Emmanuel, 145, 173
sylphide, Le (1832 ballet), 126–27
symbols: balances as, 103; Humboldt's geophysical instruments, 78–81, 121; Leroux's theory of, 228–29, 242, 245; steam engines of the Saint-Simonians, 191–92
Symphonie Fantastique (Berlioz), 141

Talbot, William Fox, 112
tâtonnement (tactile testing), 39, 54, 165–66
technaesthetics, 125–31, 295, 345 n 7; Ampère and "la technesthétique," 57, 127–28, 146, 295, 331 n 108, 345 n 7, 345 n 8; Bachelard's *phénoménotechniques*, 125, 343 n 92; and Comte's new philosophy of science, 125, 295; contrasted to German "spiritualist aesthetics," 128–29; a labor theory of aesthetics, 128, 130; and new romantic conception of the imagination, 130–31; and Robertson's *Fantasmagoria*, 130–31. *See also* fantastic spectacles (visual and auditory illusions)
technological animals, humans as, 308–9; Comte's theory of, 255, 257, 273
"technonomie" (Christian's concept of technology), 97–98, 331 n 108, 345 n 8
Teilhard de Chardin, Pierre, 309, 348–49 n 76
Tenniel, John, 176
teratology, *163*, 163–64, 238–39
terrestrial magnetism, 23, 41, 48, 336 n 60
thermodynamics, laws of, 210, 358–59 n 52
Thierry, Augustin, 198
Thom, René, 352 n 47
Thomas à Kempis, *The Imitation of Christ*, 35, 50
Thompson, E. P., 117
Thomson, William, 18–19, 210, 358–59 n 52
Todorov, Tzvetan, 345 n 5
transcendental anatomy. *See* Geoffroy Saint-Hilaire, Etienne
transcendentalism (American philosophical romanticism), 345–46 n 10
transmutationism: and the Cuvier-Geoffroy debate, 156–58, 161, 162; Lamarckian, 156–58, 161–62, 362–63 n 45
Tristan, Flora, 185, 289
Turner, J. M. W., 137

"unity of composition" (unity of type), 162–63, 178–79, 226, 234, 238–39; and the animal series, 162–63, 238–39, 351 n 24; and Grandville's *Un autre monde*, 178–79
Ure, Andrew, 98

utopias and utopian visions: Arago on utopias, 111; Bonaparte's limits on, 300–301; Comte's "intendancies," 278–81, 285–86; and freedom-in-connection, 284–85, 286, 372 n 93; intentional communities, 309–10, 376 n 43; Second Republic, 291, 292, 300–301. *See also* Comte's Religion of Humanity; Leroux, Pierre; Saint-Simonians (religion of)

Valson, Claude-Alphonse, 326–27 n 24
Varela, Francisco, 372 n 93
Vatin, François, 99
Vaucanson, Jacques de, 97
Vestiges of the Natural History of Creation, 268, 361 n 21
Vicq d'Azyr, Félix, 195–96, 236–137, 356 n 8
Vienna Circle (logical positivism), 255–56
Vigny, Alfred de, 131
virtuality: and Geoffroy Saint-Hilaire's transcendental anatomy, 162–63, 236, 239, 242–43; Leibnizian language of, 162–63, 236, 242, 362 n 40; and Leroux's ideal of virtual humanity, 226, 239–41, 242–43; and Leroux's theory of society as a milieu, 242–43, 244, 247–48
vitalism, 158–59, 195
Volquin, Suzanne, 185
Volta, Alessandro, 36–37
Vuillaume, Jean-Baptiste, 141, 144

Wagner, Richard, 143, 373 n 12
Watt, James, 110–11, 342 n 78
Watt engine and indicator diagram, 208–11
Weber, Max, 14, 321 n 36
Weber, Wilhelm, 7, 144
Wellek, René, 318 n 5
Whitehead, Alfred North: *Science and the Modern World*, 1
Wiener, Norbert, 332 n 110
Williams, L. P., 38–39
Winckelmann, Johann Joachim, 72
Winter, Alison, 143
Wise, Norton, 103, 321 n 29, 339 n 37, 370 n 54
Woolf, Arthur, 208
work, science of: and the "engineer-scientists" of the Ecole Polytechnique/CNAM, 95–97, 206–8; *forces vives* and "potential energy," 12, 217
workers' revolution of 1848. *See* Revolution of 1848 (workers' revolution)

Young and Delcambre's composition machine, 223